Development Geography

Xiangzheng Deng · Malin Song

Development Geography

Springer

Xiangzheng Deng
Beijing Technology and Business University
Beijing, China

University of Chinese Academy of Sciences
Beijing, China

Institute of Geographic Sciences and Natural Resources Research, Chinese Academy of Sciences
Beijing, China

Malin Song
Anhui University of Finance and Economics
Bengbu, Anhui, China

ISBN 978-981-96-6916-5 ISBN 978-981-96-6917-2 (eBook)
https://doi.org/10.1007/978-981-96-6917-2

This work was supported by Textbook Publishing Center of University of Chinese Academy of Sciences.

This Springer imprint is published by the registered company Springer Nature Singapore Pte Ltd.
The registered company address is: 152 Beach Road, #21-01/04 Gateway East, Singapore 189721, Singapore

Acknowledgements

This work was supported by the Textbook Publishing Center of University of Chinese Academy of Sciences, the Science Fund for Creative Research Groups of the National Natural Science Foundation of China (Grant No. 72221002).

Development Geography

Textbook Series of University of Chinese Academy of Sciences (Graduate Level) (YJC0705012).

Contents

Chapter 1
Disciplines of Development Geography

Abstract In response to the unprecedented challenges posed by globalization, such as development imbalances, regional poverty, and ecological degradation, the field of development geography faces critical adjustments in its research priorities and methodologies. The United Nations' 2030 Agenda for Sustainable Development, launched in 2016, outlines 17 Sustainable Development Goals aimed at addressing these global issues across social, economic, and environmental dimensions. This agenda builds upon previous frameworks such as the Millennium Development Goals, highlighting the urgent need to enhance human livelihood sustainability amidst environmental fragility and resource constraints. Despite global efforts, disparities in Millennium Development Goal implementation persist. Moreover, the theory of attachment underscores the evolving dynamics between developed and developing nations in a globalized context, influencing efforts toward a more equitable international order. As research in development geography intensifies globally, China's contributions are significant but reveal a notable disparity compared to international standards. Thus, strengthening research efforts within China is crucial not only for advancing geographical sciences but also for enhancing its global academic influence. The objectives of this chapter are as follows: 1. To understand the current classification of geography; master the concepts of physical, human, economic, and political geography; and understand the developmental stages of these disciplines. 2. To understand the development stages of geography. 3. To understand professional geography terms. 4. To learn to use geographic information systems, remote sensing technology, mathematical modeling, and other geographic tools to discover and analyze certain geographic phenomena.

X. Deng and M. Song, *Development Geography*,
https://doi.org/10.1007/978-981-96-6917-2_1

1.1 Connotation of Development Geography

1.1.1 Definition of Development Geography

In the middle of the twentieth century, following the Cold War between the United States (US) and the Soviet Union, the US government proposed the European Recovery Program and the Point Four Program successively to attract Western Europe and third-world countries. In doing so, they pioneered the use of the term "underdevelopment" to redivide the world pattern (Rist, 2019). As a result, a new interdisciplinary field emerged: development research (Currie-Alder, 2016). With the rise of development research, the same scholars also review the local, national, regional, and global scales of development. Development geography has since become an important branch of human geography, exploring the development of developing countries and regions, human welfare, convergence issues.

Considering the developed lineage (Escobar, 1995; Unwin & Potter, 1992; Potter et al., 2019; United Nations, 2000), the social and ideological history of the term and concept of "development" dates back to the emergence of the term "underdeveloped areas" in the 1940s. During this period, the concepts of development economics and development sociology began to grow rapidly. However, influenced by the metrological revolution in the 1950s and 1960s and the geographical cultural shift in the 1980s, development geography has developed very slowly. It was not until the 1990s, when geographers strengthened their study of regional sustainable development and global change, that development geography began to rise.

Development geography refers to the study of the level of human development and the quality of life, focusing on the related issues in the process of development convergence or improvement of human well-being in both underdeveloped and developed areas, under environmental constraints. In the context of development research, development geography is the combination of development research and geography, serving as both the theoretical basis for development research and the foundation of geography spatial and space research. Currently, there is no uniform definition of development geography; however, from its development course, its main meaning includes two aspects: narrow sense and broad sense. In a narrow sense, development geography focuses on development issues such as poverty reduction, inequality, and equity in developing countries and regions, under a specific geographical environment. In a broad sense, it extends to the global development perspective, focusing on the spatial and temporal changes, characteristics, mechanisms, and decision-making of developing countries and regions in terms of globalization, transnationalism, geopolitics and economy, climate change, environmental pollution, trade, and tariffs. In conclusion, development geography can be regarded as a comprehensive geography subdiscipline for the study of developing countries and regions and global development.

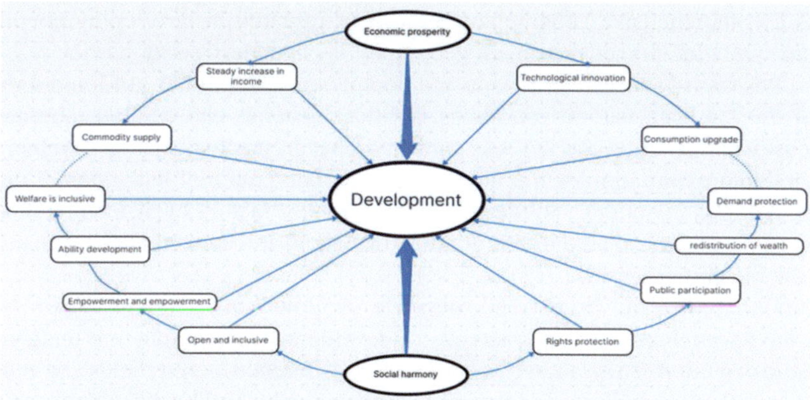

Fig. 1.1 Connotation of the concept of "development"

Development geography draws on systems science methods, integrates the perspectives of management, economics, and ecology, and focuses on the ecological, social, economic, and cultural factors influencing regional development differences. This field focuses on the level of human development and quality of life, especially the convergence of development in less developed and developed areas under environmental constraints, as well as on the joint efforts to improve human well-being. As one of the newer branches of geography, development geography has interdisciplinary characteristics, whose core is the study of the root causes of global inequality and mass poverty (Dong et al., 2021) and the associations among development, the economy, politics, and society. The purpose of development is to reduce poverty and global inequality, improving human well-being and quality of life. In essence, development must reflect the process of human social development with multi-dimensional interests of different subjects, different levels, and different social and economic backgrounds. Clarifying the criteria of the key objectives of the whole process in the definition of development is an important challenge for future geography research (Fig. 1.1).

1.1.2 Characteristics of Development Geography

As a comprehensive branch of geography for developing countries, regions, and global development, development geography is committed to discussing the development problems in specific spatial and temporal areas. It integrates social development trends with the frontier fields of other disciplines, making its research increasingly rich and interdisciplinary.

(i) **Research fields are becoming increasingly rich**

As a comprehensive geographical branch of research in developing countries, regions, and globally, development geography is characterized by its rich research fields. Research in this field is wide and complex, including but not limited to the social environment, natural conditions, political systems, and cultural inheritance. It focuses on revealing the key elements and change mechanisms of development and in-depth exploration of the development models, internal and external differences; cultural, economic, political, social, and other factors of different countries and regions; and the spatial differentiation, diffusion, and convergence mechanisms of these factors.

Currently, with the intertwined influence of global changes and human activities, the research content and methods of development geography are undergoing dynamic evolution. To accurately capture subtle differences across fields and explore core research directions, it is necessary to systematically build a diversified research system that encompasses multiple fields, scales, and research subjects. This approach helps to innovate within the geography discipline system in China and supports the goal of sustainable development.

(ii) **Interdisciplinary characteristics**

Development geography is characterized by its comprehensive and interdisciplinary nature, skillfully integrating methods and tools from management, economics, ecology, sociology, and other disciplines. By drawing on the rich heritage of social sciences, natural sciences, and the humanities, it explores critical issues such as balanced regional development, poverty, and inequality, thereby promoting the integration and advancement of various disciplines. In addition, development geography is committed to building a set of comprehensive theoretical knowledge and method systems, as well as realizing the deep intersection and integration between disciplines to better adapt to the trends and frontier challenges in the development of various disciplines. In its early stages, development research benefited from contributions from "divorced" economics, which led to the rise of the sub-discipline now known as "development economics" (Potter et al., 2012). Similarly, other mainstream social science disciplines, such as political science and sociology, have given rise to development politics and development sociology, respectively. Studies such as demography, international relations, anthropology, history, and regional planning have also contributed to development studies. Deepening the theoretical knowledge and method systems of development geography helps explore the internal connections between different disciplines and promotes the development of geography to a new peak. Development geography will provide solid theoretical support and practical guidance for economic globalization and regional sustainable development.

(iii) **Dynamics**

Development geography is dedicated to revealing the geographical phenomena and the evolution of geographic systems. It focuses not only on geographic states at a certain time but also on the continuous changes of these states and their characteristics in the time series. Through long-term observation, historical data analysis, and model construction, development geography can depict the development trajectories and trends of geographical phenomena, providing an important basis for understanding their internal mechanisms and development laws. Development geography also focuses on the dynamic interactions among elements within geographic systems. A geographic system is a complex network composed of multiple interrelated elements with mutual influence. These elements may include natural elements (such as terrain and climate), social and economic elements (such as population and industry), and cultural elements. Development geography focuses on the interactive relationships between these elements to reveal how they jointly shape the development process of geographical phenomena and predict future trends. In addition, the dynamic characteristics of development geography are also reflected in its ability to describe the spatial changes of geographical phenomena. These phenomena not only change with time but also exhibit complex distribution and evolution patterns in space. With the help of modern geographic information technology and remote sensing technology, development geography allows for high-precision monitoring and quantitative analysis of the spatial changes. Through spatial analysis and visual expression, it reveals the spatial distribution characteristics, evolution trends, and regional differences in geographic phenomena, providing scientific support for regional planning and sustainable development.

1.1.3 Relationship Between Development Geography and Related Disciplines

Building on the strong foundation of geography and incorporating the methods and means of management, economics, ecology, sociology, and other disciplines, development geography has gradually proven its significance in addressing the imbalances in regional development. Its research practice dates back to the 1940s, emerging from development practices and related concepts, and presents distinct interdisciplinary characteristics and a rich form of social science. In this process, governments, international agencies, non-governmental organizations, and community organizations have played active roles in promoting the research process of policies related to development practices (Willis, 2011). Mainstream social science disciplines such as political science, sociology, demography, international relations, anthropology, history, and regional planning have also made indispensable contributions to development research (Fig. 1.2) (Potter et al., 2012). Current geographers, with their strengths in spatial and multi-factor comprehensive research, have gradually surpassed the role of economists in development research. They actively participate in this field and have contributed to the rapid growth of "development geography", which now holds a significance comparable to that of economics (Todaro, 1977). Compared to other subdisciplines of geography, development geography has unique connotations. It not only focuses on the development convergence of less-developed countries or regions but also explores the processes of improving of the quality of life and human well-being in more developed countries or regions (De Haan & Zoomers,

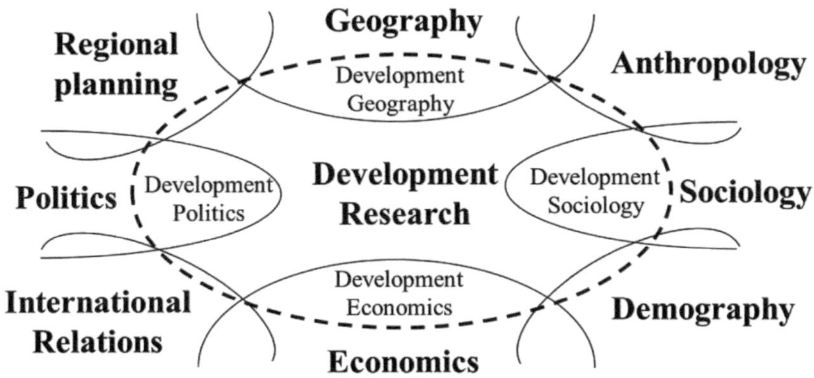

Fig. 1.2 Development geography is closely related to related disciplines and research fields

2003). In the context of global change, economic integration, and an increasingly complex geopolitical environment, all regions face severe challenges, such as poverty, regional development imbalance, and ecological degradation, all of which threaten human well-being. Therefore, using the knowledge of development geography to promote transformative development, alleviate resource and environmental pressures, and achieve the Sustainable Development Goals (SDGs) has become the focus of development geography research.

1.2 Research Progress in Related Disciplines

The study of development geography has a long history, dating back to colonial studies in the 1930s. In the late 1940s, its development focus shifted to wartime intelligence services and the reconstruction of post-war society, and research during this period was closely linked to the concept of "less developed areas." By the 1960s, the rise of Marxist theory had promoted profound change in the concept and practice of development, and the research focus of development geography had shifted from simple economic growth to an in-depth discussion of social welfare and freedom (Lawson, 2007). In this context, in the 1970s, the University of East Anglia took the lead in offering development geography as an undergraduate course, which greatly promoted the development of the discipline and the training of professional talents. With the wave of globalization in the 1980s, the research topic of development geography has gradually focused on the balanced development of less developed and developed areas, as well as the improvement of the quality of life in developed areas (De Haan & Zoomers, 2003). Focusing on the core issue of sustainable development in less developed areas, the researchers analyzed the development mechanisms and strategies in depth and formed multiple theoretical frameworks such as modernization theory, attachment theory, post-modernization theory, neoclassical theory, new Marxism, and neoliberalism (Hodder, 2005). In recent years, the development geography has constantly absorbed new technologies and new methods and comprehensively used new information technology and spatial economics to measure the spatial pattern, heterogeneous characteristics, diffusion state, and convergence mode of national or regional development. This not only provides solid technical support for regional sustainable development but also greatly promotes the rapid development of research related to development geography.

1.2.1 Early Modernization

The study of early modernization theory can be traced back to the 1950s, which marked the period of the formation and construction of modernization theory, gradually becoming the mainstream school leading development theory. Its core concept was widely recognized by both Western and non-Western societies. Early modernization theory mainly focused on the transmission of Western European and North American seventeenth-century systems and values to other parts of the world, especially Asia and third-world countries. To achieve national liberation and independence, these countries started to participate in the development of industrialization and further apply modernization theory.

From the 1940s to the late 1950s, scholars in the development geography field proposed ideas to promote national economic growth and social transformation, which can be regarded as the initial germination of development of geography. At that time, scholars mainly relied on the theory of modernization for theoretical discussions, but this theory exhibited characteristics of multidisciplinary integration at both the construction and ideological levels. To some extent, the study of modernization theory has played a positive role in promoting the spread of the Western capitalist economic model and the adoption of modern innovation and technology in less developed countries. However, by the late 1950s, the foundations of modernization theory began to be questioned. Myrdal (1957) proposed that industrialization could break the cycle of poverty and emphasized that the realization of the well-being and freedom of human society were far better than the process of modernization itself. This view introduced new dimensions of thought into development geography and guided the discipline toward a wider range of social and humanistic fields.

1.2.2 Attachment Theory

Attachment theory advocates the attachment of developing countries to developed countries in politics, economy, and culture, revealing the imbalance of power and resource allocation under the background of globalization. This theory provides theoretical support for understanding the status and challenges faced by developing countries within the global system and for seeking a more just and equitable international order. The study of attachment theory evolved from the pessimistic "classical attachment theory" to the optimistic "attachment development theory." Classical attachment theory emphasizes the dependence of developing countries on developed nations and their exploitation by them, whereas attachment development theory focuses more on the possibility of economic growth and development within the attachment relationship. This change reflects a deeper understanding of the complexity of the attachment relationship and indicates the adaptation and change of attachment theory in coping with the challenges of globalization. With continuous development, related research began to focus on the internal economic structure and social relations in developing countries. Gradually, attachment theory integrated other theoretical frameworks such as new Marxism and world-systems theory, which contributed to its development and helped explain contemporary international political and economic phenomena, providing new perspectives and tools.

In the 1960s, development geography focused on "tropical geography", "colonial geography", and "military geography", aiming to provide intelligence support for wartime military conditions and post-war reconstruction. At that time, radical political views emerged within the mainstream science of society, laying the groundwork for the birth of attachment theory. The dependency theory originated from development practices in Latin America and the Caribbean, suggesting that the Western-led global development model actually maintains the status quo of poor countries and regions, rather than promoting their social progress and poverty eradication (Frank, 1978). Humanistic theories such as postmodernism and post-structuralism also gained prominence during this period and were widely used in interdisciplinary research. Harvey (1975) further expanded upon Marx's ideas and developed a comprehensive set of theories to explain the geographical mechanism of capital accumulation. His core concept was the "spatial way out" of capital, which was gradually accepted by the Institute of Political Economy and Structuralism in Juman Geography in the mid-1970s. Additionally, a bottom-up development model has been proposed to address the development needs of less developed countries and regions, promoting sustainable development that aligns with local conditions.

1.2.3 Neoclassical Theory

The development of neoclassical theory is a process of continuous evolution and deepening, covering multiple fields including economics, geography, and organization theory. After the 1980s, western countries, with the US's Reagan and Thatcher governments as models, abandoned Keynesian policies in favor of Hayek's neoliberal ideas. Under this ideology, the government significantly reduced economic intervention, promoted the privatization of state-owned enterprises, and worked to liberalize investment and trade (Hudson, 2016). Meanwhile, under the early structuralist development economics theory, many countries once relied on planning, nationalization, and industrialization to accelerate their capital accumulation. However, these countries encountered several problems and contradictions in their economic operation.

By contrast, those less developed countries and regions with high degrees of openness, focusing on market forces and implementing export-oriented strategies, made remarkable breakthroughs in the economic field. The economic reform of Latin American countries has further consolidated the dominant position of neoliberal economic thought in economic theory (Edwards, 1995). This stage was mainly characterized by the rise of the "new rights doctrine" in Europe and "new conservatism" in the US. During this period, the British geographer A. Gilbert proposed the concept of development geography for the first time, marking the basic formation of tropical geography being gradually replaced by development geography (Gilbert, 1971). This stage was mainly characterized by the rise of "new authoritarianism" in Europe and "new conservatism" in the United States.

1.2.4 Multi-dimensional Research

Since the beginning of the twenty-first century, the developing geography research field has gradually shown vigorous vitality. Scholars have shifted their attention from the macro-scale category to the more specific and detailed small and medium-scale development process and explored the core issues related to development convergence, such as the basic movement, justice and democracy, civil sovereignty, globalization conflict, and environmental change. Meanwhile, postmodernism emerged at the beginning of this century and became an emerging paradigm in the social sciences. In this context, the new growth theory has also attracted extensive attention, which emphasizes the key role of government policies in promoting long-term social and economic growth and development (Gibson, 2002). Power (2004) reconsiders the discourse power and geopolitical nature of development geography, extensively analyzes the spatial dimensions of development and practice emphasized by development geography, and provides a new perspective to understand its complexity. Lawson (2007) further highlighted that future studies of development geography should be based on the close relationship between the improvement of human material life and securing discourse power, especially in the context of the neoliberal economy, where the influence of the discourse power, and policy on development cannot be ignored. Potter et al. (2012) suggests that tracing the evolution process of development geography research, along with its theoretical and practical applications, provides a new multi-dimensional approach to explore the development relationship from the perspective of politics, culture, economy, society, and other dimensions. These studies not only enrich the theoretical connotation of development geography but also provide strong theoretical support for us to understand and address the development problems in the real world.

1.3 Development of Related Disciplines

1.3.1 Discipline Construction and Development

In 1966, the University of Sussex established a research group dedicated to discussing the development of geography, marking the beginning of research in this field. Then, in 1973, the University of East England started its first undergraduate teaching course on the development of geography. It was followed by Oxford University, Bath University, Manchester University, London School of Economics, University College of London, and the University of Birmingham, successively, which strengthened the course domestically and laid a solid foundation.

The 21st-century world is facing multiple challenges, such as climate change, environmental degradation, regional poverty and inequality, the reshaping of national trade barriers, and geopolitical risks. In this context, the discipline construction and progress of development geography has gradually attracted extensive attention from academia and decision-makers. The international academic community has reached a consensus on the status of development geography as an independent discipline. In 2009, the International Encyclopedia of Human Geography subdivided human geography, including development geography as one of 17 branches covering 66 research entries, fully demonstrating the universality and depth of the field (Kitchin & Thrift, 2009).

Currently, the United Kingdom (UK), the United States, Sweden, Australia, Singapore, and other countries have established development geography as a formal academic discipline, with well-developed education and research systems. Although China entered this field later, its development has been rapid. In 2015, the Geographical Society of China, together with other relevant academic groups, established a working group on industrial policy and development geography and held several academic seminars on "China's Industrial Policy and Development Geography" in Beijing, Wuhan, Jinan, and Yinchuan. These conferences have provided a valuable communication platform for development geography scholars locally and abroad and have greatly promoted the rapid development of development geography in China. Since 2000, issues related to the development of geography have gradually become a research hotspot, and the number of published papers has increased by approximately 18%, highlighting the increasing attention within academic circles. Meanwhile, development geography research has also been reflected in the literature of related disciplines, such as the aggregation pattern and characteristics of economic geography, the optimal allocation of resources in economic transformation, and the policy guarantee mechanism of innovation and transformation. Relevant studies have shown an obvious upward trend since 2008.

However, domestic development geography studies still lag behind related international studies. A search for terms like "development geography", "development economics", and other keywords in the ScienceDirect literature database confirms this discrepancy. To some extent, this shows that China still has potential and room for growth in the field of development geography, requiring more scholars and research institutions to invest in it to collectively promote the progress and development of this field.

According to the ScienceDirect literature database, international research on development geography started early and showed a trend of continuous growth. As shown in Fig. 1.3, the number of papers on related topics was not large. Bus as shown in Fig. 1.4, since 1996, the number of papers on related topics has gradually increased from approximately 600 to more than 8400 in 2018, and the growth rate in the last five years is as high as 17%, which demonstrates the increasing attention and importance of development geography in the international academia.

However, compared with international research, although thc progress of development geography research in China has shown a rapid development momentum in recent years, there is still an obvious gap. As shown in Fig. 1.5, although the

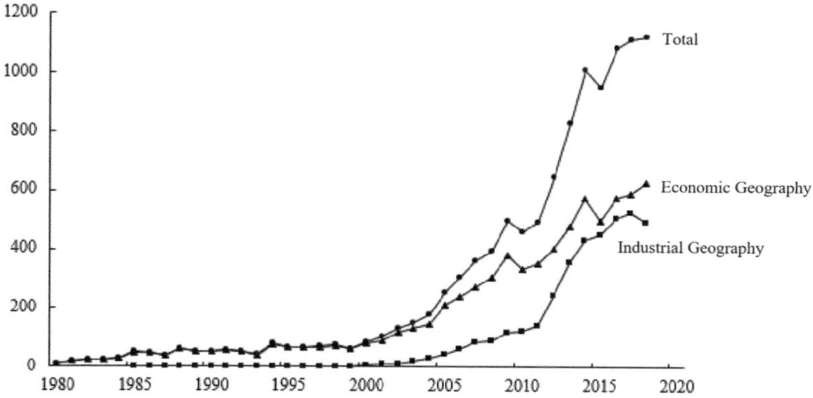

Fig. 1.3 Statistics of papers related to development geography

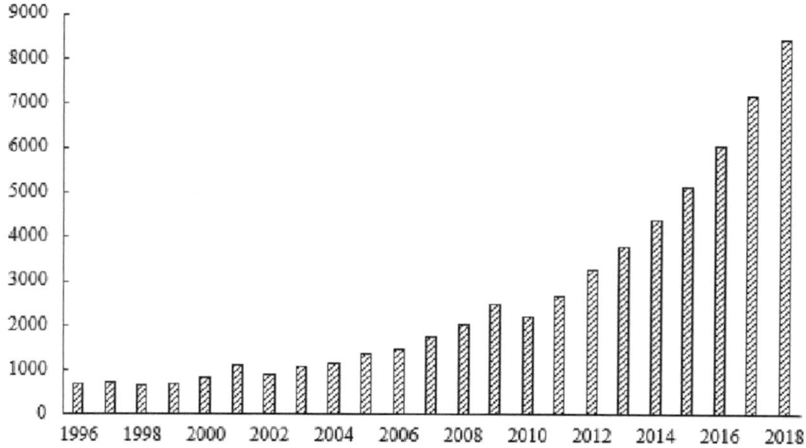

Fig. 1.4 ScienceDirect statistics of papers related to development geography

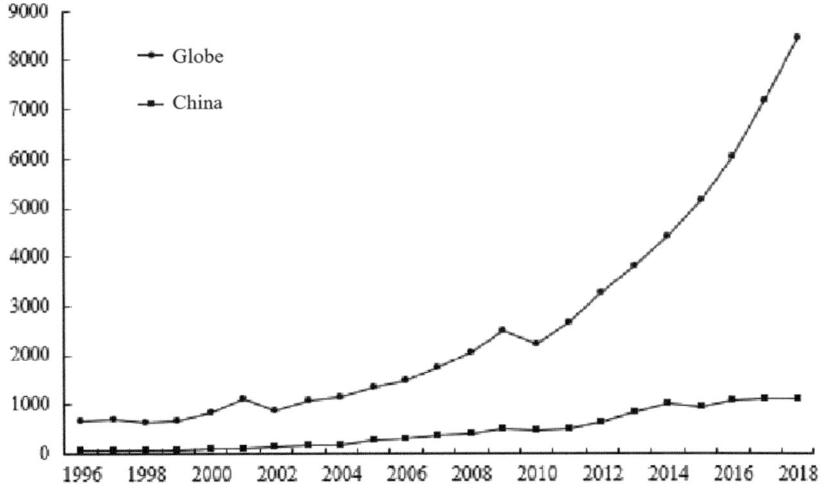

Fig. 1.5 Comparison of the number of papers published on development geography in China and internationally

number of papers published on issues related to development geography in China is increasing, there were still only 1110 papers in 2018, far lower than the 8,449 international papers. This data comparison highlights that China's research level and international influence in the field of development geography need to be improved.

Therefore, there is a need to strengthen research in the field of development geography in China, increase investment, and improve the research level to narrow the gap with the international literature. This is not only an important measure to promote the development of China's geography discipline but also the key to enhancing China's status and influence in the international academic community. We expect more scholars and research institutions to join the development geography research team to jointly promote the progress and development of this field.

1.3.2 Research Indicators and Development

Development geography aims to assess the degree of development convergence of a country or region through regional development indicators, using both qualitative and quantitative methods (Table 1.1). In the mid-twentieth century, qualitative methods such as comparative analysis, observational description, and theoretical discussion, as well as quantitative analysis methods such as data analysis, model interpretation, and systematic diagnosis were widely used and continuously developed. In the 1960s, quantitative indicators such as per capita gross domestic product (GDP) and unemployment rate began to be widely used to measure the level of economic and social development. Gould (1963) emphasized that the cognitive process is the objective result of individuals after environmental stimulation and proposed the "zero-sum game" strategy to minimize loss. In the 1970s, the Council on Overseas Development proposed the Physical Quality of Life Index (PQLI), which aims to assess the material welfare status in less developed countries and regions. Subsequently, in the 1980s, influenced by the humanistic trend, development geography began to introduce psychological analysis methods, giving rise to new research trends (Couclelis & Golledge, 1983). The United Nations Development Program (UNDP) introduced the Human Development Index (HDI) in 1989 as an important indicator of the level of national or regional development. In the 1990s, with the deepening of subject research, scholars in the developing geography field gradually realized the close connection between non-economic factors and the development process. Quantitative research methods gradually increased, including the application of system science, operational research science, metrology, etc., to reveal the spatial connection and interaction law among humanistic elements. The hierarchical analysis method proposed by American operations Chiologist Saaty (1990) has effectively reduced the influence of subjective factors by combining qualitative and quantitative methods and has been widely used in related fields. In addition, the development concept of "people-oriented" has gradually become an important means to realize development, which not only promotes the development of feminism but also makes the study of development geography more humanized. Moreover, the advent of remote sensing technology has greatly improved the accuracy of the collection and processing of developmental characterization indicators. The Scale-Invariant Feature Transformation method (SIFT) proposed by Lowe (2004) has significantly enhanced the geographical registration technology through continuous improvement. These technological advances and theoretical developments jointly promote continuous progress and improvement in the development geography field.

Table 1.1 Relevant indicators of regional development convergence

Metric		Use
Quantitative indicators	Economic indicators: gini coefficient, per capita gross national product (PC GNP), unemployment rate, energy consumption, and proportion of primary industry	Used to measure the degree and balance of the social and economic development of a country
	Social indicators: clean water resources, sanitation facilities, and adult literacy rates	Used to measure whether China meets people's needs
	Population indicators: birth rate, mortality rate, and fertility rate	Used to measure the industrialization of a country
	Composite indicators: Gross domestic product (GDP), Physical Quality of Life Index (PQLI), Human Poverty Index (HPI), and gender-related development index (GDI)	A comprehensive measure of whether China improves the quality of life and standards of people
Qualitative indicators	Including freedom, rights, security and other non-material interest indicators	Use to analyze properties that may not be numerically calculated or measured easily

Development geography has a wide range of research fields, and the need for data is correspondingly huge. To ensure that we can deeply reveal the objective laws and provide solid support for decision-making, technical support such as accurate data analysis, model building, and intuitive visual expression is particularly important. In recent years, with the rapid progress and wide application of night lighting data, statistical analysis, and big data technology, geographic information system (GIS) and remote sensing technology have provided strong technical backing for the development of geographic spatial analysis and planning. We comprehensively apply qualitative and quantitative methods to deepen the understanding of the sustainable development problems in less developed countries or regions, explore the solutions for balanced regional development, and deepen the understanding of the development model. Currently, many new technologies and methods have emerged in the development of geography research, including the use of GIS technology for spatial analysis and model simulation, evaluation, and prediction, as well as cloud computing, big data, Internet of Things, artificial intelligence, social network analysis method, input–output method, Computable General Equilibrium (CGE) model, etc. The application of these methods opens up new research perspectives and promotes the continuous progress of development geography. Spatial analysis, model simulation and visual expression using GIS technology greatly enhance our ability to dynamically analyze the evolution process of spatial patterns. Recently, 3S technologies (the global positioning system (GPS), remote sensing (RS), and GIS as the core) have been widely used in various fields, such as the evaluation of regional environmental carrying capacity, the optimization of industrial development and layout, and the evaluation of the vulnerability of social and economic systems in countries along the "Belt and Road." The application of these technologies guarantees solid technical support for the green development of less developed countries or regions. In today's rapid development of the Internet of Things and technology, big data, visualization, virtual reality, and other technologies provide key technical support for the development of geography research. With the help of big data and the Internet of Things, we can more accurately identify poor areas, innovate poverty alleviation models, and establish a dynamic poverty alleviation management and supervision mechanism. In addition, the identification of hotspots of population spatial flow can provide a scientific basis for the optimization and planning of industrial layout to further promote the innovation and development of the research paradigm, methods, and content of development geography. In summary, the research indicators of development geography are constantly improving and developing to adapt to the trends and the needs of regional development. Future research should pay more attention to sustainable development, coordinated regional development, and interdisciplinary research integration to provide scientific guidance and support for regional development.

1.4 Relevance to Sustainable Development Goals

1.4.1 Sustainable Development

Sustainable development is designed to meet both the local and global needs of modern people. Its research content includes general sustainable development theory, sectoral sustainable development, regional sustainable development, and global sustainable development. In this process, geography provides an important theoretical and scientific basis for realizing the SDG. Sustainable development is the core issue of developing geography research. With the comprehensive perspective of development geography and the advantages of multidisciplinary development, we can more comprehensively solve the regional social, economic, political, environmental, and other multidimensional development challenges, which is greatly significant for realizing SDGs.

The concept of sustainable development is gradually gaining popularity and becoming an important supplement to the traditional concept of economic development. One of the important tasks of social development is to address various challenges encountered in sustainable development. Due to its comprehensive characteristics, geography science plays an essential role in promoting sustainable development in regions. Scientific research on developing geography has a profound impact not only on China but also on the sustainable development of all countries worldwide.

The concept of sustainable development developed through three main stages. First, in the pre-Stockholm period (before 1972), during which people began to pay more attention to the survival issue, the issues of environmental constraints and carrying capacity gradually entered the public eye, and the environmental movement began to enter the political stage. This was followed by the Stockholm to the World Commission on Environment and Development (WCED) (1972–1987), when the United Nations Conference on the Human Environment was held in Stockholm in 1972, emphasizing the importance of the environment. In 1987, the WCED released a report titled “Our Common Future”, which proposed the classic definition of sustainable development, referring to the ability to meet contemporary needs without compromising the needs of future generations, which was widely accepted but also caused much discussion. Finally, in the post-WCED period (1987–present), the United Nations Conference on Environment and Development in 1992 marked a new milestone; however, the debate on “sustainability” has not subsided.

Over the past three decades, the United Nations has held several summits dedicated to the design of a system of SDGs at the global level, including Agenda 21, Millennium Development Goals (MDGs), and SDGs. However, huge gaps remain in the global implementation of these goals, with uneven progress and new challenges. To promote cooperation among the international community on major issues of human development, the United Nations launched the 2030 Agenda for Sustainable Development in 2016 on the basis of the progress made in MDGs, establishing 17 SDGs covering three major areas, namely social, economic, and environmental development, aiming to end global poverty and build a dignified society (Table 1.2).

Table 1.2 United Nations' 17 SDGs

	Sustainable development goals
Goal 1	Eliminate all forms of poverty in the world
Goal 2	Eliminate hunger, achieve food security, improve nutritional status, and promote sustainable agriculture
Goal 3	Ensure a healthy lifestyle and promote the well-being of people of all ages
Goal 4	Ensure inclusive and equitable quality education so that all people can enjoy lifelong learning opportunities
Goal 5	Realize gender equality and empower all women and girls
Goal 6	Water and sanitation and sustainable management for all
Goal 7	Ensure access to affordable, reliable, and sustainable modern energy sources for all
Goal 8	Promote sustained, inclusive, and sustainable economic growth and promote adequate productive employment and decent jobs for all
Goal 9	Build infrastructure able to withstand disasters, promote inclusive and sustainable industrialization, and promote innovation
Goal 10	Reduce inequalities within and between countries
Goal 11	Build inclusive, safe, resilient, and sustainable urban and human settlements
Goal 12	Adopt sustainable consumption and production patterns
Goal 13	Take urgent action to address climate change and its impacts
Goal 14	Ensure conservation and sustainable use of marine and marine resources to promote sustainable development
Goal 15	Protect, restore, and promote the sustainable use of terrestrial ecosystems, enable the sustainable management of forests, combat desertification, stop and reverse land degradation, and curb the loss of biodiversity
Goal 16	Create a peaceful, inclusive society to promote sustainable development, provide access to justice for all people, and establish effective, responsible, and inclusive institutions at all levels
Goal 17	Strengthen implementation tools to revive funding for the Global Partnership for Sustainable Development

1.4.2 Millennium Development Goals

In the context of global change, the development of human society is facing unprecedented challenges, especially in areas with fragile environments, limited resources, and serious poverty problems; therefore, improving the sustainable livelihood of human beings has become an urgent task. Since the 1990s, in response to the problem of sustainable development, the United Nations, through several meetings, established the "21 agenda", "Millennium Development Goals", and "we expect the future" as the core target system framework for the future development of the global economy and society to provide a clear blueprint and goal orientation. The Johannesburg Declaration on Sustainable Development, building upon the spirit of the 2000 United Nations Millennium Declaration and incorporating the consensus of the Doha and Monterrey Conferences, expanded the concept of poverty beyond mere economic growth to encompass broader aspects such as improvements in quality of life and the assurance of well-being. It also expressed a strong commitment to sustainable development goals, addressing the current state of development across various sectors. The Future We Want, launched at the "Rio + 20" Conference, is profound from a global perspective and calls on all countries across the world to take concerted action to protect our common future.

MDGs have successfully integrated the concept of development with the indicators of poverty and the level of development, injecting strong impetus into the global development process. However, the data in Table 1.3 indicates that the implementation of MDGs still needs to be promoted globally, as the progress has been extremely uneven and new challenges are gradually emerging. To promote extensive cooperation among countries around the world on major issues of human survival and development, the United Nations adopted the SDGs, included in the 2030 Agenda for Sustainable Development, in 2016. This agenda is based on the Global SDGs, and a series of institutional mechanisms to promote international development financing and technology transfer have been developed, aimed at restructuring the current global sustainable development governance system. This agenda has a profound impact on the development space and domestic development guidelines of different countries and regions; thus, it has received wide attention from the international community.

Although the MDGs have achieved some results globally, the implementation is not balanced. Some developed and developing countries have made remarkable progress in achieving these goals; however, many least developed countries (LDCs) still face enormous challenges. In addition, MDGs lasted until 2015 and were later replaced by the more comprehensive SDGs to continue to guide global sustainable development efforts.

1.5 Summary

Research in development geography has advanced rapidly in China in recent years, yet there remains a significant disparity compared to global studies. According to Science Direct statistics, there has been a noticeable increase in the number of globally published papers on development geography, underscoring its importance. However, China's research output in this domain is still lagging, suggesting the need for greater investment and research endeavors. This chapter highlights the necessity of bolstering research efforts in development geography within China, aiming to enhance its global impact and contribute to the discipline's advancement. Given the interdisciplinary nature of development geography, which integrates methodologies

Table 1.3 Millennium development goals and achievement of development geography

Target	Main progress
Goal 1: eliminate extreme poverty and hunger	The number of people living in extreme poverty decreased from 1.9 billion in 1990 to 836 million in 2015 and the extreme poverty rate in developing countries decreased from 47% in 1990 to 14%
Goal 2: popularize primary school education	From 2000 to 2015, the net primary school enrollment in developing countries and regions increased from 83 to 91%; from 1990 to 2015, the global literacy rate of young people aged 15 to 24 increased from 83 to 91%
Goal 3: promote gender equality and enhance women's rights	By 2015, developing countries and regions will basically eliminate gender gaps in primary, secondary, and higher education
Goal 4: reduce child mortality rate	From 1990 to 2015, the under-5 mortality rate decreased sharply globally, from 90 to 43 deaths per 1000 live births
Goal 5: improve maternal healthcare	From 1990 to 2013, the number of maternal deaths per 100,000 live births fell from 380 to 210
Goal 6: fight HIV/AIDS, malaria, and other diseases	From 2000 to 2013, the number of new HIV infections decreased by approximately 40%
Goal 7: ensure the sustainability of the environment	Ozone-depleting substances were almost eliminated; from 96% of improved drinking water in 1990 to 91% in 2015; and the proportion of people living in slums decreased from 39.4% in 2000 to 29.7% in 2014
Goal 8: ensure global cooperation for development	From 2000 to 2014, the actual value of official aid from developed countries rose from $81 billion to $135.2 billion

Note According to the relevant contents of the UN Millennium development goals progress report 2015

from diverse fields, it plays a pivotal role in addressing regional development indicators and promoting sustainable development objectives. In summary, the document advocates for further expansion and progression in China's development geography landscape, encouraging collaboration among scholars and research institutions to foster growth in this field.

Exercise

1. How do you understand the connotation of development geography and its evolution in the course of history? Please analyze the major changes in conceptual definition, scope of research, and methodology of development geography in the light of theoretical research results in different periods, and explore how

these changes have influenced the maturation and improvement of development geography as a discipline.

2. What are the important advances made by development geography in related theoretical research? Please select a few representative theoretical frameworks or models and analyze their core ideas, research methodologies, and contributions to the study of development geography. In addition, please reflect on the limitations and challenges in the practical application of these theoretical frameworks or models, as well as the future direction of development.
3. How can development geography be interrelated and mutually reinforced with the SDGs? Please analyze the roles and functions of development geography in promoting the achievement of the SDGs and explore how development geography can provide theoretical support and practical guidance for sustainable development. In addition, reflect on how development geography research should adjust its focus and methodology in addressing development challenges in the context of globalization, such as development imbalances, regional poverty, and ecological damage to better serve the SDGs.

Extended Reading

Babcicky, P. (2013). Rethinking the foundations of sustainability measurement: The limitations of the environmental sustainability index (ESI). *Social Indicators Research, 113*(1), 133–157.

Fujita, M., Krugman, P. R., & Venables, A. J. (2011). *The spatial economy: Cities, regions, and international trade.* MIT Press.

Gibson graham, J. K. (2002). Intervenciones posestructurales. *Revista Colombiana De Antropologia, 38*(1), 261–286.

Hudson, R. (2016). Rising powers and the drivers of uneven global development. *Area Development and Policy, 1*(3), 279–294.

Jin, G., Deng, X., Chu, X., Li, Z., & Wang, Y. (2017). Optimization of land-use management for ecosystem service improvement: A review. *Physics and Chemistry of the Earth, Parts A/B/C, 101*, 70–77. https://doi.org/10.1016/j.pce.2017.03.003.

Jin, G., Wang, P., Zhao, T., Bai, Y., Zhao, C., & Chen, D. (2015). Reviews on land use change induced effects on regional hydrological ecosystem services for integrated water resources management. *Physics and Chemistry of the Earth, Parts A/B/C, 89–90*, 33–39. https://doi.org/10.1016/j.pce.2015.10.011.

Sharma, R. C., Tateishi, R., Hara, K., Gharechelou, S., & Iizuka, K. (2016). Global mapping of urban built-up areas of year 2014 by combining MODIS multispectral data with VIIRS nighttime light data. *International Journal of Digital Earth, 9*(10), 1004–1020. https://doi.org/10.1080/17538947.2016.1168879.

The United Nation. (2015). *The millennium development goals.*

References

Couclelis, H., & Golledge, R. (1983). Analytic research, positivism, and behavioral geography. *Annals of the Association of American Geographers., 73*(3), 331–339.

Currie-Alder, B. (2016). The state of development studies: Origins, evolution and prospects. *Canadian Journal of Development Studies., 37*(1), 5–26.

De Haan, L., & Zoomers, A. (2003). Development geography at the crossroads of livelihood and globalization. *Tijdschrift Voor Economische En Sociale Geografie, 94*(3), 350–362.

Dong, Y., Jin, G., Deng, X., & Wu, F. (2021). Multidimensional measurement of poverty and its spatio-temporal dynamics in China from the perspective of development geography. *Journal of Geographical Sciences, 31*(1), 130–148.

Edwards, S. (1995). *Crisis and reform in Latin America: From despair to hope.* Oxford University Press.

Escobar, A. (1995). *Encountering development: The making and unmaking of the third world[M].* Princeton University Press.

Frank, A. G. (1978). *Dependent accumulation and underdevelopment.* https://doi.org/10.1007/978-1-349-16014-3

Gibson graham J.K. (2002). Intervenciones posestructurales. *Revista Colombiana DeAntropologia, 38*(1), 261–286.

Gilbert, A. L. A. N. (1997). Some thoughts on the new geography and the study of development. *Area, 3*(2), 123–128.

Gould, P. R. (1963). Man against his environment: A game theoretic framework. *Annals of the Association of American Geographers, 53*(3), 290–297.

Harvey, D. (1975). *The geography of capitalist accumulation: A reconstruction of the Marxian theory. Antipode, 7*(2), 9–21. https://doi.org/10.1111/j.1467-8330.1975.tb00616.x.

Hodder, R. (2005). *Development geography.* Routledge.

Hudson, R. (2016). Rising powers and the drivers of uneven global development. *Area Development and Policy, 1*(3), 279–294.

Kitchin, R., & Thrift, N. (2009). *International encyclopedia of human geography.* Elsevier Ltd.

Lawson, V. A. (2007). *Making development geography.* Routledge.

Lowe, D. G. (2004). Distinctive image features from scale in variant key points. *International Journal of Computer Vision, 60*(2), 91–110.

Myrdal, G. (1957). *Economic theory and underdeveloped regions.* London: Gerald Duckworth.

Potter, R., Conway, D., Evans, R., et al. (2012). *Key concepts in development geography.* Sage Publications.

Potter, R., Binns, T., Elliott, J., Nel, E., & Smith, D. (2019). *Geographies of development: An introduction to development studies* (4th ed.). London and New York: Routledge. https://doi.org/10.4324/9781315759319

Power, M. (2004). *Rethinking development geographies.* London and New York: Routledge.

Rist, G. (2019). *The history of development: From western origins to global faith* (5th ed.). Zed Books.

Saaty, T. L. (1990). How to make a decision: The analytic hierarchy process. *European Journal of Operational Research., 48*(1), 9–26.

Todaro, M. P. (1977). Economics for a developing world; An introduction to principles, problems and policies for development.

United Nations. (2000). *United Nations millennium declaration.* United Nations.

Unwin, T., & Potter, R. (1992). Undergraduate and postgraduate teaching on the geography of the third world. *Area, 24*(1), 56–62.

Willis, K. (2011). *Theories and practices of development.* Routledge.

Chapter 2
Theories and Methodologies of Development Geography

Abstract In recent decades, China has undergone rapid economic transformation, marked by unprecedented growth and urbanization. However, this trajectory has highlighted critical challenges related to environmental sustainability, regional economic disparities, and resource management. As the world's most populous nation, China's development strategies not only impact its own citizens but also resonate globally, influencing discussions on climate change, economic globalization, and sustainable development alongside initiatives such as the Belt and Road. This section explores these complex dynamics, emphasizing the urgent need for balanced growth that harmonizes economic advancement with environmental preservation and social stability. Through the lens of regional development convergence and the theory of resource and environmental carrying capacity, this chapter examines how China navigates these intricate issues to achieve sustainable and equitable development in the pursuit of societal well-being and environmental resilience. The objectives of this chapter are as follows: 1. To understand the patterns and mechanisms of regional development in China, focusing on the evolution and convergence dynamics between developed and underdeveloped regions. 2. To analyze the impact of globalization, climate change, and the Belt and Road Initiative on China's regional economic development. 3. To evaluate strategies for achieving sustainable development by balancing economic growth with environmental protection in different regions in China. 4. To apply theoretical frameworks such as new economic geography and sustainable development theory to analyze regional economic disparities and propose policy recommendations for achieving balanced development.

2.1 Economic Growth Theory

2.1.1 Background

The background of economic growth theory can be traced back to the Industrial Revolution in the eighteenth century. Before that time, economic development mainly depended on the development of natural resources and the growth of the labor force.

X. Deng and M. Song, *Development Geography*,
https://doi.org/10.1007/978-981-96-6917-2_2

However, with the rise of industrialization, the economic development of European countries led by Britain was rapid, and the theory of economic growth also began to emerge in several European countries. The growth of residents' income is closely related to the welfare of the people. Many scholars have attempted to uncover the mystery of economic growth and devoted themselves to the theoretical research of economic growth. The Industrial Revolution triggered a fundamental economic paradigm shift, which catalyzed scholarly exploration into growth theories aimed at illustrating economic expansion mechanisms and enhancing public well-being.

The theory of economic growth has been developed over more than 200 years, which can be roughly divided into three stages: classical economic growth theory, new classical economic growth theory, and new economic growth theory. Classical economic growth theory dates back to the Industrial Revolution in the late eighteenth and early nineteenth centuries. In the development process of classical economic growth theory, classical economists emphasize the importance of labor force, capital accumulation, and laissez-faire market mechanism to economic growth (Smith, 1776; Ricardo, 1817). They believe that through the division of labor and free competition, the market can achieve effective allocation of resources and promote economic growth. In the 1950s, economists such as Robert Solow (1956) and Samuelson (1947) proposed the neoclassical economic growth theory to overcome the shortcomings of the Harold-Domar model (Harrod, 1939; Domar, 1946). Neoclassical economic growth theory holds that stable economic growth can be realized by adjusting the input ratio of two production factors, namely labor and capital. However, long-term economic growth still needs to rely on technological progress. Since the 1980s, economists such as Cass (1965) and Romer (1986) have proposed new growth theories. They believe that factors such as innovation, entrepreneurship, human capital, and the institutional environment have an important impact on economic growth. In addition, sustainable development has become an increasingly important issue in new growth theory. The emergence of new economic growth theory marks the integration of neoclassical economic development theory into economic development theory and provides a theoretical basis for long-term economic growth.

2.1.2 Overview

The theory of economic growth explains the law of economic growth and its influencing factors. Generally, economic growth refers to the growth of the capacity of a country or region to produce goods and services. Economic growth also includes welfare growth per capita. According to American economist Simon Kuznets, a country's economic growth can be defined as its growing ability to provide residents with various economic products over a long time, and the growing ability is based on advanced technology, systems, ideology, and corresponding adjustments.

Simon Kuznets summarized six characteristics of economic growth according to historical data: (1) High growth rate of output by population and high growth rate of population. The most remarkable characteristic of economic growth is that the output growth rate, population growth rate, and per capita output growth rate are all quite high. (2) High growth rate of productivity. Productivity improvement is a sign of technological progress. (3) High speed of economic structural change. (4) Rapidly reforming social and ideological structures. (5) Rapid economic growth across the world. (6) Balanced economic growth across the world. Economic growth is generally defined as an increase in production, where output can be expressed either as the total output of the economy or as output per capita. The extent of economic growth can be described in the growth rate. The definition of economic growth has three aspects.

First, economic growth is mainly reflected in the increase in commodity output and total labor services, that is, the sustained growth of GNP. Considering population changes and price fluctuations, it can be defined as an increase in GDP per capita. Therefore, the most accurate definition of economic growth is the sustained growth of GNP.

Second, the development of science and technology is one of the essential conditions for achieving economic growth. Only through the development of science and technology and technological innovation can the economy achieve growth. Among the many factors that promote economic growth, scientific and technological progress is the most important. The course of the development of science and technology is the process of economic growth.

Third, the sufficient condition for realizing economic growth lies in the adaptive adjustment of system and ideology. Only when the social and economic system and ideology meet the needs for economic growth can science and technology play a role and economic growth be truly realized. Therefore, a reasonable social system and ideology are the prerequisites for promoting economic growth.

Thus, economic growth is defined as the continuously increasing production of material products and services in a country and region. This indicates the expansion of the economic scale and production capacity, reflecting the growth of the economic strength of a country and a region. Currently, China mainly uses GNP and GDP to measure economic growth.

Economists have different views regarding the source of economic growth. Marx, Engels, and Schumpeter argue that technological progress and innovation can improve productivity and promote capital accumulation and labor productivity, thus viewing it as the main driver of economic growth (Marx, 1867; Schumpeter, 1942). This view is widely supported in neoclassical growth theory (Solow, 1956). Other economists see capital accumulation as a major source of economic growth. They believe that increasing production capital through investment can expand the scale and improve production efficiency, thus achieving economic growth. David Ricardo has repeatedly emphasized the importance of capital accumulation in his economic theory(Ricardo, 1817; Harrod, 1939; Domar, 1946). Other scholars, such as Becker and Schultz, believe that human capital and education can improve the skills and knowledge of labor producers, improve productivity and innovation capacity, and promote economic growth (Schultz, 1961; Becker, 1964). In new economic growth theory, Romer and Lucas focus on endogenous growth, especially the role of increasing return for scale, in economic growth, and it focuses on endogenous technological innovation (Romer, 1986; Lucas, 1988). Some economists also focus on the impact of an economic system and policy environment on economic growth. They believe that good institutional arrangements, transparent market rules, and effective macroeconomic policies can improve the efficiency of resource allocation and innovative activities, thus promoting economic growth (North, 1990; Acemoglu et al., 2001). In sum, there are four main sources of economic growth: human resources, natural resources, technology, and capital. The source of growth can be studied according to the total production function: Y = AF (L, K, R), where Y represents total production, A represents technology, L represents labor, K represents capital, and R represents natural resources. It can be seen from the total production function that the source of economic growth is the improvement of natural conditions, technological progress, the continuous accumulation of capital, and the improvement of labor quality.

2.1.3 *The Main Model*

(i) **The neoclassical economic growth theory model**

The neoclassical growth model was proposed in the 1950s by Robert Merton Solow and others. It is called the neoclassical growth model because its basic assumptions and analytical methods follow the ideas of neoclassical economics, The most representative theory is Solow's economic growth model, which puts forward five basic assumptions: diminishing marginal return to capital, technological progress is exogenous, population growth rate n is fixed, $d\%$ of capital needs to be reset every year, and savings is a fixed percentage $s\%$ of income (Solow, 1956).

The Solow model is similar to the Cobb–Douglas production function:

$$Y(t) = K(t)^{\alpha} H(t)^{\beta} (A(t)L(t))^{1-\alpha-\beta} \tag{2.1}$$

The relationship between output (Y) and physical (K), human (H) capital level, quantity of labor (L) and technological development level (A) is reflected in this function, where α represents the income share of physical capital and β represents the income share of human capital. Assuming that the rate growth of labor and technical level is n and g, respectively, then:

$$L(t) = L(0)e^{nt} \tag{2.2}$$

$$A(t) = L(0)e^{gt} \tag{2.3}$$

Thus, the effective labor increases at a rate of $n + g$.

The core view of the Solow model reflects the impact of savings on the actual output level. When economic development is in a stable state, the per capita income can be expressed as follows:

$$\ln(y(t)) = \ln(A(0)) + gt + \frac{\alpha + \beta}{1 - \alpha - \beta} \ln(n + g + \delta) + \frac{\alpha}{1 - \alpha - \beta} \ln(S_k) + \frac{\beta}{1 - \alpha - \beta} ln(S_h) \quad (2.4)$$

$$y(t) = Y(t)/L(t) \quad (2.5)$$

where $y(t)$ represents the per capita income level, the depreciation rate is expressed by δ, S_k represents the proportion of physical capital in the income, and S_h represents the proportion of human capital in the income.

This model only describes the situation when the economy is in a stable state; however, in real life, economic activity is always in an unbalanced state, which is a dynamic process of transformation from imbalance to equilibrium. Order $y(t)$ is the income of effective labor in the unit, namely:

$$y^\wedge(t) = \frac{Y(t)}{A(t)L(t)} \quad (2.6)$$

where $y^\wedge(t)$ means the steady-state income level in the unit's effective labor. In addition, the Solow model states that in the case of stable economic growth, the convergence mode of economic growth can be expressed as follows:

$$\frac{dlny^\wedge(t)}{dt} = \lambda\left[ln\left(y^{\wedge*}\right) - lny^\wedge(t)\right] \quad (2.7)$$

$$\lambda = (n + g + \delta)(1 - \alpha) \quad (2.8)$$

where λ represents the rate of economic convergence, and the final equation can be expressed as:

$$lny^\wedge(t) = \left(1 - e^{-\lambda}\right)ln\left(y^{\wedge*}\right) + e^{-\lambda}lny^\wedge(t - 1) \quad (2.9)$$

In the neoclassical growth theory, capital accumulation is the core and the decreasing law of capital accumulation mechanism is the basic assumption, with the neoclassical production function and the capital accumulation equation. The basic conclusion is: the law of diminishing capital returns (from the neoclassical production function) leads to the gradual reduction of capital accumulation power; unless there is exogenous population growth or technological progress, the economy cannot achieve sustained growth; government policy has only a horizontal effect and no growth effect. The growth rate of effective labor in the neoclassical model is exogenous. Thus, the neoclassical model does not provide any economic explanation for this difference. In short, although neoclassical growth theory logically fits into these empirical facts, its explanation is inadequate: exogenous technological progress does not reveal the underlying mechanisms of economic growth.

(ii) Theoretical model of new economic growth

New economic growth theory refers to economic theory that is based on the modification of neoclassical economic growth theory, also known as endogenous growth theory, proposed by Robert Merton Solow and others in the 1950s. It aims to explain the fundamental causes and driving forces of economic growth. Compared with traditional economic growth theory, new economic growth theory emphasizes the importance of technological innovation, knowledge-intensive industries, and human capital. The theory holds that if a country or region wants to achieve long-term economic growth, it must ensure the constant or increasing income of scale and the endogenous accumulation of material, human, knowledge, and other factors rising from spillover effects. Solow's study shows that exogenous technological advances are the only driver to maintain the economy's growth in the long term. Arrow believes that the externalities of "learning while doing" and knowledge spillover can have a positive impact on the overall labor productivity level of the economy. Based on previous studies, Romer innovatively proposed endogenous growth theory by improving the previous economic growth model. Endogenous growth theory has the following assumptions:

The marginal return of capital remains unchanged (that is, the production function is a straight line);

Changes in capital K will cause changes in technology A;

If the population growth rate n, and the capital stock ratio of depreciation are both 0, the output growth rate is related to the savings rate s.

Technological progress was regarded as an endogenous variable of the endogenous growth model mainly considering the four inputs: L (labor), K (capital), A (technology), H (human capital). In addition, in economic activity, the theory in is divided into three types: the final product production, the production of intermediate products, and technology research and development (Romer, 1986). Assuming that the supply of labor and the stock of human capital are constant, the function relationship between Y (output) and labor and capital can be obtained as follows:

$$Y(t) = H_{\gamma}^{\alpha} L^{\beta} \sum_{i=1}^{\infty} X_i^{1-\alpha-\beta} \tag{2.10}$$

where X represents several capital goods broken down from the capital serving for production activities. The production function described above assumes that each specific capital product acts separately for the final output, that is, there is no complete substitution between the capital products.

Intermediate goods refer to capital goods used in the production of final goods. Firms producing intermediate goods must possess the technology to transform 1 unit of intermediate goods into n units of final goods. In the model, it is assumed that each type of capital good i is exclusively sold by a single manufacturer. Based on this assumption, the demand function for final goods can be derived:

$$K = n \times \sum_{i=1}^{\infty} X_i \tag{2.11}$$

In technology research, to obtain new technologies, a certain amount of human capital and the existing knowledge stock must be invested. Considering that knowledge can be shared, the technical level obtained after adding up all the research input is as follows:

$$\dot{A} = \delta H_A A \tag{2.12}$$

where H_A represents the human capital invested in the process of technological research, and δ represents the productivity coefficient.

2.1.4 Research and Applications

Economic growth theory studies the medium- and long-term growth trends and growth mechanisms of economic systems. It attempts to explain the growth in indicators of long-term economic output, employment, and the standard of living in a country or region, as well as the factors driving this growth. Economic growth theory has two main categories: traditional economic growth theory and new economic growth theory.

Traditional economic growth theory mainly focuses on the accumulation of production factors and the growth of productivity to explain economic growth. Among these factors, the production factors include labor force, capital, and natural resources. The representative models of traditional economic growth theory are the Harold-Domar economic growth model and the Solow economic growth model, which emphasize the impact of capital investment, labor force growth, and technological progress on economic growth. The Harold-Domar economic growth model was proposed by Roy Harold and Evans-Pierre Thomas in the late 1930s to early 1940s. The model aims to explain the economic growth trends and factors in the medium and long term. It is based on Keynes's "effective requirements principle." The Harold-Domar model mainly aims to highlight the elements needed for sustained economic growth, the root causes of economic fluctuations, and how the economy can achieve long-term growth. As one of the representatives of traditional economic growth theory, this model laid the foundation for the development of economic growth theory, introduced the concept of investment and capital accumulation, and emphasizes the importance of capital to economic growth. However, the model also has some limitations, namely disproportionate assumptions, and overly strict realization of stable economic growth. American economist Solow compared the very narrow equilibrium growth path assumed by the Harold-Domar model to "edged" growth.

New economic growth theory emphasizes the importance of new elements such as technological innovation, knowledge economy, and human capital to economic growth. It proposes that technological innovation and the application of knowledge can improve productivity, create new markets, and drive continuous economic development. The representative models of this theory are the neoclassical economic growth model and the endogenous economic growth model. Economic growth theory of the modern neoclassical school proposed the neoclassical economic growth model, which was expounded by American scholar Solow in A Contribution to the Theory of Economic Growth (Solow, 1956). Later, British scholar Mead systematically explained a neoclassical Theory of Economic Growth (Meade, 1961). This model differs from the Harold-Domar model in the following ways. First, the basic assumptions are different. The new classical economic growth model is based on assumptions such as capital accumulation, technological progress, and population growth, and assumes that long-term economic growth is mainly driven by technological progress and capital accumulation. In contrast, the Harold-Domar model puts more emphasis on the impact of effective demand on economic growth and advocates adjusting effective demand to achieve economic growth. Second, the equilibrium path is different. The neoclassical economic growth model assumes that there is a stable equilibrium path and that the economy will follow the path to achieve sustained and steady growth. However, the Harold-Domar model assumes that economic growth can be influenced by factors such as cyclical depression and recovery, and economic growth will not always follow a stable path.

Endogenous growth theory is a new economic growth theory that was gradually developed in the late 1980s and early 1990s. Compared with traditional exogenous growth theory, it puts more emphasis on the influence of endogenous factors such as technological progress, human capital accumulation, and institutional environment on economic growth. Arrow's (1962) model and the increasing growth model have a significant impact on theory of economic growth. In Arrow's model, output is not only tangible input but also a process of learning and accumulating experience. Increased capital not only increases the output through commodity production but also promotes technological progress and the development of new ideas in the production process; therefore, the contribution of capital is greater than the traditional contribution. However, in this model, the technology remains exogenous, and it changes with the endogenous capital stock. Romer's model (Romer, 1986), emphasizes the end nature of technological progress; that is, technological progress is not exogenous, but can be stimulated and promoted through a series of economic activities and policies. Romer's model proposes that technological innovation and knowledge accumulation are the main drivers of economic growth, and enterprises should take steps to promote technological innovation, intellectual property protection, and investment in education to drive economic growth. This new economic growth theory based on technological progress pays attention to the role of knowledge accumulation in the process of capital accumulation and promotes the development of the high-tech revolution and the arrival of the era of knowledge economy.

2.2 Convergence Theory of Regional Development

2.2.1 Background

Since the twentieth century, people have begun to notice the imbalance of economic growth and development levels between different regions. Some regions may have achieved rapid economic growth due to resource abundance, advanced technology, or policy support, whereas others are relatively lagging. This imbalance has drawn much attention to the regional development gap. In recent years, the convergence problem of regional development has gradually become the focus of academic discussion. Relevant studies include areas such as theory of regional development convergence, convergence reasons, and convergence test methods. With the widening income gap globally, narrowing the gap between rich and poor regions and promoting regional development convergence has gradually become a topic of common concern to policymakers and scholars across the world. The main objectives of academic discussions on convergence are: first, to compare the ability of new growth theory and neoclassical growth theory to describe the real economic phenomena; second, to provide a reference for the development policies of relevant countries or regions; and third, to explore whether developing countries are growing faster or whether the gap between the rich and the poor is widening to solve the fairness problem. With the development of economic statistics and econometrics methods, scholars began to empirically analyze economic growth in different regions. They attempt to test whether the convergence hypothesis holds and what the convergence rate and mechanism are through the analysis of historical data.

2.2.2 Overview

The theory of regional development convergence proposes that the economic development level between different regions tends to converge or diverge under certain conditions. The central idea of the theory is that areas with poorer or lower economic development will grow at a faster rate, gradually narrowing the economic gap with affluent areas. This phenomenon of convergent development may involve many factors, including resource allocation, technological progress, human capital accumulation, and market mechanism.

One of the important assumptions of the convergence theory of regional development is the diminishing marginal return. This means that in poorer areas, the same investment or resource allocation may bring higher marginal gains, thereby contributing to economic growth. Meanwhile, the wealthier regions are already at relatively high economic levels, and the same investment may only have smaller growth effects. Another important hypothesis is the technical spillover effect. This refers to the dissemination and sharing of technology, knowledge, and experience among different regions. Technological innovation and experience in more developed areas can be transmitted to less developed areas through various channels, thus promoting the economic development of the latter. The research methods of the convergence theory of regional development usually involve the comparative analysis of the economic growth rate, income level, and employment rate in different regions, as well as the empirical study of the factors affecting regional economic development. Although the convergence theory of regional development presents a potential economic development path, in practice, the differences in regional development are affected by several factors, including policies, institutions, and geographical location. Thus, although convergence may occur, there may be some limitations or obstacles.

The convergence of regional development can be theoretically explained in three ways, namely, neoclassical growth theory, new growth theory, and system-growth theory. The early neoclassical growth model reveals the convergence mechanism of economic growth theoretically, and its basic idea is that the return of scale is unchanged in production, and the marginal output of factor input decreases. The convergence hypothesis based on neoclassical growth theory partly reveals the general mechanism of convergence in regional development, which is the most widely used convergence theory. In the 1980s, a new form of economic growth theory called endogenous growth theory gradually emerged. This theory explains the change in technical level and sustainable growth through factors such as dry learning theory, knowledge spillover, and purposeful research and development. Endogenous growth theory mainly focuses on the internal driving mechanism of economic growth and emphasizes the role of two mechanisms: diminishing efficiency of technological innovation and technological diffusion. These two mechanisms together lead to non-monotonic changes in the cost of imitation, which makes less developed countries differentiate, some countries gradually approach, and other countries gradually move away from developed countries. Therefore, the endogenous technological progress model can also explain the occurrence of club convergence. According to institutional economics, the possibility of regional development convergence depends on the differences in the institutional environments of different regions. In areas with relatively stable and good institutional environments, economic growth may accelerate, leading to the convergence of regional development; whereas in areas with poor or unstable institutional environments, economic growth may be limited, leading to the continuation or widening of regional differences in development. Based on political economy and philosophy, Marx first integrated the system into economic theory, analyzed the interactive relationship between productive forces and the relations of production, the economic foundation, and the superstructure in the process of human development, and questioned how and why the system develops. Early institutional economics often stayed on the "ideological experiment" and lacked a theoretical connection with the growth convergence, and it was difficult to directly analyze the influence of the system on the economic growth differences in different countries or regions. However, in recent years, people have gradually realized the two-way relationship between a system and economic development and understood the role of institutional change in economic growth in the transition economy, and institutional economics and growth theory have led to the trend of integrated development. Some scholars also explain regional growth differences based on the economic growth model, including endogenous institutional change.

2.2.3 Theoretical Framework

Regional development convergence refers to the gradual convergence of the development levels of different regions. Simply put, this implies that regions with relatively lower levels of development tend to experience faster growth compared to more advanced regions, eventually leading to the economic levels of different regions gradually converging. This phenomenon, known as convergence, is a characteristic observed in the process of regional development. Many empirical studies on regional convergence use α convergence, β convergence, and club convergence. α convergence refers to the degree of dispersion of an economic variable in the economic system among economies (Barro & Sala-i-Martin, 1997). β convergence refers to the process in which the economic development level of economic entities with different per capita output levels in the initial stage tends to converge (Baumol, 1986). Club convergence means that the per capita income level of a group of countries or regions with similar economic characteristics approaches a steady state (Baumol et al., 1988). Overall, α convergence and β convergence are mainly used to describe the convergence properties of the algorithm, whereas club convergence is used to describe the convergence phenomenon of individuals in social groups. The following is a brief introduction and analysis of the basic model of the above three concepts.

(i) **α convergence and analysis**

In the existing empirical studies, per capita GDP is generally used to reflect the convergence of regional economies. Some scholars suggest that α convergence is a measure of the average degree of the per capita income level gap among countries at different points (Barro & Sala-i-Martin, 1997). Due to the strict condition constraints, α convergence is also called strong convergence. The coefficient of variation, Gini coefficient, and σ coefficient can be used to reflect the distribution state of α convergence.

$$\textit{Coefficient of variation}: CV = \sqrt{\frac{1}{n}\sum_{i=1}^{n}\left(x_i - \overline{X}\right)^2/\overline{X}} \tag{2.13}$$

$$\text{Gink coefficient: } G = f(X_1, X_2, \ldots, X_n; P_1, P_2, \ldots, P_n) \tag{2.14}$$

Based on the nature of the research object, the standard deviation can also be considered the standard to measure convergence. For example, considering the difference in the per capita output levels of different regions, the standard deviation is used to construct the test formula of α convergence, as follows:

$$\sigma = \sqrt{\frac{1}{n}\sum_{i=1}^{n}\left[y_i - \frac{1}{n}\sum_{i=1}^{n} y_i\right]^2} \tag{2.15}$$

where y_i represents the per capita output of the ith region, and σ represents the standard deviation of the actual total output of the n regions.

(ii) **β convergence and analysis**

β convergence is the most commonly used method in convergence models, including absolute β convergence and conditional β convergence. Absolute β convergence is a trend of convergence among regions without considering the influence of external factors, whereas conditional β convergence is a trend of convergence among regions after taking external factors into account. In the field of economic growth, the former means the relative convergence of economic development. It shows that although the initial economic level of a region is low, its future economic growth rate is high; therefore, the economic level of the region will catch up with the high economic level. However, the latter argues that the economic growth rate of a region does not only depend on its initial level but should also consider the socioeconomic characteristics of the region. In other words, different regions will have different steady-state levels to some extent due to their own social and economic characteristics, and their development speed also depends to some extent on the differences between their initial development level and steady-state development level. It is unreasonable to simply assume that regions with low initial economic levels develop relatively fast, as the steady-state level of the region itself may be low. Therefore, the initial level should not only consider the factor but also consider the characteristics of the corresponding region. If regions tend to stabilize to a certain steady-state level only after controlling for these characteristic factors that lead to regional initial-level differences, then this phenomenon is called conditional β convergence.

The economic growth of the closed economic system is mainly in a log linear form, namely:

$$\frac{1}{T-t}\log\left(\frac{y_{iT}}{y_{it}}\right) = x_i^* + \frac{1-e^{-\beta(T-t)}}{T-t}\log\left(\frac{y_i^*}{y_{it}^0}\right) + u_{it} \tag{2.16}$$

Including i said i economic units, t and T respectively at the beginning and the end of economic development, y_{it}^0 for the per capita real income, y_i^* said the state of economic development stable per capita income level, x_i^* for the economic development into the steady state of per capita income growth rate, y_{it} and y_{iT} respectively for the regional economic development, the end of per capita income, coefficient β said the rate of economic growth convergence, u_{it} as the error item. Thus, the β convergence coefficient substantially reflects the speed at which y_{it}^0 approaches y_i^*. The larger the value is, the faster the regional economy develops into a stable state. The critical value of the regional economic growth trend is 0. If β > 0, the regional economic growth tends to converge. Otherwise, it indicates that the regional economic growth shows a trend of divergence. If it is constant, the formula can be expressed as: $x_i^* y_i^*$

$$\frac{1}{T-t}\log\left(\frac{y_{iT}}{y_{it}}\right) = B + \frac{1-e^{-\beta(T-t)}}{T-t}\log y_{it} + u_{it} \tag{2.17}$$

where B is a constant, and the regression analysis using the above equation will show that the β value is only related to the per capita income at the beginning, that is, the absolute β convergence.

(iii) **Club convergence and analysis**

Club convergence means that different economies with similar initial conditions will converge to the same steady-state level in the long term. It shows that the income or productivity levels of less developed and developed economies tend to converge internally but not between the less developed and developed economies. Conditional convergence refers to the convergence of the income or productivity of a single economy to its own steady-state level, whereas club convergence refers to the convergence of income or productivity of a class of economies with similar initial conditions to a certain steady-state level. Club convergence focuses on considering the spatial characteristics of the convergence phenomenon. Therefore, club convergence can explain better the coexistence of rich and poor areas in the process of regional economic development. Using per capita GDP as the core variable to determine the convergence of the club, the calculation formula is as follows:

$$y_{it} = \alpha + \beta lnY_{i0} + \varepsilon_{it} \tag{2.18}$$

$$y_{it} = \left(\frac{1}{T}\right) ln\left(\frac{Y_{iT}}{Y_{i0}}\right) \tag{2.19}$$

$$\beta = -\frac{1-(1-\lambda)^T}{T} \tag{2.20}$$

where Y_{i0}, Y_{iT} represent region i in the beginning ($t = 0$) and end ($t = T$) of per capita GDP, y_{it} represents the average growth rate of per capita GDP during this period. λ represents growth convergence rate, ε_{it} is the error. β represents the convergence coefficient of per capita GDP at the base time. If the value is positive, the regional economic development level tends to diverge; otherwise, it represents convergence.

2.2.4 Research and Applications

The convergence theory of regional development proposes that the economic growth rate of poorer areas will exceed that of rich areas, thus reducing the economic gap between different regions, which is one of the important theories of economic geography. The theory is based on the observation of long-term economic growth and development, proposing that the relatively poor areas will experience faster growth under certain conditions, thus narrowing the gap with the rich areas, and eventually reaching a state of convergence. This convergence can be due to the more investment opportunities and low-cost resources in poorer areas, which can attract more inflow of capital and human resources and accelerate economic growth. At the same time, rich areas may face problems such as resource depletion and rising costs, leading to slower growth rates. Therefore, under certain conditions, the difference in the economic growth rate between such regions will gradually decrease, and even reach a relatively balanced state. However, the difference may be affected by many factors, such as policy system, technological progress, and resource allocation. Therefore, regional development convergence is not inevitable; there is still a certain degree of regional development differences. Some studies also point out that although there may be a convergence trend in a certain period, in the long run, regional development differences may be affected by more factors and show greater volatility.

Both national and international scholars have explored the theory of regional development convergence. International scholars have mainly focused on the influencing factors of growth convergence and regional economic convergence. Based on the experience of less developed and developed countries around the world, some scholars have studied the convergence problem from an empirical perspective and concluded that for industrialized countries, innovation and investment have the property of public goods and positive externalities for other countries, thus promoting convergence between different countries (Baumol, 1986). In addition, many economics historians also attach great importance to the technological progress in economic catch up and think the technology gap between less developed and developed countries is the main cause of economic growth convergence (Gerschenkron, 1962). Moreover, some scholars, based on the long history of developed countries, proved that economic globalization and immigration have a decisive influence on growth convergence (Williamson, 1996). Many economists have extended the analysis of transnational convergence regression under the framework of neoclassical growth theory, and Barro and Sala-i-Martin (1991, 1992). Based on the Summers and other databases, many economists have extended the transnational convergence regression analysis under the framework of neoclassical growth theory (Barro & Sala-i-Martin, 1992). The Summers database covers most countries or regions around the world. Some scholars found that there was significant conditional convergence in the Summers transnational data from 1960 to 1985, and the empirical results of the elastic convergence rate of factor output were highly consistent with the theoretical model and thus believed that the extended Solow model could well explain the economic differences and growth convergence of various countries (Mankiw et al., 1992). Kim conducted a unit root test of the panel data of 17 East Asian countries and found that the conditions converged; however, the steady-state growth trend of different countries was different. Kim believed that this supported the neoclassical growth theory rather than the endogenous growth theory. In China, the applied research on development convergence theory starts from the empirical statistical analysis of σ convergence. In terms of total description, many scholars have decomposed the regional differences. With the continuous improvement of data and the introduction of new analysis tools, many scholars have extended the analysis of the growth convergence in China based on the classical methods of β convergence and σ convergence and made some new achievements. Some scholars have begun to use kernel density estimation and other methods to describe the trend of convergence in China more comprehensively. From the measurement and description of regional gaps to trying to explain the causes of regional differences, domestic research on the convergence of economic growth has made great progress.

2.3 Resource and Environmental Carrying Capacity

2.3.1 Background

With the rapid development of industrialization and urbanization, the impact of human activities on the natural environment, such as pollution, ecosystem degradation, and natural resource depletion, has attracted much attention. In recent years, China has achieved rapid economic growth. However, the economic development mode of high investment and high consumption leads to a series of problems such as excessive resource consumption, deterioration of the ecological environment, and low production efficiency. Facing the pressures of resource scarcity and environmental deterioration, there is an urgent need for a scientific approach to evaluate and manage resource utilization and environmental protection to achieve the goals of sustainable development.

The theory of resource and environmental carrying capacity refers to the theoretical framework of evaluating and exploring the pressure and load that resources and environmental systems can bear. It involves concepts of resource utilization, environmental quality, ecosystem health, and quantification and analysis to assess the upper limit of human activity stress that a particular region or system can withstand. Resource and environmental carrying capacity has now become an important content of ecology, geography, resource science, environmental science, and other disciplines. The proposal of different carrying capacity concepts and related evaluation theories and methods is highly important for further understanding the interaction between population, resources, environment, and regional sustainable development and its coupling mechanism.

2.3.2 Overview

The basic concept of bearing capacity was first proposed by Hadwen and Palmer (1922) in an article on the management of pastoral land. The term "carrying capacity" originated in ecology, and it refers to the maximum amount of inventory of an individual organism in a certain environment. In 1953, Odum defined the concept of carrying capacity in his Principles of Ecology as the "upper limit of population growth" (Odum, 1971; Price, 1999). The study of bearing capacity mainly focuses on theoretical discussion, mainly on the population growth law of organisms outside of human beings and the population problem under food restriction. In the 1960s and 1970s, a global resource and environmental crisis broke out. The study of carrying capacity was no longer limited to simple theoretical research but widely used in the study of human economic and social development under the restriction of resources and environment. In the 1980s, with the determination of the concept of resource and environmental carrying capacity by the United Nations Educational, Scientific and Cultural Organization (UNESCO), the demand for land and water resources increased in social and economic development, and large-scale research on land and water resources carrying capacity was conducted in China (UNESCO, 1985). Under the guidance of the general theory of sustainable development, local scholars have proposed various definitions of resource and environmental carrying capacity, and as the basic standard to evaluate the coordination of resources, the environment, and social economy, carrying capacity has been widely recognized, and has become an important means to reflect the interaction mechanism between human beings and the environment.

Resource and environmental carrying capacity refers to the maximum value of regional resources (natural resources, science and technology, economy, etc.) that can be used to support human activities or economic development in a certain period. Environmental carrying capacity evolves from environmental science. It is more commonly understood as "the maximum ability of regional resources and environment to support human activities or economic development in a certain region and time", which is generally measured by population and land area. The carrying capacity of resources and the environment involves several subsystems that influence and restrict each other, such as resources, environment, society, and economy, and is composed of three elements: carrier, carrying object, and environmental carrying rate. The carrier includes the natural environment carrier composed of life; material; and material production support systems such as air, soil, and mineral resources; and the artificial environment carrier composed of public facilities and economic foundation. The bearing object refers to the capacity of resources and the environment to bear pollutants, population size, human production, and living activities. The environmental carrying capacity refers to the reflection of the resource and environment system in a certain range of human social and economic activities of time and space. Since the late twentieth century, research on resource and environmental carrying capacity, focusing on population, resources, environment, and development, has gradually become a key area of study in disciplines such as geography, ecology, and environmental economics. It has gained increasing attention in fields such as regional planning, ecosystem service assessment, global environmental status and development trends, and sustainable development.

2.3.3 Research and Applications

The carrying capacity of resources and the environment is an important category of environmental science research. It is an important index to measure environmental quality and environmental capacity tolerance, as well as a comprehensive capacity value reflecting various factors such as economic and social aggregate. The increasingly serious global environmental problems have attracted global attention, and taking the path of sustainable development has become the common choice of the international community. Serious resource and environmental problems require countries to change the traditional extensive mode of economic development, low input, low consumption, and high return of economic model. Therefore, the process of economic development must first break through the "bottleneck" of the theory, and directly guide the economic practice of standard operation to solve the rising resource and environment bearing capacity problems and constantly improve economic development. The correct understanding of the sustainable development of economic environmental capacity is the prerequisite to solving the problem of environmental resources. Research has been conducted on the multifunctional comprehensive carrying capacity of society, economy, and ecology of different elements (land, water, ecological environment, energy, mineral resources, etc.).

(i) **Carrying capacity of land resources and the environment**

The carrying capacity of land resources and environment refers to the ability of land resources and environment to withstand and meet the needs of human social and economic development in a certain period. This concept involves factors such as the quantity and quality of land resources and the stability and sustainability of the environment. Land resources are limited, and the demand for land in human economic activities is constantly increasing. Emphasizing the environmental carrying capacity of land resources can remind people to pay attention to the scarcity of land resources, avoid excessive exploitation and waste, and ensure the sustainable use of land resources. With the deepening of the urbanization process, the contradiction between economic development, urban construction, and ecological environment is increasingly prominent. Scientific research on land carrying capacity is critical in developing and utilizing land resources efficiently and intensively and optimizing the spatial structure and layout of different land use types.

The carrying capacity of land resources, including land resources, land resources quality, environmental load, environmental capacity, and sustainable development, is mainly evaluated using the experience model method, remote sensing method, and system dynamics method based on the specific food consumption level, that is, largest population the region can feed.

(ii) **Environmental carrying capacity of water resources**

Water resource environmental carrying capacity refers to the ability of human activities and natural factors to pressure and load water resources and the environment over a certain period. With the expansion of human activities and the increasingly global climate problems, environmental problems such as water pollution, water shortage, and ecological balance destruction have gradually become the focus of human society. Owing to the uneven spatial and temporal distribution of water resources in China, the scarcity of per capita water resources, water pollution, and other environmental problems have been increasingly troubling China's industry and agriculture in recent years, and the water resource problem has become the main constraint to the social and economic development of some regions in China. Therefore, examining regional water resource environmental carrying capacity, optimizing the spatial allocation of regional water resources, and coordinating water resource usage are critical to realizing regional sustainable development. The environmental bearing capacity of water resources mainly considers the balance, water environment quality, water resource utilization efficiency, ecological environment protection, regional ecosystem, and social economy maximum carrying capacity and scale research. Research methods mainly include trend analysis, comprehensive evaluation, principal component analysis, system dynamics method, and multi-objective decision analysis. The construction of the evaluation index system is the basis of water resource carrying capacity research, and the commonly used indicators can be roughly divided into the water resource system index, social system index, economic system index, and ecological environment system index.

(iii) **Ecological carrying capacity**

Ecological carrying capacity refers to the ability of an ecosystem to accommodate and maintain the number and activities of specific organisms within a certain period. In other words, it describes the extent of biodiversity and ecological function that an ecosystem can support and its resistance to external disturbances. The assessment of ecological carrying capacity usually involves many aspects, such as biodiversity, ecological function, environmental quality, and natural resource utilization. It not only considers the sustainability of natural resources and the ecological environment but also pays attention to the coordinated relationship between human economic and social development and ecological and environmental resources. Given challenges such as the excessive consumption of resources and serious damage to the ecological environment, research on ecological carrying capacity based on ecological service function and facing regional sustainable development has gradually become a new research hotspot, which provides a strong theoretical basis for promoting the healthy development of regional economy and ecological civilization.

Scholars have expanded the perspective of ecological carrying capacity research from the social-ecological-economic benefits of single-dimensional elements (land, water, forest) to the comprehensive evaluation of the impact of multiple factors (human and natural factors) on the ecological environment. Related contents involve the improvement of ecosystem service function, construction of an ecological compensation mechanism, construction of ecological security protection pattern, comprehensive evaluation and policy design of fragile and typical areas, evaluation of sustainable utilization based on ecological footprint, and optimization of territorial space development. The research methods mainly include the comprehensive evaluation model, state space method, ecological footprint method, resource and demand difference method, and system dynamics model. In addition, modern technical methods such as GIS and remote sensing provide a full range of dynamic monitoring data for ecological environment-related research and technical support for the comprehensive assessment of the sustainable development capacity of the ecological environment.

(iv) **Carrying capacity of energy and minerals**

Due to the over-exploitation and use of energy and minerals, especially fossil energy, the contradiction between the sustainable development of society and the carrying capacity of energy and minerals and between extensive production mode and the shortage of energy and minerals are increasingly prominent, and the negative effect of the sustainable development of human economy and society is increasingly intensifying. The carrying capacity of energy and minerals refers to the capacity of specific areas to accommodate and maintain the development and utilization of energy and mineral resources within a certain range of time and space. The evaluation result of the carrying capacity of energy and minerals is an important indicator to measure the sustainable development of a region.

In recent years, many scholars have conducted extensive research on energy and mineral resources from different perspectives, mainly including the development status and evaluation of energy and mineral resources and the spatial and temporal differentiation and driving mechanism of energy and mineral resources. Under the background of the current ecological civilization construction, the economic and ecological benefits of energy and mineral development and utilization have also gradually drawn the attention of scholars and opened up new fields for the development of a green economy and the realization of emission reduction targets. In addition, energy and mineral resources, as an important strategic resource for the development of the national economy, research on mineral energy security and strategic research, and the development and utilization of new energy, have become the focus of research.

2.4 Sustainable Development Assessment

2.4.1 Background

Since the Industrial Revolution, the human economy has developed rapidly; however, it has also brought serious environmental problems, such as pollution, resource depletion, and ecosystem destruction. In the 1960s and 1970s, the rise of the environmental movement made human beings begin to realize that economic growth could not erode natural resources and the ecological environment without restriction. In the second half of the twentieth century, a series of global environmental problems, such as climate change, biodiversity loss, and land degradation, have attracted global attention. These issues highlight the impact of human activities on the earth system, creating an urgent need for sustainable development in the future. In 1972, the United Nations held the first Global Environment Conference and issued the Declaration on the Human Environment, which emphasized the importance of environmental protection and sustainable development and marked the environment as the beginning of the global agenda. In 1987, the United Nations established the "World Council on Environment and Development", led by Norwegian Prime Minister Holland Brandand. The committee published Our Future: The World Council Environment and Development Report, defining the concept of "sustainable development" as "meeting the needs of current generations without compromising the ability to meet the needs of future generations." In 1992, the United Nations Conference on Sustainable Development was held in Rio de Janeiro, Brazil, where the Rio Declaration and Agenda 21 were issued, setting out the principles and direction of action for sustainable development and the comprehensive development goals of reducing poverty, promoting social justice, and protecting the environment. In 2015, the United Nations adopted the 2030 Agenda for Sustainable Development, which sets out 17 Sustainable Development Goals (SDGs) covering three areas—economic, social, and environmental—to achieve global sustainable development by 2030.

2.4.2 Connotation

Sustainable development thought can be traced back to 1789, with political economist Malthus's principle of demography, where he comprehensively discusses population growth and the contradictions in resource limitations, pointing out that only between coordinated and sustainable development can there be social progress (in 1859, Darwin's origin of species, from the perspective of biological evolution to sustainable development ideas). In 1987, the United Nations World Commission on Environment and Development published a study entitled "Our Common Future", which first proposed the concept of "sustainable development." Sustainable development is based on three basic principles: fairness, sustainability, and justice, the core of which means "to meet the current needs without affecting the survival needs of future generations." After continuous improvement, revision, and promotion, the concept of sustainable development has been recognized by most countries in the world.

Sustainable development refers to meeting the needs of the contemporary people without damaging the environment under the premise without damaging the needs of the later generation. Its core purpose is development, but to achieve economic development on the basis of maintaining the sustainable use of resources and the environment. With the development and progress of society, Chinese people have gradually recognized the strategic goal of sustainable development, and the connection between sustainable development and the national economy has become a hot spot of common concern and recognition for society. The theory of sustainable development requires that social development matches the carrying capacity of natural resources.

Because resources are limited, particularly natural resources, it requires coordinated common environmental and economic development. The implementation of sustainable development requires constant overall probing of the social, resource, economic, and environmental aspects of planning. The connotation of the economy and environment of sustainable coordinated development, including common, coordinated, fair, efficient, multidimensional, and sustainable development, is the overall coordinated development of the economy, society and the environment. However, the development we pursue is the development of contemporary people, which cannot be sustainable at the cost of the development ability of one region or that of other regions. Sustainable development should be efficient, multi-dimensional, and multi-mode. Since the theory of sustainable development conforms to China's national conditions, China has been applying the theory of sustainable development in its social and economic development. Local and international scholars have conducted rich research on the theory of sustainable development. International scholars have expounded the concept and connotation of sustainable development from the perspective of intergenerational fairness, sustainable utilization of resources, and the coordinated development of economic society and ecological environment. Homberg et al. (1992) believe that the concept of "sustainable development" is still developing and changing, and needs to be improved according to different spaces and times. Escowar (1995) believes that sustainable development is a new theoretical construction, aiming to transfer the problem of natural environment protection to the social field.

2.4.3 Research and Applications

Sustainable development theory aims to achieve coordinated economic, social, and environmental development to meet the needs of current and future generations. The sustainable development of "development geography" takes the sustainable development of the world, countries, or regions as the research object, the harmony between man and land as the main research line, the system science as the basic research method, and the system sustainable development as the research goal, which has strong theoretical and application value.

The applied research is mainly reflected in the systematic analysis of the basic elements of regional sustainable development, and the diagnosis and early warning of the environmental situation and development of each stage in the region. First, study the current situation of the regional population, formulate a regional population development strategy, develop countermeasures for sustainable utilization of regional human resources and sustainable development of livelihood, get rid of regional poverty, and build a regional security mechanism. Second, based on the research on the carrying capacity and suitability of resources and environment in different regions, analyze the correlation between social and economic structure and industrial structure; then, analyze the countermeasures and suggestions of rational allocation of regional resources, optimization, and adjustment of regional industrial structure and sustainable development of regional ecology. Third, analyze the current situation of regional development. On the one hand, explore the evolution pattern and restriction mechanism of underdeveloped regions, on the other hand, identify the aggregation and convergence development mechanism of developed regions, and explore the path of regional equilibrium and sustainable development. Fourth, clarify the opportunities and challenges facing global development from the three aspects of climate change, economic and social globalization, and the regions along the Belt and Road routes, and study the establishment of a global coordinated development mechanism and sustainable development approaches.

As one of the most populous countries in the world, China is facing severe environmental and resource challenges. With its rapid economic growth, China also faces environmental problems such as air, water, and soil pollution, and the limited and uneven distribution of energy resources. How to protect the environment; how to manage resources; how to maintain sustained and steady economic growth, and how to seek stable social development have increasingly become the core issues of our future development. Therefore, China should urgently implement the concept of sustainable development and promote the coordinated development of economic growth and environmental protection to realize the sustainable development of man and nature.

2.5 New Economic Geography

2.5.1 Background

Globalization refers to the phenomenon of deepening economic, social, and cultural ties and exchanges among countries. In the context of globalization, the contacts and interactions between countries in the world are increasingly frequent. This is mainly due to economic, technological, political, and cultural factors. Globalization has accelerated the flow of goods, capital, information, and people; broken national boundaries; and promoted economic complementarity and cooperation among countries. The rise and development of multinational corporations, the increase of international trade, and the formation of a global production network are all economic factors of globalization. Meanwhile, the rapid development of science and technology makes the dissemination of information more rapid and convenient and promotes global exchanges and cooperation in various fields. The popularization and application of the Internet, e-commerce, mobile communication, and other technologies have changed the way people live and work and promoted the process of globalization. In addition, cultural communication and communication in general are also an important part of globalization. Large-scale immigration, the prosperity of tourism, and the popularization of media communication have all promoted communication and integration among different cultures, forming a diversified cultural pattern. The connotation of globalization has not formed a unified concept in academia. The

IMF defines globalization as a typical representative of economic globalization. The organization noted that "globalization refers to the increase in the scale and form of transnational transactions of goods and services and international capital circulation, as well as the wide and rapid spread of technology to enhance the interdependence of the economies around the world" (International Monetary Fund, 1997). Anthony Giddens believes that globalization is not simply an economic process but also has many social, political, and cultural elements (Anthony Giddens, 2000). Different scholars hold different views regarding the impact of globalization. Martin Wolf believes that globalization refers to the world economy, science, and technology; therefore, globalization is the inevitable trend in human society (Martin Wolf, 2008). Peter Martin and others believe that globalization will expand inequality among countries, thus increasing the global "Matthew effect" (Peter and Lalt, 1998).

2.5.2 Development

Globalization is a trend of the deepening and extensive integration of the world economy, culture, politics, and society since the end of the twentieth century, and it is regarded as one of the main features of the contemporary world. Globalization has increased contacts and interactions among countries and resource and information flow across national boundaries, forming a highly interdependent global system.

Economic globalization is one of the core contents of globalization, which was brought about by the rise of a free-market economy and international trade. With the progress of science and technology and transportation convenience, goods, capital, services, and labor can be exchanged more quickly across national borders. The formation of a global production network, the rise of multinational corporations, and the deepening of the international division of labor are all characteristics of economic globalization. Economic globalization has accelerated the economic complementarity and cooperation among countries and promoted the growth and development of the global economy. However, economic globalization has also brought about many problems, such as the widening gap between the rich and the poor, the unbalanced use of resources, and the spread of financial risks. Cultural globalization is also another important aspect of globalization. With the development of information technology and the popularity of large-scale communication media, the frequency of communication and integration among different cultures is increasing. Global pop culture, the increasing influence of multinational media companies, and the increase in immigration and tourism are all characteristics of cultural globalization. Cultural globalization promotes the spread and communication of cultural diversity and enriches people's life experiences; however, it also causes concerns about cultural homogenization and cultural conflict. In addition, political globalization is reflected in the change of the international political pattern and the rise of international organizations. With the deepening of international cooperation and the development of international organizations, political ties between countries have become increasingly close. Political globalization has brought about the formulation and implementation of international rules and strengthened the capacity for international governance; however, it has also brought about problems such as the weakened power of sovereign states and the rise of nationalism. In addition, globalization also includes social globalization, which involves the flow of people, the spread of social values, and the renationalization of social issues. With the increasing tide of immigrants and refugees, personnel communication between different countries is more frequent. Social globalization has intensified the connection and interaction among different societies and promoted the dissemination and identification of common human values, but it has also brought challenges such as cultural conflicts and identity issues. To sum up, globalization is a complex and diversified phenomenon. In the process of globalization, countries need to strengthen cooperation to jointly tackle various challenges in the economic, cultural, political, and social fields and achieve common development and prosperity.

2.5.3 Research and Applications

The formation of new economic geography can be traced back to the late 1980s and early 1990s, when the process of globalization, technological innovation, and the challenges of local economic development jointly promoted the rise of this school of thought. Globalization is one of the important backgrounds for the formation of new economic geography. During this period, with the rapid development of information technology and the acceleration of international trade, global economic ties became increasingly close, and the rise of multinational corporations and the deepening of international division of labor changed the geographical pattern of economic activities. It is difficult to fully explain these new phenomena and trends; therefore, scholars begin to seek new theoretical and analytical tools to better understand the geospatial economic activities in the context of globalization. With the increasing development of globalization and the continuous development of regional integration, mainstream economics increasingly encountered, unexplained phenomena when explaining the existing regional economic differences. Western economists represented by Krugman tend to discuss the reasons for the development differences of different regions from the perspective of geospatial. In 1991, Krugman introduced the space transportation cost to the trade model and combined the monopoly competition model with the iceberg transportation cost model to discuss the development of urban agglomeration and industrial agglomeration under the condition of integration, thus creating new economic geography.

Neoclassical economics assumes complete competition and constant return on scale; therefore, economic development will eventually converge, even after departing from equilibrium; however, the assumptions are too harsh to explain the general phenomenon. Growth and cumulative effect theory proposes that agglomeration in economic development is self-reinforcing; therefore the initial balance will constantly change. Growth theory can explain economic phenomena but lacks strict logical reasoning. The new economic geography combines the advantages of the one to make up for the shortcomings of the other, both strict mathematical reasoning and clever explanation of the general phenomenon of real economy. In 1996, scholars introduced ecological environment quality and other factors into the study of new economic geography for the first time. In 2004, the annual conference of Cologne University proposed to include environmental factors in the research system of new economic geography. In general, the new economic geography is currently in the early growth stage, and the frontier research topics mainly focus on three aspects: theory, policy, and demonstration. Scholars discuss the ecological civilization construction under the environmental externality of economic space optimization and regional coordinated development problem from the perspective of human and natural life community, the human and natural coupling system, based on the new economic geography theory in the urban agglomeration and industrial transformation multidimensional integration research.

After decades of development, new economic geography research has made continuous progress. Among them, central-peripheral theory, as one of the basic theories of new economic geography, further explains the internal operation mechanism of economic geography agglomeration, pointing out that two adjacent areas will develop into the core and surrounding areas. The theory assumes that there are two regions, and the industries in these two regions are composed of two sectors, namely agriculture and manufacturing. When the crowding effect reaches a certain level, it will stimulate enterprises to migrate to the marginal areas with a small number of competitors, thus breaking the center-periphery equilibrium. The basic equations of the theory are as follows:

$$Y_j = (1 - \mu)\varphi_i + \mu\lambda_i\varpi_j \tag{2.21}$$

$$T_j = \left[\sum_k \lambda_k(\varpi_k D_{ik})^{1-\sigma}\right]^{\frac{1}{1-\sigma}} \tag{2.22}$$

$$\varpi_j = \left[\sum_k Y_k T_k^{\sigma-1} D_{ik}^{1-\sigma}\right]^{\frac{1}{\sigma}} \tag{2.23}$$

$$\varpi_j^* = \varpi_j T_J^{-\mu} \tag{2.24}$$

where Y represents regional income, μ and $1-\mu$ indicate the number of workers and farmers in the region, respectively. φ_i represents the share of agricultural labor as an exogenous variable, and λ_i represents the region's share of the manufacturing labor force at any time. Assuming that agricultural transportation does not have to pay the costs, the transportation cost of the manufacturing industry is reflected as "iceberg transportation"; that is, when a unit of product is transported from region i to region k, only the $\frac{1}{D_{ik}}$ units are accessible. T is the price index. The lower the price index, the lower transportation costs the region maintains with region i, and the larger the share of its manufacturing industry in region i. $\varpi_j and \varpi_j^*$ indicate the nominal and actual wages for the district, respectively.

2.6 Summary

China faces serious environmental and resource challenges and should implement the concept of sustainable development, promote the coordinated development of economic growth and environmental protection, and realize the sustainable development of human beings and nature. Research has shown that progress has been made in describing and explaining the convergence trend of China's economic growth. The theory of resource and environmental carrying capacity involves the concepts of resource utilization, environmental quality, and ecosystem health and aims to assess and explore the upper limit of the pressure that resource and environmental systems can withstand. Regional development convergence theory states that the levels of economic development of different regions converge under certain conditions, with poorer regions growing faster and gradually narrowing the economic gap with richer regions. The SDGs aim to achieve global sustainable development by 2030, including the economic, social, and environmental spheres. The theory and research of sustainable development are dedicated to achieving economic development while maintaining the sustainable use of resources and the environment.

Exercise

1. The environmental and resource challenges facing China and grasp the importance of implementing the concept of sustainable development and promoting the harmonization of economic growth and environmental protection.
2. The description and explanation of the convergence trend of China's economic growth and understand the assessment framework of the theory of resource and environmental carrying capacity and its application in sustainable development.
3. The theory of convergence of regional development and understand the trend of economic development levels in different regions and the mechanism of narrowing the economic gap between poor and rich regions.
4. The importance of the SDGs and familiarize with the content and meaning of the 2030 Global SDGs, including economic, social, and environmental aspects.
5. The content and research history of the theory of sustainable development and grasp the core principles and methods for realizing economic development while maintaining the sustainable use of resources and the environment.

Extended Reading

Todaro, M. P., & Smith, S. C. (2014). *Development economics* (Nie Q., Cheng J., & Wang X., et al., Trans.; 11th ed.). Machinery Industry Press.
Wang, X. (2015). *A study of Singapore's great power balance diplomacy* (1965–2014). Doctoral dissertation. Yunnan University.

References

Arrow, K. J. (1962). The economic implications of learning by doing. *The Review of Economic Studies, 29*(3), 155–173. https://doi.org/10.2307/2295952

Acemoglu, D., Johnson, S., & Robinson, J. A. (2001). The colonial origins of comparative development: An empirical investigation. *American Economic Review, 91*(5), 1369–1401. https://doi.org/10.1257/aer.91.5.1369

Barro, R. J., & Sala-i-Martin, X. (1991). Convergence across states and regions. *Journal of Political Economy, 100*(2), 223–251.

Barro, R. J., & Sala-i-Martin, X. (1992). *Convergence.* NBER Working Papers, 3419. National Bureau of Economic Research.

Barro, R., & Sala-i-Martin, X. (1997). *Technological diffusion, convergence, and growth.*

Baumol, W. J. (1986). *Productivity growth, convergence, and welfare: What the long-run.*

Becker, G. S. (1964). *Human capital: A theoretical and empirical analysis.* Columbia University Press.

Cass, D. (1965). Optimum growth in an aggregative model of capital accumulation. *The Review of Economic Studies, 32*(3), 233–240.

Domar, E. D. (1946). Capital expansion, rate of growth, and employment. *Econometrica, 14*(2), 137–147.

Gerschenkron, A. (1962). *Economic backwardness in historical perspective: A book of essays.* Cambridge, MA: Belknap Press of Harvard University Press.

Harrod, R. F. (1939). An essay in dynamic theory. *The Economic Journal, 49*(193), 14–33.

Lucas, R. E. (1988). On the mechanics of economic development. *Journal of Monetary Economics, 22*(1), 3–42.

Mankiw, N. G., Romer, D., & Weil, D. N. (1992). *A contribution to the empirics of economic.*

Marx, K. (1867). Capital: A critique of political economy (Vol. 1). *Verlag von Otto Meisner.* (Original work published 1867)

Meade, J. E. (1961). *A neo-classical theory of economic growth.* Oxford

North, D. C. (1990). *Institutions, institutional change and economic performance.* Cambridge University Press.

Odum, E. P. (1971). *Fundamentals of ecology.* W. B. Saunders.

Ricardo, D. (1817). *On the principles of political economy and taxation.* John Murray.

Romer, P. M. (1986). Increasing returns and long-run growth. *Journal of Political Economy, 94*(5), 1002–1037.

Samuelson, P. A. (1947). *Foundations of economic analysis.* Harvard University Press.

Schumpeter, J. A. (1942). *Capitalism, socialism and democracy.* Harper & Brothers.

Schultz, T. W. (1961). Investment in human capital. *The American Economic Review, 51*(1), 1–17.

Smith, A. (1776). *An inquiry into the nature and causes of the wealth of nations.* W. Strahan and T. Cadell.

Solow, R. M. (1956). A contribution to the theory of economic growth. *The Quarterly Journal of Economics, 70*(1), 65–94.

Venables, A. J. (1996). Trade policy, cumulative causation, and industrial development. *Journal of Development Economics, 49*(1), 179–197.

Chapter 3
Geographic Differences in Regional Development

Abstract Differences in regional geographic development among countries and regions are one of the research focuses of development geography and a special feature of development geography research. Various indicators for measuring regional geographic development differences can reflect the development differences between different regions within countries and between different countries from different perspectives, and then help analyze the causes and theoretical origins of the regional differences between the western and eastern regions, and between developing and developed countries. The objectives of this chapter are as follows: 1. To learn how to use various indicators such as economy, population, culture, education, and ecological resources to measure geographic development differences between countries and regions. 2. To understand the geographic development differences and their causes by studying the current development status in the economic, population, education, and cultural aspects of representative developing and developed countries. 3. To study the role of resources and environment in regional sustainable development.

3.1 Regional Disparities in Geographic Development

Differences in the geographical development of countries and regions are one of the popular topics of research in today's world, and also one of the distinctive features of the differences in the development of countries and regions in the world. In this chapter, we first study the differences in geographical development among regions in China, and second, we study the current differences in geographical development among developing and developed countries. To quantitatively compare the differences in geographic development between countries and regions, and to illustrate how to measure the differences in geographic development, we select corresponding indicators in terms of the economy, income, population, culture, education, ecological resources, and environment and elaborate on them in detail.

X. Deng and M. Song, *Development Geography*,
https://doi.org/10.1007/978-981-96-6917-2_3

3.1.1 Economic Development

The most intuitive way to measure geographic development differences between different regions is by comparing the economic development of different regions. Gross Domestic Product (GDP) is the sum of the value of all final goods and services produced by a country or region in a given period of economic activity and can effectively measure the economic level of a country.

Gross Domestic Product per capita (GDP per capita) is calculated based on the ratio of the GDP of a country or region in a certain period to the resident population in that period. GDP reflects the overall economic development of a country, whereas GDP per capita is an important symbol to measure the degree of development of a country and people's standard of living. It is an effective tool for people to understand the macroeconomic operation of a country or region.

The total savings rate is the percentage of GDP (or national income) that the total increase in the savings rate of China's residents accounts for over a certain period. The total savings rate is a potential resource for economic growth and is an important indicator reflecting the relative level of savings development in a country or region. When analyzing macroeconomic issues such as investment, consumption, growth, and social security in a country, the total savings rate is an important indicator of concern (Skufina et al., 2015), as well as an important reflection of the economic development of a country or region.

Consumption expenditure refers to all final consumption expenditures on goods and services by resident households over a certain period. In 2021, consumption re-emerged as the first driving force of China's economic growth, with not only the total retail sales of consumer goods growing rapidly but also the proportion of service consumption increasing. There is a more significant positive relationship between per capita consumption expenditure and the variable of heightened industrial structure in the eastern, central, and western regions of China, and per capita consumption expenditure in all regions can promote the heightened and rationalized industrial structure (Egbunike & Emudainohwo, 2018). The indicator of the level of consumption of the population can effectively reflect the economic development of China.

3.1.2 Population Development

As the main body of economic activity, human beings are the main source of labor and vitality for economic and social development. The size of a population can, to a certain extent, reflect the economic development of a region, and the quality of a population can also reflect the development of education in a country. Human resources are a combination of the size and quality of a population, and human capital is the engine of economic and social development of a stronger power and is longer lasting, so that the differences in geographic development in terms of population development can be fully embodied.

The total resident population refers to the population regularly residing in the local area, whether or not they are at home at the standard time of the population survey, and is quantitatively equal to the number of local people registered under the household registration system. The total resident population reflects the economic development and livability of a region or country, and an increase in the resident population in a region or country means that the region has a certain attraction for the population analyzing the differences in the total resident population, the impact of the total population on the differences in geographic development can be derived (Teixeira & Queirós, 2016). By analyzing the differences in the total resident population, the impact of the total population on the differences in geographic development can be derived.

The natural rate of population growth is the ratio of the natural increase in the population over a given period (usually one year) to the average number of people over that period, expressed as a percentage in thousands. The natural population growth rate reflects the trend and speed of population growth, and the study of changes in the natural population growth rate in different regions has an important significance for the consumption structure and economic development, as well as facilitating the study of geographic development differences in various regions.

3.1.3 Cultural Development

Cultural development varies according to economic, social, ethnic, and geographical conditions. Cultural qualities and patterns are different in different countries and regions, and the differences in spiritual culture are particularly significant, which results in the specificity of the cultures of different countries and regions; therefore, through the study of the state of cultural development (Canestrino et al., 2020), it is possible to reflect the geographical development of the differences.

The number and loans of public library books and periodicals can reflect the level of the cultural development of a region; the greater the demand for books by local residents, the higher the level of cultural development of a region.

The number of patent applications filed by residents is an important indicator of whether a country's cultural and technological development activities are active and is also an important indicator of a country's cultural and technological development level.

3.1.4 Resource and Environmental Development

Resources and the environment together constitute the background of regional development, and the resource and environmental carrying capacity is an important carrier for the coordinated development of socio-economic and ecological environment and other composite systems, which is important for the sustainable development of human society (Yadav et al., 2021). To a certain extent, it reflects the differences in geography, and the overall geographic differences are a comprehensive concept.

Total water resources refer to the total amount of surface and subsurface water produced from precipitation. Water resources are fundamental natural resources, as well as strategic economic and social resources (Jiang et al., 2024), supporting the water needs of cities and towns, agriculture, industrial development, residential life, and the maintenance of ecosystems, and are an important foundation for nature and economic and social development (Chen, 2024). Without sufficient water resources as support, the production development of any industry will be stagnant.

Afforestation area refers to the afforestation area on all lands that can be afforested during the reporting period, in which mature trees and shrubs are planted by artificial seeding, seedling planting, aircraft seeding, and other methods, and in which the survival rate reaches 85% or more after inspection and acceptance in compliance with relevant requirements and regulations. In social development, forest resources have become an important ecological resource for environmental protection and regional sustainable development and play an irreplaceable ecological function in maintaining soil and water, preventing wind and fixing sand, and climate regulation (Zhao et al., 2021), which can be measured by the total afforestation area.

Wastewater discharge is mostly harmful water, which not only pollutes water resources and harms aquatic organisms but also destroys our ecosystem. Sulfur dioxide (SO_2) is mainly produced by the combustion of coal in production and life in excess, this exhaust can cause serious threats to people and animals as well as nature, reduce life expectancy, increase mortality, increase the probability of respiratory diseases, and reduce the firm productivity, and it also widens the income gap through brain drain (Raimi et al., 2021). Sulfur dioxide is one of the most important atmospheric pollutants and an important indicator of national or regional ecological development.

3.2 Regional Development Differences in China

The division of China's regional geography can be divided in different ways, taking into account various considerations such as research focus, comprehensiveness, and comparability. For the division between the east and west of China, it is a division based on economic factors on the policy side. The eastern part is the region with a higher level of economic development, the central part is the sub-developed region, whereas the western part is the less developed region. The division began in 1986 with the official announcement of the Seventh Five-Year Plan adopted by the National People's Congress at its sixth and fourth sessions, and after revisions in 1997 and 2000, the eastern region consists of Liaoning, Beijing, Tianjin, Hebei, Shandong, Jiangsu, Shanghai, Zhejiang, Fujian, Guangdong, and Hainan. The central region comprises Shanxi, Jilin, Heilongjiang, Anhui, Jiangxi, Henan, Hubei, and Hunan. The western region comprises Shaanxi, Gansu, Qinghai, Ningxia, Xinjiang, Sichuan,

Chongqing, Yunnan, Guizhou, Tibet, Guangxi, and Inner Mongolia. The above division does not include Hong Kong, Macao, and Taiwan. For the current status of development differences between east and west China, we start from the indicators that measure the geographical development differences and detail the current status of development differences between east and west China through comparative analyses and coefficients of variation.

3.2.1 Differences in Economic Development

To explore the current status of economic development differences in China, we choose two indicators (GDP per capita and consumption level) that are more representative of the economic level to provide a detailed description of the economic development differences between the eastern and western regions of China in recent years.

Because of the limitations of GDP as an indicator of economic level, we use the GDP per capita indicator to provide a more detailed description of the current situation of economic development differences in China. GDP per capita can illustrate the productivity level of a region, the level of labor productivity, and the economic situation of a region's development, which helps to understand the macroeconomic operation of a particular region.

As the economy continues to develop, the differences in GDP per capita between regions in China are gradually decreasing. In 2007, the coefficient of variation of GDP per capita by region was 0.53, and by 2018, it had dropped to 0.38, illustrating that with the aid of a series of policies, the differences in the economic development of the regions have been on a markedly narrowing trend (Table 3.1).

The pulling effect of consumption on China's economic growth is generally weak, whereas consumption demand directly affects the improvement of the economic level. The ultimate goal of any production activity is to realize consumption, which is the ultimate force driving economic growth, and the transformation of consumption potential into actual consumption capacity is the goal and destination of national economic growth.

If income and consumption can grow in harmony, it will promote the healthy and stable development of the economy; therefore, in the economic background of the "new normal", improving consumption, especially the level of consumption of

Table 3.1 Variation coefficient of GDP per capita from 2007 to 2018

Year	Variation coefficient	Year	Variation coefficient	Year	Variation coefficient
2007	0.53	2011	0.41	2015	0.38
2008	0.49	2012	0.39	2016	0.38
2009	0.48	2013	0.38	2017	0.39
2010	0.45	2014	0.37	2018	0.38

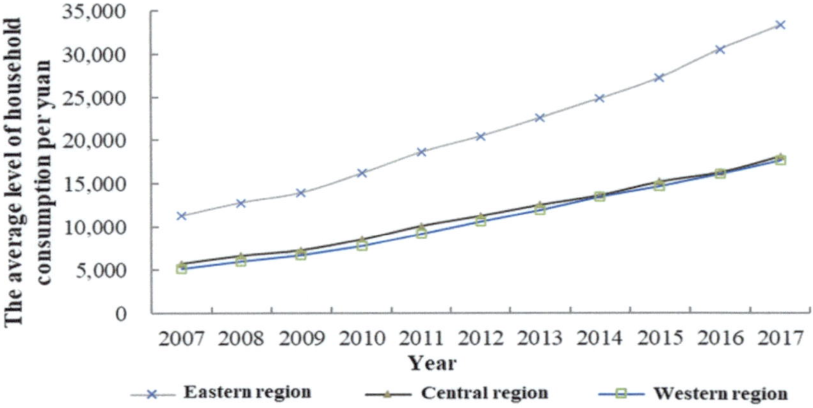

Fig. 3.1 Average consumption level of the eastern, central, and western regions

residents, can effectively promote the stable growth of the economy; therefore, in studying the differences in the development of the domestic economy, the cited resident consumption level indicator can effectively reflect the economic development situation.

According to the National Bureau of Statistics and the China Statistical Yearbook, the average value of consumer water in the central and eastern regions from 2007 to 2017 is calculated and plotted as a line graph (Fig. 3.1).

Overall, the consumption level of residents in the eastern region is higher than that in the central and western regions, and the average value of consumer water in both the central and western regions shows a rising trend. The average value of consumer water rose at different rates in each region. The growth rate of the consumption level of residents in the eastern region is slightly lower than that in the central and western regions, the growth rate of the consumption level of residents in 2007–2017 showed a downward trend in each region. The central and western regions have a slower rate of decline than that of the eastern region, and the difference between the consumption level of the eastern region and the central and western regions is narrowing.

The analysis shows that the economic level of the eastern region is ahead of the central and western regions; however, the economic development of the central and western regions has gradually accelerated significantly, and the economic difference with the eastern region is gradually narrowing.

3.2.2 Differences in Population Development

As the mainstay of economic activity, human beings are the main source of labor and vitality for economic and social development. To a certain extent, the size of the population can reflect the economic development of a region, thus reflecting the differences in geographical development.

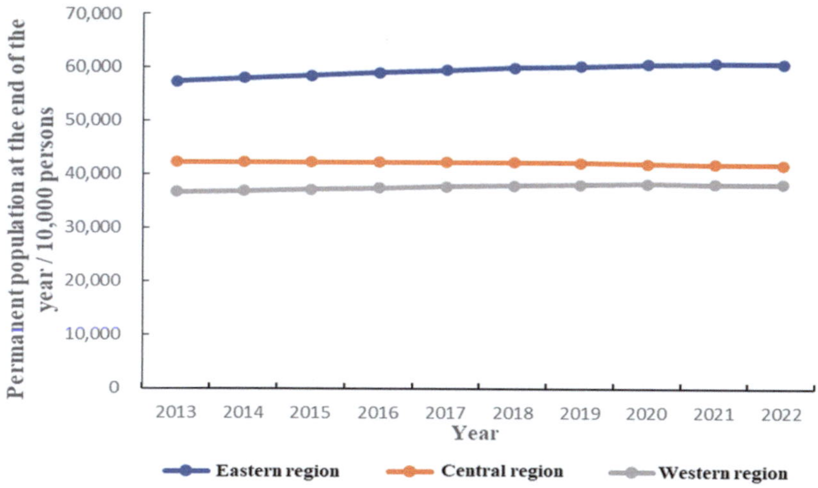

Fig. 3.2 Permanent resident population at the end of the year 2013–2022 in the eastern, central, and western regions

Based on relevant data from the National Bureau of Statistics and the China Statistical Yearbook, we collated and obtained the total year-end resident population of the central and eastern regions from 2013 to 2022 and drew a line graph of the total year-end resident population (Fig. 3.2).

Overall, the total number of permanent residents in the eastern region is higher than that in the central and western regions, with the number of permanent residents in the eastern region showing an overall upward trend between 2013 and 2022, the number of permanent residents in the central region showing a downward trend in the same period, and the number of permanent residents in the western region showing an upward trend in the same period.

Figure 3.3 compares the year-end resident population numbers of selected regions of China in 2013 and 2022, and the results show that the data for the two years basically overlap, with no significant change in the resident population numbers of most areas.

The natural population growth rate refers to the ratio of the natural increase in population over a certain period (usually one year) to the average number of people in that period, expressed as a thousandth of a cent, and it reflects the trend and speed of population growth.

As the mainstay of the economic structure, it affects people, consumption, and the economy. A rise in the natural rate of population growth leads to an increase in consumption at the micro level. Therefore, studying the changes in the natural population growth rate in different regions is of great significance to the consumption structure and economic development, as well as facilitating the study of geographic development differences in various regions.

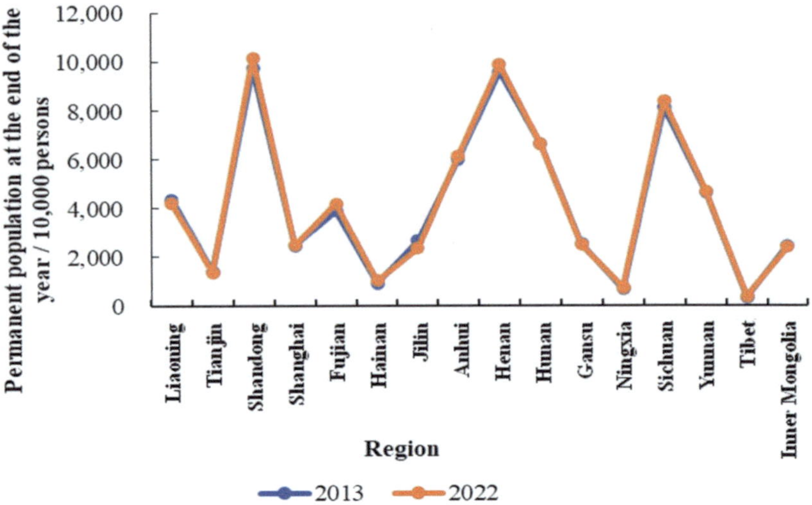

Fig. 3.3 Permanent resident population at the end of 2013 and 2022 by region

Based on relevant data from the National Bureau of Statistics and the China Statistical Yearbook, we compiled the average value of the natural population growth rate in the central and eastern regions for the period 2008–2022 (Table 3.2).

The comparison found that in the period 2008–2019, the average natural population growth rate of the western region is much higher than that of the central and eastern regions, fluctuating above and below 7‰, which is more stable, whereas the average natural population growth rate of the central and eastern regions is not similar, fluctuating above and below 4.5‰. In the period 2021–2022, the average value of the natural population growth rate of the central and eastern regions declines; however, the western region's average natural growth rate is still much higher than that of the central and eastern regions. In the period 2021–2022, the average natural population growth rate in the central and eastern regions decreases; however, the average

Table 3.2 Eastern, central, and western regions' 2008–2022 average rate of natural population growth (‰)

Year	Eastern region	Central region	Western region	Year	Eastern region	Central region	Western region
2008	4.59	4.57	6.86	2015	4.27	4.42	6.51
2009	4.68	4.73	6.78	2016	5.40	4.55	6.83
2010	4.54	4.97	6.64	2017	5.56	4.89	7.09
2011	4.33	4.58	6.49	2018	4.28	4.04	6.12
2012	4.77	4.66	6.61	2019	3.85	3.19	5.57
2013	4.42	4.63	6.49	2021	0.42	−1.05	2.11
2014	4.94	4.70	6.61	2022	−0.58	−1.98	1.25

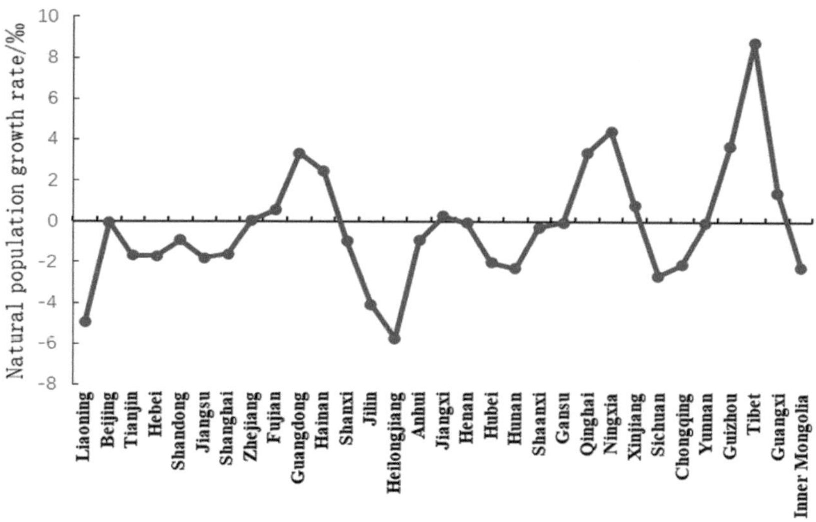

Fig. 3.4 Natural population growth rate in selected regions in 2022

natural population growth rate in the western region is still much higher than that in the central and eastern regions. In recent years, the natural population growth rate in the eastern region has been slightly higher than that in the central region. A line graph of the natural population growth rate in 2022 for each region of China is shown in Fig. 3.4.

A comparison of the natural population growth rate charts of various regions in China shows that the population growth rate of the central and eastern regions is smaller than that of the western regions, with the natural growth rate of most of the regions concentrating around 0 per thousand. The natural population growth rate in the western region is generally concentrated between 2 and 5%, with the natural growth rate in Tibet reaching exceeding 8%.

The high level of economic development and inherent geographical advantages of the eastern region have caused part of the population to migrate to the eastern region, resulting in a much higher number of permanent residents in the eastern region than that in the central and western regions, whereas the natural growth rate of the population in the western region exceeds that in the central and eastern regions.

3.2.3 Differences in Cultural Development

Culture is the foundation of a country's development. To explore the current situation of the differences in cultural development in China, we chose two indicators that are more representative of the level of cultural development (the number of public library collections per capita and the frequency of lending of public library books and periodicals), to describe in detail the differences in cultural development between the eastern and western parts of China in recent years.

To study the differences in the cultural development of various regions in China, we first looked up the indicator of the per capita ownership of public library collections, using data from the National Bureau of Statistics and the China Statistical Yearbook, and since the data before 2011 were not available, we collected the data for the period 2011–2022, and collated to obtain the average value of the indicator of per capita ownership of public library collections in the central, eastern and western regions for the period 2011–2022 (Table 3.3).

A comparison showed that the average value of public library collections per capita in the eastern region is much higher than that in the central and western regions. The average value of collections in the eastern region showed a fluctuating upward trend, with a fall in 2012–2013, and basically showed a straight upward trend after 2013. The average per capita ownership of public library collections in the central and western regions was closer, and the development was more stable, with a slow rate of increase; however, the western region was more advantageous than the central region. The per capita library collection in the eastern region is about twice as large as that in the central region, and by 2017, the growth rate of the central and western regions exceeded that of the eastern region, and the difference in library collections between regions was narrowing.

To explore the differences in the mean value of per capita owned public library collections across regions, the concept of coefficient of variation is introduced. The calculated difference in the mean value of public library collections per capita in each region shows a trend of increasing and then decreasing, and the difference in library collections between the central and western regions is gradually decreasing (Fig. 3.5).

The number of public library books and periodicals issued can reflect reading culture. To compare the reading culture in different regions, data on the number of books and periodicals borrowed from libraries in various regions across China were collected, and the average value of the frequency of borrowing books and periodicals

Table 3.3 Average number of public library collections per capita (volumes/person), eastern, central, and western regions, 2011–2022

Year	Eastern region	Central region	Western region	Year	Eastern region	Central region	Western region
2011	0.86	0.38	0.45	2017	1.09	0.51	0.60
2012	0.98	0.42	0.48	2018	1.16	0.54	0.61
2013	0.89	0.40	0.48	2019	1.23	0.57	0.64
2014	0.93	0.41	0.52	2020	1.26	0.65	0.68
2015	0.98	0.44	0.54	2021	1.33	0.71	0.72
2016	1.04	0.47	0.57	2022	1.41	0.77	0.77

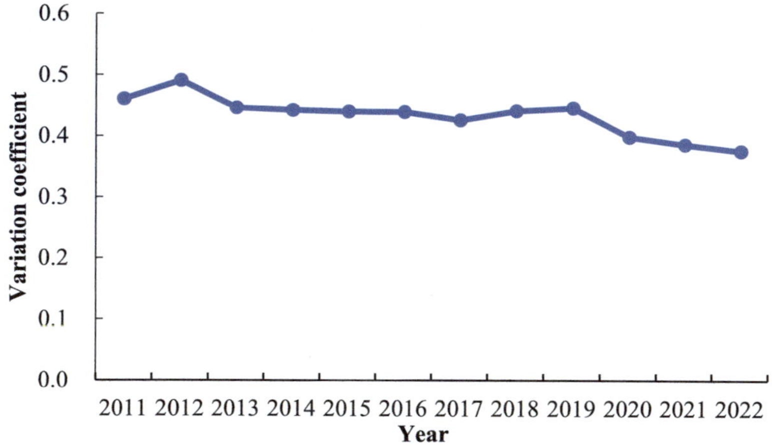

Fig. 3.5 Variation coefficient of per capita collections in the central and western regions, 2011–2022

from public libraries and its growth rate in the central and eastern regions from 2008 to 2022 were compiled (Table 3.4).

The table shows that the frequency of literature lending in the eastern region is much higher than that in the central and western regions. From 2008 to 2017, the frequency of book and periodical lending in the central and eastern regions doubled and 2017 to 2022, the frequency decreased first and then increased. A comparison shows that the frequency of book and periodical lending in Jiangsu in the eastern region is the highest, and it increases annually, followed by the lending frequency in Shandong, Zhejiang, and Guangdong (approximately 20 million times). In the central region, the frequency of book and periodical lending in Anhui, Henan, and Hubei is higher (more than 10 million times), close to the average level of the eastern region. The lending frequency in Shanxi, Jilin, and Heilongjiang is low, and is to be further developed. In the western region, in addition to Sichuan, the frequency of book and periodical lending in Sichuan has decreased and then increased. In the western region—except for Sichuan, which has a high frequency of book and periodical lending—the lending frequency in other regions is relatively low, and the western region should urgently strengthen its cultural development and narrow the gap with the eastern region.

3.2.4 *Differences in Educational Development*

As a common development goal pursued by countries worldwide, education has an important influence on comprehensively improving the national quality and the scientific and technological levels, as well as enhancing China's international competitiveness and influence. The development of education can also differ because of

Table 3.4 Average number of borrowings of books and other documents from public libraries and their growth rates in the eastern, central, and western regions, 2008–2022

Year	Eastern region		Central region		Western region	
	Number of borrowings of books and periodicals from public libraries per 10,000 persons	Annual growth rate/per cent	Number of borrowings of books and periodicals from public libraries per 10,000 persons	Annual growth rate/per cent	Number of borrowings of books and periodicals from public libraries per 10,000 persons	Annual growth rate/per cent
2008	653.44	–	358.55	–	177.96	–
2009	706.79	8.17	360.42	0.52	211.31	18.74
2010	714.48	1.09	418.83	16.21	220.26	4.23
2011	780.19	9.20	471.34	12.54	244.36	10.94
2012	855.11	9.60	576.69	22.35	281.87	15.35
2013	1068.43	24.95	626.46	8.63	315.62	11.97
2014	1188.72	11.26	667.35	6.53	359.14	13.79
2015	1215.63	2.26	689.27	3.28	346.26	−3.59
2016	1333.18	9.67	724.31	5.08	366.17	5.75
2017	1277.29	−4.19	827.33	14.22	399.51	9.11
2018	1263.90	−1.05	891.625	7.77	395.16	−1.09
2019	1295.90	2.53	951.50	6.72	392.25	−0.74
2020	888.81	−31.41	585.50	−38.47	249.83	−36.31
2021	1207.00	35.80	875.71	49.57	316.91	26.85
2022	1256.54	4.11	896.12	2.33	323.67	2.13

geographical differences, and the differences in geographical development can be reflected by studying the status of educational development in different regions. To explore the current situation of the differences in the development of education in China, two indicators that are more representative of the level of education development (investment in education and years of education per capita) are chosen to describe in detail the differences in the development of education in the eastern and western regions of China in recent years.

Education investment is the most fundamental material guarantee of educational development. Therefore, the basic situation of education development in China can also be reflected by studying the investment in education in China in recent years. To study the differences in education development in each region, we select the data on education investment in each region in China from 2007 to 2021, obtained from the National Bureau of Statistics and the China Statistical Yearbook. The data for 2013 is replaced by the average value of education investment in 2012 and 2014, and the data on education investment in the central and eastern regions from 2007 to 2021 are obtained through the collation of education funding input and annual growth rate data (Table 3.5).

Table 3.5 Investment in education and annual growth rates in the eastern, central and western regions, 2007–2021

Year	Eastern region		Central region		Western region	
	Education expenditure input (billion)	Annual growth rate (%)	Education expenditure input (billion)	Annual growth rate (%)	Education expenditure input (billion)	Annual growth rate (%)
2007	4596	–	2157	–	1900	–
2008	5577	21.33	2720	26.13	2489	30.95
2009	6413	14.99	3329	22.40	3190	28.17
2010	7096	10.66	3744	12.46	3852	20.75
2011	8433	18.84	4397	17.43	4574	18.75
2012	10,348	22.70	5576	26.83	5610	22.66
2013	11,761	13.66	6410	14.95	6514	16.11
2014	13,174	12.02	7244	13.01	7418	13.88
2015	14,115	7.14	7560	4.37	7921	6.78
2016	15,253	8.06	8384	10.89	9019	13.87
2017	16,450	7.85	9024	7.63	9792	8.56
2018	19,809	20.42	10,696	18.53	11,471	17.15
2019	22,001	11.07	11,617	8.61	12,394	8.05
2020	22,308	1.40	12,261	5.54	13,300	7.31
2021	25,395	13.84	12,782	4.25	13,854	4.17

The data in Table 3.5 show that the overall annual investment in education funding in the eastern region is significantly higher than that in the central and western regions, and investment in education funding in the central and western regions is generally roughly the same, and the investment in education funding in both the central and eastern regions shows an upward trend.

Although the input of education expenditure in the central and western regions is rising every year, the data on the annual growth rate show a trend of slowing down in general, except for the annual growth rate of education expenditure in 2011, 2012, 2016, and 2018, which shows an increase compared with the previous year. The annual growth rate of education expenditure in the rest of the years is lower than that in the previous year. In addition, the calculation of the average annual growth rates of the central and eastern regions by the geometric mean method reveals that the average annual growth rates of the eastern, central, and western regions are 12.99%, 13.55%, and 15.25%, respectively, with the western region having the highest average annual growth rate.

As reflected in the education expenditure of each region, the level of education development in the eastern region is superior to that in the central and western regions; however, the annual growth rate of education expenditure in the central and western regions is relatively fast, and the level of education development in the central and western regions is rising, gradually narrowing the gap between them and the superior eastern region.

The number of years of education per capita refers to the number of years of academic education per capita of a certain population group, which is measured by weighting and summing the corresponding education levels and dividing by the corresponding total population. This indicator has been used by many scholars to measure the overall education level of a region; therefore, by studying the number of years of education per capita in each region of China, we can also understand the differences in the level of education development in different regions.

Figure 3.6 shows the years of education per capita in the central and eastern regions and the whole country from 2007 to 2022. As the figure shows, the years of education per capita in the eastern region are generally higher than those in the central and western regions and the national average, whereas the years of education per capita in the central region are roughly the same as the national average, and those in the western region are the lowest. Overall, the number of years of education per capita has risen in both the central and western regions. Having achieved nine-year compulsory education, the per capita number of years of education in the western region has risen from 7.397 to 9.774, meeting the requirement for nine-year compulsory education, and the per capita number of years of education nationwide has risen from 8.186 in 2007 to 10.143 in 2022. Nine-year compulsory education has been achieved nationwide. In terms of the disparity and trend of change in the number of years of education per capita by region, the gap between the eastern and the central regions has narrowed, the gap between the central and the western regions has decreased, and, similarly, the gap between the eastern and western regions has gradually narrowed; therefore, roughly the same development trend of the gap narrowing is observed in the three regions.

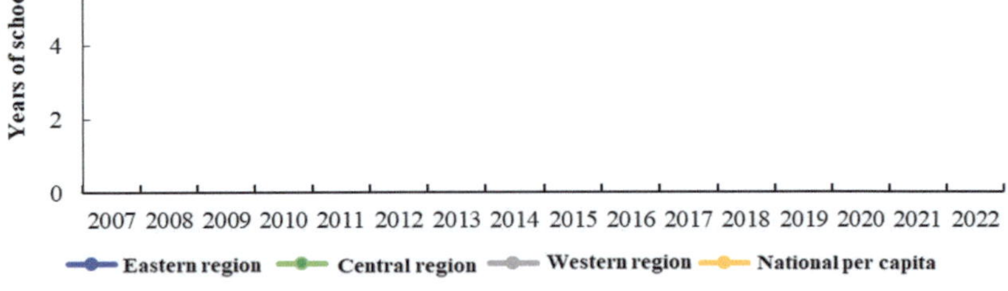

Fig. 3.6 Years of schooling per capita at the national level and in the eastern, central, and western regions, 2007–2022

As the differences in the level of education development between regions show, as reflected in the number of years of education per capita in each region, the level of education development in the eastern region is still higher than that in the central and western regions; however, the differences are gradually narrowing.

The above analysis reveals that the level of education development in the eastern region is greater than that in the central and western regions. However, over time and with the support of national policies, the pace of education development in the central and western regions has accelerated significantly, and the difference in education development between the eastern region and the central and western regions is gradually diminishing.

3.2.5 *Environmental Resources*

Environmental resources are highly important for the sustainable development of human society. The development of environmental resources also varies according to geography, and the study of the development of environmental resources in different regions can also reflect the differences in geographic development. To investigate the current situation of the differences in the development of environmental resources in China, two indicators (per capita water resources and total afforestation area), which are more representative of the level of development of environmental resources, are selected to analyze the differences in the development of environmental resources in the central and eastern regions of China in recent years.

As stated earlier, the total amount of water resources in all regions of China is relatively abundant in general; however, China's population base is large, and the per capita ownership is very small, which is less than a quarter of the world's average; therefore, we introduced the indicator of per capita water resource ownership to analyze the differences in ecological resources reflected by the differences in the water resources of each region. We have compiled data on water resources per capita for the whole country and the central and eastern regions from 2007 to 2022 (Table 3.6).

A comparison showed that in the period 2007–2022, the per capita water resource ownership in the western region is much higher than that in the central and eastern regions and in the whole country, with a high of 8355.91 m^3 per capita in 2010 and a low of 6146.46 m^3 per capita in 2022, while the per capita water resource ownership in the central and eastern regions basically stabilizes at 1000–3000 m^3 per capita, which is less than one-third of that in the western region. Meanwhile, the development trend of per capita water resource ownership in the east-central region is more or less the same. To further illustrate the development trend of per capita water resource ownership in the central and east-central regions, we calculated the annual growth rate of per capita water resource ownership in the central and east-central regions for the period 2008–2022 and plotted a line graph (Fig. 3.7).

The per capita water resource ownership in the eastern and central regions showed negative growth in several years, with the lowest annual growth rates of −32.52%

Table 3.6 Water resource per capita (m^3/person) for the whole country and the central, eastern, and western regions, 2007–2022

Year	Eastern region	Central region	Western region	Total	Year	Eastern region	Central region	Western region	Total
2007	1304.59	1265.99	7810.62	1916.34	2015	1598.10	1606.41	6674.59	2039.20
2008	1532.77	1304.76	8212.19	2071.05	2016	1988.98	1938.16	7825.99	2354.90
2009	1408.21	1251.32	7426.17	1816.18	2017	1517.96	1385.50	7620.14	2074.50
2010	1911.00	1893.58	8355.91	2310.41	2018	1062.69	1542.98	7369.21	1971.85
2011	1289.54	1116.27	7455.45	1730.20	2019	1073.32	1957.52	6997.14	2077.75
2012	1764.01	1717.39	7667.26	2186.20	2020	903.75	2450.89	7057.84	2239.80
2013	1780.89	1652.88	7675.17	2059.70	2021	1074.12	2044.68	6677.79	2098.49
2014	1539.12	1461.88	7520.55	1998.64	2022	1110.51	1841.56	6146.46	1918.18

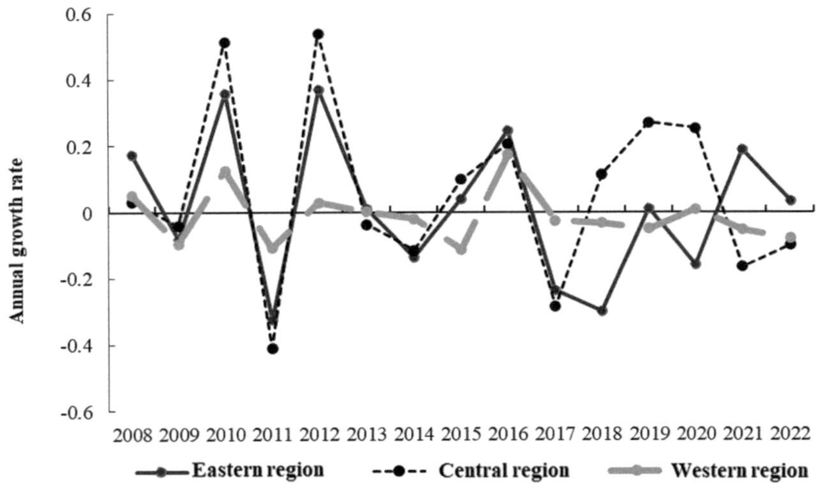

Fig. 3.7 Annual growth rate of water resources per capita in the eastern, central, and western regions, 2008–2022

and −41.05%, respectively, in 2011, and the lowest growth rate of per capita water resource ownership of −11.25% in the western region in 2015. In addition, the geometric mean method was used to obtain the average annual growth rates of −1.07%, 2.53%, and −1.58% for the three regions, respectively.

The inter-regional differences in ecological resources development level reflected by water resources in each region indicate that the ecological resources development level of the western region is superior to that of the central and eastern regions; however, the annual growth rate of water resources in the central and eastern regions has been faster in recent years; therefore, the gap between the ecological resources development level of the central and eastern regions and that of the advantaged regions in the west is gradually narrowing.

Table 3.7 Total afforestation area and annual growth rate, 2007–2022, eastern, central, and western regions

Year	Eastern region		Central region		Western region	
	Total afforestation area/10^3hm^2	Annual growth rate (%)	Total afforestation area/10^3hm^2	Annual growth rate (%)	Total afforestation area/10^3hm^2	Annual growth rate (%)
2007	634.7	–	811.6	–	2361.4	–
2008	798.5	25.81	1300.8	60.28	3175.1	34.46
2009	837.8	4.92	1557.8	19.76	3840.4	20.95
2010	946.9	13.02	1485.58	−4.64	3427.7	−10.75
2011	1228	29.69	1504.87	1.30	3245.74	−5.31
2012	1123.81	−8.48	1506.87	0.13	2951.79	−9.06
2013	1188.83	5.79	1711.37	13.57	3179.88	7.73
2014	1125.61	−5.32	1698.56	−0.75	2712.12	−14.71
2015	1629.97	44.81	2156.16	26.94	3877.89	42.98
2016	1539.65	−5.54	1833.15	−14.98	3794.17	−2.16
2017	1420.61	−7.73	2125.85	15.97	4110.73	8.34
2018	1539.58	8.37	2094.65	−1.47	3641.72	−11.41
2019	1496.58	−2.79	2220.79	6.02	3648.53	0.19
2020	1452.29	−2.96	1982.82	−10.72	3471.12	−4.86
2021	471.24	−67.55	1289.19	−34.98	1965.51	−43.38
2022	599.35	27.19	1455.33	12.89	2134.84	8.62

To analyze the difference in ecological resource development reflected by the total afforestation area, we compiled data on the total afforestation area and its annual growth rate in the eastern, central, and western regions from 2007 to 2022, using raw data from the National Bureau of Statistics and the China Statistical Yearbook (Table 3.7).

The comparison revealed that the total area of afforestation in the eastern and central regions is much smaller than that in the western region, and the total area of afforestation in the central and eastern regions, overall, shows an upward and then a downward trend, and the growth rate of each region fluctuates and changes over time.

3.2.6 *Differences in Environmental Resources*

The differences in geographical development can be understood by studying the environmental conditions of different regions. To explore the status quo of the differences in environmental resources between regions in China, we chose two indicators, total wastewater discharge and SO_2 emission, to describe the differences between the environment of the eastern and western regions in recent years.

Wastewater not only pollutes water resources and harms aquatic organisms but also destroys our ecosystem. We selected the data of total wastewater discharges in each region of China from 2007 to 2017, which were obtained from the National Bureau of Statistics and the China Statistical Yearbook, and calculated the average value of total wastewater discharges and the annual growth rate of the central and eastern regions from 2007 to 2017 (Table 3.8).

The comparison shows that the total annual wastewater discharge in the eastern region is significantly higher than that in the central and western regions, whereas the total wastewater discharge in both the central and eastern regions shows an upward trend. Overall, the growth rate of wastewater emissions in the central and western regions is slower, and the wastewater emissions in the eastern region are under control compared with the central and western regions.

Sulfur dioxide, as one of the most important atmospheric pollutants, causes serious harm to humans and animals, as well as to nature. To study the differences in SO_2 emissions between the three regions in China, we calculated the average value of SO_2 emissions in the eastern, central, and western regions and plotted it as a line graph (Fig. 3.8).

As shown in the figure, the difference in SO_2 emissions between the central and eastern regions is relatively small, with emissions in the central region slightly higher than those in the eastern and western regions. The emissions in both the central and eastern regions show a decreasing trend over time. The eastern and central regions

Table 3.8 Average value and growth rate of total wastewater discharges in eastern, central, and western China, 2007–2017

Year	Eastern region		Central region		Western region	
	Total wastewater discharge /$10^4 t$	Annual growth rate (%)	Total wastewater discharge /$10^4 t$	Annual growth rate (%)	Total wastewater discharge /$10^4 t$	Annual growth rate (%)
2007	269,660.68	–	177,892.68	–	98,257.10	–
2008	273,977.73	1.60	181,458.13	2.00	104,281.75	6.13
2009	285,287.09	4.13	189,111.88	4.22	103,318.68	−0.92
2010	301,678.36	5.75	199,276.25	5.37	104,991.00	1.62
2011	321,027.01	6.41	221,414.00	11.11	107,442.78	2.34
2012	328,823.79	2.43	233,749.73	5.57	113,379.39	5.53
2013	332,996.01	1.27	237,007.14	1.39	116,284.96	2.56
2014	342,981.19	3.00	241,493.31	1.89	121,417.58	4.41
2015	351,095.27	2.37	248,249.05	2.80	125,432.20	3.31
2016	342,741.38	−2.38	226,507.02	−8.76	127,395.21	1.56
2017	329,180.04	−3.96	225,025.87	−0.65	131,285.21	3.05

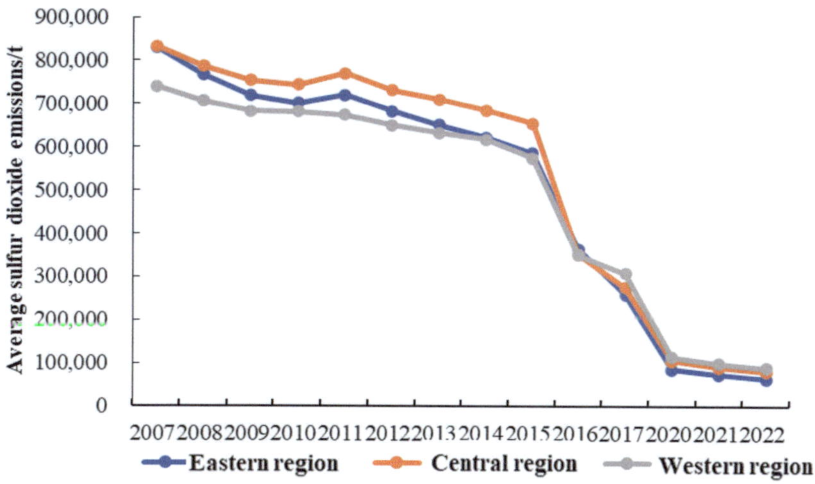

Fig. 3.8 Average SO_2 emissions in the central and eastern parts of China, 2007–2022

are better than the western region in controlling SO_2 emissions, and the difference between the three regions is further reduced.

Further analysis of the 2022 SO_2 emissions from various regions in China (Fig. 3.9) reveals that the SO_2 emissions from the eastern regions of Beijing, Tianjin, Shanghai, Hainan, and Tibet are much less compared to those from other cities, probably due to policy control. The SO_2 emissions from Hebei, Liaoning, Shandong, Shanxi, Yunnan, and Inner Mongolia are very high, and the environmental pollution in these regions is more serious and the policy aspect should be prioritized to exercise some control over the environment to achieve sustainable development.

3.3 Differences in Development Between Developing and Developed Countries

The United Nations has broadly classified the countries of the world into three categories based on a combination of economic, social, scientific, and technological factors: least developed countries (LDCs), developing countries, and developed countries. A study of the geographical developmental differences between developing and developed countries helps to understand the differences in their development and find ways to reduce them.

To this end, we selected several developed and developing countries from the seven continents of the world at different latitudes to illustrate the differences in various aspects of development between developed and developing countries. The selected countries are: developed countries (Japan, Germany, the United Kingdom,

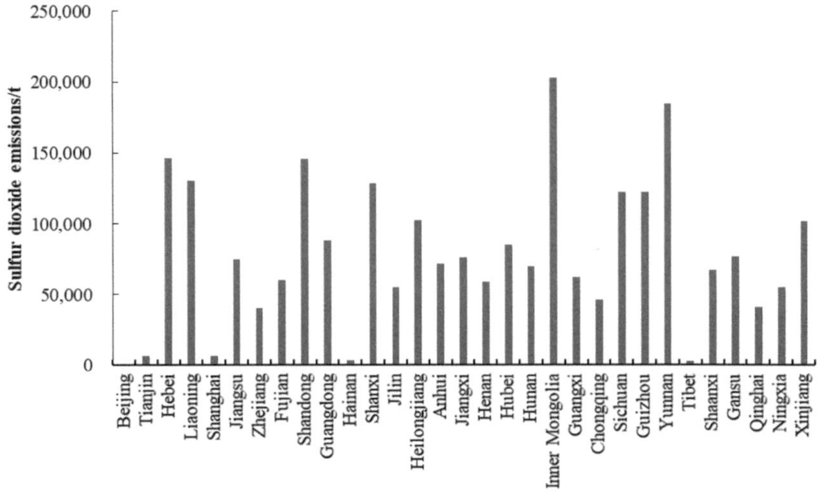

Fig. 3.9 Sulfur dioxide emissions in China in 2022 by region

France, the United States, Canada and Australia), and developing countries (China, India, Malaysia, Russia, Mexico, Brazil, Egypt, and South Africa).

3.3.1 Differences in Economic Development

The most distinctive feature of the global economy is the great diversity of economies among countries, and in the case of developing and developed countries, the diversity of economic development is reflected in the two dimensions of GDP and the aggregate savings rate.

GDP refers to the total outcome of productive activities and is a measure of the total value of all final goods produced by both domestic and non-domestic residents of a country. To compare the differences in economic development as measured by GDP between developing and developed countries in the current period, we used data from selected developed and developing countries, sourced from the International Statistical Yearbook for the period 2018–2022, measured the level of development of developing and developed countries in the current period using the five-year average of GDP, and collated the data as shown in Table 3.9.

In recent years, the economies of developed and developing countries have varied considerably, with the US having the highest current value of GDP in numerical terms, accounting for 24.3% of the world's GDP, followed by China with 17.7% of the world's GDP. Overall, the level of GDP of developed countries is higher than that of developing countries. In the case of the selected representative countries, most of the developed countries, except for the US, have a present value of GDP in the range of 1 trillion to 5 trillion USD, whereas the developing countries, except for the

Table 3.9 Average GDP of developing and developed countries, 2018–2022 ($ billion)

Region	GDP	Developed country	GDP	Developing country	GDP
World	915,251.1	Japan	48,684.3	China	161,756.4
High-income countries	572,174.5	Germany	40,046.0	India	29,929.6
Middle-income countries	335,350.2	United Kingdom	29,415.4	Malaysia	3635.3
Middle-lower-income countries	71,764.7	France	27,761.5	Russia	17,520.4
Middle-upper-income countries	263,588.5	United States	222,649.5	Mexico	12,575.6
Lower middle-income countries	339,948.3	Canada	18,515.1	Brazil	17,454.5
Low-income countries	4589.2	Australia	15,370.9	Egypt	3536.6
				South Africa	3908.3

higher GDP of China, have a slightly higher level of GDP in India, Russia, Mexico, and Brazil, and the rest of the world has a lower level of GDP. This shows that there are some differences among developed countries and among developing countries.

Comparing the GDP values of China and the US in recent years, although the total volume of the US is much larger than that of China, the gap between China and the US is gradually narrowing, and the GDP and annual growth rates of the two are shown in Table 3.10.

While a high savings rate has always been considered to lead to high economic growth, Keynes put forward the famous "paradox of thrift", that is, a high savings rate may lead to a slowdown in international economic development. Solow's neoclassical growth theory suggests that changes in the savings rate would only temporarily affect economic growth; however, in Solow's model, the conversion of savings into investment is the single most important precondition for the contribution of high savings to economic growth.

Table 3.10 Present value and annual growth rates of GDP in China and the US, 2018–2022

Year	United States		China	
	GDP ($ billion)	Annual growth rate (%)	GDP ($ billion)	Annual growth rate (%)
2018	205,331	–	143,119	–
2019	213,810	4.13	146,982	2.70
2020	210,605	−1.50	150,583	2.45
2021	233,151	10.71	181,332	20.42
2022	250,352	7.38	186,766	3.00

Note GDP data in the table are in nominal terms

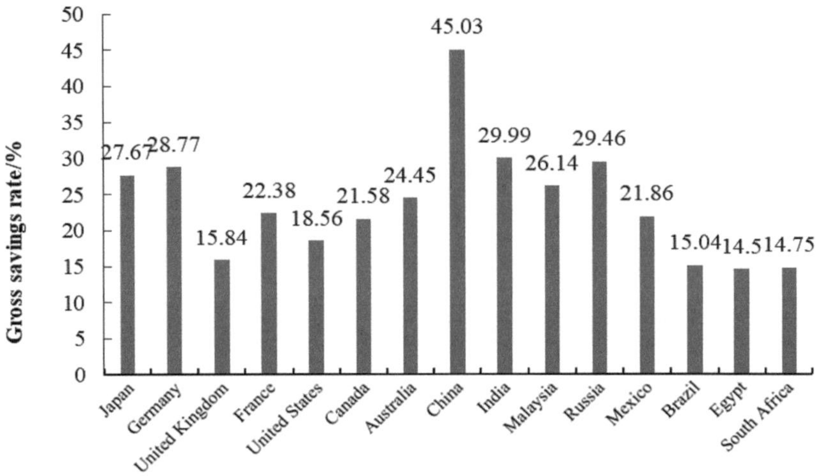

Fig. 3.10 Current gross savings rates in developing and developed countries

To compare the differences in economic development as measured by the gross savings rate between developing and developed countries in the current period, the average gross savings rate for selected developed and developing countries for the five-year period 2018–2022 was calculated and plotted as a bar chart (Fig. 3.10), using data from the World Bank.

The comparison reveals that the difference between the current gross savings rate of developing countries and that of developed countries is relatively small. Overall, the total savings rate in developing countries is slightly higher than that in developed countries, as shown by the fact that China's total savings rate reaches 45.03%, which is more than twice that of the United States and about three times that of the United Kingdom.

Further comparing the development of the total savings rate between China and the United States in 2018–2022 (Table 3.11), China's total savings rate has been consistently high, whereas that of the United States is less than 20%, and China's total savings rate is more than twice as high as that of the United States. The savings rate of developed countries such as the United States has been kept at a very low level, probably because these countries had a rather long period of capital accumulation after the industrial revolution and had enough capital to sustain their development, whereas China's capital accumulation mainly relies on savings, and only by maintaining a high savings rate can it ensure investment and thus promote economic growth.

A huge difference still exists between the stage of economic development of China and the United States, and comparing the economic indicators of the two countries in recent years does not have much value. Therefore, we selected the economic data of the United States at the stage of accelerated industrialization and compared it

Table 3.11 Total savings rate and annual growth rate of China and the United States (2018–2022)

Year	China		United States	
	Gross saving rate (%)	Annual growth rate (%)	Gross saving rate (%)	Annual growth rate (%)
2018	44.68	–	19.02	–
2019	43.90	−1.76	19.29	1.42
2020	44.28	0.88	18.48	−4.21
2021	45.79	3.41	17.80	−3.68
2022	46.52	1.58	18.20	2.22

with China's current stage of development. We believe that this will provide useful insights to China's current macroeconomic research.

The modernization process of the United States can be divided into classical and neo-modernization, with the industrialization period roughly spanning a century from 1814 to 1913. Since the industrialization of the United States was too early for detailed information to be available, the stage of industrialization in the United States was divided into two important phases based on the relevant literature: 1714–1865 (the beginning stage of industrialization) and 1865–1913 (the accelerated stage of industrialization).

In terms of per capita income, China entered the second stage of industrialization around 2000; in terms of industrial structure, China began to move toward the late stage of industrialization in 2005. In 1865, the American Civil War ended, and from then on until 1913, it was the accelerated stage of industrialization in the United States. The United States' GDP data from 1960 to 1975 are shown in Table 3.12. In 1960, United States' per capita GDP was about US$3007.123, and by 1975, the per capita GDP grew to US$7801.457. The annual average per capita GDP growth rate fluctuated and increased after the industrialization of the United States.

The gross national income (GNI) and the proportion of the three major industries during the industrialization period in the United States are shown in Tables 3.12 and 3.13, respectively, and the data of the indicators in China for the same period are shown in Table 3.14.

The data show that the economic development of the United States in the late industrialization period is similar to that of China, and China only started to reach the economic level of the United States in 1971 in 2000 in terms of GDP/GNI. By 2005, China had already surpassed the level of the United States' GNI in 1975, which indicates that China's economic development was faster compared to that of the United States during the same period. In 2000, China's per capita GDP was US$959.373, less than that of the United States in 1960. By 2022, China's GDP has achieved a 13-fold growth, surpassing the per capita GDP level of the United States in 1975, indicating that China's economic development is faster than that of the United States during the same period. In addition, China's industrial structure is changing, and its tertiary industry is gradually becoming the leading industry.

Table 3.12 United States indicators, 1960–1975

Year	Gross national income ($ billion)	GDP ($ billion)	GDP per capita ($)	Annual growth rate of GDP per capita (%)	Year	Gross National Income ($ billion)	GDP ($ billion)	GDP per capita ($)	Annual growth rate of GDP per capita (%)
1960	5464	5433	3007.123	–	1968	9486	9425	4695.923	8.29
1961	5668	5633	3066.563	1.98	1969	10,260	10,199	5032.145	7.16
1962	6092	6051	3243.843	5.78	1970	10,743.6	10,733	5234.297	4.02
1963	6431	6386	3374.515	4.03	1971	11,629.3	11,649	5609.383	7.17
1964	6907	6858	3573.941	5.91	1972	12,805.1	12,791	6094.018	8.64
1965	7490	7437	3827.527	7.10	1973	14,318.5	14,254	6726.359	10.38
1966	8201	8150	4146.317	8.33	1974	15,533	15,452	7225.691	7.42
1967	8671	8617	4336.427	4.59	1975	16,845.5	16,849	7801.457	7.97

Table 3.13 National income and the share of the three major industries and the government during the industrialization period in the United States

Year	National income ($ billion)	Primary sector (%)	Secondary sector (%)	Tertiary sector (%)	Government (%)
1869	68.27	22.2	32.7	40.9	4.2
1879	72.27	14.5	37.7	43.3	4.5
1889	107.01	14.2	38.2	42.4	5.2
1899–1903	173.13	18.2	36.2	39.6	6.0
1910–1913	291.11	18.9	38.5	37.1	5.5

Table 3.14 China on indicators, 2000–2022

Year	National income ($ billion)	GDP ($ billion)	GDP per capita ($)	Annual growth rate of GDP per capita (%)	Primary sector (%)	Secondary sector (%)	Tertiary sector (%)
2000	11,966.8	12,113.5	959.4	7.64	4.2	59.6	36.2
2005	22,698.6	22,859.7	1753.4	10.74	5.2	50.5	44.3
2010	60,610.9	608,71.6	4550.6	10.10	3.6	57.4	39.0
2015	109,737.5	110,155.4	8033.4	6.36	4.5	42.5	53.0
2018	135,568.4	136,081.5	9770.9	6.12	4.2	36.1	59.7
2022	177,708.5	179,632.0	12,720.0	2.28	10.5	47.7	41.8

3.3.2 *Differences in Population Development*

In 2009, the world's population reached 6.8 billion, and it is estimated that the world's total population will reach 9.5 billion by 2050, and the ever-increasing population will have a huge impact on the world economy. The development of the world is closely related to the trend of population development, and the change in the world's population is reflected differently among countries. We can reflect the development of the world by studying and analyzing the differences between the populations of developing and developed countries, which in turn reflects the differences in geographical development.

Table 3.15 shows the total population and the average rate of population growth for selected developed and developing countries for 2018–2022. The population difference between developed and developing countries is large, and the total population of most developing countries is higher than that of developed countries. Among developed countries, the United States has the largest total population, with an average total population of 330.4 million in recent years, followed by Japan, with an average total population of 126.1 million, but its average population growth rate in recent years has shown a negative value of −11.38%. The rest of the developed countries have a total population of less than 100 million; however, their population growth rates are higher, especially in Canada and Australia, whose average population growth rates are 80.48% and 86.48%, respectively.

Table 3.15 Average of total population and growth rate of selected developed and developing countries, 2018–2022

Developed country	Population (billion)	Growth rate (%)	Developing country	Population/billion	Growth rate (%)
Japan	1.261	−11.38	China	14.092	21.42
Germany	0.832	12.35	India	13.947	61.10
United Kingdom	0.669	30.77	Malaysia	0.334	79.84
France	0.676	13.72	Russia	1.443	−5.78
United States	3.304	39.11	Mexico	1.259	65.84
Canada	0.380	80.48	Brazil	2.130	45.00
Australia	0.255	86.48	Egypt	1.074	119.25
			South Africa	0.587	79.37

3.3.3 Differences in the Development of Education and Culture

In today's world of social development dominated by the knowledge economy, education determines the global influence of a country. With the rapid development of the world economy, the process of globalization is further accelerated, and with culture as an important element of national exchanges, each country is striving to explore the diversity of cultures in order to seek closer ties in the future. We compare the differences in educational and cultural development between developed and developing countries in terms of the gross enrolment rate in tertiary education and the number of resident patent applications.

The gross enrolment rate in tertiary education is the ratio of the number of persons enrolled in tertiary education to the population of the appropriate age. It is generally considered that when the gross enrolment ratio in higher education is below 15%, it belongs to the elite education stage, when it is 15 to 50% is the stage of mass higher education, and more than 50% is the stage of the popularization of higher education. To analyze the differences between the educational development of developed and developing countries, we selected the data on the gross enrolment rate in higher education in some developed and developing countries, and used the data of 2014 for the statistical analysis. For missing data of some countries, we use the data of the near year, obtained from the International Statistical Yearbook and the World Bank (Table 3.16).

The comparison reveals that among the selected developed countries, the highest gross enrolment ratio in tertiary education is in Australia, at 90.31%, and the lowest is in the United Kingdom, at 56.48%. The gross enrolment ratios in tertiary education in the selected developed countries are at the stage of tertiary education universalization. Among the selected developing countries, Russia has the highest gross enrolment ratio in tertiary education, at 78.65%, which is at the stage of tertiary education popularization, with South Africa having the lowest, at 19.38%. The selected developing

Table 3.16 Gross enrolment ratio in tertiary education in selected developed and developing countries

Country	Enrolment ratio in tertiary education (%)	Year	Country	Enrolment ratio in tertiary education (%)	Year
Japan	63.36	2014	China	43.39	2015
Germany	64.18	2014	India	25.54	2014
United Kingdom	56.48	2014	Malaysia	36.87	2014
France	64.39	2014	Russia	78.65	2014
United States	86.88	2014	Mexico	29.94	2014
Canada	75.70	2019	Brazil	49.28	2014
Australia	90.31	2014	Egypt	30.77	2014
			South Africa	19.38	2014

countries, except Russia, are all at the stage of tertiary education popularization. Overall, the gross enrolment rate of tertiary education in developed countries is generally higher than that in developing countries, and the popularization of tertiary education in developed countries is higher.

The number of resident patent applications is an important reflection of whether a country's cultural and technological activities are active. Analyzing the number of resident patent applications in each country helps compare the differences between the cultural and technological development of developed and developing countries. Based on International Statistical Yearbook and World Bank data, we compiled the number of resident patent applications in some developed and developing countries from 2005 to 2020 (Table 3.17).

A comparison of the number of resident patent applications between developed and developing countries at five-year intervals from 2005 to 2020 reveals that Japan and the United States of America have a huge number of patent applications, all of which are over 2 million, indicating that Japan and the United States are in a leading position in the world in terms of science and technology and culture. In developing countries, the number of patents is higher than in China and Russia, which is at the general level of the developed countries.

We also find that in Japan, Germany, and the United Kingdom, the number of patents is decreasing annually, whereas in the United States and other countries, the patents are increasing. The gap between the scientific and technological development of the developed countries is narrowing. Developing countries show a year-on-year trend of growth in the number of patents.

In terms of educational culture, China and Japan stand out again in comparison with the rest of the world. Geographically, the northeastern part of China is separated from Japan by the sea, and a large part of Japanese culture has been handed down from China, thus, there are many similarities, and the development of education in the two

Table 3.17 Number of patent applications filed by residents of selected developed and developing countries (in 10,000s)

Developed country	Year				Developing country	Year			
	2005	2010	2015	2020		2005	2010	2015	2020
Japan	367.96	290.08	258.84	227.35	China	93.49	293.07	968.25	1497.16
Germany	48.37	47.05	47.38	42.26	India	4.52	8.85	12.58	23.14
United Kingdom	17.83	15.49	14.87	11.99	Malaysia	0.52	1.23	1.27	0.99
France	14.33	14.75	14.31	12.77	Russia	23.64	28.72	29.27	23.76
United States	207.87	241.98	288.34	269.59	Mexico	0.58	0.95	1.36	1.13
Canada	5.18	4.55	4.28	4.45	Brazil	3.91	4.23	4.64	5.28
Australia	2.56	2.41	2.29	2.37	Egypt	0.43	0.61	0.72	0.98
					South Africa	1.00	0.82	0.89	0.54

Table 3.18 Enrolment in higher education in China and Japan

China		Japan	
Year	Enrolment in tertiary institutions (%)	Year	Enrolment in tertiary institutions (%)
2018	54.01	1971	17.34
2019	57.28	1972	19.02
2020	62.24	1973	20.31
2021	67.39	1974	22.04
2022	71.98	1975	23.94

countries is even more unique and twisted. Therefore, to better reflect the differences in the development of education culture between developing and developed countries in the same period, we choose China as a representative of developing countries and Japan as a representative of developed countries and compare the differences in education culture in terms of enrolment in tertiary institutions. Since China is now in the late stage of industrialization and Japan's corresponding historical development period is roughly in the 1970s, we collate relevant data from 2018–2022 for China and 1971–1975 for Japan (Table 3.18).

Both China and Japan show an upward trend in tertiary enrolment, with China's enrolment increasing from 54.01% in 2018 to 71.98% in 2022 and that of Japan during the same period increasing from 17.34% in 1971 to 23.94% in 1975. Thus, the difference in the level of education and culture reflected in the enrolment rate in higher education shows that China's level of education and culture is higher than that of Japan during the same period.

3.3.4 Differences in Environmental Resources

To analyze the differences in ecological resources and environmental development between countries, we compared land area, renewable freshwater resources per capita, and CO_2 emissions.

To understand the differences in land resources among countries, we collected data on the current land area of developing and developed countries from the World Bank and plotted them in a bar chart (Fig. 3.11).

There are large differences in land area between countries. Among developed countries, Canada, the United States, and Australia have large land areas, while France, the United Kingdom, Germany, and Japan have very sparse land areas and are in short supply of land resources. Among developing countries, Russia has the largest land area of 17,098,000 km^2, and China's land area is second only to Russia's, which is rich in resources and has a great deal of land and resources. Countries such as South Africa, Egypt, and Malaysia have relatively small land areas, and there are large differences in land resources among developed countries and among developing countries.

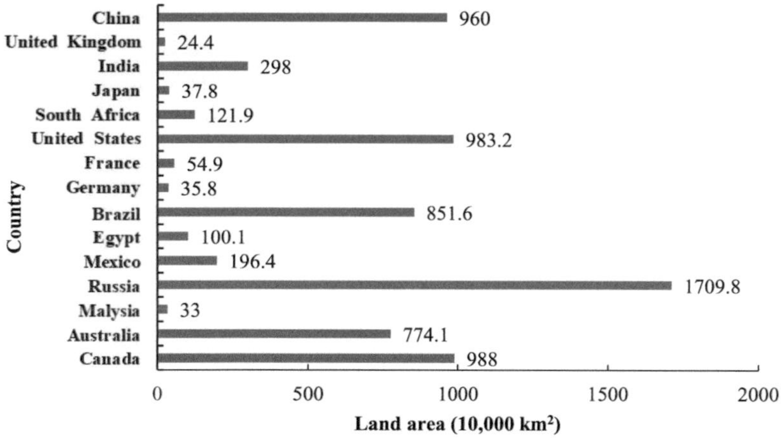

Fig. 3.11 Land area in developing and developed countries

In Asian countries the land area and resource reserves of Japan are much lower than in China; however, Japan's economy is higher than China's. This could be related to advanced science and technology. Although resource-rich and populous, China's overall scientific and technological level compared to that of Japan and other developed countries is still behind. This gap in scientific and technological development accounts for the economic differences. Figure 3.12 reflects the developmental differences between countries' water resources, using data from the World Bank. The comparison reveals a relative lack of renewable freshwater resources per capita in most developed countries. There are differences in renewable water resources per capita in developing countries, and water scarcity can be detrimental to all aspects of human life, and if the scarcity is severe, it can affect relations between countries and lead to conflict over water resources.

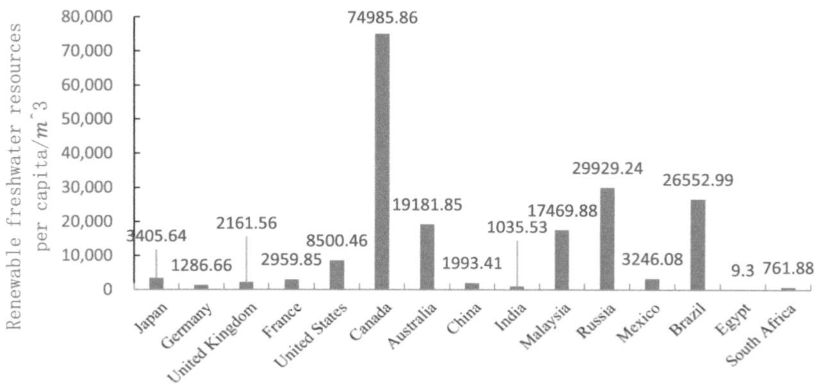

Fig. 3.12 Renewable freshwater resources per capita in developing and developed countries

Carbon dioxide emissions are an important indicator of environmental quality (Li et al., 2018; Wang et al., 2019). Table 3.19 shows the data of CO_2 emissions in selected developing and developed countries in 2020, obtained from the World Bank.

The comparison reveals that developed countries, with the exception of the United States, and developing countries, with the exception of China, have relatively low CO_2 emissions. Both China and the United States have relatively high GDP and high CO_2 emissions.

Analyzing the resource environment of developing and developed countries in the current period reveals that Canada and Russia are both relatively resource-rich countries, with Russia possessing a land area of 17,098,000 km^2 and Canada possessing a land area of 9,880,000 km^2. Russia and Canada are separated by the sea and have similar geographical latitudes; however, their economic development is not the same. Below, we explore the differences in the resource environment between them.

Canada is rich in natural resources, providing the most basic material conditions for its development. Canada attaches great importance to environmental protection and sustainable development in the process of economic development. Russia is also very rich in natural resources, Russia's early economic strength is very strong, with a strong industrial and scientific and technological base.

Resources are the foundation of a country's development, and environmental issues are of great significance to economic development. Table 3.20 shows the CO_2 emissions of Canada and Russia, which are both in the late industrialization stage.

A comparison reveals that Canada's CO_2 emissions in the late industrialization period were only about 200 million tons, whereas those of Russia were about 1.7 billion tons, eight to nine times that of Canada during the same period, a figure that is sufficient to show that the differences between Canada and Russia in terms of environmental pollution are too significant to be ignored. Thankfully, Russia's CO_2 emissions have shown a downward trend in recent years, and the environment is moving in a positive direction.

Table 3.19 Carbon dioxide emissions from selected developing and developed countries in 2020 (10^6t)

Developed country	Carbon dioxide emission	Developing country	Carbon dioxide emission
Japan	1014.07	China	10,944.69
Germany	603.35	India	2200.84
United Kingdom	308.65	Malaysia	245.14
France	267.16	Russia	1618.27
United States	4320.53	Mexico	383.13
Canada	516.87	Brazil	414.14
Australia	379.00	Egypt	210.75
		South Africa	393.24

Table 3.20 Canada and Russia's CO_2 emissions for the same period in history

Canada-year	Carbon dioxide emissions (million tons)	Russia-year	Carbon dioxide emissions (million tons)
1960	192.90	2010	1670.53
1961	194.18	2011	1763.41
1962	207.19	2012	1830.83
1963	211.11	2013	1778.56
1964	237.80	2014	1705.34

3.4 Reasons for Regional Development Differences

We have extensively discussed the differences in the development of each region in the previous section. The causes and theoretical basis of these differences are summarized in Fig. 3.13.

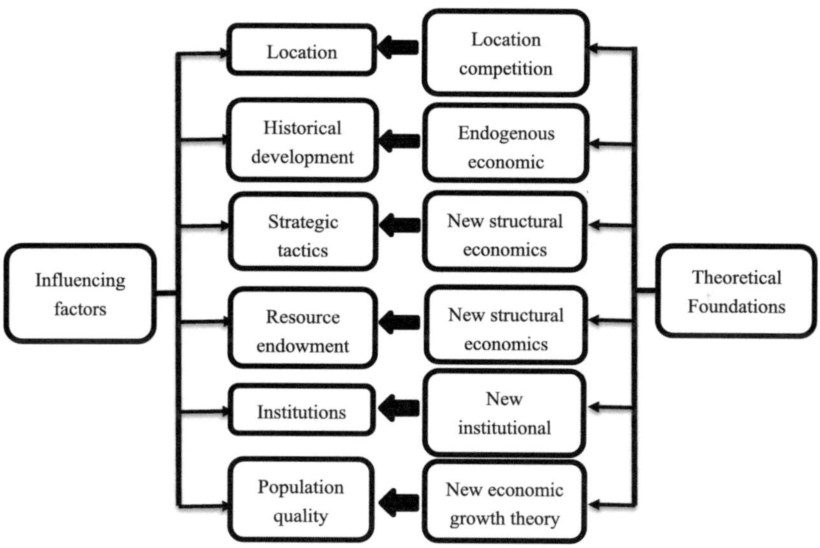

Fig. 3.13 Factors affecting regional development disparities and theoretical origins

3.4.1 *Reasons for Differences in Economic Development*

This section discusses the causes of geo-economic development disparities in each of the six areas of location, historical development foundations, policies and development strategies, differences in resource endowments, institutional factors, and demographic factors.

3.4.1.1 Location Factors

Modern location theory has been studied in terms of natural resources, transport costs, institutional structure, and government policies. Geography is the foundation of economic development, and although it cannot directly determine economic development, it plays an important role in economic development. Most parts of China's eastern region are in the coastal areas, and the potential opportunities for regional development are much more compared to the central and western regions, whereas most of the central and western regions are inland and underdeveloped, and geographically, the inconvenient traffic and few foreign exchange activities are not conducive to economic development, which leads to the differences in the economy between the eastern region and the central and western regions.

3.4.1.2 Historical Development Basis

New economic geography theory proposes that changes in the space of economic activities are completed with the flow of factors, and the flow of factors in space will make various factors gather in a certain region, thus leading to the imbalance of economic growth among regions. The eastern region of China has a superior geographical location and convenient transport system compared with the central and western regions, which will attract more production factors, thus shifting the production factors of less developed regions to the eastern region. This leads to the economic differences between the eastern region and the central and western regions, and over time, this difference will have a cumulative effect and may further expand the gap.

In his study on endogenous economic growth, Paul Romer, winner of the 2018 Nobel Prize in Economics, identified economics and technology as endogenous variables of economic growth and suggested that capital and (unskilled) labor, as well as human capital (as measured by years of education) and new ideas (as measured by patents, with an emphasis on innovation), are the four main elements of current economic growth. The end of the nineteenth century and beginning of the twentieth century witnessed large numbers of international immigrants. Between 1850 and 1914, more than 60 million people immigrated to the United States. The massive population transfer that brought an abundant labor force to the developed world, as well as the impetus for economic growth. Basic scientific research and technological development play an important role in the economic growth of contemporary developed countries, the advancement of knowledge leads to technological innovation, the accumulation of surplus wealth further strengthens the knowledge reserve, the two interact with each other, and the cycle leads to further economic growth and maintains the high economic growth rate of developed countries.

3.4.1.3 Policies and Development Strategies

New structural economics emphasizes the role of the market in resource allocation and suggest that the government should assist in solving the problem of externalities and coordinating development and help enterprises carry out industrial upgrading. The formulation of policy strategies plays a crucial role in economic development, and national policies can influence the direction of the development of a region at the macro level. Since the reform and opening up, China has proposed a strategic layout that gives priority to the development of the eastern coastal region, which has led to the rapid development of the eastern region's economy, and then, in accordance with the changing situation, it has put forward a timely development strategy for the development of the western region, which has accelerated the rate of economic growth in the central and western regions, and has gradually narrowed the economic disparities between the regions.

3.4.1.4 Differences in Resource Endowments

New structural economics suggests that the economic outcomes of an economy are endogenous to the structure of its factor endowments. There are large differences in resource endowments among the regions of China, with the central and western regions rich in energy resources, mineral resources, climate resources, land resources, etc., and the eastern region relatively deficient in these resources, but with a high level of culture, education, science and technology, a well-qualified labor force, abundant capital, and strong capacity for technological innovation. The eastern region is the main production base of consumer goods in China, whereas the central and western regions are important energy and raw material export bases.

Contemporary developing countries are relatively weak compared to developed countries at the beginning of the natural resource endowment. In most of the less developed countries, particularly in Asia, oil, minerals, and other resources are scarce. Paul Rohmer points out that the technology gap between economically poor countries and economically developed countries is reflected in the physical gap and the concept of the gap, and that developing countries are poor mainly because they do not grasp the concept of creating economic value in industrialized countries. To reduce the difference, it is necessary to understand the concept of creating economic value and formulate a specific plan.

3.4.1.5 Institutional Factors

American economist Douglas North suggested that institutional change is an important factor affecting economic growth. North stressed that technological innovation certainly injects vitality into economic growth; however, if people do not have the impulse of institutional innovation and institutional change and consolidate the results of technological innovation through a series of institutional (including property rights system, legal system, etc.) construction, then, the long-term economic growth and social development of human society is unthinkable. According to North, institutions play a determining role in a country's economic growth and social development.

In the early stage of economic development, many developed countries provided a wide range of entrepreneurial opportunities for people with entrepreneurial motives, and after a long period of practice, the economy developed rapidly. North believes that the economic system is the rule of the game for economic development and plays a crucial role in the development of the economy. In recent years, the international political environment has been relatively stable, and many developing countries have actively formulated relevant economic policies to achieve rapid economic development.

3.4.1.6 Demographic Factors

New economic geography holds that the market size of a region affects economic development and is often reflected in the population size. The difference in population between the eastern, central, and western regions also indirectly leads to the difference in economic development. The demographic factor is not only reflected in the size of the population but also in the cultivation and introduction of talents, and the difference in the quality of the population may lead to the difference in the economic growth of the region.

Population issues are also closely related to natural resources, which are gradually diminishing with people's social activities, further affecting the economy and threatening the survival of human beings. The large population base of developing countries and the relative scarcity of per capita resources are important factors contributing to the economic disparities between developing and developed countries.

3.4.2 Reasons for Differences in Population Development

People are the mainstay of economic activity, and population development is related to economic development. The following sections discuss the reasons for the differences in population development.

3.4.2.1 Location Factors

Natural geography is the basis of human existence, providing the most basic living space for human beings, and human beings are the product of natural evolution and natural enhancement, and regional differences in natural conditions further lead to regional differences in population distribution. The eastern region is more attractive to people than the western region because of its better agricultural conditions in the eastern region, its geographical location near the sea, and its warm and humid climate, which is suitable for people to live and develop agriculture, resulting in the difference in population size between the regions.

3.4.2.2 Economic Factors

Economic development has an increasing impact on the regional population distribution. Economic development drives the rapid development of a population. Generally, the more developed the economy is and the better the social security function is, the lower the willingness of people to raise children becomes. Most developing countries have a low level of economic development, most families improve their lives through the increase in the size of the population, raising children is regarded as a kind of capital investment, and the differences in economic development and ideological concepts cause the population development differences between countries.

3.4.2.3 Ideology

In recent years, with economic growth, the natural population growth rate in eastern China has been lower than in the western region, partly because of the lower birth rate in the eastern region. The main reason for the lower birth rate is the change in people's ideological concepts, coupled with the fact that the eastern region is economically developed and the pressure on people's lives is becoming greater. Thus, a combination of factors has led to the lowering of the birth rate in the eastern region. As for the central and western regions, some people's concepts are still stuck on raising children for old age, resulting in a higher natural population growth rate in those regions.

3.4.2.4 Policy Factors

Rapid population growth has brought about a lack of resources, and some developing countries have changed from encouraging fertility to birth control. Before the 1970s, the Mexican government had adopted a policy of encouraging fertility. In 1972, Mexico switched to a population control policy centered on birth control in order to control the accelerated growth of the population and rationalize the distribution of the population and the layout of socio-economic development. In some countries with small populations, governments have established fertility incentives to encourage population development. The different fertility policies of countries have also resulted in differences in population development.

3.4.3 *Reasons for Differences in Education and Culture*

The development of education and culture can be reflected in the quality of the population to a certain extent, and the following reasons may explain the regional differences in the development of education and culture.

3.4.3.1 Location Factors

The eastern region of China is close to the sea, and the traffic and communication systems are more developed, which is conducive to the dissemination of culture. In addition, because of the absorption of advanced education concepts, education flourishes, whereas in the central and western regions, due to the topography and natural conditions, are not conducive to the dissemination of culture, which indirectly leads to the slow development of education, hence the difference in the development of education and culture between the eastern region and the central and western regions.

3.4.3.2 Ideological Differences

The eastern region is more innovative and creative due to the absorption of advanced educational concepts, whereas the western region is relatively more traditional in its way of thinking, which further leads to the differences in educational culture between the regions.

3.4.3.3 Economic Factors

The income gap between developed and developing countries creates differences in their educational development. Developing countries have relatively lower incomes and their economies are more backward than those of developed countries. There is also an imbalance in economic development within developing countries, and this imbalance leads to a more backward development of basic education. In developed countries, although there is also a gap in income, with the governments' macro-control, basic education is well developed, and the education penetration rate is high, which creates a difference between the educational development of developing and developed countries.

3.4.3.4 Policy Factors

National investment in education plays an important role in the development of education in a region. Compared to the eastern region, the investment in education is significantly higher than the central and western regions, and the state has preferential policies for education in the eastern region. As the difference became increasingly obvious, the state began to take relevant measures to support education in the western region by implementing the eastern region's "two projects" and other measures, to narrow the gap in the development of education between the central and western regions and the eastern region.

Each country has its own education development policy. The education provided in developed countries is better than that provided in developing countries, and most developed countries encourage scientific and technological talents, cultivate a large number of excellent talents, and improve the overall education level and comprehensive quality of China. The development of education in developing countries is relatively short, and the education system is not mature enough; therefore, the policy may not be flexible.

3.4.4 Reasons for Differences in Environmental Resources

Environmental problems not only affect economic development but also have a significant impact on the survival of human beings. We explain the differences in the development of environmental resources in various regions below.

3.4.4.1 Demographic Factor

American biologist Paul Ehrlich proposed the famous population explosion theory, which holds that the rapid increase and excess in population has exceeded the carrying capacity of the earth's ecological environment and is threatening the survival of human beings.

The uneven distribution of resources in China is determined by geographic location, the eastern region has a relative lack of resources; however, funds and talents are concentrated in that region. The central and western regions are relatively rich in resources; however, due to insufficient development, the pace of economic development is slower. The original uneven distribution of resources between the eastern region and the central and western regions due to the movement of the population caused the per capita difference in resources to further increase.

3.4.4.2 Economic Factors

The eastern region has better economic development, as the developed areas in this region are mainly industrial, whereas the economically underdeveloped areas in the central and western regions are mainly agricultural. The process of developing industry will inevitably have an impact on the environment, resulting in environmental pollution. In recent years, with the growth of the economy and the relevant policies of the state, the western region began to develop its, further increasing environmental pollution.

The better economic development in the eastern region also brings environmental pollution to the region.

Environmental problems are divided into two categories: one due to industrial production, traffic, and living sewage caused by environmental pollution; the other due to the irrational exploitation of natural resources, resulting in environmental pollution. Generally, developed countries, due to the faster industrialization process, produced the first type of environmental pollution. Conversely, developing countries have weaker economic development, industrial underdevelopment, and thus experience the second type of environmental pollution. Therefore, economic differences arise between developing and developed countries due to resource and environmental differences.

3.4.4.3 Industrial Policy Differences

In recent years, China has launched several policies to solve the problem of environmental pollution and narrow the differences in the environmental development of the eastern and western regions. China has also implemented the "West–East Gas Pipeline" and other development strategies to balance the distribution of resources and reduce the development gap between the eastern region and the central and western regions.

The damage to the global environment has led countries to take corresponding measures to alleviate the problems caused by environmental pollution. Most developed countries have adopted the practice of relocating polluting enterprises to economically disadvantaged developing countries to reduce environmental pollution in their own countries. At the same time, they also treasure the resources of those developing countries, and most of them have adopted the practice of importing the resources of economically underdeveloped countries to achieve the protection of their own resources. Some developing countries have exploited resources in large quantities to develop their economies, destroying the environment, and the difference in the development of environmental resources between developing and developed countries has increased.

3.5 Development in Less Developed Countries or Regions

Most of the world's poor people are concentrated in less developed countries or regions. Industrial capitalism, economic globalization, and the acceleration of the world's urbanization process have changed the economic structure of these regions, and a series of problems such as a low level of economic development, backward infrastructure, uneven distribution of social wealth, intricate social conflicts, low level of education of the people, and difficulties in the effective use of energy resources have made the governance of poverty in these countries and regions a challenge. Development geography focuses on poverty reduction in less developed countries and regions. This chapter describes the adjustment problems faced by less developed countries and regions, focusing on the economic development of the region from the perspectives of environmental vulnerability and regional development, demographic imbalance and regional division of labor, cultural plurality and regional differences, as well as regional poverty and poverty reduction.

3.5.1 Environmental Vulnerability and Regional Development

With the rapid development of the economy, the process of urbanization has accelerated. Strongly influenced by human activities, the impact of climate change, environmental damage, and other phenomena in less-developed countries and regions has become increasingly prominent, and industrial pollution and ecological damage are intensifying and presenting regional and comprehensive characteristics. Development geography scholars have found that economic growth, policies and regulations, international trade, financial and tax system, economic structure, and the degree of urbanization are relatively important factors affecting regional carbon emissions (Brajer et al., 2011). The state, cities, and enterprises account for the majority of this study. The research on cities or urban agglomerations focuses on identifying spatial and temporal characteristics of regional carbon emissions, the relationship between the development of cities or urban agglomerations and the level of carbon emissions, and assessing the interaction between public policies and carbon emissions (Lutsey & Sperling, 2008). The theoretical mechanism of convergence of regional development, such as technology diffusion and factor flow, has gradually become a topic of interest in development geography; however, the environmental vulnerability of regional development requires the attention of theoretical research and relevant management departments, and the results of geographic research and analysis will be inaccurate if regional development is neglected. Given the differences in the resource and environmental endowments of less developed countries and regions, development geographers have analyzed the measurement and spatial and temporal patterns of regional economic development and ecological and environmental coordination and found that environmental regulation can significantly improve the level of synergy between the economy and the environment.

3.5.2 Population Imbalances and Regional Division of Labor

Population is a fundamental element of economic and social development, and the level of population quality is more important than other measures of economic development when comparing levels of development among countries and regions, particularly the indicators of health, education, demographics, and human rights. Development geography focuses on the study of socially marginalized groups, notably child labor, diversified childhood, and transition pathways to adulthood (Arora & Puranik, 2004); the impact of indicators of adolescents' mental health, self-adjustment, and social behaviors and the constraints to their academic achievement (Cui et al., 2022); the status of women, the relationship between gender inequality and development, and the reversal of gender differences in education (Jayachandran, 2015); the right to survival and development of people with disabilities (Andrea Wood, 2013); education and health inputs and the formation of human capital (Schultz, 1961); the gender division of labor and the relationships among families (Fraser, 2017); and the economic effect of population agglomeration (Combes et al., 2011).

3.5.3 Cultural Pluralism and Regional Differences

Culture as an endogenous factor is integrated with economic development and plays an irreplaceable role in economic and social development. Differences in regional cultures directly or indirectly lead to differences in regional patterns, and economic activities are always rooted in the network of local socio-cultural interpersonal relationships. Due to the abstract nature of culture and the hidden nature of its mechanism of action, the measurement of culture and the quantification of the relationship between culture and the economy have become the key to deepening the study of the geography of development. As cultural pluralism varies in regional development, policy formulation should be integrated with the culture and value system of a particular region to fully realize people's basic political and cultural rights. Many development geography scholars have studied this: the mutual influencing relationship between culture and economic development has been analyzed, and the theoretical basis, operating mechanism, and strategic support for the integration of culture and economic development have been explored (Ferrara et al., 2014). As a kind of informal system, culture affects the operating efficiency of various factors and various formal systems and acts on economic growth. The cultural environment can play a role in improving labor productivity, increasing labor supply, and promoting technological innovation (Shi et al., 2014). A rights-based approach to development is constructed, pointing out the relationship between human rights and the potential challenges and development of society (Wise et al., 2013). Finally, the link between the expansion of women's economy, political rights, and economic development is analyzed (Razavi, 2016).

3.5.4 Regional Poverty Reduction

Regional poverty is a core component of development geography research and one of the major topics that development geographers have closely followed and studied for a long time. Poverty is a major global social and practical problem, and rural poverty and agricultural issues continue to dominate the regional development agenda. Development geography scholars have explored and analyzed social poverty from different perspectives. Poverty has multidimensional, regional, and dynamic characteristics, and the academic community's perception of poverty has gone through a process of transformation from a single dimension of income/consumption poverty to multidimensional poverty (including income, education, health, employment, and living conditions) (Zhou, 2018). The theory of basic needs from the perspective of economics has been proposed, social exclusion theory from the sociological perspective, ability poverty and multidimensional poverty theory from the developmental perspective, and rights poverty and class theory from the political science perspective. Regional multidimensional poverty measurement dimensions and measurement methods have been explored, including income/consumption criterion method, Human Development Index (HDI), and Multidimensional Poverty Index (MPI) (Alkire & Foster, 2011).

3.6 Development in Developed Countries or Regions

In the current complex and changing global economic environment, developed countries or regions are facing more serious development problems. Even the most developed economies are plagued by ecological environmental and internal development imbalance and other development problems. Exploring their quality of life in the process of improving the issue of convergence is an important area of research in development geography. Therefore, there is an urgent need to analyze and reflect on the exercise of political and economic power conducting quality of life convergence studies in developed countries or regions from a developmental perspective. This chapter details the main challenges of environmental degradation, global environmental change, and regional development approaches faced by developed countries or regions, with a view to exploring the issue of convergence in the process of improving the quality of life in developed countries and regions and searching for new models and processes to achieve more sustainable development.

3.6.1 Environmental Sustainability

The concept of ecological carrying capacity shows the mutual constraints between regional natural resources and biological survival and development, and ecological carrying capacity is dynamic and reversible; when the carrying capacity of regional resources exceeds its limit, the natural ecological environment can be repaired through timely intervention using human measures, thus providing new ideas for the sustainable development of resources and the environment (Zhang & Zhang, 2019). Neoclassical economics suggests that the limited supply of the public environment and the lack of control over free-riding behavior (free-riding behavior is speculative behavior that enjoys the benefits of others without paying the cost) have become the root causes of the current poor climate, water pollution, and gradual deterioration of other public resources in some countries (Franck et al., 2011). The field focuses on ecosystem services and their linkages to human well-being and the impact of changes in ecosystem services on material needs, security, health, social relationships, and rights to freedom and choice (Leemans & Groot, 2003). More systematic results have focused on changing perceptions and practices in environmentalism and development thinking, exploring the competing nature of sustainable development (Martin & Rice, 2014); identifying constraints to environmental sustainability and proposing models for achieving environmental sustainability and growth (Nissing & Blottnitz, 2010); optimizing land-use management to improve ecosystem services and determining the Environmental Sustainability Index (ESI), which assesses the ability of countries to protect the environment in the future (Babcicky, 2013). In addition, from the perspective of land use, development geography scholars have proposed that a comprehensive evaluation of the degree of coordinated development of the urban land use system and the selection of a reasonable urban land use model to regulate the way human beings utilize land resources is an important means of improving the efficiency of intensive land use and ensuring sustainable development.

3.6.2 Transformation of Regional Development Mode

In recent years, society has become increasingly concerned about whether the development of developed countries and regions is converging, and the research on the coordinated development of inter-regional and rural agriculture, and the balanced development of industry and cities has become a hotspot of recent development geography research. In addition, research on regional development mode and development policy has also achieved remarkable results, and the relevant research topics are mainly reflected in the coordinated development of regional agriculture and rural areas and the coordinated development of industry and cities. (1) Interregional rural, agricultural, and development convergence: unlike the urban–rural dichotomy in the original social policy, the regional type of rural policy constructs a new type of urban–rural partnership in practice through development cooperation that breaks through administrative and territorial boundaries (Wiest, 2016); the impact of innovative agricultural production methods on rural development and farmers' well-being (Dawson et al., 2016); the role of urban–rural industrial complementarities in guiding the direction of rural industrial structure development (Nazzaro & Marotta, 2016); and the framework of sustainable livelihoods, the new paradigm of rural development, and coordinated rural development (Nesar et al., 2010). (2) Urban, industrial, and development convergence: accelerating the implementation of regional coordinated development strategy, major regional strategy, main functional area strategy, new urbanization strategy, optimizing the layout of major productive forces, promoting the deep integration and development of city-regional innovation centers and regional industrial systems, and strengthening the synergistic integration of the regions (Feenstra et al., 2015); local regional policies on employment and economic growth (Cappelen et al., 2003); synergistic development of urban–rural regional integration based on spatial governance perspectives (Durkheim, 2013); dissecting the impact of urban geometrical forms on urban wages, rents, population growth, traffic, etc., and stressing the need for cities to carry out orderly horizontal expansion under compact forms efficiently promote urban development (Li et al., 2019); extended metropolitan areas and peri-urban development (Kelly, 2010); and the contribution of changes in urban management practices to sustainable urbanization.

3.7 Coordinated Development in Developing Countries or Regions

With the rapid development of globalization and integration, developing countries or regions are faced with new opportunities and challenges in the process of globalization. On the one hand, due to the uneven impact of globalization on the level of development and internal spatial differences of developing countries or regions, the degree of their integration into globalization shows regional differences, whereas the imbalance in internal spatial development is gradually widened by the impact of globalization. On the other hand, the development of globalization has led to several global problems that require both developed and developing countries to share the responsibility of governance, and the difference in the level of development between the two is also an important basis for sharing responsibility. Overall, in the wave of globalization, developing countries or regions should pay more attention to spatial equilibrium and coordinated development in space and time, such as regional differences and convergence under globalization, global modernization governance under the new situation, ecological civilization system construction in the new era, and mitigation and adaptation strategies under climate change.

3.7.1 Regional Differences and Convergence Under Globalization

Currently, the globalization of society, economy, culture, technology, and ecology shows the trend of difference and diversity, which is a concentrated manifestation of the response of neo-liberal capitalism to globalization. The world landscape has produced complex and profound changes, with the deep-rooted impacts of the international economic crisis continuing to emerge, and the investment and trade patterns and multilateral investment and trade rules constantly being adjusted, making the development problems faced by less developed countries and regions even more severe. In recent years, research in this area has focused on regional cooperation and regional economic integration, the impact of transnational corporations and foreign investment on industrial agglomeration and the pattern of regional development, structural contradictions, and the debt crisis, and has achieved important research results (Amin, 2010). In the process of globalization, the globalization of capital causes the spatial contradiction of capital accumulation and the globalization of the ecological crisis, which constantly intensifies the occupation and deprivation of ecological space in less developed countries or regions (Zhang, 2014). The globalization of trade and the economy leads to the increase in inequality in less developed countries (Milanovic & Squire, 2007); and under the globalization of transnational corporations, debt, and other globalization products, the contradiction between the economic and sustainable development of less developed countries or regions is further expanded (Shin et al., 2017). Relevant studies are mainly related to the impact of globalization on regional trade and regional patterns in less developed countries, and the trade control of developed countries' multinational corporations on less developed countries and regions under the globalization of monopoly capital (Foster, 2015); the evolutionary trend of the debt crisis and its impact on less developed countries or regions (Aizenman et al., 2016); the shaping of globalization on the industrial development and the regional community resource patterns and modes of co-operative development, among others.

The globalization of the world economy has been accelerating, corresponding to the widening of the development gap between rich and poor economies. In addition, inequality within some countries has also been widening, thus advancing the study of regional inequality. The rise of state-centered approaches has been attributed to the central government's coastal development policies, whereas the globalization school has argued for the importance of foreign investment in development, and others have emphasized reform and development, such as the "Western Development Strategy" launched in 1999, where the development divide between the coastal regions and the hinterland continues to widen. The study of regional inequality has been further developed by Chinese geographers, who view regions as geographic scales best suited to constructing the interplay of complex social processes in the era of globalization, arguing that regional inequality is shaped by the complex interplay of globalization, institutions, and region-specific factors. The spatial heterogeneity/connections and development trajectories of regions have a great impact on regional inequality, that regional inequality in China is sensitive to geographic agglomeration and clustering, and that changes in the trajectories and destinies of leading or lagging regions have a great impact on regional inequality.

3.7.2 Modernized Global Governance in the New Situation

The world today is undergoing complex and profound changes, and changes in the global political landscape are affecting the territorial, economic, resource, and ecological security of all countries. The deep-rooted impacts of the international financial crisis continue to emerge, and the investment and trade patterns and multilateral investment and trade rules continue to be adjusted, making the development problems faced by less developed countries and regions particularly severe. Scholars of development geography have conducted research on the external challenges of the global governance system such as a weak sense of democracy, the failure of the global governance system, and the incoherence of the plurality of global governance subjects (Leite et al., 2015). The main research area re-articulates the transformation of the concept of global governance, stresses the importance of promoting the transformation of global governance subjects, and highlights the construction of a new type of great power relationship as an important way of accelerating the reform of the global governance mechanism to maximize the interests of all countries in the context of cooperation (Mawdsley, 2012).

As the world's largest developing country, China's globalization strategy has gradually become an international consensus, and the concept of a community of human destiny is China's new program and proposition in today's context of globalization and diversification. Compared with other governance programs, the idea of building a community of human destiny is pragmatic, systematic, inclusive, and universal, and, in this way, seeks new connotations for the common interests and values of mankind. Major studies in this area cover regional trade and development, the world free trade area, trade protectionism, and the World Trade Organization (WTO). Relevant studies have also explored various forms of aid such as government aid, aid from international organizations, and humanitarian aid (Essex, 2013); the complexity of the aid landscape, emerging countries, and global development aid patterns and challenges (Kanbur & Sumner, 2012); and the heterogeneous impacts of international aid on the economic growth of recipient countries (Dreher & Langlotz, 2020). In recent years, the "Belt and Road" strategy has become a new mode of international regional cooperation and a new platform for exploring global governance to promote the in-depth development of economic globalization, as well as an effective way to promote the building of a community of human destiny. Relevant scholars have stressed the role of the "Belt and Road" strategy in the reform of the international trade system (Juma et al., 2001); and pointed out that the development of the "Belt and Road" trade may impact global trade governance (Ran, 2017).

3.7.3 Ecological Civilization Construction in the New Era

To promote the construction of ecological civilization, according to the wholeness and systemic nature of ecosystems as well as the inherent laws, the overall protection, systematic restoration, and comprehensive management of ecosystems are promoted. Relevant research mainly focuses on solving key and difficult problems in ecosystem protection and governance, embedding green and low-carbon transformation measures into economic and social development, maintaining the strategic determination to strengthen the construction of ecological civilization under high-quality development, high-level security, and high-level protection, and searching for the optimal emission reduction path suitable for China, considering both long-term and short-term goals.

Since 2013, precise poverty alleviation has become an important strategic task for China to eliminate poverty and build a moderately prosperous society in all aspects. It is worth noting that China's poor population is mainly concentrated in underdeveloped regions, which often highly overlap with China's ecological security barrier zones, ecologically fragile zones, and resource and environmentally sensitive zones, and are facing the double predicament of insufficient economic development momentum and ecological environment deterioration. The rough exploitation and utilization of the development mode has led to and exacerbated the continuous deterioration of the ecological environment, whereas the continuous deterioration of the ecological environment has induced a lack of motivation for economic development, forming a vicious circle of poverty and returning to poverty, and forming a pattern of ecological poverty. Therefore, ensuring the coordinated development of the two national strategies of poverty alleviation and ecological civilization construction is important to the smooth implementation of poverty alleviation and ecological environmental protection.

Related research in this field focuses on precision poverty alleviation or ecological compensation and on the theoretical connotation of poverty alleviation, working mechanism, implementation path, development model, effect assessment, development strategy, etc. Specific models and case studies in tourism, finance, agricultural poverty alleviation, etc. have achieved remarkable results. Among them, ecological compensation is an important aspect of research by scholars local and international, and foreign research is mainly reflected in the scope of the main body of ecological compensation, ecological compensation mechanism, and ecological compensation sustainability. Domestic ecological compensation mechanism, standards and methods, systems, impact on regional development, and other areas have made important progress. As China's precise poverty alleviation strategy is in its final stage, it is important to explore how precise poverty alleviation and eco-compensation can be reconciled to promote a win–win situation of ecological protection and precise poverty alleviation in less developed regions.

3.7.4 Mitigation and Adaptation Strategies to Climate Change

Global climate change is one of the biggest challenges to human survival and development, and the extreme meteorological disasters, sea level rise, species extinction, ecosystem destruction, and other consequences triggered by it are causing far-reaching negative impacts on the global economy, society, and natural ecology (Barreca & Schaller, 2020; Khatana et al., 2023). Global warming, extreme weather events and their impacts and mitigation strategies triggered by carbon emissions have received widespread attention from all sectors of society, and the success or failure of climate change mitigation and adaptation actions will have a significant impact on the future development of the world. The carbon emission data published by the World Bank show that the global emission is in an increasing trend from 1960 to 2013, and after 2013 the global carbon emission is gradually decreasing, which is mainly concentrated in a few countries or regions, showing the characteristics of unbalanced distribution (Fig. 3.14).

To mitigate the adverse impacts of climate change, the Paris Agreement specifies a long-term goal of global average warming by the end of the twenty-first century of no more than 2 °C above the pre-industrial level and endeavoring to keep it within 1.5 °C. Countries have formulated autonomous contribution goals (Intended Nationally Determined Contributions [INDCs]) for action on climate change according to their own circumstances and adopted a series of climate mitigation and adaptation strategies. Developing countries are actively responding to climate change and have formulated corresponding mitigation and adaptation measures covering agriculture, water resources, risk hazards, land use, social well-being, and other aspects. Existing research mainly focuses on the development challenges and response mechanisms of developing countries and LDCs or regional areas. In terms of development practice, China, as a responsible carbon emitting country, has promulgated a series of policies

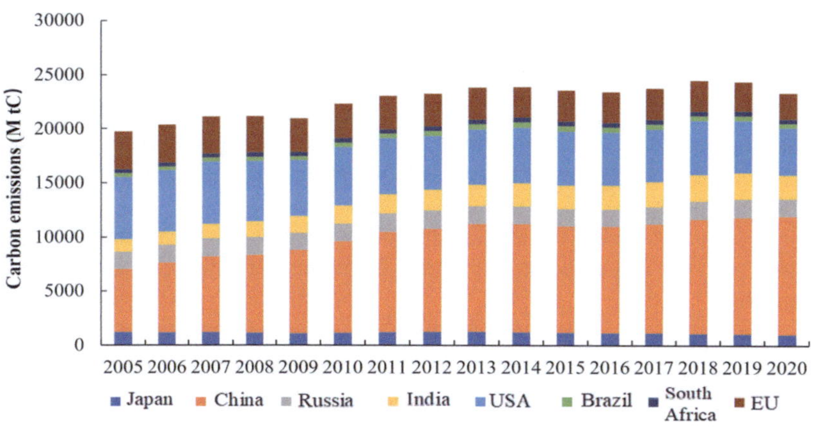

Fig. 3.14 Carbon emissions in selected developed and developing countries, 2005–2020

and regulations such as the National Plan for Addressing Climate Change (2014–2020) and the 13th Five-Year Comprehensive Work Program on Energy Conservation and Emission Reduction, and has formulated a plan to reach the peak of carbon emissions by 2030, achieve carbon neutrality by 2060, and complete the carbon emission mitigation and adaptation measures. In addition, to strengthen the risk assessment of the impact of climate change, China has set up a series of policies and regulations, such as the "Comprehensive Work Program on Energy Conservation and Emission Reduction in the 13th Five-Year Plan", the China Climate Risk Assessment Report 2023, Technical Guidelines on Adaptation to Climate Change—Vulnerability, Impacts and Risks Assessment, the Outline for High-Quality Development of Meteorology (2022–2035), and the National Strategy for Climate Change 2035, among others. The Program for Quality Development of Meteorology (2022–2035) and the National Strategy for Adaptation to Climate Change 2035 were issued. As a member of the "Basic four" countries in the international negotiations on climate change, Brazil has formed a legal system mainly based on the National Climate Change Policy Act and the National Plan on Climate Change. As a large developing biodiversity country, India has formulated eight action plans, namely Water Resources, Sustainable Himalayan Ecosystems, Green India, National Plan on Climate Change, Himalayan ecosystem, Green India, sustainable agricultural development, and strategic research on climate change to improve the ability of China and its people to adapt to climate change. As a country that is more vulnerable to the negative impacts of climate change, Egypt has strengthened its top-level design from the standpoint of a developing country, integrated climate governance into the overall development strategy of China, and established specialized agencies to form a professional and diversified governmental climate governance network, and it promotes industrial transformation and the implementation of green development projects.

Exercise

1. Explain the importance of economy, population, culture, education, and ecological resources in measuring regional geographic development differences.
2. What is the significance of total population and natural growth rate when studying regional geographic development differences? Please provide examples.
3. Explain the role of resources and the environment in regional sustainable development and discuss the differences in water resources and afforestation areas among different regions in China.
4. What are the main differences between developing and developed countries in terms of GDP and gross savings rate? Compare these with data from specific countries?
5. Compare the educational and cultural development levels of developing and developed countries using the gross enrolment rate in higher education and the number of resident patent applications. What do you think are the impacts of these differences on national development?

Extended Reading

Kiani, V., Ashrafi, A., & Alishahpoor, F. (2024). Identifying and prioritizing political-economic factors affecting Baloch ethnic challenges inside the political geography of Iran. *Human Geography Research.*

Li, R., Yu, Y., Cai, W., Liu, Y., & Li, Y. (2024). Exploring the gradient impact of climate and economic geographical factors on city-level building carbon emissions in China: Characteristics and enlightenments. *Sustainable Cities and Society.*

Ma, Y., Wang, X., & Zhong, C. (2024). Spatial and temporal differences and influencing factors of eco-efficiency of cultivated land use in main grain-producing areas of China. *Sustainability.*

Todaro, M. P., & Smith, S. C. (2014). *Development economics* (11th ed., Translated by Qiaoping Nie, Jingrong Cheng, Xiaowen Wang, et al.). Machinery Industry Press.

Wang, X. (2015). *A study of Singapore's great power balance diplomacy (1965–2014) (Doctoral dissertation).* Yunnan University.

You, C., Qu, H., Feng, C. C., & Guo, L. (2024). Evaluating the match between natural ecosystem service supply and cultural ecosystem service demand: Perspectives on spatiotemporal heterogeneity. *Environmental Impact Assessment Review.*

References

Aizenman, J., Jinjarak, Y., Lee, M., et al. (2016). Developing countries' financial vulnerability to the eurozone crisis: an event study of equity and bond markets. *Journal of Economic Policy Reform, 19*(1), 1–19.

Alkire, S., & Foster, J. (2011). Counting and multidimensional poverty measurement. *Journal of Public Economics, 95*(7–8), 476–487.

Amin, A. (2010). An institutionalist perspective on regional economic development. *International Journal of Urban and Regional Research, 23*(2), 365–378.

Arora, B., & Puranik, R. (2004). A review of corporate social responsibility in India. *Development, 47*(3), 93–100.

Babcicky, P. (2013). Rethinking the foundations of sustainability measurement: the limitations of the environmental sustainability index (ESI). *Social Indicators Research, 113*(1), 133–157.

Barreca, A., & Schaller, J. (2020). The impact of high ambient temperatures on delivery timing and gestational lengths. *Nature Climate Change, 10*(1), 77–82. https://doi.org/10.1038/s41558-019-0632-4.

Brajer, V., Mead, R. W., & Xiao, F. (2011). Searching for an environmental Kuznets curve in China's air pollution. *China Economic Review, 22*(3), 383–397.

Canestrino, R., Ćwiklicki, M., & Magliocca, P. (2020). Understanding social entrepreneurship: A cultural perspective in business research. *Journal of Business Research.*

Cappelen, A., Castellacci, F., Fagerberg, J., et al. (2003). The impact of EU regional support on growth and convergence in the European union. *Journal of Common Market Studies, 41*(4), 621–644.

Chen, P. (2024). Unlocking policy effects: Water resources management plans and urban water pollution. Journal of Environmental Management.

Combes, P., Duranton, G., & Gobillon, L. (2011). The identification of agglomeration economies. *Journal of Economic Geography, 11*(2), 253–266.

Cui, M., Hong, P., & Jiao, C. (2022). Overparenting and emerging adult development: a systematic review. *Emerging Adulthood, 10*(5), 1076–1094.

Dawson, N., Martin, A., & Sikor, T. (2016). Green revolution in Sub-Saharan Africa: Implications of imposed innovation for the wellbeing of rural smallholders. *World Development, 78*(2), 204–218.

Dreher, A., & Langlotz, S. (2020). Aid and growth: New evidence using an excludable instrument. *Canadian Journal of Economics/revue Canadienne D'économique, 53*(3), 1162–1198.

Durkheim, E. (2013). De la division du travail social. *History of Economic Thought Books, 30*(10), 297–301.

Egbunike, F. C., & Emudainohwo, O. B. (2018). Tax revenue and economic growth: A study of Nigeria and Ghana. *Signifikan: Jurnal Ilmu Ekonomi.*

Essex, J. (2013). *Development, security, and aid: Geopolitics and geoeconomics at the US agency for international development*. University of Georgia Press.

Evidence: Globalization and Poverty. (2006). University of Chicago Press NBER.

Feenstra, R. C., Inklaar, R., et al. (2015). The next generation of the Penn world table. *American Economic Review, 105*(10).

Ferrara, A., Salvati, L., Sabbi, A., et al. (2014). Soil resources, land cover changes and rural areas: Towards a spatial mismatch. *Science of the Total Environment, 478*(8), 116–122.

Foster, J. B. (2015). The new imperialism of globalized monopoly-finance capital: An introduction. *Monthly Review, 67*(3), 1–22.

Franck, S., von Bloh, W., Muller, C., Bondeau, A., & Sakschewski, B. (2011). Harvesting the sun: New estimations of the maximum population of planet earth. *Environmental Sciences and Ecology, 222*(12), 2019–2026.

Fraser, N. (2017). After the family wage: A postindustrial thought experiment. In B. Hobson (Ed.), *Gender and citizenship in transition.* New York: Routledge.

Jayachandran, S. (2015). The roots of gender inequality in developing countries. *Economics, 7*(1): 63–88.

Jiang, S., Li, E., Wei, Y., Yan, X., He, R., & Banny, E. T. (2024). Measurement and influencing factors of carbon emission efficiency based on the dual perspectives of water pollution and carbon neutrality. *Science of The Total Environment.*

Juma, C., Fang, K., Honca, D., Huete-Perez, J., Konde, V., Lee, S. H., Arenas, J., Ivinson, A., Robinson, H., & Singh, S. (2001). Global governance of technology: Meeting the needs of developing countries. *International Journal of Technology Management, 22*(7–8), 629–655. https://doi.org/10.1504/IJTM.2001.002982.

Juma, C., Fang, K., Honca, D., Huete-Perez, J., Konde, V., Lee, S. H., et al. (2001). Global governance of technology: meeting the needs of developing countries. *International Journal of Technology Management, 22*(7/8), 629–655.

Kanbur, R., & Sumner, A. (2012). Poor countries or poor people? Development assistance and the new geography of global poverty. *Journal of International Development, 24*(6), 686–695.

Khatana, S. A. M., Eberly, L. A., Nathan, A. S., & Groeneveld, P. W. (2023). Projected change in the burden of excess cardiovascular deaths associated With extreme heat by midcentury (2036–2065) in the contiguous United States. *Circulation, 148*(20), 1559–1569. https://doi.org/10.1161/CIRCULATIONAHA.123.066017

Kelly, P. F. (2010). Everyday urbanization: the social dynamics of development in Manila's extended metropolitan region. *International Journal of Urban and Regional Research, 23*(2), 283–303.

Leemans, R., & De Groot, S. R. (2003). *Millennium ecosystem assessment: Ecosystems and human well-being: A framework for assessment*. London: Island Press.

Leite, I. C., Pomeroy, M., & Suyama, B. (2015). Brazilian south–south development cooperation: The case of the ministry of social development in Africa. *Journal of International Development, 27*(8), 1446–1461.

Li, Y., Xiong, W., & Wang, X. (2019). Does polycentric and compact development alleviate urban traffic congestion? A case study of 98 Chinese cities. *Cities, 88*, 100–111.

Li, Y., Chui, Y., & Lu, L. (2018). Regional energy, CO_2, and economic and air quality index performances in china: A meta-frontier approach. *Energies, 11*(8), 2119.

Lutsey, N., & Sperling, D. (2008). America's bottom-up climate change mitigation policy. *Energy Policy, 36*(2), 673–685.

Martin, N., & Rice, J. (2014). Sustainable development pathways: determining socially constructed visions for cities. *Sustainable Development, 22*(6), 391–403.

Mawdsley, E. (2012). The changing geographies of foreign aid and development cooperation: Contributions from gift theory. *Transactions of the Institute of British Geographers, 37*(2), 256–272.

Milanovic, B., & Squire, L. (2007). Does tariff liberalization increase wage inequality? Some empirical evidence. In A. Harrison (Ed.), *Globalization and poverty*. University of Chicago Press. https://doi.org/10.7208/chicago/9780226318004.003.0005

Nazzaro, C., & Marotta, G. (2016). The common agricultural policy 2014–2020: Scenarios for the European agricultural and rural systems. *Agricultural and Food Economics, 4*(16).

Nesar, A., Edwardh, A., & Jamesf, M. (2010). Using the sustainable livelihoods framework to identify constraints and opportunities to the development of freshwater prawn farming in Southwest Bangladesh. *Journal of the World Aquaculture Society, 39*(5), 598–611.

Nissing, C., & Blottnitz, V. H. (2010). Renewable energy for sustainable urban development: A review of the effectiveness of policies in developing countries.

Raimi, M. O., Abiola, I., & Alima, O. (2021). Exploring how human activities disturb the balance of biogeochemical cycles: Evidence from the carbon, nitrogen and hydrologic cycles. *Nitrogen and Hydrologic Cycles.*

Ran, L. (2017). An empirical study on the promotion of contemporary international trade innovation and network trade to China the Belt and road development strategy. *Boletín Técnico, 55*, 335–343.

Razavi, S. (2016). The 2030 agenda: challenges of implementation to attain gender equality and women's rights. *Gender and Development, 24*(1), 25–41.

Schultz, W. T. (1961). Investment in human capital. *The American Economic Review, 51*(1), 1–17.

Shi, S. X., Huang, K. M., Ye, D. Z., & Yu, L. H. (2014). Culture and regional economic development: Evidence from China. *Papers in Regional Science, 93*(2), 281–299.

Shin, J., Mendoza, X., Hawkins, M. A. et al. (2017). The relationship between multinationality and performance: Knowledge-intensive versus capital-intensive service micro-multinational enterprises. *International Business Review, 26*(5).

Skufina, T., Baranov, S., & Samarina, V. (2015). Production functions in identifying the specifics of producing gross regional product of Russian Federation. *Social and Economic Journal.*

Teixeira, A. A. C., & Queirós, A. S. S. (2016). Economic growth, human capital and structural change: A dynamic panel data analysis. *Research Policy, 45*(8), 1470–1488.

Wang, X. (2017). *The measurement of poverty: Theories and methods* (2nd ed.). Beijing: Social Sciences Academic Press.

Wang, Y., et al. (2019). Does environmental regulation affect CO_2 emissions? Analysis based on threshold effect model. *Clean Technologies and Environmental Policy, 21*(3), 565–577.

Wiest, K. (2016). Migration and everyday discourses: Peripheralisation in rural Saxony-Anhalt: From a gender perspective. *Journal of Rural Studies, 43*, 280–290.

Wise, R. D., Covarrubias, M. H., & Puentes, R. (2013). Reframing the debate on migration, development and human rights. *International Journal of Population Geography, 19*(4), 430–443.

Wood, A. (2013). Drawing disability in Japanese manga: visual politics, embodied masculinity, and wheelchair basketball in Inoue Takehiko's real. *Culture, Medicine and Psychiatry, 37*(4), 637–655.

Yadav, S., Lama, S., & Kapley, A. (2021). Urban environmental issues and mitigation by applying ecological and ecosystem engineering. In *Handbook of ecological and environmental engineering*. Wiley Online Library.

Zhang, J., & Zhang, C. L. (2019). Quantitative evaluation and optimized utilization of water resources-water environment carrying capacity based on nature-based solutions. *Journal of Hydrology*, 568.

Zhang, K. H. (2014). Does foreign direct investment promote economic growth? Evidence from East Asia and Latin America. *Contemporary Economic Policy, 19*(1), 175–185.

Zhao, Y., Wang, Y., & Wang, Y. (2021). Comprehensive evaluation and influencing factors of urban agglomeration water resources carrying capacity. *Journal of Cleaner Production.*

Zhou, J. (2018). Why have almost all efforts to reduce poverty in sub-Saharan Africa failed? *Chinese Science Bulletin, 63*(7), 606–610.

Chapter 4
Spatial Patterns of Development in Less Developed Regions

Abstract This Chapter explores the evolution of regional development in China's less developed areas, focusing on spatial and temporal patterns. It reviews historical strategies such as the prioritization of the eastern seaboard, the Western Development Strategy, and the rise of central China, addressing regional disparities. Key drivers include rapid economic growth, urbanization, ecological civilization, and social improvements. The chapter analyzes economic and ecological efficiency, highlighting their roles in sustainable development. Finally, it examines social development and anti-poverty measures, using empirical models to identify influencing factors. The chapter provides a concise overview of the challenges and achievements in balancing regional development in China. The objectives of this chapter are as follows: 1. To examine the economic disparities between developed and less developed regions, analyze the main factors contributing to these imbalances, and explore the drivers of regional development, such as industrial structure, technological progress, and levels of informatization. 2. To analyze the spatial and temporal patterns of regional development during the process of rapid urbanization, study the impact of urbanization on economic growth, resource allocation, and the ecological environment, and evaluate urbanization's role in promoting regional development. 3. To evaluate the ecological efficiency of different regions, study the role of sustainable development policies and measures in promoting the balance between economic growth and ecological conservation, and explore pathways to achieve high-quality development. 4. To investigate the main drivers of regional development, including economic growth, industrial transformation, ecological civilization construction, and social and livelihood improvements, and analyze how these factors collectively promote the development of less developed regions.

X. Deng and M. Song, *Development Geography*,
https://doi.org/10.1007/978-981-96-6917-2_4

4.1 Development in Less Developed Regions

The report of the 19th Communist Party of China (CPC) National Congress points out that the main contradiction of Chinese socialism has changed into "the contradiction between the people's growing needs for a better life and the imbalance and insufficiency of development (Wen et al., 2023). The imbalance of development presents the typical characteristic of spatial heterogeneity, and this imbalance is expanding due to the development of industrial transformation, technological progress, and informatization (Brewer, 2006; Tulenty & Riabinina., 2022). In terms of the level of economic development, there is a large discrepancy between the developed and underdeveloped regions, and the level of regional well-being and the ability to develop also shows an expanding trend; therefore, the development of less developed regions has become an important topic (Jiang et al., 2014; Wang et al., 2021). The 40 years of reform and opening up have provided an important foundation for China's high-quality development, and the new China has undergone extensive changes in the past 70 years from its establishment in 1949 to 2019. China's economic development has even embarked on the road of rapid economic development. In 2007, China's total economic output succeeded in being the third-largest economic entity, surpassing Germany, and from 2009, China has surpassed Japan, becoming the second largest economy. In 2018, China's total gross national product (GNP) exceeded 90 trillion CNY, accounting for 16.7% of the global share, which is equivalent to an average annual growth of 9.4 percentage points since the reform and opening up in 1978.

In terms of economic growth rate, over the past 40 years of reform and opening up, China's nominal GDP growth rate has averaged as high as 14.5% per annum, which is still very high after excluding the annual inflation rate of 4.8%. In terms of the structure of economic growth, China's three-industry structure has become increasingly reasonable, and in 2018, the proportion of China's three industries was 4.4:38.9:56.5, and the tertiary industry has gradually become an important engine for pulling China's economic growth. As China's economic development has entered a new normal, China has shifted from high-speed development to high-quality development. At the 19th National Congress held in 2017, President Xi Jinping put forward the economic development of "promoting economic transformation and upgrading of high quality, and building a modern economic system", which has become the main theme of China's economic development in the next stage. China's embarking on a sustainable and high-quality development road, transforming the mode of economic growth, and moving toward the green direction have become an inevitable requirement for development under the historical conditions.

4.1.1 Development History of Less Developed Regions in China

The development practice of 40 years of reform and opening up has become important in Chinese development history. Since the Third Plenary Session of the Eleventh Central Committee of the Party, the development of China's reform and opening up has gone through a 40-year-long development practice. In the past 40 years, China has introduced different development policies to support fast-developing regions (Fig. 4.1), including policies to prioritize the development of coastal regions, implement the strategy of western development, implement the revitalization of the old northeastern industrial bases, promote the rise of central China, and accelerate the development of "One Belt, One Road", which has resulted in the emergence of different fast-developing regions.

4.1.1.1 Priority Development of the Eastern Seaboard

Prioritizing the development of the eastern coastal region is a development path explored on the basis of summarizing the socialist construction economy in the 30 years of the founding of New China to address the weak state of China's overall development, the total number of the rural population, and reduce the number of the population living in poverty, and efforts are being made to solve the difficult problem of the large development gap between regions. Over the past 40 years of reform and opening up, the eastern coast, as a pilot and early adopter region, has come up with a reform path of efficient government functions, rapid financial development, reasonable total volume and structure of trade services, expanding foreign investment, and optimizing tax policies and business environment, which has provided valuable experience for the establishment and improvement of China's socialist market economic system (Wang et al., 2023a, 2023b). In addition, the eastern coastal region has also played an important role in China's use of foreign investment, especially under the two markets locally and internationally, giving full play to China's comparative advantages, and through the use of foreign capital, technology, and management technology, etc., which has enabled China's enterprises to gradually integrate into

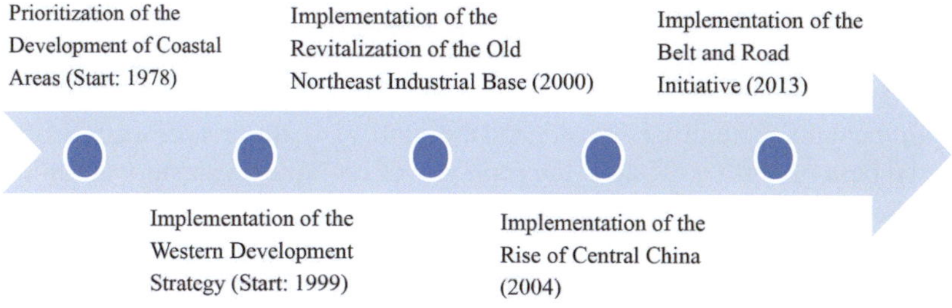

Fig. 4.1 Major regional development strategies implemented in China

the global trade market, and become an important part of the division of labor in the global trade. Finally, the eastern coastal region has also become an important carrier of technological innovation and industrial agglomeration in China. With the emergence of new technologies, methods, modes, and forms of business, the eastern region has become the "leader" of China's technological innovation, and as an important platform for opening up to the outside world, it also demonstrates China's "creation" to the world.

4.1.1.2 Western Development Strategy

The Western Development Strategy was proposed in response to the contemporary development, and it also provides an important foundation for China's poverty alleviation. The strategy of western development has made a qualitative leap since it was proposed in 1999. From a series of major implementation projects, the successive completion of the West–East Gas Transmission and the West–East Electricity Transmission have provided important support for China's economic development (Wang et al., 2023a, 2023b). In addition, with the implementation of poverty eradication strategies, China's western region has achieved significant results, with oil roads to counties, electricity delivery to townships, and radio and television broadcasting to villages basically realized; rural drinking water, rural energy, and ecological migration projects actively promoted; and the construction of public health facilities strengthened, thus reducing the population of the poor in rural areas annually. In the western development strategy, China adheres to the principle of giving equal importance to development and protection and has implemented a series of major projects for ecological environmental protection, which has led to a fundamental improvement in the ecologically fragile environment.

4.1.1.3 Revitalizing the Old Industrial Base in the Northeast

The revitalization of the old industrial base in the northeast was proposed in 2002, as a typical representative of resource-based regions, the development of the northeast region is affected by various factors and lags, but after the introduction of the policy of revitalizing the old industry, the economic aggregate of the northeastern region has been rising annually, and the economic structure has been continuously optimized; especially as the grain silo of China, the three eastern provinces have already become the agricultural production, processing, and distribution center. Meanwhile, the equipment manufacturing industry and the quality of enterprise products in the old industrial base of northeast China, the economy of multiple ownership, the ability of independent innovation, the infrastructure conditions have improved significantly, the emerging industries are steadily developing, and the key aspects of people's livelihoods have been protected in a focused manner.

4.1.1.4 Rise of Central China

The rise of central China is an important development geography practice that began in 2004. Since its introduction, it has achieved remarkable results, specifically, the central region's economic development momentum is good, the economic strength has significantly improved, and as China's important food and energy base, the central region has begun to accelerate the transition to undertake the eastern coastal equipment manufacturing industry, focusing on the development of high-tech industrial bases and new heights of comprehensive logistics hubs, and the construction of resource conservation and environmental protection in the central region has achieved initial results. In addition, the ecological environment in the central region has been protected.

4.1.2 Drivers of Development in Less Developed Regions

The extension of less developed regions has been expanding (Fig. 4.2), from the early stage of the national construction, which was centered on economic construction, gradually expanding the radius of development. With the gradual development of China's economy, there was a period of rapid urbanization, which was most typically characterized by the urbanization rate as the main measure of urban development. With the 18th National Congress writing ecological civilization into the Party Constitution for the first time, ecological civilization and economic development have taken an equally important position, and weighing the development relationship between the two has become an important metric for fast-developing regions. The rural revitalization proposed in the report of the 19th National Congress of the Communist Party of China (CPC) has provided an important directional guideline for China's agricultural and rural reform and development, and in particular in 2025 and 2030, it will become the rapid rural revitalization area's core component.

From the perspective of development geography, the idea of underdeveloped regions is dynamic and spatial-temporally heterogeneous, for this reason, correctly identifying the underdeveloped regions of the current development of China should be scientifically identified in terms of economic efficiency and eco-efficiency as the key elemental items, and analyzing their state of flux and influencing factors is important.

4.1.2.1 Rapid Economic Growth

Rapid economic growth is a development strategy centered on economic construction that was clarified by the state at the beginning of the establishment of new China. During the period 1978–2017, China's economy experienced relatively rapid growth, and from 2000–2017, China's total GDP increased rapidly, with a growth rate significantly higher than that of 1990–2000.

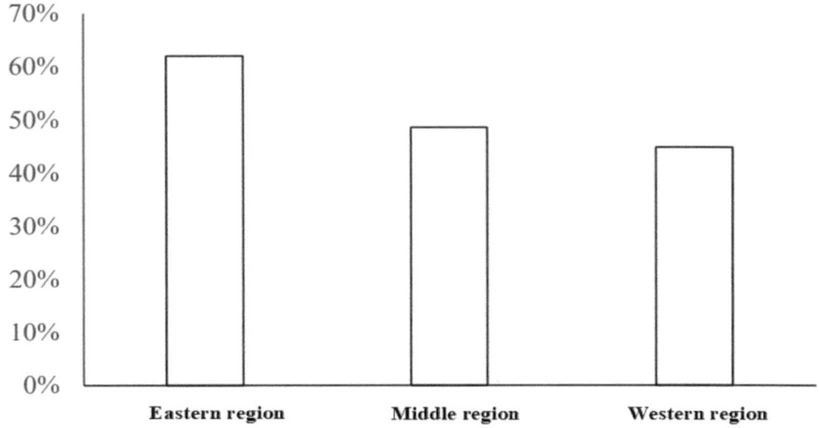

Fig. 4.2 Levels of urbanization rates in China's eastern, central and western regions, 2017

However, in the process of development, regional development also emerged, but it created the problem of uneven development in some areas. The total GDP for 2018 in the southeast provinces and redundant northwestern provinces was analyzed. Out of the top ten provinces, only Shandong province belongs to the northern provinces, whereas the rest are all southern provinces. The total GDP of the province with the highest GDP is 70 times higher than that of the province with the lowest GDP. The Guizhou and Tibet provinces have a GDP growth rate of more than 9%, whereas 12 provinces have a growth rate lower than the national average.

In the past 40 years of reform and opening up, China's economy has been at the stage of crossing the river by touching the stones, the economic system has been continuously improved, and the supporting measures for reform tend to be clear. After the initial growth, China's economy has experienced radical change China's and has transformed from a planned economy to a market economy. The first stage of the economic system reform was mainly focused on the rural areas of China. Due to the mismatch between productive forces and production relations, the land contract responsibility system was implemented, whereby farmers signed long-term land contracts with the government, had the right to use the land under collective ownership, and effectively increased the efficiency of agricultural production through the establishment of non-farming enterprises. Moreover, by encouraging rural credit and increasing rural savings, the production and living conditions in China side improved rapidly (Wei & Liefner, 2012; Wu et al., 2020), and there was surplus labor movement between different sectors.

The second stage of economic reform was the reform of state-owned enterprises (SOEs), on which SOEs were gradually given greater autonomy in development, restrictions limiting the efficiency of SOEs were cancelled, and the participation of township and village enterprises began to form a pattern of international and domestic competition.

The third stage of economic development came at the end of the Southern Tour speech. China's economy began to move toward openness and sharing, with the acceleration of factor mobility between urban and rural areas, administrative and regulatory reforms in the banking system, the tax system, foreign trade, and foreign investment, removing various binding constraints on growth, and the rapid development of the urban economy.

Overall, China's economy is undergoing a shift from high-speed growth to high-quality growth, and achieving a rational allocation of resources and promoting high-quality economic growth by relying on the labor force and productivity levels is the next step in the direction of progress.

4.1.2.2 Rapid Urbanization

As an inevitable stage of China's development, between 1978 and 2017, China's urbanization has made world-renowned achievements. Its urbanization rate was 17.9% in 1978 and 58.5% in 2017, China's an increase by 40 percentage points in the past 40 years of reform and opening up, and the resident urban population has increased from 172 million in 1978 to 813 million in 2017. The urban population has gradually increased, and China has successfully transformed from a country dominated by an agricultural population to one dominated by an urban population.

Urbanization has followed the development of China's industrial economy to achieve the agglomeration of non-agricultural industries. Concentration toward towns and cities throughout the agglomeration process is an inevitable condition of historical development, the objective law of human development, and an important symbol of a country's modernization. The realization of urbanization is important for the orderly advancement of China's overall economic development, particularly in the primary stage of socialist construction (Luo et al., 2016). Urbanization is the path that China must go through to achieve modernization, and with the development of the economy, together with industrialization, information networking, and agricultural modernization, urbanization has become an important part of China's sustainable development. These aspects complement each other. Industrialization is the precursor to development and the driving force for the overall development. Agricultural modernization is the bedrock of development and the source of service to industrialization and urbanization. Information modernization injects new vitality into urbanization.

In addition, urbanization injects a new engine for sustained and healthy economic development. An important pillar of economic development is the release of domestic demand to stimulate stable economic growth with domestic demand, and an important part of the release of domestic demand (Fan et al., 2012). Although China's urbanization rate has steadily increased after years of development, the household population remains at much lower level compared with that in developed countries. Urbanization results in the transfer of surplus rural labor to urban settings, and citizens will enjoy better public services, which are enjoyed in towns and cities, thus releasing new energy, providing activities, and enhancing the healthy development of the economy.

Urbanization is also the main factor in achieving high-quality development of industry. The optimization and upgrading of industrial structure is an important task for the transformation of the economic growth mode, and accelerating the development of the service industry is also an important direction for the optimization of China's industrial structure. Currently, there is still a discrepancy between the level of China's service industry and that of developed countries, especially in the proportion of the total output value of the service industry to GDP. Achieving urbanization is closely related to increasing the proportion of the service industry, and the level of the service industry, particularly the concentration and convergence of factors in the production process, changes in lifestyles, and changes in the level of production will expand the demand for the service industry. Urbanization will bring about the reconfiguration and distribution of urban and rural factor resources, which is conducive to promoting the further development of the service industry.

Urbanization is a powerful tool for promoting coordinated regional development. Since the reform and opening up, China's eastern coastal areas have taken the lead in opening up and development, forming a number of urban agglomerations such as Beijing-Tianjin-Hebei, the Yangtze River Delta, the Pearl River Delta, etc., which have strongly promoted the rapid development of the eastern region and become an important growth pole of the national economy. However, the development of the central and western regions is relatively lagging behind, mainly because the urbanization process is unbalanced, and the development of cities in the central and western regions is insufficient. Currently, the urbanization rate of the resident population in the eastern region has reached 62.2%, whereas that of the central and western regions is only 48.5% and 44.8%, respectively (Fig. 4.2). With the in-depth promotion of the strategy for the development of the western region and the rise of central China, and the acceleration of the transfer of industries from the eastern coastal region, it will be beneficial to accelerate the process of urbanization and cultivate the formation of new poles of growth in the central and western regions of China, which have a strong carrying capacity in terms of resources and the environment, to promote the graded expansion of economic growth and market space from east to west, and from south to north, and to push for a more reasonable population and economic layout and a more coordinated regional development.

Urbanization is an inevitable requirement for the promotion of overall social progress. As a product of the progress of human civilization, urbanization can not only improve the efficiency of production activities but also enrich farmers improve people's quality of life. With the prosperity of the urban economy, the improvement of urban functions, the level of public services and the quality of the ecological environment, people's material lives will be more substantial and abundant, and their spiritual lives will be more enjoyable. With the gradual breaking up of the urban–rural dualistic system and the gradual dissolution of the contradictions of the dualistic structure within the city, the whole population will share the fruits of modern civilization. This is conducive to maintaining social justice and eliminating hidden social risks, as well as promoting comprehensive human development and harmonious social progress.

4.1.2.3 Ecological Civilization Construction

Since the report of the 18th National Congress of the Communist Party of China (CPC) included ecological civilization in the Party's constitution, China's ecological civilization construction has entered the fast lane. The Congress, which was held in November 2012, raised the construction of ecological civilization to an unprecedented strategic height, as the overall layout of the construction of socialist undertakings with Chinese characteristics, and compared it with the economic construction, political construction, cultural construction, and social construction. The report clearly states that "the construction of ecological civilization is the most important part of the socialist cause with Chinese characteristics." The report clearly states that "building an ecological civilization is a long-term plan for the well-being of people and the future of the nation. In the face of the severe situation of tightening resource constraints, serious environmental pollution, and degradation of ecosystems, it is necessary to establish the concept of ecological civilization of respecting nature, adapting to nature, and protecting nature, and to place the construction of ecological civilization in a prominent position, integrating it into all aspects and the whole process of economic, political, cultural, and social construction, striving to build a beautiful China, and realizing the perpetual development of the Chinese nation. The continuous development of socialist construction with Chinese characteristics from the "three-in-one" to the "four-in-one" to the "five-in-one" is a summary of the practice of socialist construction. It marks the Party's more precise understanding of the essence of socialism with Chinese characteristics in both theory and practice and the Party's deepening understanding of the laws of socialist construction, communist party rule, and human social development.

The third session of the 18th CPC Central Committee, held in November 2013, made further deployment of ecological civilization construction, explicitly stating that it would deepen the reform of the ecological civilization system closely around the construction of a beautiful China, accelerate the establishment of a systematic and complete ecological civilization institutional system, and improve the institutional mechanism for the development of territorial space, the economic use of resources, and ecological environmental protection to push forward the formation of a new pattern for the construction of modernization that harmonizes the development of human beings and nature. The Outline of the 13th Five-Year Plan, adopted at the Fourth Session of the 12th National People's Congress in March 2016, includes a section on "Accelerating the Improvement of Ecological and Environmental Problems", which points out that improving environmental quality is the core of the plan and focuses on solving the outstanding problems in the field of ecological and environmental protection, increasing ecological environmental protection, and improving the quality of the ecological environment, and concertedly promoting the prosperity of people, the wealth of China, and the beauty of China.

4.1.2.4 Social and Livelihood Improvement Demands

The shift from precise poverty eradication to comprehensive well-off social marks a fundamental transformation of underdeveloped regions. The report of the 19th National Congress points out that the main contradiction in China's society has been transformed into the contradiction between the growing need for a better life and the unbalanced and inadequate development, especially with the gradual improvement of the people's living standards, the demand for external conditions is not only the improvement of the material and cultural needs but also the enhancement of the demand for democracy, the rule of law, fairness, justice, as well as security and the environment, and other topics (Liu et al., 2014). 2020 was the year of building a moderately prosperous society in all respects, and also the year of the final sprint out of poverty. As of 2019, about 5% of China's remaining poor people and about 10% of China's poor counties have not been lifted out of poverty. China aimed to complete the lifting of the remaining poor people and poor counties out of poverty by 2020 in the underdeveloped regions to achieve leapfrogging development. Precision poverty eradication not only realizes the economic development of underdeveloped regions but also improves the industrial system, production status, and sustainability framework, and the welfare of people in poor regions.

4.1.3 Evaluation of Development in Less Developed Regions

4.1.3.1 Evaluation Development Levels Based on the Principal Component Analysis Method

For the multi-dimensional time series multi-indicator comprehensive evaluation method, researchers have used methods, which can be mainly divided into subjective and objective methods for determining the weight of the indicators. For example, hierarchical analysis is based on the evaluation of subjective experts to determine the importance of the indicators. This method tends to exaggerate or diminish the weight of these different indicators. In contrast to the subjective evaluation method, objective methods of empowerment, such as the principal component analysis and entropy value methods are not influenced by external factors. The principal component analysis method is an important method to achieve the indicators of dimensionality reduction, based on the quality of the data. The method can help determine the quality of economic growth weights in order to reflect the quality of China's economic growth in terms of the dimensions of the contribution of the weights of the different indicators of the degree of magnitude. For this reason, this study mainly uses the main components of the analysis method to assess the quality of economic growth of the 19 indicators of the measurements.

The key equation of the principal component analysis method is as follows:

$$\begin{cases} x_1 = \alpha_{11}F_1 + \alpha_{12}F_2 + \ldots + \alpha_{1m}F_m + \varepsilon_1 \\ x_2 = \alpha_{21}F_1 + \alpha_{22}F_2 + \ldots + \alpha_{2m}F_m + \varepsilon_2 \\ \ldots \\ x_n = \alpha_{n1}F_1 + \alpha_{n2}F_2 + \ldots + \alpha_{nm}F_m + \varepsilon_n \end{cases} \tag{4.1}$$

where, $x_1, x_2, \ldots, x_n$ are the n variables, $F_1, F_2, \ldots, F_m$ are the m principal components extracted for this study, and $\varepsilon_1, \varepsilon_2, \ldots, \varepsilon_n$ are the error terms.

After simplification, the above equation can be reduced to

$$\begin{bmatrix} x_1 \\ x_2 \\ \ldots \\ x_n \end{bmatrix} = \begin{bmatrix} \alpha_{11} & \alpha_{12} & \ldots & \alpha_{1m} \\ \alpha_{21} & \alpha_{22} & \ldots & \alpha_{2m} \\ \ldots & \ldots & \ldots & \ldots \\ \alpha_{n1} & \alpha_{n2} & \ldots & \alpha_{nm} \end{bmatrix} \begin{bmatrix} F_1 \\ F_2 \\ \ldots \\ F_n \end{bmatrix} + \begin{bmatrix} \varepsilon_1 \\ \varepsilon_2 \\ \ldots \\ \varepsilon_n \end{bmatrix} \tag{4.2}$$

Further abbreviated as:

$$X = AF + \varepsilon \tag{4.3}$$

where the matrix A is the component coefficient matrix in principal component analysis, α_{ij} is the variable in the coefficient matrix, x_i is the factor loading on the principal component F_j, and the larger the value of the factor loading, the higher the correlation between it and the principal component.

Before analyzing the data, it is necessary to carry out a series of data processing takes for principal component analysis, including the stages of standardization of data, applicability test of principal components, and extraction of principal components.

Specifically, the first step is to standardize the data. In the process of collecting the indicators of the quality of China's economic growth in this study, different indicator data are collected; however, the quantitative outline of these data is inconsistent. Therefore, before analyzing the data, it is necessary to eliminate the influence of the quantitative outline, standardize the data, which can be trended using the inverse method of the indicators, and then the sampled the data. The data are standardized to obtain the standard sample matrix:

$$Z_{ij} = \frac{x_{ij} - \overline{x}_j}{S_j} \tag{4.4}$$

where x_{ij} is the original data before processing, $\overline{x}_j$ is the data mean, S_j indicates the data standard deviation, and Z_{ij} is the data value after processing.

Second, the KMO test and Barlett's sphericity test are carried out on the data, which is mainly to test whether the data after standardization meet the requirements of principal component analysis, and the values of the two test indexes range from 0 to 1. The calculation formula is as follows:

$$KMO = \frac{\sum \sum \gamma_{ij}^2}{\sum \sum \gamma_{ij}^2 + \sum \sum p_{ij}^2} \quad (4.5)$$

where γ_{ij}^2 characterizes the correlation coefficient of variables i and j, and p_{ij}^2 characterizes the partial correlation coefficient between variables i and j.

According to the test, in general, the KMO value is greater than 0.5 to characterize the applicability of principal component analysis. A value less than 0.5 means that it is not applicable.

Finally, the principal component analysis is extracted. Generally, in the process of principal component analysis, it is not necessary to select all the variables; therefore, only the principal components are selected. There are two methods of selection. The first is to select the factor with an eigenvalue greater than 1 as a component factor of the principal component analysis. This method can lead to missing information. The other method is to select the cumulative contribution rate of the variance that is more than 85%. This method reduces the problem of missing information in the process of principal component analysis.

Based on the eigenvalues, the eigenvectors $\lambda_1, \lambda_2, \ldots, \lambda_p$ can be obtained, where λ_i is the variance of the principal component F_i, and the larger the value, the larger the contribution to the principal component.

The variance contribution of the ith principal component can be further obtained as follows:

$$e_i = \frac{\lambda_i}{\sum_{k=1}^{p} \lambda_k} \quad (4.6)$$

Further calculation of the cumulative contribution of the first q principal components can be expressed as follows:

$$\sum_{i=1}^{q} e_i = \frac{\sum_{k=1}^{q} \lambda_k}{\sum_{k=1}^{p} \lambda_k} \quad (4.7)$$

Generally, the selected cumulative contribution is more than 85%.

Based on the above steps, the number of principal components can be determined, and then based on the number of principal components, the linear combination of the original variables, also known as the principal component score, is derived, with the expression:

$$F_j = \beta_{j1}x_1 + \beta_{j2}x_2 + \ldots + \beta_{jp}x_p \quad (4.8)$$

where $x_1, x_2, \ldots, x_n$ are the data after further normalization.

4.1.3.2 Development Correlation Analysis Based on Spatial Statistics

Exploratory spatial data analysis is an important method to analyze spatial interrelationships, and the spatial weight matrix should be determined before carrying out the analysis.

Generally, the n*n-order spatial weight matrix includes the spatial interrelationships between the ith spatial unit and the jth unit; therefore, the determination of the matrix should be all exogenous. The diagonal of the spatial weight matrix has, in general, a spatial distance of 0. The size of the other elements is confirmed based on the specific spatial cell relationship, and the spatial weight matrix can be expressed as follows:

$$W = \begin{pmatrix} W_{11} & W_{12} & \dots & W_{1n} \\ W_{21} & W_{22} & \dots & W_{2n} \\ \dots & \dots & \dots & \dots \\ W_{n1} & W_{n2} & \dots & W_{nn} \end{pmatrix} \tag{4.9}$$

Considering the universality and applicability of the spatial weight matrix, this study mainly adopts the binary to determine the spatial weight relationship, that is, the method of determining the weight matrix by using the "car" collocation method, which is denoted as $W_{ij} = 1$ if the spatial units i and j are adjacent to each other. If the spatial units are not adjacent to each other, it is denoted as $W_{ij} = 0$, that is:

$$W_{ij} = \begin{cases} 1 \text{ } space\ cell\ i\ is\ adjacent\ to\ j \\ 0 \text{ } i = j\ or\ space\ cells\ i\ and\ j\ are\ not\ adjacent \end{cases} \tag{4.10}$$

Rook contiguity adjacency matrix was chosen for this study, that is, collocation is represented by 1 and non-collocation by 0.

Exploratory spatial data analysis is a spatial data analysis method with identification function, which is mainly used to explore the non-randomness or spatial autocorrelation of the spatial distribution. The method not only reflects the spatial distribution characteristics of the data but also serves as a basis for empirical research using spatial econometrics. Spatial autocorrelation can reflect the spatial association and clustering of variables and data. There are two specific methods: global spatial autocorrelation and local spatial autocorrelation.

An important indicator of global spatial autocorrelation is Moran's I index, which can reflect the average degree of clustering of spatial unit variables and test whether the variables are similar, different, or independent in the whole region. The value of Moran's I index ranges from −1 to 1. Values greater than 0 mean that high values are neighboring high values or low values are neighboring low values, and values less than 0 mean that high values are neighboring low values or low values are neighboring high values. Values equal to 0 mean that the variables are independent. In the presence of spatial autocorrelation, the larger the absolute value, the higher the spatial correlation of the variables. Moran's I index is calculated using the following formula:

$$I = \sum_{i=1}^{n}\sum_{j=1}^{n} W_{ij}(Y_i - \overline{Y})(Y_j - \overline{Y}) / \sum_{i=1}^{n}(Y_i - \overline{Y})^2 \tag{4.11}$$

Local spatial autocorrelation is used to test for clustering of variables in local space, and two commonly used statistics are the Moran scatterplot and LISA clustering plot. If the variables are positively correlated in space, the scatter falls in quadrants one and three, and if the variables are negatively correlated in space, the scatter falls in quadrants two and four. A scatter in the first quadrant indicates that both the spatial unit itself and the surrounding spatial units are high. A scatter in the third quadrant indicates that both the spatial unit itself and the surrounding spatial units are low. A scatter in the second quadrant indicates that the spatial unit itself is low but the surrounding spatial units are high. A scatter in the fourth quadrant indicates that the spatial unit itself is high but the surrounding spatial units are low. The scatter in the fourth quadrant indicates that the spatial unit itself is high, but the surrounding spatial units are low.

The LISA clustering map enables the observations of variables to be visualized on a map, with different colors used to indicate that high values are adjacent to high values, high values are adjacent to low values, low values are adjacent to low values, and low values are adjacent to high values. Where spatial clustering is not evident, it is not marked on the map, making it a very intuitive tool for representing clustering effects.

4.1.3.3 Measurement of Development Efficiency Based on Stochastic Frontier Functions

Stochastic Frontier Analysis (SFA) has been applied to multifaceted efficiency measurement and estimation since it was proposed in the 1970s and adapted to the study of multiple-input but output problems. Before applying SFA, it is necessary to make assumptions about the distribution of the error term, and the basic model assumptions are as follows:

$$y_{it} = f(x_{it}; \beta)\exp(\upsilon_{it} - \mu_{it}) \tag{4.12}$$

$$\mu_{it} = \mu_i \exp[-\eta(t - T)] \tag{4.13}$$

$$TE_{it} = \exp(-\mu_{it}) \tag{4.14}$$

where outputs and inputs are denoted as y_{it} and x_{it}, respectively, the parameter to be estimated is β, and the error term is $(\upsilon_{it} - \mu_{it})$, $\upsilon_{it} \in iid$ (independently and identically distributed) and obeys the $N(0, \sigma_v^2)$ distribution, which characterizes the measurement error as well as the non-estimable stochastic factor, and the second part of μ_{it} characterizes the stochastic error for region i in time period t and represents the inefficiency term of the production process, and the term obeys a truncated nontrivial distribution. The key influence of the time factor on the inefficiency term μ_{it} is shown in Eq. (4.13), where η is the attenuation coefficient and μ_{it} represents the level of technical inefficiency in region i that does not change over time. $TE_{it} = \exp(-\mu_{it})$ represents the technical efficiency, and when $\mu_{it} = 0$, it indicates that the individual is at the decision-making frontier, and when $\mu_{it} > 0$, it means that the individual is below the production or frontier level, that is, in a state of non-technical efficiency.

This study applies the Cobb–Douglas production function to establish a stochastic frontier model to empirically test China's economic growth efficiency and influencing factors using 2000–2017 panel data as a sample. The following empirical model is used:

$$ln(y_{it}) = \beta_0 + \beta_1 ln(L_{it}) + \beta_2 ln(K_{it}) + \upsilon_{it} - \mu_{it} \tag{4.15}$$

$$\begin{aligned} m_{it} = \delta_0 + \delta_1 (FDI/GDP)_{it} + \delta_2 (TRA/GDP)_{it} \\ + \delta_3 (TIAV/GDP)_{it} + \delta_4 [(DEP + LOA)/GDP]_{it} \end{aligned} \tag{4.16}$$

$$TE_{it} = \frac{E[f(x_{it}; \beta)\exp(\upsilon_{it} - \mu_{it})]}{E[f(x_{it}; \beta)\exp(\upsilon_{it} - \mu_{it})|\mu_{it} = 0]} = \exp(-\mu_{it}) \tag{4.17}$$

$$\mu_{it} = \mu_i \exp[-\eta(t - T)] \tag{4.18}$$

$$\gamma = \sigma_\mu^2 / (\sigma_\upsilon^2 + \sigma_\mu^2) \tag{4.19}$$

where β_0 is the intercept term, β_1 and β_2 are the parameters to be estimated, and the error stochastic term consists of υ_{it} and μ_{it} together. L_{it} is the number of people employed in the whole society (in 10,000 people), and K_{it} is the fixed capital stock (in billion yuan), which is depreciated using the perpetual inventory method. m_{it} is the technological inefficiency term, δ_0 is the constant term, and δ_1 to δ_4 are the parameter terms to be estimated. The positive and negative effects of the parameters are used to estimate the impact of the parameters on technical efficiency. Drawing on the results of previous research, this study focuses on the impact of the following factors on the technical efficiency of economic growth: $(FDI/GDP)_{it}$ represents the level of openness to the outside world, reflecting the scale of foreign direct investment absorbed by each province; $(TRA/GDP)_{it}$ represents the degree of foreign trade dependence, using the current year's export and import rates as the basis. *DEP*, expressed as the ratio of total import and export trade to GDP in that year, mainly indicates the development of foreign trade in that year. $(TIAV/GDP)_{it}$ represents the industrial structure, expressed as the industrial structure and as the proportion of the added value of the tertiary industry in each province to the GDP in that year, responding to the development of the industry. $[(DEP + LOA)/GDP]_{it}$ represents financial development and is characterized by the ratio of deposit and loan balances to the current year's GDP for each province, indicating the financial development status of different provinces.

4.1.3.4 Evaluation of Eco-efficiency Convergence Based on Spatial Club Indicators

The measurement of the convergence model was initially mainly applied to measuring whether there is convergence or divergence in the income gap between countries, and has gradually been used in the fields of consumption, trade, and environment. In the study of convergence, the β-convergence method is the most commonly used detection method, and its theoretical basis is mainly derived from the assumption that diminishing capital gains will cause economic growth to be regionally consistent. In regions with differences in income levels, β-convergence refers to the phenomenon that in the process of economic development, the development of regions with a lower economic development status is initially faster than that of those with a higher economic development status, and over time, the economic development levels of the two regions eventually become consistent. β-convergence includes absolute β-convergence and conditional β-convergence. Absolute β-convergence is used to depict the convergence between different regions by excluding external factors, and conditional β-convergence is a test of the convergence between the regions taking external factors into consideration. Based on the convergence model of Barro and Sala-i-Martin, the absolute β-convergence model is as follows:

$$\frac{1}{T}ln\left(\frac{y_{i,t+T}}{y_{i,t}}\right) = \alpha + \beta ln\left(y_{i,t}\right) + \varepsilon_{i,t} \quad \varepsilon \sim N\left(0, \delta^2\right) \tag{4.20}$$

where $y_{i,t+T}$ and $y_{i,t}$ represent the level of economic efficiency in period $t + T$ and period t, respectively, and α is a constant term with $\beta = -\left(1 - e^{-\theta T}\right)$, where θ denotes the rate of convergence between regions. After regression, if $\beta < 0$, it means that there is a converging absolute β over the time period under study, that is, the economic efficiency of the lagging regions is significantly faster than that of the more developed regions.

Conditional β-convergence is developed on the basis of the absolute β-convergence mentioned above, and it is calculated as follows:

$$\frac{1}{T}ln\left(\frac{y_{i,t+T}}{y_{i,t}}\right) = \alpha + \beta_1 ln\left(y_{i,t}\right) + \beta_2 X_{i,t} + \varepsilon_{i,t} \quad \varepsilon \sim N\left(0, \delta^2\right) \tag{4.21}$$

From the traditional β-convergence test, it can be seen that the spatial effect is not included in the test, but in reality, the economic efficiency of a region is affected by the technological level of the neighboring region and the type of production Based on this, this study includes the spatial effect in the ordinary convergence model to establish the spatial autoregression (SAR) model, which is based on the spatial effect of a region. The SAR model and the Spatial Error Model (SEM) are established for convergence analysis.

4.1.3.5 Analysis of Factors Affecting Regional Development Based on Spatial Autoregressive Model

The SAR model considers the spatial correlation, and in the classical regression model, the attribute data and location data of the spatial relationship is combined to analyze the spatial interrelationships between geographic factors. After adding the spatial autoregressive effect, the equation of the spatial regression model with absolute β-convergence and conditional β-convergence is deformed as:

$$\frac{1}{T}ln\left(\frac{y_{i,t+T}}{y_{i,t}}\right) = \alpha + \beta ln\left(y_{i,t}\right) + \rho W ln\left(\frac{y_{i,t+T}}{y_{i,t}}\right) + \varepsilon_{i,t} \quad \varepsilon \sim N\left(0, \delta^2\right) \tag{4.22}$$

$$\frac{1}{T}ln\left(\frac{y_{i,t+T}}{y_{i,t}}\right) = \alpha + \beta_1 ln\left(y_{i,t}\right) + \beta_2 X_{i,t} + \rho W ln\left(\frac{y_{i,t+T}}{y_{i,t}}\right) + \varepsilon_{i,t} \quad \varepsilon \sim N\left(0, \delta^2\right) \tag{4.23}$$

where ρ is the spatial regression coefficient representing the magnitude of the spatial forces in each province, W is the spatial weight matrix, and ε is the random error term.

4.1.3.6 Evaluation of Regional Development Influencing Factors Based on Spatial Error Convergence Model

The spatial error convergence model focuses on analyzing the relationship between regional spillover effects and random shocks, and it is used to portray the differences in interactions between different regions because of their different locations. After adding the spatial error term, the equation of the spatial regression model with absolute β-convergence and conditional β-convergence is deformed as:

$$\frac{1}{T}ln\left(\frac{y_{i,t+T}}{y_{i,t}}\right) = \alpha + \beta ln\left(y_{i,t}\right) + \varepsilon_{i,t}\varepsilon_{i,t} = \lambda W + \mu, \mu \sim N\left(0, \delta^2 I\right) \tag{4.24}$$

$$\frac{1}{T}ln\left(\frac{y_{i,t+T}}{y_{i,t}}\right) = \alpha + \beta_1 ln\left(y_{i,t}\right) + \beta_2 X_{i,t} + \varepsilon_{i,t}\varepsilon_{i,t} = \lambda W + \mu, \mu \sim N\left(0, \delta^2 I\right) \tag{4.25}$$

where λ represents the regression residual correlation coefficient and μ is a spatially uncorrelated disturbance term.

4.1.3.7 Analysis of Factors Affecting Regional Poverty Based on Panel Regression Models

Panel-data vector autoregression (PVAR) was first proposed by Holtz-Eakin et al. (1988), and has been continuously improved by McCoskey and Kao, Joakim Westerlund, and other scholars. It has now become an important tool for the analysis and research of panel data and time series data in the field of modern econometrics. Compared with the VAR model, PVAR is flexible in terms of the time length of the data and is more applicable. When T is the length of the time series and m is the lag order, the parameters of the equation can be estimated as long as they satisfy $T \geq m + 3$; and when T satisfies $T \geq 2m + 2$, the parameters of the lagged term can be estimated in the steady state. PVAR combines the heterogeneity of individuals in the panel data model characteristics and the characteristics of interaction between variables included in VAR, which effectively solves the system endogeneity problem and truly reflects the dynamic response relationship between variables, and it has been widely used in the analysis of economic problems. Therefore, this study establishes the following PVAR model:

$$\delta_{i,t} = \lambda_0 + \sum_{j=1}^{k} \lambda_j \delta_{i,t-j} + \mu_i + o_i + \varepsilon_{i,t} \tag{4.26}$$

where $\delta_{i,t}$ and $\delta\_\delta_{i,t-j}$ denote the matrix of explanatory variables consisting of the matrix of endogenous variables and the lagged terms of the endogenous variables, respectively; λ_j is the estimation matrix of the jth order of lagging; μ_i is a fixed-effects term that captures inter-provincial differences; o_i denotes a time effect to account for the characteristics of trends in the systematic variables; and $\varepsilon_{i,t}$ is assumed to be a random perturbation term obeying a normal distribution.

4.2 Spatial and Temporal Patterns of Regional Economic Efficiency

The development of regional economy contains several aspects. In addition to the total GDP and GDP growth rate, it includes the efficiency indicators of economic development, particularly in the process of China's high-speed economic development, achieving high-quality development of the economy, and efficiency as an important consideration indicator.

Economic efficiency is a key indicator linking the turning point between high economic growth and high-quality economic development. High-quality development is an important expression for China's economic development in the report of the 19th National Congress of the Communist Party of China (CPC), which shows that China's economy has shifted from the stage of high-speed economic growth to the stage of high-quality growth and that realizing a turnaround in the mode of economic growth is not only a necessity for economic development but also where the objective economic law lies (Wang et al., 2021). In the process of the turnaround, with the economic efficiency drive as the core, defining the economic growth efficiency and quality of the connotative growth path is the key to achieve high-quality development, enhance the ability of the market to allocate resources, optimize the industrial structure, and focus on industrial transformation and upgrading. In 2016, in the government's report, the concept of the economic growth interval was introduced for the first time, with an expected economic growth interval of 6.5–7% as the development goal. Economic efficiency has not received sufficient attention; therefore, identifying the current status of economic efficiency and conducting a convergence test for different provinces in China is important.

Spatial (spillover) effects become an important path for regional (provincial) units to interact in terms of efficiency (Bai et al., 2012). The spatial effect refers to the mutual-influence relationship between different units within a neighborhood unit, and due to the existence of the spatial (spillover) effect, it will make the economic efficiency of different regions converge with each other. The economic efficiency of a spatial unit does not only depend on the level of resource endowment of the economy but is also affected by the economic efficiency of neighboring economies, and similarly, inextricably linked within an economy for various reasons. Therefore, it is of great scientific value to conduct research on regional economic efficiency in the context of spatial effects.

Spatial club convergence is based on the concept of spatial heterogeneity to explain the interconnectedness of regions. Related research around regional economic growth convergence has become a topic of great interest in economics research (Wang & Wang, 2019), and the "absolute convergence hypothesis" and "conditional convergence hypothesis" have been proposed based on traditional neoclassical economic growth theory and new economic growth theory. The absolute convergence hypothesis argues that, based on the assumption of constant returns to scale and diminishing marginal returns to capital in the production function, different economies have higher economic efficiency due to the difference in the capital stock in the growth process, so that the economy with a smaller capital stock has higher economic efficiency, which ultimately leads to the same growth efficiency in different economies (Ding et al., 2019). However, because of the lack of an empirical verification process, the key problem is unable to explain the problem of sustained economic growth. The conditional convergence hypothesis posits that convergence is based on one's own resource endowment status, that is, "conditional" convergence, which does not incorporate the concept of space.

4.2.1 Regional Economic Efficiency and Spatial Club Convergence

The study of efficiency can be traced back to Adam Smith; however, the study of economic efficiency began after Cobb–Douglas production theory (Laffont, 1975). With regard to the role of efficiency in economic growth, scholars are mainly concerned about the spatial and temporal evolution of efficiency changes. Specifically, scholars in China have pointed out that accelerating the process of marketization and improving the efficiency of resource allocation can effectively enhance economic efficiency and that China's economic efficiency is mainly manifested in the transformation of labor resources from the agricultural sector to the non-agricultural sector, as well as the reallocation of resources and the enhancement of production efficiency within the industrial sector and the enterprise. The improvement of production efficiency (Fan et al., 2012). Local and international scholars mainly adopt the production function method, growth accounting method, stochastic frontier analysis method, and data envelopment method to measure economic efficiency. The literature on China's economic efficiency points out that the total factor productivity of the economy fluctuated from 1952 to 1998, and since 1978 onward, China's economic efficiency has improved significantly. Based on the growth accounting method, studies have found that the difference of technical efficiency between industries and the gradual increase is the key reason for the improvement of economic efficiency (Reinhard et al., 1999). Other scholars have used the production function method and the stochastic frontier model to measure regional growth efficiency, asserting that China's economic growth is mainly capital-driven, with a small degree of efficiency contribution, and that there are obvious differences between the eastern, central, and western regions.

Spatial spillovers are expressed in various forms, specifically in domestic and international studies. The manifestations of spatial spillovers are mainly in the form of spatial externalities, geographical spillovers, spatial heterogeneity and spatial dependence (Lin et al., 2017). Spatial externality focuses on the interaction mechanism under the non-market role, and spatial externality was initially used in the research between micro-economies; however, with the depth of the research, the externality also exists across the border of the economy, and in particular, when there is a strong similarity between the neighboring economies, the spatial spillovers between the regions will be closer. Geographical spillover is a concept of the new economic geography, which is aimed at explaining the spatial differences in economic activities from a micro-economic perspective to elucidate the spatial agglomeration mechanism of economic activities. Spatial heterogeneity portrays the unequal nature of the relationship between observational units. Spatial dependence characterizes the coupling between similar variables and locations, where the spatial agglomeration of intervals with similar variables shows positive spatial autocorrelation results, whereas intervals with spatial disparities show negative spatial autocorrelation states.

This section extends the existing literature in two ways. First, several studies have explored economic efficiency; however, there is no unified definition and measurement of the concept. This study explores the use of the stochastic frontier production function (SFA) method to study economic efficiency. Second, the existing literature is more mature for the study of spatial convergence of economic growth, whereas little research has been conducted on the spatial spillovers of economic efficiency, and the existing literature does consider economic efficiency a key indicator. (Kontodimopoulos et al., 2011). This study explores the spatial mutual agglomeration and dispersion of economic efficiency and its influencing factors.

4.2.2 Data, Variables, and Research Design

4.2.2.1 Variables and Data

According to the literature review, the measurement model for studying economic efficiency is divided into two main kinds; one is the stochastic frontier model and the other is the data envelopment analysis model, based on the characteristic that SFA can directly estimate the influencing factors. This study adopts SFA to estimate the economic efficiency and its influencing factors in China (Table 4.1). Drawing on the previous research results, in this study, $(FDI/GDP)_{it}$ represents the level of openness to foreign investment, focusing on the impact of various factors on the technical efficiency of economic growth. $(TRA/GDP)_{it}$ indicates the degree of dependence on foreign trade, calculated as the ratio of the total import and export trade volume to GDP for each province in a given year. This reflects the development of foreign trade during that year. $(TIAV/GDP)_{it}$ represents industrial structure, expressed as the proportion of the tertiary industry's added value to GDP in each province, capturing the state of industrial development. $[(DEP + LOA)/GDP]_{it}$ measures financial development, represented by the ratio of deposits and loans to GDP for each province, indicating the level of financial development across different regions.

Based on the above research methods and panel data, this study analyzed China's economic efficiency and its influencing factors using the analytical software Frontier 4.1 and obtained the economic efficiency indices of different provinces, and the coefficients were estimated.

Absolute β-convergence and conditional-convergence checking models containing spatial effects were established, and the models were set up as follows:

Absolute-convergence:

$$gEE_{i,t} = \alpha_i + \phi_t + \beta ln\left(EE_{i,t-1}\right) + \delta \sum_j W_{ij}\left(gEE_{i,t}\right) + \mu_{i,t} \tag{4.27}$$

Conditional β-convergence:

$$gEE_{i,t} = \alpha_i + \phi_t + \beta_1 ln\left(EE_{i,t-1}\right) + \beta_2 X_{i,t} + \delta \sum_j W_{ij}\left(gEE_{i,t}\right) + \mu_{i,t}$$

Table 4.1 Descriptive statistics of relevant variables

Variable	Meaning	Number	Mean	Standard Error	Minimum	Maximum
Y	Total output (in billions)	660	16,582.37	733.7979	263.68	124,719.5
L	Labor input (million people)	660	2473.962	62.91973	275.5	7072
K	Material capital stock	660	35,649.4	1327.392	1569.7	197,446
FDI/GDP	Level of opening to the outside world	660	0.024154	0.000906	0.00006	0.163485
TIAV/GDP	Industrial structure	660	0.457767	0.003531	0.296445	0.837317
(DEO + LOA)/GDP	Financial development	660	2.987666	0.041744	1.381581	7.578346
TRA/GDP	Trade dependence	660	0.306026	0.014082	0.007627	1.711295

$$\mu_{i,t} = \lambda \sum_{j} W_{ij} * \mu_{i,t} + \varepsilon_{i,t} \tag{4.28}$$

where $gEE_{i,t}$ denotes the economic efficiency growth rate of the i province $ln\left(EE_{i,t-1}\right)$ at the beginning of the period. The positive or negative coefficient of the variable indicates the existence of regional economic efficiency convergence, if the coefficient is negative, it means that the economic efficiency growth has a significant tendency to converge.

4.2.3 *Regional Economic Efficiency and Influencing Factors*

Considering the availability of data and information, this study takes China's provincial areas (provinces, municipalities directly under the central government, and autonomous regions) as the evaluation unit (excluding Hong Kong, Macao, Taiwan, and Tibet, because of the lack of data). The sample data are panel data of 30 provinces in China from 2000 to 2021, from national statistical yearbooks, provincial statistical yearbooks, statistical bulletins, and the China Economic website.

The estimation of influential factors based on SFA and the significance test (Table 4.2) indicate that the likelihood ratio (LR) passes the test with a significance level of 1%, which indicates that SFA is appropriate for the regression of influential factors.

Table 4.2 Factors affecting economic efficiency

Variable	Estimated coefficient	Standard error	t-value
Constant	−1.72135***	0.06800	−25.3133
Ln(L)	0.19189***	0.02827	6.7873
Ln(K)	0.99282***	0.01459	68.0301
FDI/GDP	1.38138**	0.78009	1.7708
TIAV/GDP	−0.88111***	0.23844	−3.6952
(DEO + LOA)/GDP	−0.00004**	0.00002	−1.7256
TRA/GDP	−0.19022***	0.05866	−3.2427
σ^2	0.06300***	0.00510	12.3530
λ	0.99999****	0.00003	38,240.4350
Log-likelihood function (log)	25.22***		
Likelihood ratio test (LR)	99.61***		

Note *, **, and *** represent significance at the 10%, 5%, and 1% significance levels, respectively

The coefficient of labor force is 0.19189, which means that a 1% change in labor force increases China's GDP by 0.1919%, whereas the coefficient of capital is 0.9928, which means that a 1% change in capital increases China's GDP by 0.9928%. This shows that the current development of China's economy is dominated by the drive of capital. In addition, the sum of the elasticity of labor output and capital output is 1.1847, which is close to 1. In line with the law of constant returns to scale, it is necessary to fundamentally improve the economic output through technological progress, and the current Chinese economy should achieve high-quality transformation to curb the blind scale of the change.

The level of foreign direct investment (FDI) is an indicator of the level of openness of a region's foreign economy and the degree of modernization, thus reflecting the proportion of different regions in the international market. As shown in the table, the parameter value of FDI/GDP is 1.38138, indicating the impact of opening up to the outside world on China's economic efficiency; that is, for every 1% increase in the scale of opening up to the outside world, economic efficiency will increase by 1.38138%; therefore, the impact of the scale of opening up to the outside world on the economic efficiency shows a positive impact.

The industrial structure indicates the reasonableness of the economic development of a region and is an important reference for economic efficiency. After regression, the coefficient of the proportion of the tertiary industry in GDP is −0.8811, which shows that the development of the tertiary industry has a negative effect on China's economic efficiency, which may be closely related to the stage of transformation and upgrading of the industrial organizations in different regions.

The level of financial development can indicate the optimal allocation of capital in the region and to sectors with higher production efficiency. China's financial development level has a negative impact on China's economic efficiency. This is consistent with the estimation results of studies by Chinese scholars, probably because the financial market is the most important means to stimulate economic efficiency. China's financial development is still lagging behind the demand for economic efficiency; therefore, China's level of financial development has a negative impact on its economic efficiency.

Trade scale is an indicator that reflects the active participation of regions in international trade, with a parameter value of −0.1902, indicating that the trade scale has a negative impact on China's economic efficiency, and that for every 1 percentage point rise in the scale of import and export, the economic efficiency will be reduced by −0.1902 percentage points (Li & Wu, 2012). The degree of foreign trade dependence is mainly related to the local economic development situation. Currently, China's economy is highly dependent on foreign countries, which has resulted in inflated economic output but low economic efficiency in some regions.

4.2.4 Modeling Test of Spatial Club Convergence

The test revealed that there is a spatial correlation of economic efficiency. Therefore, this study uses the spatial measurement method to simulate and analyze the measurement model and obtains the SAR model and the SEM, and after simulation, it is found that the results of the spatial measurements in the absolute β-convergence and the conditional β-convergence models (Table 4.3), all show that the Log-L value of the SEM results is greater than the Log-L value of the SAR model. Therefore, after comparison, it is found that the SEM model is the optimal choice for the spatial measurement model.

After analysis, it can be concluded that in the absolute β-convergence model, the coefficient of lnEEt-1 is negative and passes the significance level test of 1%, which indicates that there is a significant absolute β-convergence of China's regional economic efficiency, and after calculating the β-convergence of the neoclassical economic growth model, it is concluded that the absolute rate of China's regional economic efficiency convergence in the sample period is 6.96%. Conditional β-convergence rate is 6.75%, which indicates that the convergence speed is slightly slower than the absolute β-convergence after controlling the influencing factors. As control variables, the degree of openness to the outside world and industrial structure has a significant negative impact on economic efficiency convergence and have passed the significance test of 1% and 5%, respectively. Trade dependence has a significant positive effect on economic efficiency convergence, whereas financial development has a nonsignificant effect on economic efficiency convergence.

In this section, from China's economic efficiency, the exploration of spatial spillovers, and regional economic efficiency convergence is calculated, and the economic efficiency derived from the calculation of the model using the SFA is used as an input term into the spatial club convergence test.

Table 4.3 Two-way fixed effects spatial econometric model estimation and test results

Variable	Absolute β-convergence		Relative β-convergence	
	SAR	SEM	SAR	SEM
$\ln EE_{t-1}$	−0.05836*** (−5.97)	−0.06720*** (−6.29)	−0.05806*** (−6.06)	−0.06523*** (−6.19)
FDI/GDP			−0.42181*** (−4.20)	−0.40240*** (−4.09)
TIAV/GDP			−0.11566** (−2.37)	−0.10023** (−2.07)
(DEO + LOA)/GDP			0.000001 (1.01)	0.000001 (1.00)
TRA/GDP			0.03346*** (2.91)	0.03988*** (3.35)
λ	0.26145*** (5.37)	0.27944*** (5.68)	0.27341*** (5.68)	0.30709*** (6.14)
σ^2	0.00096*** (17.62)	0.00095*** (17.61)	0.00092*** (17.61)	0.00091*** (17.56)
Log-L	1289.2204	1290.8354	1302.4563	1304.7351

Note *, **, and *** represent significance at the 10%, 5%, and 1% significance levels, respectively

4.3 Spatial and Temporal Patterns of Regional Ecological Efficiency

Over the past 40 years, China's rapid urbanization and industrialization have brought about huge social progress and regional economic prosperity, but they have also brought about a series of problems such as resource consumption, environmental pollution, and destruction. In 2014, the Central Economic Work Conference pointed out that "the carrying capacity of China's environment has reached or is close to the upper limit", and that resource shortage and environmental degradation have become important bottlenecks restricting sustainable economic development. Environmental degradation has become an important bottleneck restricting sustainable economic development. Obtaining maximum economic and social benefits with minimum ecological loss, that is, the improvement of ecological efficiency, is an important issue facing China's economic development in the context of the "new normal" and the promotion of the construction of ecological civilization. With the rapid advancement of urbanization and industrialization in China, what is the status of eco-efficiency in each region? What are the evolving characteristics of regional eco-efficiency patterns? What are the driving factors of regional eco-efficiency?

The foreign concept of eco-efficiency was first proposed by German scholars Schaltegger and Sturm (1990), defining it as the ratio of the value added of economic activities to the role of the environment. Eco-efficiency involves both the ecological environment and the economy and requires the participation of multiple subjects such as enterprises, regions, and the state. Research on the theoretical level of eco-efficiency mostly focuses on conceptual extension and evaluation method exploration. Evaluation methods mainly include the ratio method, indicator system method, and modeling method (Ye et al., 2017). The ratio method usually measures the ratio of desired economic output and non-desired environmental impact, in which the economic output is expressed by the enterprise value added or profit, and the environmental impact is expressed by resource consumption and pollution emission. The indicator system method has become the main method of environmental reporting due to its comprehensiveness, systematicity, and unity. For example, Finland's ecological environment project uses a participatory approach to construct an ecological efficiency indicator system covering economic, social, and ecological aspects, so that decision makers can not only detect the changes in the economic-environmental ratio but also obtain information on regional socio-cultural changes. The data envelopment analysis (DEA) method has been widely used due to its objectivity and comprehensiveness, and has been continuously extended to include models such as U-SBM, SSBM, US-SBM. At the empirical level, this study focuses on the measurement of eco-efficiency and the analysis of influencing factors of enterprises, industries, and industrial sectors at the micro-level and places more emphasis on the operability of eco-efficiency.

In China, studies on eco-efficiency started late, mainly focusing on three aspects: (1) Definition of the connotation of eco-efficiency. Some scholars suggest that the essence of eco-efficiency is the reduction of the materialization of circular economy and that the goal of circular economy is the improvement of eco-efficiency, and they emphasize that the social development should consider environmental benefits while pursuing economic benefits. (2) Evaluation methods of eco-efficiency. These methods mainly include the single ratio method, index system method, and model method. The single ratio method aims to introduce the green GDP and ecological footprint algorithm into the traditional economic/environmental ratio method to measure regional eco-efficiency. The indicator system method evaluates regional eco-efficiency by constructing a comprehensive indicator system, such as a composite eco-efficiency evaluation indicator system of resource input economic benefit-environmental impact to evaluate the overall level of eco-efficiency of the counties in the Beijing-Tianjin-Hebei Urban Agglomeration. The modeling method uses the DEA model and its improvement model to measure eco-efficiency. For example, the three-stage DEA model, super-efficiency DEA model, non-desired output DEA model, non-desired output SBM model, etc. are used to measure the eco-efficiency of China's provinces and China reveals a gradient distribution pattern of high in the eastern region and low in the western region. (3) Spatial differentiation characteristics and influencing factors of eco-efficiency. Research on the characteristics of spatial differentiation of eco-efficiency from the perspective of industry has found that the overall eco-efficiency of Chinese agriculture has a "W" structure, and the eco-efficiency of various industrial sectors in terms of energy consumption and pollutant emissions has an obvious industry imbalance.

Some scholars have further explored the driving factors of regional eco-efficiency on the basis of its spatial pattern and predicted that the center of gravity of the spatial distribution pattern of eco-efficiency in mainland China will move toward the northeast in the next 10–20 years, the spatial distribution pattern will show a contraction trend, and the driving factors are mainly related to urbanization, industrial structure, policy support, economic efficiency, the use of foreign investment, human resources, education level, and environmental regulation.

Throughout the domestic and foreign research on eco-efficiency, the research content has changed from defining the concept of eco-efficiency and its spatial distribution to exploring the driving factors, the research method has changed from the traditional single ratio method and indicator system method to the complex model method, and the fields involved have become increasingly extensive. Foreign research on eco-efficiency pays more attention to the guidance of eco-efficiency to the environmental protection behavior of enterprises, and there are relatively few studies at the national level. Currently, the method of measuring regional eco-efficiency has become more mature; however, the modeling method adopted at this stage for regional scale eco-efficiency research is less likely to consider further comparisons of efficient regions and cannot fully reflect the differences in regional eco-efficiency in China.

4.3.1 Model, Indicator, and Data Sources

This study first measures China's provincial eco-efficiency using the panel data of 30 provinces (Hong Kong, Macao, and Taiwan, and Tibet are temporarily excluded from the study area due to incomplete data) from 2005 to 2016 as the basic decision-making unit, and takes the provinces' non-desired output (CO_2 emissions) and real GDP as the output variables. The primary energy variable is also added to the traditional production function, which means that capital, labor, and energy are used as the output variables (Sueyoshi & Yuan, 2015). Meanwhile, the primary energy variable is added to the traditional production function. In the calculation, the output variable is the effective output of real GDP converted to the price of the base period of 2005, and the capital input variable is obtained using the method of perpetual inventory.

$$K_t = I_t + (1 - \delta)K_{t-1} \tag{4.29}$$

where K_t denotes the capital stock in period t, δ denotes the depreciation rate, and I_t denotes the investment amount in period t.

The labor input indicator takes the number of people employed at the end of the year as the labor input indicator, whereas the energy input indicator variable uses the discounting method to convert the primary energy of coal, oil, and natural gas in each province to standard coal in tons of standard coal. The non-desired output indicator takes CO_2 as the bad output, and this study draws on the previous measurements and calculates the formula as:

$$C = \sum E_j \times K_j \tag{4.30}$$

Table 4.4 Carbon emission factors for major energy sources

Energy type	Carbon emission factor (t/t standard coal)	Energy type	Carbon emission factor (t/t standard coal)
Raw coal	0.7559	Fuel oil	0.6185
Coking coal	0.855	Other petroleum by-products	0.5857
Clean coal	0.7559	Liquefied petroleum	0.5054
Other coking products	0.649	Natural gas	0.4483
Crude oil	0.5857	Coke oven gas	0.3548
Gasoline	0.5538	Refinery dry gas	0.4602
Kerosene	0.5714	Other gas	0.3548
Diesel fuel	0.5921		

Note According to IPCC (2014)

where C represents the CO_2 emissions of a certain energy source, E_j is the energy consumption, K_j is the energy carbon emission coefficient, and this study adopts the coefficients published by the United Nations Intergovernmental Panel on Climate Change (IPCC) to calculate carbon emissions (Table 4.4).

The data used in this study are mainly from the China Energy Statistical Yearbook (2006–2017), the China Statistical Yearbook (2006–2017), and the statistical yearbooks of each province (2006–2017), and the data have been largely organized and calculated.

4.3.2 *Spatial and Temporal Patterns of Spatial Eco-efficiency*

The changes in eco-efficiency values in each province are mixed. Overall, the eco-efficiency shows a slow rise (Table 4.5). In particular, since 2010, the eco-efficiency has made more progress in all provinces. Taking Inner Mongolia as an example, the eco-efficiency was 0.836 in 2010 and 0.889 in 2016, with an increase ratio of 6.33%. Sub-regionally, the eco-efficiency generally decreases in the order of eastern, central and western regions, and the eco-efficiency values of some provinces in the west also show special high values in different years, which is inseparable from the low total amount of energy used locally. The imbalance of eco-efficiency is closely related to China's economic development and the strength of policy inclination.

Exploring the spatial differentiation characteristics of the data after spatializing the eco-efficiency in 2005, 2010, and 2015, the different eco-efficiencies show a spatial pattern of gradual decrease in the "East, Central, and West" and symmetrical distribution in the "South, Central, and North." In 2016, the eco-efficiency regions are mainly distributed in Shanghai, Beijing, Jiangsu, Hainan, Inner Mongolia, and

Ningxia. Ningxia and Inner Mongolia are economically underdeveloped provinces; however, their eco-efficiency value can reach a higher level, which is closely related to the water and blue sky projects in these provinces over the years, and the development of the tertiary industry in these regions, that is, the tourism industry, has continued to increase, which has further contributed to the improvement of eco-efficiency (Song & Chen, 2019). Provinces with eco-efficiency values above 0.8 include Shanxi, Anhui, Zhejiang, Shandong, and Xinjiang, which share the common characteristic of resource transformation, with Shanxi, for example, being subject to various preferential policies by the Chinese government to ensure high-quality development. Regions with eco-efficiency less than 0.6 mainly include Heilongjiang, Sichuan, and Yunnan. These provinces are at a critical stage of transformation, where the long-term accumulation of industrial structure and the lack of reform of institutional mechanisms have long existed, which have partially squeezed the development of the tertiary industry and created economic bottlenecks.

4.3.2.1 Spatial Distribution

Spatial distribution refers to the distribution of geographical things in China or globally according to a certain indicator or geographical location. In the macro sense, the concept of spatial distribution generally refers to a global scale, at least at a level such as that of oceans or continents. In the micro sense, it refers to regional and local or smaller geographic spaces, which are subject to the uniform constraints of macroscopic geographic distribution laws.

4.3.2.2 Clustering

The term agglomeration is mainly used in the context of the agglomeration effect. The agglomeration effect is understood from an economic perspective as the economic effect produced by the concentration of various industries and economic activities in space and the centripetal force that attracts economic activities to a certain area, which is an important factor leading to the formation and continuous expansion of cities. The agglomeration effect is a common economic effect, and typical examples include Silicon Valley in the United States and the small commodity and garment industries in Zhejiang.

4.3.3 Analysis of Factors Influencing Regional Eco-efficiency

This study further analyzed the influencing factors of eco-efficiency, and after comparing the SLM results with the SEM results, it was found that both the ρ value of the SLM regression and the λ value of the SEM regression were greater than 0,

Table 4.5 Changes in eco-efficiency in China's provinces, 2005–2016

Province	2005	2010	2015	2016
Heilongjiang	0.544	0.614	0.583	0.600
Inner Mongolia	0.722	0.836	0.888	0.889
Xinjiang	0.51	0.578	0.761	0.803
Jilin	0.62	0.783	0.593	0.614
Hebei	0.738	0.733	0.789	0.767
Gansu	0.695	0.729	0.647	0.601
Beijing	1.000	1.000	1.000	1.000
Shanxi	0.754	0.836	0.888	0.889
Tianjin	0.914	0.821	0.835	0.797
Qinghai	1.000	0.689	0.72	0.674
Shan'xi	0.637	0.655	0.743	0.731
Ningxia	0.745	1.000	1.000	1.000
Shandong	0.651	0.787	0.806	0.826
Henan	0.617	0.724	0.763	0.735
Jiagnsu	0.826	1.000	1.000	1.000
Anhui	0.707	0.742	0.864	0.877
Sichuan	0.517	0.504	0.55	0.548
Hubei	0.597	0.615	0.683	0.643
Shanghai	1.000	0.973	1.000	1.000
Chongqing	0.624	0.621	0.661	0.636
Zhejiang	0.943	1.000	0.954	0.828
Jiangxi	0.611	0.626	0.681	0.659
Hunan	0.557	0.607	0.644	0.613
Yunnan	0.587	0.618	0.561	0.511
Guizhou	1.000	0.746	0.715	0.666
Fujian	0.739	0.825	0.82	0.798
Guangxi	0.608	0.593	0.697	0.657
Guangdong	0.741	0.78	0.723	0.704
Hainan	0.543	0.704	0.902	1.000
Liaoning	0.684	0.759	0.71	0.717

which means that eco-efficiency has a significant positive correlation in space, and the SLM regression is more appropriate than SEM regression in R2 (Table 4.6).

The regression reveals that the energy structure passes the significance test and the estimated coefficients are negative, the increase in the proportion of coal in the energy structure will have a hindering effect on the improvement of eco-efficiency, the level of technology and the rate of urbanization have a positive contribution to the improvement of eco-efficiency, high-quality urbanization will have a further

Table 4.6 Spatial constraints affecting eco-efficiency

	SLM (Fixed effect)	SEM (Fixed effect)
Energy structure	−0.398***	−0.402***
	(10.23)	(10.54)
Industrial structure	0.112	0.115
	(1.95)	(1.92)
Technology level	0.893**	0.899**
	(21.1)	(21.5)
Urbanization rate	0.362***	0.366***
	(6.42)	(6.56)
GDP	0.12	0.14
	(1.94)	(1.93)
R2	0.816	0.804
λ		0.205** (−2.35)
ρ	0.312*** (−5.27)	

Note *, **, and *** represent significance at the 10%, 5%, and 1% significance levels, respectively

inhibitory effect on the eco-improvement of China, and the further increase in the level of technology, such as the use of more energy-saving and environmentally friendly equipment and the use of low-carbon technology, will improve the regional eco-efficiency.

The spatial characteristics of eco-efficiency is a result of multiple factors, especially the level of technology, which plays a key role. This study found that eco-efficiency is an issue and concept that crosses provincial boundaries, breaks through the concept of boundaries, breaks the existing principle of each province serving its own, influences the cooperation and enhancement program between regions, establishes the platform of resource sharing, and helps achieve interregional governance. Therefore, it should be addressed from a multi-objective decision-making perspective, through data cooperation, innovative cooperation, and intelligent construction cooperation between regions.

4.4 Spatial and Temporal Patterns of Social Development

4.4.1 Social Development and Anti-poverty

Reducing and escaping poverty has become a common goal of human development and a global problem. The 17 Sustainable Development Goals (SDGs) established by the "2030 Agenda for Sustainable Development" aim to further eradicate all forms of poverty and collectively improve the well-being of all human beings. As a responsible developing country, China will organically combine the poverty reduction goals of the SDGs with its historic mission to eradicate poverty and accelerate the implementation of the basic strategy of precise poverty eradication (Feng et al., 2019). How to further effectively innovate the poverty alleviation model and its countermeasures to enhance the quality and effectiveness of poverty alleviation efforts has become a major challenge. As an important foundation of innovative poverty alleviation ideas, correctly screening the sensitive factors affecting social poverty and studying the mechanism of each factor's effect on social poverty will be an issue worthy of research efforts.

In terms of theoretical exploration, early international scholars have researched and analyzed the issue of social poverty from the perspectives of disciplines such as economics, sociology, and political science and focused on the detailed discussion of poverty standards, poverty measurement, poverty mapping, and anti-poverty policy research Since the Nobel laureate Amartya Sen introduced the idea of "mentioning feasible ability", the multidimensional poverty research method has provided more powerful tools and perspectives for identifying the poor population and formulating pro-poor policies. Multidimensional poverty measurement dimensions, measurement methods, and indicator weights have attracted the attention of many scholars internationally, and many results have been achieved in theory and empirical evidence.

Social poverty research in China started a little later, and most of them are the follow-up studies of international studies. Due to the limitations of data acquisition and poverty alleviation mechanisms, most scholars construct multidimensional poverty measurement models based on international research methods, combined with their own data collection, to carry out comprehensive analysis of poverty. Other scholars use GIS technology and mathematical and statistical methods to explore the spatial distribution pattern and evolutionary characteristics of poverty and quantitatively analyze its influencing factors from the dimensions of social development and geographical conditions. In terms of research objects, relevant studies mainly focus on special poverty groups such as migrant workers, migrant relocated farmers, people with disabilities, and the elderly. In China, the existing studies on social poverty influencing factors are mostly qualitative description. Quantitative analysis of the research result is fewer and is mainly for a specific point in time for the regional multidimensional poverty identification and measurement of the follow-up analysis. The study of the spatial–temporal dynamics of the social poverty is insufficient. Poverty and non-poverty are relative concepts, often in a continuous process of change. The data used in the current study are mostly cross-sectional data, unable to grasp the dynamic characteristics of its longitudinal fluctuation and change, which may affect the causal judgement between external factors and social poverty, leading to errors in pro-poor policy formulation and decision-making.

In this context, based on the provincial panel data of China from 2010 to 2021, this study constructs a PVAR model with income poverty, health poverty, and cultural poverty as the kernel of social poverty, identifies the key factors influencing social poverty through generalized method of moments (GMM) estimation and impulse response and variance decomposition analysis. It systematically reveals the dynamic response process among the factors and their contribution to the evolution of social poverty, with a view to more accurately and comprehensively grasping the long-term equilibrium relationship between the factors and social poverty and their dynamic effects, so as to provide a guide for poverty alleviation policies and decision-making and a theoretical basis for relevant research.

According to the current overall perception and evaluation of social poverty internationally, social poverty not only refers to economic poverty, but also includes health deterioration, educational deprivation, and spiritual emptiness. Specifically, social poverty can be summarized as income poverty, health poverty, and cultural poverty.

In the latter sections of this paper, we establish PVAR models based on the above three aspects to portray the fluctuation transmission mechanism between each variable and social poverty and the extent of the contribution of each variable to each aspect of social poverty, and to identify the key influencing factors of social poverty. Referring to previous research results, this study chooses the three indicators of annual savings per capita, epidemiological incidence rate, and illiteracy rate to characterize income poverty, health poverty, and cultural poverty, respectively. Meanwhile, GDP per capita, the proportion of non-agricultural industry, per capita social security and employment expenditure, total social dependency ratio, and crop damage ratio are selected as the explanatory variables of income poverty in terms of the level of economic development, industrial structure, social security, population burden, and natural disasters, per capita public healthcare expenditure, and number of hospitals per 10,000 people. In terms of healthcare investment, healthcare resources accessibility, urbanization development, healthcare security and environmental quality, public healthcare expenditure per capita, hospitals per 10,000 people, urbanization rate, urban residents' healthcare insurance participation rate, and forest coverage rate are chosen as the explanatory variables for health poverty. In terms of education investment, higher education development, education resources allocation, science and technology investment and mass culture construction, education expenditure per capita, number of higher education institutions, student–teacher ratio, R&D expenditure per capita, and per capita expenditure on mass cultural undertakings are chosen as the explanatory variables for cultural poverty. The definitions of the indicators and their data descriptions are shown in Table 4.7.

In this study, the panel data of 31 provinces (autonomous regions and municipalities directly under the central government) in China from 2010 to 2021 are selected as the sample interval. The data are mainly obtained from China Statistical Yearbook, China Health and Family Planning Statistical Yearbook, China Environmental Statistical Yearbook, China Culture and Cultural Relics Statistical Yearbook, and the website of National Bureau of Statistics from 2011 to 2022. To avoid the problem of heteroscedasticity among variables, the original variables are logarithmized before modeling.

Table 4.7 Key variable definitions and descriptive statistics

Variable notation	Variable definition	Unit	Mean	Std. Dev	Min	Max
lnPSD	Annual savings per capita	Yuan/person	10.54	0.564	9.093	12.31
lnIDI	Incidence of infectious diseases	1/10 million	5.406	0.360	4.392	6.492
lnIR	Illiteracy rate	% of the total	1.481	0.677	−0.236	3.718
lnPGDP	GDP per capita	Yuan/person	10.77	0.479	9.464	12.14
lnIS	Share of non-agricultural industries	% of the total number of people	−0.106	0.0595	−0.299	−0.002
lnPSSE	Per capita social security expenditure	Yuan/person	7.327	0.577	5.937	8.748
lnGDR	Total social dependency ratio	% of the total number of people	3.614	0.202	2.959	4.057
lnFCA	Crop damage ratio	%	−2.230	0.897	−5.236	−0.363
lnPPHE	Public health expenditure per capita	Yuan/person	8.049	0.510	6.853	9.637
lnPHOS	Number of hospitals per 10,000 people	Institutions/ten thousand people	−2.606	0.735	−4.595	−1.327
lnUR	Urbanization rate	%	4.033	0.239	3.121	4.495
lnUHI	Urban residents' medical insurance participation rate	%	−0.0385	0.466	−0.825	1.053
lnFCR	Forest coverage rate	%	3.290	0.722	1.391	4.202
lnPEF	Education expenditure per capita	Yuan/person	10.27	0.370	9.339	11.32
lnHEDU	Number of high schools	Institutions	4.219	0.760	1.792	5.118
lnSTR	Student–teacher ratio	–	2.867	0.0730	2.634	3.122
lnPRD	R&D expenditure per capita	Yuan/person	12.76	0.323	11.62	13.56
lnPMCE	Per capita expenditure on mass culture	Yuan/person	3.961	0.665	2.280	5.736

4.4.2 *Empirical Tests and Analysis*

4.4.2.1 Model Validation

Before estimating by the PVAR model, to avoid the phenomenon of "pseudo-regression" due to the unstable data, it is necessary to carry out the unit root test on the data first. To improve the effectiveness of the test, this study uses the LLC, IPS, and PP-F tests. According to the results of the variable test, some variables are smooth, and the rest of the variables do not pass the test. For the non-stationary panel series, after the first-order difference processing, the results of the test of the variables are all presented as smoothness (Table 4.8), indicating that all the variables belong to the first-order monotone series. The later empirical analysis is based on the first-order difference series of the above variables, and the letter "d" is used in front of the variables to indicate that they are first-order difference series.

Constructing the PVAR model requires choosing the appropriate lag order. A lag order that is too short can easily lead to unreliable test results, whereas a lag order that is too long can affect the overall quality of data due to impaired degrees of freedom. In this study, MAIC, MBIC, and MQIC information criteria are adopted to select the lag order of the model. The results of the indicator values of the information criteria are given in Table 4.9, and the minimum lag given by each evaluation indicator is marked with "*" to show that the optimal lag order of PVAR is determined according to the principle of majority when inconsistency in the test results occurs. As shown in the results, the optimal lag order for the PVAR model for income poverty, health poverty, and cultural poverty is lag 1.

Table 4.8 Panel unit root test

Method	dlnPSD	dlnIDI	dlnIR	dlnPGDP	dlnIS	dlnPSSE
LLC	−10.6205***	−5.0429***	−9.8600***	−5.0115***	−9.9409***	−3.3048***
IPS	−6.8463***	−8.7772***	−8.0750***	−4.6712***	−6.0164***	−9.0570***
PP-F	215.2481***	427.6032***	303.1160***	339.7729***	178.0942***	496.8096***
Method	dlnGDR	dlnFCA	dlnPPHE	dlnPHOS	dlnUR	dlnUHI
LLC	−9.6178***	−12.3125***	−7.2762***	−4.9655***	−3.1708***	−100.00***
IPS	−8.7947***	−8.9662***	−8.0013***	−6.4490***	−3.8069***	−6.3065***
PP-F	366.7313***	590.8625***	359.0193***	163.0389***	110.9938***	233.3609***
Method	dlnFCR	dlnPEF	dlnHEDU	dlnSTR	dlnPRD	dlnPMCE
LLC	−9.1710***	−19.6367***	−3.7981***	−5.1604***	−5.1006***	−10.3749***
IPS	−6.2206***	−7.6950***	−8.7856***	−6.7712***	−6.3646***	−7.6028***
PP-F	112.0468***	466.2609***	843.1591***	266.9403***	296.7365***	316.7595***

Note *, **, and *** represent 10%, 5%, and 1% significance levels, respectively

Table 4.9 Lag order judgement results

Order	Income poverty			Health poverty			Cultural poverty		
	MAIC	MBIC	MQIC	MAIC	MBIC	MQIC	MAIC	MBIC	MQIC
1	−88.7*	−437.1*	−229.9*	−72.3*	−362.6*	−190.0*	−104.4*	−452.8*	−245.6*
2	−58.3	−290.6	−152.4	−43.7	−237.3	−122.1	−63.7	−296.0	−157.9
3	−27.1	−143.2	−74.2	−5.6	−102.6	−45.0	−20.7	−136.8	−67.7

4.4.2.2 Analysis of Factors Influencing Social Poverty Based on PVAR Models

This study uses GMM to determine whether each variable has a significant effect on different aspects of social poverty and whether the direction of the effect is positive or negative. GMM estimation of panel data using the PVAR model requires the removal of fixed effects from the sample. In this study, we draw on Arellano and Bovver (1995) to eliminate fixed effects by using forward-differenced Hermlet transformations, and individual time effects are treated by within-group mean differences to avoid biased coefficient estimates due to individual effects and correlation with regressors. The specific results of the GMM estimation are shown in Table 4.10.

Table 4.10 shows that the lag 1 period of GDP per capita (dlnPGDP) and the proportion of the non-agricultural industry (dlnIS) have significant positive influence on the current annual savings per capita (dlnPSD), which indicates that the regional macroeconomic operation has a favorable trend and an obvious promotion effect on the improvement of residents' income level. The lag 1 period of per capita social security expenditure (dlnPSSE) has a significant positive effect on per capita annual savings because the former is targeted to the key factors affecting employment,

Table 4.10 GMM estimation results for the PVAR model

Variable	h_dlnPSD		h-dlnIDI		h_dlnIR
L.h_dlnPSD	0.0755*** (3.77)	L.h_dlnIDI	−0.2347*** (−5.20)	L.h_dlnIR	0.1271*** (3.59)
L.h_dlnPGDP	0.2797*** (6.00)	L.h_dlnPPHE	0.0615 (0.94)	L.h_dlnPEF	−0.4822*** (−8.68)
L.h_dlnIS	0.8244*** (3.93)	L.h_dlnPHOS	−0.5936*** (−7.21)	L.h_dlnHEDU	3.1540*** (15.83)
L.h_dlnPSSE	0.0446** (2.88)	L.h_dlnUR	4.4100*** (13.93)	L.h_dlnSTR	−1.1600*** (5.12)
L.h_dlnGDR	0.2572*** (8.29)	L.h_dlnUHI	0.0411*** (6.13)	L.h_dlnPRD	0.8568*** (11.41)
L.h_dlnFCA	−0.0036** (−1.72)	L.h_dlnFCR	0.0386 (1.08)	L.h_dlnPMCE	−0.1116* (−1.74)

Note L. denotes lag 1 period and ***, **, and * denote significance at the 1%, 5%, and 10% significance levels, respectively such as illness, work injury, and unemployment, and plays an important role in guaranteeing the basic survival of an individual or a family, dispersing risks, and enhancing the level of welfare. The total social dependency ratio (dlnGDR) with a lag of 1 period has a significant positive impact on the current period's annual per capita savings, indicating that the former will, to a certain extent, promote the latter, due to the enhancement of China's modern medical and healthcare technology, the aging of the population, and the continuous reduction of the working-age population, which increases the burden of old-age pension. The lag 1 period of the crop damage ratio (dlnFCA) has a negative effect on the current per capita annual savings at the 5% significance level, and the frequent occurrence of natural disasters has produced manifold damages to the human economy and social system, which has brought about a greater impact on the social production and the residents' income increase.

The lag 1 period of the number of hospitals per 10,000 people (dlnPHOS) has a significant negative impact on the incidence of infectious diseases, due to the realization of equal allocation of basic medical and healthcare resources, guaranteeing the accessibility of medical and healthcare services in underdeveloped regions, and effectively enhancing the level of regional healthcare. The urban residents' health insurance participation rate (dlnUHI) with a lag of 1 period has a significant positive effect on the incidence of infectious diseases. This is possibly because, on the one hand, comprehensive health insurance coverage may reduce the incidence of infectious diseases by raising public health awareness and promoting disease prevention; on the other hand, problems in the health insurance system, such as higher out-of-pocket medical expenses and pressure on the operation of the health insurance fund, may indirectly increase the incidence of infectious diseases. Therefore, increasing health insurance coverage should be accompanied by further optimization of the health insurance system to ensure that it can effectively respond to the challenges posed by infectious diseases. The urbanization rate (dlnUR) with a lag of 1 period has a significantly positive effect on the incidence of infectious diseases. This suggests that urban environments with high population density concentration during the urbanization process facilitate the spread of infectious diseases to a certain extent.

The lag of 1 period of per capita education expenditure (dlnPEF) has a significant negative correlation with the current illiteracy rate (dlnIR), which indicates that the illiteracy rate can be curbed only after a certain level of government investment in education. The 1-period lag of the student–teacher ratio (dlnSTR) is significantly negatively related to the current illiteracy rate (dlnIR). A decrease in the student–teacher ratio usually means an increase in the number of teachers or a decrease in the number of students, which can improve the quality of education to some extent. The lag 1 period of R&D expenditure per capita (dlnPRD) has a significant positive effect on the illiteracy rate. This is possibly because R&D expenditure per capita has declined in recent years; however, the overall amount of R&D expenditure has increased, and the illiteracy rate has significantly declined. The decline in the illiteracy rate and the change in per capita R&D expenditures reflect China's efforts to improve national education and scientific and technological R&D capabilities. The per capita expenditure on mass cultural undertakings (dlnPMCE) with a lag of 1 period has a negative effect on the illiteracy rate of the current period at the 10% significance level, indicating that increasing investment in mass cultural undertakings, providing diversified public cultural services, and promoting the development of mass cultural undertakings is an effective way to improve the cultural level of the population.

4.4.2.3 Impulse Response Analysis

The impulse response function is used to measure the impact of a one-standard-deviation shock from a randomly perturbed term on the current and future values of other variables, and it can relatively intuitively portray the dynamic interaction effects among variables. To further examine the dynamic process of the impact of each variable on social poverty, this study obtains the impulse response plots of each variable with social poverty using 500 Monte-Carlo simulations.

Figure 4.3 represents the shock response of annual savings per capita (dlnPSD) to itself and to each of the other variables. In particular, 1 standard deviation annual savings per capita (dlnPSD) has a largely positive and progressively weaker effect on itself. For a one-unit standard deviation change in GDP per capita (dlnPGDP), the share of non-agricultural industries (dlnIS), and per capita social security expenditures (dlnPSSE), annual savings per capita is statistically insignificant, although it exhibits a first-rising-then-falling shock effect. For a one-unit change in the total social dependency ratio (dlnGDR), annual savings per capita shows a significant positive impulse response in the first period and reaches a peak, and then its impact diminishes from the second period onward and eventually converges to 0. For a one-unit change in the crop damage ratio (dlnFCA), the impact effect of a one-unit change in the annual savings per capita shows a downward and then an upward movement, but it is not statistically significant.

Figure 4.4 show the shock response of the incidence of infectious diseases (lnIDI) to itself and to each of the other variables. A one-unit standard deviation changes in public health expenditure per capita (dlnPPHE) exhibits a gradual but statistically nonsignificant shock effect on infectious disease incidence. A one-unit standard deviation change in hospitals per 10,000 population (dlnPHOS) shows a significant negative impulse response with a peak in the first period, and a statistically nonsignificant increase in the second period. For a one-unit change in the urbanization rate (dlnUR), the infectious disease incidence rate shows a significant positive impulse response and peaks in the first period, and then declines from the second period onward, but still has a positive effect. For a one-unit standard deviation change in urban residents' health insurance participation rate (dlnUHI), the incidence of infectious diseases shows a significant positive impulse response and peaks in the first period, then declines and becomes nonsignificant in the second period, and then its effect gradually diminishes and converges to 0. For a one-unit standard deviation change in forest coverage rate (dlnFCR), the incidence of infectious diseases shows a gradual increase in the effect of the shock, but it is statistically nonsignificant.

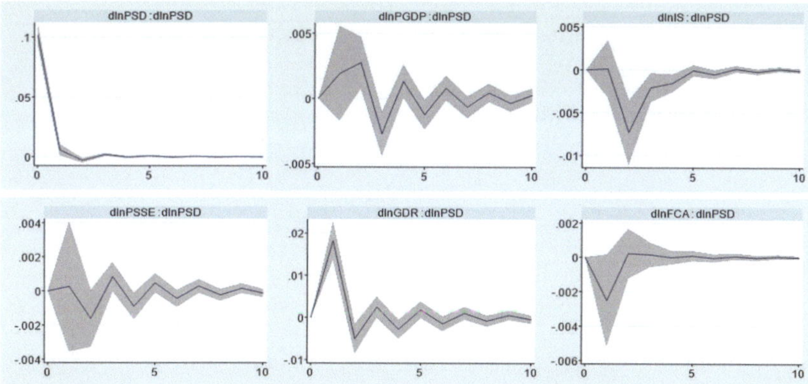

Fig. 4.3 Impulse response results for each variable and per capita annual savings *Note* The black solid line is the impulse response curve and the grey area is the 95% confidence interval obtained from 500 Monte-Carlo simulations

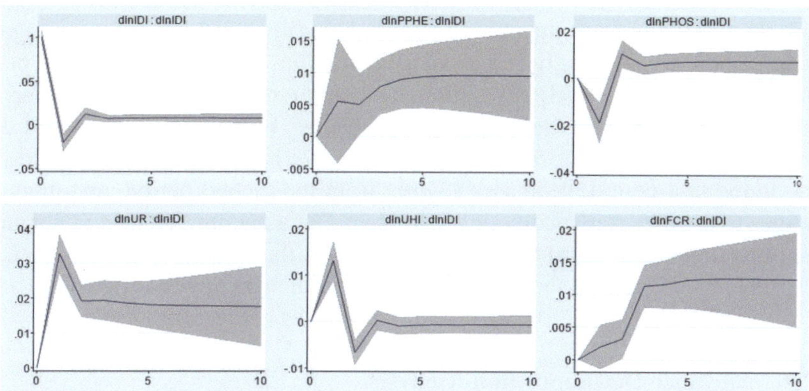

Fig. 4.4 Impulse response results for each variable and the incidence of infectious diseases *Note* The black solid line is the impulse response curve and the grey area is the 95% confidence interval plot obtained from 500 Monte-Carlo simulations

Figure 4.5 represents the shock response of the illiteracy rate (lnIR) to itself and to each of the other variables. For a one-unit change in the standard deviation of the household education expenditure (dlnPEF), the illiteracy rate shows a significant negative impulse response and peaks in the first period, rises sharply but not statistically significantly from the second period, and gradually converges to 0. For a one-unit change in the number of higher education institutions (dlnHEDU), the illiteracy rate shows a significant positive impulse response and peaks in the first period, and falls, and is statistically nonsignificant from the second period. The number of higher education institutions (dlnHEDU) changes by one unit of standard deviation, the illiteracy rate shows a significant negative impulse response and peaks in the first

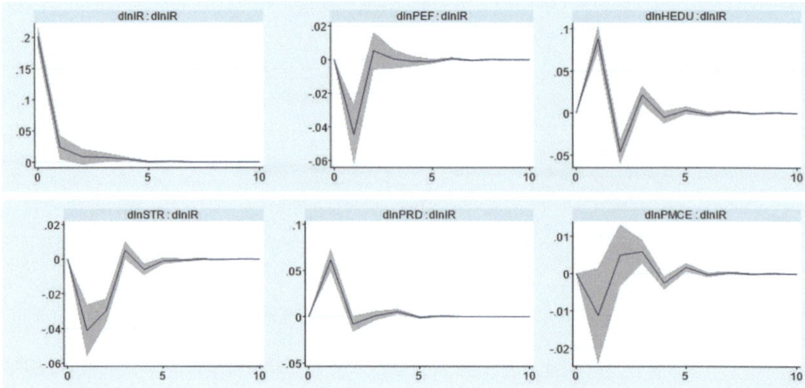

Fig. 4.5 Impulse response results for each variable and illiteracy rate *Note* The black solid line is the impulse response curve and the grey area is the 95% confidence interval plot obtained from 500 Monte-Carlo simulations

period, then declines from the second period, but still shows a significant negative impulse response, and then becomes statistically nonsignificant in the third period and converges to 0. The R&D expenditure per capita (dlnPRD) changes by one unit of standard deviation, and the illiteracy rate changes by one unit of standard deviation. The illiteracy rate shows a significant positive impulse response and reaches a peak in the first period, decreases sharply from the second period, and gradually converges to 0, but is statistically nonsignificant. For a one-unit change in per capita expenditure on mass cultural programs (dlnPMCE), the illiteracy rate is statistically nonsignificant, although it shows a downward and then an upward shock effect.

4.4.2.4 Variance Decomposition Analysis

Impulse response analysis can give the effect of a standard information shock to one variable in a system on the system and other variables; however, it cannot give the degree of contribution of each variable to the system and other variables after the shock. This study is based on variance decomposition, which calculates the degree of explanation of changes in one variable to the variance of other variables to obtain the relative contribution of each factor to the changes in social poverty and identify the key influencing factors of social poverty. The results of variance decomposition are shown in Table 4.11 and Fig. 4.6.

The results of the income poverty variance decomposition (Table 4.11, Fig. 4.6a) show that the contribution of per capita annual savings (dlnPSD) to its own variance tends to diminish over time, slowly decreasing from 100% in Period 1 to 95.65% in Period 10. The contribution of other variables to annual per capita savings over time increases annually, and the variance contribution of each variable to it gradually converges to a steady state from the 9th period onwards, with the explanatory weights

Table 4.11 Variance decomposition of the causes of social poverty

dlnPSD		dlnPSD	dlnPGDP	dlnIS	dlnPSSE	dlnGDR	dlnFCA
	1	1	0	0	0	0	0
	2	0.968155	0.000347	1.58E-06	6.64E-06	0.030901	0.000589
	3	0.960194	0.001049	0.004942	0.000248	0.032979	0.000588
	4	0.958564	0.001729	0.005325	0.000312	0.033482	0.000588
	5	0.957453	0.001892	0.005541	0.000386	0.034141	0.000588
	6	0.957001	0.002032	0.005538	0.000407	0.034434	0.000588
	7	0.956717	0.002094	0.00556	0.000424	0.034617	0.000588
	8	0.956577	0.002133	0.005561	0.000432	0.03471	0.000588
	9	0.956495	0.002153	0.005564	0.000436	0.034764	0.000588
	10	0.956452	0.002164	0.005565	0.000439	0.034793	0.000588
dlnIDI		dlnIDI	dlnPPHE	dlnPHOS	dlnUR	dlnUHI	dlnFCR
	1	100	0	0	0	0	0
	2	0.863931	0.002533	0.030457	0.088641	0.014126	0.000313
	3	0.8284	0.004376	0.037067	0.112334	0.016749	0.001075
	4	0.792732	0.008741	0.03747	0.134622	0.015961	0.010475
	5	0.759656	0.014083	0.038765	0.152727	0.015267	0.019502
	6	0.729012	0.019443	0.040273	0.167941	0.014602	0.028729
	7	0.700944	0.02452	0.041727	0.181395	0.013988	0.037426
	8	0.675296	0.029218	0.043095	0.193514	0.013426	0.045452
	9	0.651812	0.033539	0.04436	0.204548	0.012911	0.052829
	10	0.630245	0.037515	0.045527	0.214662	0.012438	0.059613
dlnIR		dlnIR	dlnPEF	dlnHEDU	dlnSTR	dlnPRD	dlnPMCE
	1	100	0	0	0	0	0
	2	0.727707	0.035219	0.137775	0.030166	0.066875	0.002258
	3	0.689686	0.033786	0.166171	0.043497	0.064316	0.002545
	4	0.683905	0.033458	0.172353	0.043491	0.063703	0.003092
	5	0.682906	0.033407	0.172439	0.044	0.064057	0.003191
	6	0.682725	0.033417	0.172567	0.044006	0.064049	0.003236
	7	0.682686	0.03342	0.172595	0.044012	0.064051	0.003237
	8	0.682664	0.033422	0.172617	0.04401	0.064049	0.003238
	9	0.682658	0.033422	0.172622	0.044012	0.064049	0.003238
	10	0.682657	0.033422	0.172624	0.044011	0.064048	0.003238

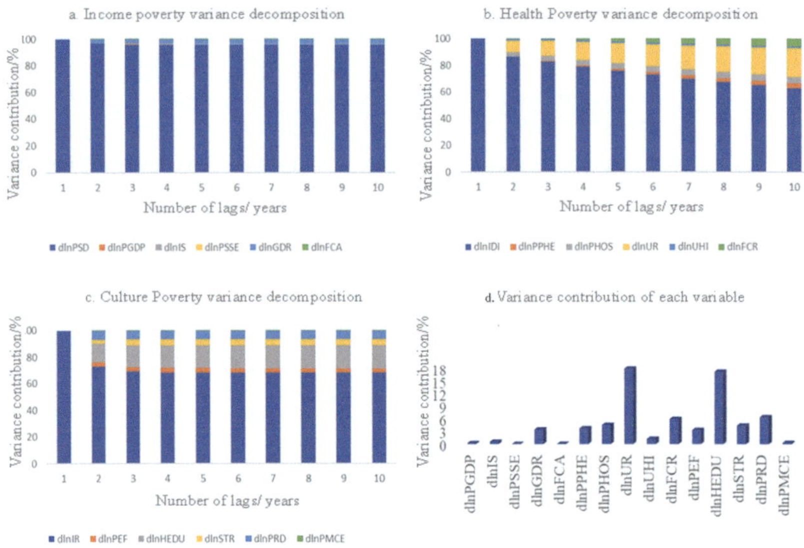

Fig. 4.6 Results of the variance decomposition of the causes of social poverty

of GDP per capita, the proportion of non-agricultural industries, per capita social security expenditures, the total social dependency ratio, and the crop damage ratio on the annual per capita savings remaining at around 0.22%, 0.56%, 0.04%, 3.5%, and 0.06%, respectively. Natural disasters, the level of regional economic development, and social security are important factors affecting long-term future regional income poverty.

Meanwhile, from the results of the variance decomposition of health poverty (Table 4.11, Fig. 4.6b), the incidence rate of infectious diseases (lnIDI) gradually declined from 100% in Period 1 to 63.02% in Period 10. Over time, the per capita public health expenditure, the number of hospitals per 10,000 people, the urbanization rate, the urban residents' health insurance participation rate and the forest coverage rate all remained at 3.75%, 4.55%, 21.47%, 1.24%, and 5.96% of the variance of the infectious disease incidence rate in the 10th period, respectively, after experiencing a period of continuous increase. This analysis shows that urbanization development explains health poverty relatively more strongly compared to other factors, and is an important means of suppressing regional health poverty.

The results of the variance decomposition of cultural poverty (Table 4.11, Fig. 4.6c) indicate that the contribution of the illiteracy rate (lnIR) to its own is all from itself in the 1st period, and the degree of contribution is maintained at the level of 68.27% after the 6th period. As the forecast period increases, the variance contribution of per capita education expenditure, the number of higher education institutions, the student–teacher ratio, per capita R&D expenditure and per capita expenditure on mass culture to the illiteracy rate increases steadily annually and stabilizes at around 3.34%, 17.26%, 4.4%, 6.4%, and 0.32%, respectively, after the

5th period. According to the long-term future trend, per capita education expenditure contributes to a greater extent than other factors beside the illiteracy rate itself, and it plays a crucial role in the future movement of cultural poverty in the region.

The results of variance decomposition get closer as the forecast period gets longer. In this paper, the variance decomposition results of the 10th period are used to illustrate the degree of influence of each variable on social poverty. It should be noted that, considering that the Human Development Index (HDI), the Human Poverty Index (HPI), and the Multidimensional Poverty Index (MPI) published by the UNDP for poverty measurement adopt the equal weight method for determining the weights of the dimensions and specific indicators, this paper follows the same method in examining the contribution of each variable to the variance of social poverty, i.e., if the weights of income poverty, health poverty and cultural poverty and their explanatory variables are the same, then, the variance decomposition results will be the same as those of income poverty, health poverty, and cultural poverty, and if their explanatory variables have the same weight, then, the variance contribution of each explanatory variable to social poverty can be directly observed and compared.

As shown in Fig. 4.6d, in descending order of relative contribution, they are urbanization rate, number of higher education institutions, R&D expenditure per capita, forest coverage rate, number of hospitals per 10,000 people, student–teacher ratio, public health expenditure per capita, total social dependency ratio, education expenditure per capita, urban residents' health insurance participation rate, proportion of non-agricultural industries, per capita expenditure on public cultural affairs, GDP per capita, crop damage ratio, and social security expenditure per capita. Based on the conclusions of the previous GMM estimation analysis, per capita public health expenditure and forest cover rate cannot be used as the explanatory variables of social poverty; thus, they are excluded.

To avoid the bias of the empirical results, this study considers the measurement results after changing the order of each variable and finds no discrepancy between the results of the main empirical analyses and the conclusions of the above econometric tests; therefore, it can be considered that the above model is robust, and that the results of the aforementioned empirical analyses have a strong explanatory power.

4.4.3 Results and Recommendations

This paper uses China's provincial panel data to construct a PVAR model with income poverty, health poverty, and cultural poverty as the core content of social poverty, and focuses on examining the dynamic mechanism between each factor and social poverty and its contribution to the evolution of social poverty through GMM estimation, impulse response, and variance decomposition analyses, and selecting the factors influencing social poverty.

In terms of income poverty, natural disasters have a significant inhibitory effect on income poverty, whereas the level of economic development, industrial structure, social security, and population burden have a positive effect. The factors affecting income poverty, in the order of variance contribution, are the total social dependency ratio, proportion of non-agricultural industries, per capita GDP, crop damage ratio, and per capita social security expenditure.

From the perspective of health poverty, the impact of healthcare expenditure and environmental quality on health poverty did not reach the level of significance. The accessibility of medical and health resources has a significant constraint on health poverty, and urbanization and medical security adversely impact the health poverty situation in a region. The order of magnitude of the impact on health poverty is the urbanization rate, the number of hospitals per 10,000 people, and the participation rate of urban residents in health insurance.

From the perspective of cultural poverty, education capital investment, education resource allocation exhibition, and mass culture construction are significantly negatively correlated with cultural poverty, whereas higher education development and science and technology investment have a significant positive impact on cultural poverty. The variables in the order of the variance contribution to cultural poverty are the number of higher education institutions, per capita R&D expenditure, student–teacher ratio, per capita education expenditure and, per capita expenditure on mass cultural undertakings.

In general, examining the variance contribution of each variable to social poverty according to the principle of equal weight, the following size ranking is obtained.

Urbanization rate, number of higher education institutions, R&D expenditure per capita, number of hospitals per 10,000 people, student–teacher ratio, total social dependency ratio, education expenditure per capita, urban residents' health insurance participation rate, proportion of non-agricultural industries, per capita expenditure on mass cultural undertakings, GDP per capita, crop-affected ratio, and per capita social security expenditure.

Exercise

1. What are the main challenges faced by less developed regions in the implementation of regional development strategies?
2. What are the specific impacts of the Yangtze River Delta integrated development strategy on the development of less developed regions?
3. What specific measures has the government taken to promote high-quality development in less developed regions?
4. What is China's future development strategies and expected goals in narrowing regional development gaps?
5. How can the economic development of less developed regions be effectively promoted through policy adjustments and coordinated regional development strategies?

Extended Reading

Brewer, E. A., et al. (2006). The challenges of technology research for developing regions. *IEEE Pervasive Computing*, 15–23.

Tulenty, D., & Riabinina, N. (2022). Searching for hidden connections between the evolution of poverty in developing countries and information technology. *International Journal of Innovation and Technology Management, 19*(06).

References

Arellano, M., & Bover, O. (1995). Another look at the instrumental variable estimation of error-components models. *Journal of Econometrics, 68*(1), 29–51. https://doi.org/10.1016/0304-4076(94)01642-D

Bai, C.-E., Ma, H., & Pan, W. (2012). Spatial spillover and regional economic growth in China. *China Economic Review.*

Brewer, E. A., et al. (2006). The challenges of technology research for developing regions. *IEEE Pervasive Computing*, 15–23.

Ding, X., Cai, Z., Xiao, Q., et al. (2019). A study on the driving factors and spatial spillover of carbon emission intensity in the Yangtze river economic belt under double control action. *International Journal of Environmental Research and Public Health.*

Fan, J., Sun, W., Zhou, K., & Chen, D. (2012). Major function oriented zone: New method of spatial regulation for reshaping regional development pattern in China. *Chinese Geographical Science, 22*, 196–209. https://doi.org/10.1007/s11769-012-0528-y

Feng, W., Liu, Y., & Qu, L. (2019). Effect of land-centered urbanization on rural development: A regional analysis in China. *Land Use Policy, 87*, Article 104072. https://doi.org/10.1016/j.landusepol.2019.104072

Holtz-Eakin, D., Newey, W., & Rosen, H. S. (1988). Estimating vector autoregressions with panel data. *Econometrica, 56*(6), 1371–1395. https://doi.org/10.2307/1913103

IPCC. (2014). Climate change 2014: Mitigation of climate change. *Contribution of Working Group III to the Fifth Assessment Report of the Intergovernmental Panel on Climate Change.* Cambridge University Press. https://www.ipcc.ch/report/ar5/wg3/

Jiang, X.-Z., Liu, T.-Y., & Su, C.-W. (2014). China's marine economy and regional development. *Marine Policy, 50*, 227–237. https://doi.org/10.1016/j.marpol.2014.06.008

Kontodimopoulos, N., Papathanasiou, N. D., Flokou, A., Tountas, Y., & Niakas, D. (2011). The impact of non-discretionary factors on DEA and SFA technical efficiency differences. *Journal of Medical Systems, 35*, 981–989. https://doi.org/10.1007/s10916-010-9521-0

Laffont, J.-J. (1975). Macroeconomic constraints, economic efficiency and ethics: An introduction to Kantian economics. *Economica.*

Li, Y., & Wu, F. (2012). The transformation of regional governance in China: The rescaling of statehood. *Progress in Planning, 78*, 55–99. https://doi.org/10.1016/j.progress.2012.03.001

Lin, J., Yu, Z., Wei, Y. D., & Wang, M. (2017). Internet access, spillover and regional development in China. *Sustainability, 9*, 946. https://doi.org/10.3390/su9060946

Liu, Y., Stillwell, J., Shen, J., & Daras, K. (2014). Interprovincial migration, regional development and state policy in China, 1985–2010. *Applied Spatial Analysis and Policy, 7*, 47–70. https://doi.org/10.1007/s12061-014-9102-6

Luo, J. M., Qiu, H., & Lam, C. F. (2016). Urbanization impacts on regional tourism development: A case study in China. *Current Issues in Tourism, 19*, 282–295. https://doi.org/10.1080/13683500.2015.1033385

Reinhard, S., Lovell, C. A. K., & Thijssen, G. (1999). Econometric estimation of technical and environmental efficiency: An application to dutch dairy farms. *American Journal of Agricultural Economics, 81*, 44–60. https://doi.org/10.2307/1244449

Schaltegger, S., & Sturm, A. (1990). Ökologische Rationalität: Ansatzpunkte zur Ausgestaltung von ökologieorientierten Managementinstrumenten. *Die Unternehmung, 44*(4), 273–290.

Song, J., & Chen, X. (2019). Eco-efficiency of grain production in China based on water footprints: A stochastic frontier approach. *Journal of Cleaner Production, 236*, Article 117685. https://doi.org/10.1016/j.jclepro.2019.117685

Sueyoshi, T., & Yuan, Y. (2015). China's regional sustainability and diversified resource allocation: DEA environmental assessment on economic development and air pollution. *Energy Economics, 49*, 239–256. https://doi.org/10.1016/j.eneco.2015.01.024

Tulenty, D., & Riabinina, N. (2022). Searching for hidden connections between the evolution of poverty in developing countries and information technology. *International Journal of Innovation and Technology Management, 19*(06).

Wang, P., Cen, C., & Lin, X. (2023a). Internet development and the spatial optimization of regional productivity: Evidence from China. *Growth and Change, 54*, 912–939. https://doi.org/10.1111/grow.12680

Wang, S., Zhang, Y., & Wen, H. (2021). Comprehensive measurement and regional imbalance of China's green development performance. *Sustainability, 13*, 1409. https://doi.org/10.3390/su13031409

Wang, X., Zhang, M., Jie, S., Zhang, M., & Zhang, Z. (2023b). Exploring the coordinated evolution mechanism of regional sustainable development and tourism in China's beautiful China initiative. *Land, 12*, 1003. https://doi.org/10.3390/land12051003

Wang, Y., & Wang, J. (2019). Does industrial agglomeration facilitate environmental performance: New evidence from urban China? *Journal of Environmental Management, 248*, Article 109244.

Wei, Y. H. D., & Liefner, I. (2012). Globalization, industrial restructuring, and regional development in China. *Applied Geography, 32*, 102–105. https://doi.org/10.1016/j.apgeog.2011.02.005

Wen, H., Liang, W., & Lee, C.-C. (2023). China's progress toward sustainable development in pursuit of carbon neutrality: Regional differences and dynamic evolution. *Environmental Impact Assessment Review, 98*, Article 106959. https://doi.org/10.1016/j.eiar.2022.106959

Wu, H., Li, Y., Hao, Y., Ren, S., & Zhang, P. (2020). Environmental decentralization, local government competition, and regional green development: Evidence from China. *Science of the Total Environment, 708*, Article 135085. https://doi.org/10.1016/j.scitotenv.2019.135085

Ye, X., Ma, L., Ye, K., Chen, J., & Xie, Q. (2017). Analysis of regional inequality from sectoral structure, spatial policy and economic development: A case study of Chongqing, China. *Sustainability, 9*, 633. https://doi.org/10.3390/su9040633

Chapter 5
Agglomeration and Convergence of Regional Development Process in Developed Regions

Abstract Agglomeration and convergence of regional development is an important component of development geography. Since the 1980s, global regional development has shown obvious agglomeration characteristics, particularly in developed areas, and has not only attracted more industries, especially high-tech industry settlement and development, resulting in high-tech industry and economic agglomeration, but also attracted more competent technical personnel, forming a talent agglomeration. China's Beijing-Tianjin-Hebei city cluster, Yangtze River Delta city cluster, and Pearl River Delta city cluster have developed rapidly over the past 30 years. The economic share of the three major city clusters had exceeded 45% by 2021, and the absorption capacity of the economic growth of the three regions has reached a record high. However, the agglomeration of regional development is not contrary to the balanced development of the region. Many developed regions adopted the development theory of "agglomeration convergence," which has become an important future direction of regional development in China's developed regions. Agglomeration and convergence theory of regional development in developed areas is currently a topic of great interest in regional development research. Since the beginning of the last century, geographers have studied the phenomenon of regional agglomeration and convergence and explored the process of industrial agglomeration. With the rise of the economic activity in large enterprises and transnational corporations, more attention is being paid to the internal structure of enterprises, which loosens the research on development agglomeration and convergence. Then, in the late 1970s, with economic globalization, geographers quickly revived the study of development agglomeration and convergence. In the 1980s, research attention turned to "new economic geography," which laid a good theoretical foundation for the study of agglomeration and convergence of development geography. The objectives of this chapter are as follows: 1. To elaborate the agglomeration process of different elements in the regional development process of developed areas from the perspectives of urban space, industry, economy, and innovative talents. 2. To analyze the convergence characteristics of different elements. 3. To reveal the main driving mechanism of the agglomeration and convergence process of different elements of urban space, industry, economy, and innovative talents.

X. Deng and M. Song, *Development Geography*,
https://doi.org/10.1007/978-981-96-6917-2_5

5.1 Convergence of Urban Spatial Expansion and Development

5.1.1 Urban Spatial Expansion

Urban spatial expansion is an important representation of urbanization, an important regional development process experienced by developed regions over the past 30 years, and an important representation of the regional development process of development geography. Urban spatial expansion includes not only the dynamic expansion process of urban land but also the optimization and adjustment of the spatial pattern of different land use structures within cities. In addition, the dynamic process of regional diffusion and promotion of urban economic, cultural, population, and other factors is also an important part of urban spatial expansion. Therefore, the process of urban expansion is characterized not only by urban land area but also by indicators such as urban primacy and urban rank-scale based on population and other factors. The process of urban expansion in developed areas has different characteristics at different stages of urbanization development. Studying the development law of urban spatial expansion in developed areas at different stages provides an important reference for the future development planning of underdeveloped areas. Urban space expansion is driven by many factors of urban development. The high density of resources, environment, population, social economy, and other factors has led to the rise of cities, which not only makes the city full of vitality but also leads to the continuous expansion of urban land and urban elements. This kind of expansion is dynamic in time and space, irreversible in time, orderly, and has phases. Spatial expansion patterns vary with different driving factors. The research object of urban spatial expansion is mainly the spatial form of urban expansion, and the research aim is mainly to explore the spatial development law of urban land use and urban elements. Urban spatial expansion theory can be divided into urban internal regional structure theory and urban regional development theory based on the research object. The research object of urban internal regional structure theory is a single city or urban internal structure, whereas that of urban regional development theory is mainly the distribution of urban agglomerations or multiple cities in the region.

5.1.1.1 Theory of Urban Internal Regional Structure

Urban internal regional structure theory can be divided into modern urban internal regional structure theory and traditional urban internal regional structure theory. The traditional regional structure of the city has produced many theories, including Burgess's concentric circle theory, Hoyt's sector theory, and Harris and Ullman's multi-core theory. According to Burgess's concentric circle theory, the distribution of non-urban land from the urban center to outside the urban built-up area is completely in accordance with land use efficiency. However, this theory assumes that cities develop uniformly, ignoring the influence of traffic or natural obstacles

(Burgess, 2015). According to Hoyt's sector theory, the urban center is the CBD (central business district), and the accessibility and connectivity of traffic routes have an important impact on the formation of various functional areas of a city. Therefore, the functional areas of a city are not differentiated by their distance from the center but are distributed according to the two sides of the traffic routes (Hoyt, 1939). According to Harris and Ullman's multi-core theory, the CBD is still the activity center of the city; however, it is not necessarily located in the geometric center of the city and will form several other sub-cores, with wholesale areas and low-pollution light industries next to the layout of the city center and low-grade residential areas around the periphery (Harris and Ullman, 1945). Although the above three theoretical models are all early research results, they depict a certain type of urban spatial expansion process from different perspectives such as location rent, urban traffic, and multiple cores. Modern urban internal regional structure theory research is conducted from three aspects: urban area, marginal area, and influence area. It includes Dickenson's three-zone model, Talosum's regional city model, and Mueller's metropolitan structure model. Dickenson's three-zone model holds that the inner regional structure of a city is composed of the central zone, the middle zone, and the outer or suburban zone (Clarke et al., 1996). According to Talosum's regional urban model, there exists a continuous unity between the urban area and the rural hinterland, including the urban core area, urban fringe area, urban influence area, and rural hinterland. Mueller's model of metropolitan regional structure consists of four parts: declining central area, inner suburbs, outer suburbs, and urban fringe area (Mueller, 1981). In addition to declining central cities, several small cities are forming in the outer suburbs, which form their own urban areas according to the natural environment, regional transportation networks (Akihiro, 2023), and internal regionalization of economic activities, and then combine these urban areas into metropolitan areas (Fig. 5.1).

5.1.1.2 Urban Regional Development Theory

The modern theory of urban regional development studies the distribution of different cities or the urban groups in a certain region. The main theories are growth pole theory, point-axis theory, core–edge theory, and metropolitan theory. According to growth pole theory, economic growth will be affected by the benefits generated by the growth pole, and the formation, development, decline, and disappearance of the growth pole will affect regional industrial development and industrial distribution in space, leading to changes in regional and urban spatial structure. Core–edge theory combines growth pole theory with other spatial systems and divides regions into core and edge regions that have the effect of intensification and diffusion. This theory comprehensively explains regional economic development and spatial structure and is, therefore, widely recognized. According to the point-axis theory, because of its central location, important transportation lines will attract the convergence of industry, population, and other factors, thus forming residential areas in advantageous locations. The urban spatial expansion pattern shows different characteristics

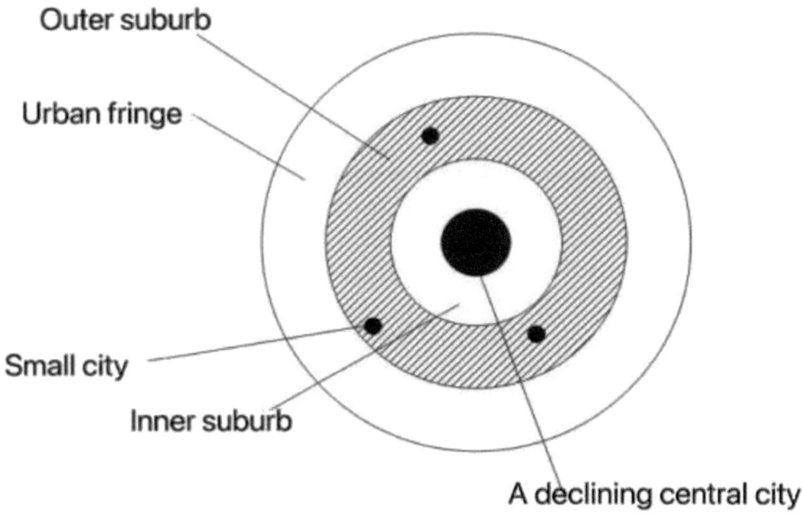

Fig. 5.1 Mueller's model of regional structure within cities (According to Mueller, 1981)

at different stages, including the single-core concentric expansion pattern, the axial belt expansion pattern, the multipolar nuclear growth expansion pattern, and the metropolitan area expansion pattern. In the single-core concentric expansion mode, due to the development of the social economy, the radiation force of the central urban area is continuously enhanced, and the original urban area is the center of the all-round outward expansion. The form of its expansion depends on the surrounding topography and geomorphic features on the one hand, and on the main influencing factors, influence intensity, and influence mode that drive the urban expansion on the other hand. The axial belt expansion mode is mainly caused by the change in traffic conditions, which leads to the change in the urban form of expansion. New development areas continue to extend along the traffic lines, which will also give rise to new growth points and evolve into the spatial structure of multiple core areas. The multipolar core growth and expansion model refers to the development of new growth points in the outer areas of the city to meet the needs of urban function adjustment or new urban functions, forming a new core area. The expansion pattern of the metropolitan area is mainly due to the blurring of the interface between the central city and the suburban area, or the emergence of sub-centers or satellite cities in the peripheral areas of big cities. The quantitative analysis of urban spatial expansion is an important prerequisite for the study of urban spatial expansion, and remote sensing technology is one of the important monitoring means of urban land spatial expansion. Since the 1970s, Landsat and MSS (Maximum Segment Size) remote sensing data have been applied to the dynamic monitoring of urban fringe settlements. In the 1990s, remote sensing images were applied to study the dynamic changes in the spatial expansion of regional metropolitan areas. In the twenty-first century, different urban spatial expansion simulation models based on remote sensing data have emerged and been widely used. Remote sensing technology began to play a key role in monitoring the evolution of urban spatial form. Then, combined with GIS technology, remote sensing technology provided a more effective method for the dynamic monitoring of urban spatial expansion. Although China's research focus on urban spatial expansion has shifted from qualitative analysis of urban spatial form to quantitative analysis of dynamic changes in urban spatial expansion and determination of urban spatial expansion boundaries and has achieved rich results, there are still many problems (Couch & Karecha, 2006). For example, China's global international metropolis, represented by Shanghai and Beijing, has entered a new stage of urban development. However, existing relevant studies have not discussed the characteristics and mechanism of urban expansion in the new development stage.

The Yangtze River Delta city cluster is centered in Shanghai. According to the 2021 Yangtze River Delta Regional Integrated Development Plan outline, the planning scope is officially designated as the whole area of Jiangsu, Zhejiang, Anhui, and Shanghai. The Yangtze River Delta city cluster is an important intersection of the "Belt and Road" and the Yangtze River Economic Belt, and is one of the regions with the best foundation for urbanization in China. In the past 20 years, it has also experienced the fastest urban space expansion process. The rapid economic development and the acceleration of industrialization and urbanization in the Yangtze River Delta urban agglomerations have led to a rapid increase in the demand for urban construction land. This case analyzes the change in urban construction land in the Yangtze River Delta urban agglomerations to measure the temporal and spatial evolution characteristics of urban development in the Yangtze River Delta urban agglomerations. The results show that from 1998 to 2017, the proportion of urban construction land area in three provinces and one city in the Yangtze River Delta has an increasing trend. The urban construction land area ratio of Shanghai and Zhejiang Province is higher than the average level of urban agglomerations, whereas that of Anhui and Jiangsu provinces is lower than the average level of Yangtze River Delta urban agglomerations (Fig. 5.2).

According to the statistical analysis of the construction land area of 26 cities in the Yangtze River Delta urban agglomeration from 1998 to 2017, the growth rates of Shanghai, Anhui, Jiangsu, and Zhejiang are 82%, 260%, 324%, and 96%, respectively, indicating that the overall urbanization process of the nine cities in Jiangsu Province is developing rapidly and the urban construction land has greatly increased. However, the urbanization process in Shanghai has become stable, and the expansion of urban construction land is the smallest. According to the comparison of urban construction land changes in different regions, there is a large gap between the construction land area of the capital cities of Anhui Province and Jiangsu Province and other cities in the provinces. Except for the capital cities of Hefei and Nanjing, the difference in the construction land area of other cities is not large; however, the construction land area of Nanjing is almost twice that of Hefei, indicating a large regional difference level. Hangzhou, the capital city of Zhejiang Province, has a small gap with other cities in the province. The construction land area of Hangzhou is between Hefei and Nanjing. The other cities except Zhoushan, have a small gap in construction land area. The change in urban primacy is an important representation of

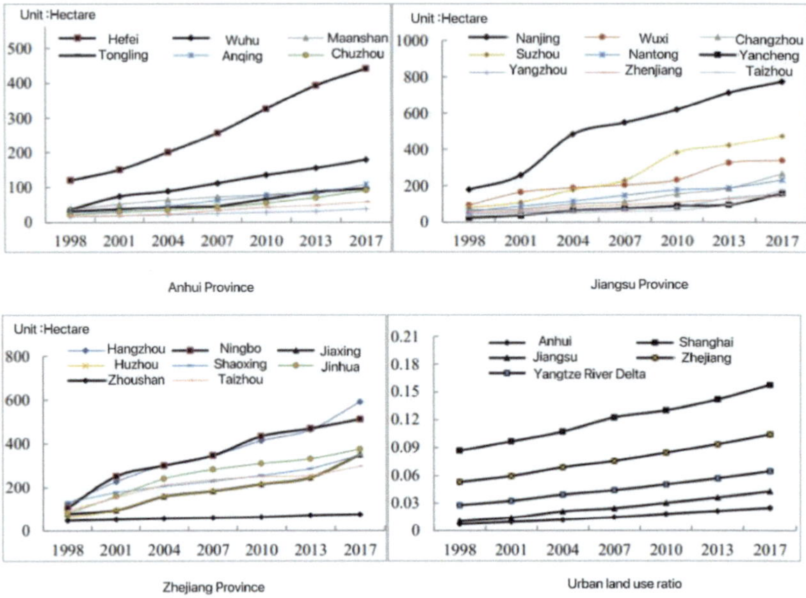

Fig. 5.2 Area and proportion of urban construction land in the Yangtze river delta urban agglomeration

the development process of regional urban spatial expansion, which is closely related to the regional development process from the perspective of development geography. City primacy refers to the importance of the largest city in a country or region. To a certain extent, the degree of city primacy represents the degree of concentration of urban development elements in the urban system in the largest cities. In the urban system, the city with the largest permanent population is the first city, the two-city index, four-city index, and eleven-city index are used to calculate the city primacy. The two-city index is the ratio of the population size of the first city to the second city in a region. The formula is $S = P1/P2$. It is generally believed that if the two-city index is less than 2, the urban system structure is normal and the concentration is appropriate. If the value is greater than 2, there is a trend of structural imbalance and excessive concentration. The more the index of two cities is greater than 2, the more unreasonable the scale structure of the urban system, the more unbalanced the development of the urban system, the more obvious the difference between the scale levels of cities, and the greater the difference between the regions of urbanization. Due to the oversimplification of the comparison between the two cities, to avoid generalization, later scholars proposed the four-city index: $S = P1/(P2 + P3 + P4)$ and the 11-city index: $S = 2P1/(P2 + P3 + ... + P11)$. According to the principle of city rank-size distribution, under the ideal urban population distribution, the two-city index of the urban system in the region should be 2, and the value of the four-city and 11-city indexes should be 1. The more the index of two cities is greater than 2, and the more the index of four cities and eleven cities is greater than 1, the more the distribution of city size belongs to the first distribution. If the index of two cities is less than 2, and the index of four cities and eleven cities is less than 1, then, the city size distribution belongs to the sequential size distribution. In this case, the two-city index, four-city index, and 11-city index are constructed based on the urban population size, and the spatial expansion process of the Yangtze River Delta urban agglomeration is analyzed. Based on the city primacy formula, the calculation results are shown in Table 5.1.

Table 5.1 shows that the two-city index, four-city index, and 11-city index of the Yangtze River Delta region show an increasing trend from 1998 to 2017, indicating that with the development of urbanization, the development speed of the first city, Shanghai, is obviously faster than that of other cities in the urban agglomeration, and Shanghai plays an increasingly leading role in the entire urban agglomeration. After 2015, the two-city, four-city, and eleven-city indexes all showed a downward trend, indicating that the concentration of the first city decreased after 2015, and other cities in the urban agglomeration developed rapidly. Rank-scale analysis is one of the important methods to analyze urban spatial expansion and development from the population perspective, and it is also an important content in the process

Table 5.1 City primacy of Yangtze river delta urban agglomeration

Year	First urban population size (10,000)	Two-city number	Four-city index	Eleven-city index
1998	1527.32	1.929	0.897	0.906
1999	1567.25	1.975	0.902	0.926
2000	1608.67	2.022	1.102	1.077
2001	1668.33	2.097	1.392	1.254
2002	1712.97	2.153	1.409	1.412
2003	1765.84	2.217	1.451	1.504
2004	1834.98	2.299	1.472	1.624
2005	1890.26	2.349	1.496	1.698
2006	1964.11	2.425	1.513	1.613
2007	2063.58	2.339	1.561	1.654
2008	2140.65	2.346	1.656	1.666
2009	2210.28	2.359	1.634	1.695
2010	2302.66	2.2	1.735	1.772
2011	2347.46	2.232	1.793	1.799
2012	2380.43	2.257	1.8	1.82
2013	2415.15	2.283	1.874	1.925
2014	2415.27	2.288	1.926	1.971
2015	2425.68	2.275	1.916	1.764
2016	2419.76	2.273	1.717	1.755
2017	2418.33	2.264	1.698	1.718

of regional development of development geography (Ficcadenti et al., 2022). The ordination-size rule means that after all cities in an urban system are ranked from large to small in terms of population, there is a regular relationship between the city size and the ordination, and this rule better describes the size distribution of an urban population. This case takes the Yangtze River Delta urban agglomeration as an example to analyze the urban rank-scale. A common expression of the bit-size rule is:

$$P_r = \frac{P}{r^q} \tag{5.1}$$

where P_r represents the population size of the city with the bit order as; P is the first urban population size; q is a constant, also known as the Zipf index.

To facilitate the calculation, the logarithm of both sides of the formula is taken to obtain the expression:

$$lgP_r = lgP_i - qlgr \tag{5.2}$$

The value of the parameter q is obtained by linear regression of the city bit order. When q is greater than 1, it indicates that the urban population distribution is more dispersed than the ideal state. When q is close to 1, it indicates that the city size distribution is close to the ideal state. When q is less than 1, it indicates that the size distribution of the urban system is more concentrated than the ideal state. After ranking 26 cities in the Yangtze River Delta urban agglomeration by scale from largest to smallest, the regular relationship between city size and the ranking in each year is calculated, and the regression equation and R^2 are obtained. The calculation results are as follows (Table 5.2).

The calculation results show that during the 20 years from 1998 to 2017, the parameter q is less than 1 in all cases, which indicates that the size distribution of the urban system is more concentrated than the ideal state, the large cities are too large, and small and medium-sized cities are relatively underdeveloped. The q value increases first and then tends to stabilize, and the value is closer to 1. After 2010, the parameter q decreased and stabilized at about 0.75, indicating that small and medium-sized cities within the urban system were developing and expanding. This is mainly due to the adjustment of national policies, which will accelerate the development of small cities as the focus of urban size structure optimization. The national "Ninth Five-Year Plan" (1996–2000) and the "2010 Vision Target Outline" clearly define the development direction of regional economic integration in the Yangtze River Delta and propose that Shanghai should give full play to its own radiating role, release resources to the surrounding areas, and drive the development of surrounding cities. The industrial transfer of Shanghai is always first transferred to other big cities in the provinces, and then transferred from the big cities to the small and medium-sized cities in the provinces. From the decline of the primacy of the Yangtze River urban agglomeration and the change in the Zipf index, it is clear that in recent years, the spatial expansion of the Yangtze River Delta urban agglomeration mainly occurs

Table 5.2 Analysis results of urban rank-scale in Yangtze river delta urban agglomerations

Year	lgPi	q	R^2	Regression equation
1998	3.280	0.695	0.715	lgPr = 3.280–0.695lgr
1999	3.280	0.697	0.717	lgPr = 3.280–0.697lgr
2000	3.289	0.702	0.721	lgPr = 3.289–0.702lgr
2001	3.296	0.706	0.724	lgPr = 3.296–0.706lgr
2002	3.301	0.708	0.708	lgPr = 3.301–0.708lgr
2003	3.307	0.713	0.712	lgPr = 3.307–0.713lgr
2004	3.315	0.718	0.733	lgPr – 3.315–0.718lgr
2005	3.323	0.722	0.734	lgPr = 3.323–0.722lgr
2006	3.350	0.733	0.729	lgPr = 3.350–0.733lgr
2007	3.368	0.746	0.740	lgPr = 3.368–0.746lgr
2008	3.379	0.754	0.754	lgPr = 3.379–0.754lgr
2009	3.389	0.760	0.747	lgPr = 3.389–0.760lgr
2010	3.412	0.778	0.760	lgPr = 3.412–0.778lgr
2011	3.398	0.739	0.739	lgPr = 3.398–0.739lgr
2012	3.402	0.741	0.746	lgPr = 3.402–0.741lgr
2013	3.405	0.743	0.747	lgPr = 3.405–0.743lgr
2014	3.408	0.745	0.746	lgPr = 3.408–0.745lg
2015	3.377	3.377	0.813	lgPr = 3.377–0.750lgr
2016	3.379	3.379	0.813	lgPr = 3.379–0.750lgr
2017	3.383	3.383	0.813	lgPr = 3.383–0.750lgr

outside the first city, and the spatial distribution of the city size develops from the centralized state to the ideal state.

5.1.2 Convergence of Urban Spatial Expansion

The convergence trend of urban spatial expansion is the result of the development of cities or urban agglomerations to a certain stage. Continuous attention paid to urban land intensification, will gradually strengthen the convergence of urban space expansion. Currently, the urban expansion in the developed coastal areas of southeast China is gradually slowing down, and the trend of regional spatiotemporal convergence is increasing. Due to the continuous industrial gradient transfer from these regions to the surrounding regions, the urban spatial expansion of urban agglomerations in the central region is also continuously enhanced; however, the regional spatio-temporal convergence is slower than that of the southeastern coastal urban agglomerations.

5.1.2.1 The Convergence Trend of Urban Spatial Expansion and Development in Developed Areas

Urban agglomerations are relatively economically developed areas in China, and are also the highest spatial organization form of urban development to the mature stage. The convergence process of urban spatial expansion and development are the geographical characteristics of the development of urban agglomerations in China. Since the 1990s, China's transportation network has been developing continuously, particularly the comprehensive and rapid development of China's high-speed rail in this century, and the cutting of transportation and time costs have caused significant changes in the mode, form, and pattern of urban development in China and also promoted the spatio-temporal convergence of China's urban agglomerations. Relevant theories and several empirical studies suggest that there are two characteristics of spatial convergence in economic geography, namely subset convergence and dynamic and long-term convergence. From the perspective of time and space, the first law reflects the relationship between space and convergence, and the second law reflects the relationship between time and convergence. With the continuous rise in urban land and housing prices in developed areas, which leads to higher economic costs, coupled with related indicators such as energy conservation and emission reduction becoming new benchmarks for regional development competition, urban space utilization efficiency in developed areas will attract more attention. Therefore, the urban spatial expansion trend of Yangtze River Delta and Pearl River Delta urban agglomeration will slow down. The construction of high-speed railway has caused the continuous transfer of industries to the rear of these regions and accelerated the development of cities in the surrounding provinces of Anhui, Hunan, and Jiangxi; thus strengthening the trend of urban spatial convergence in Yangtze River Delta and Pearl River Delta urban agglomeration. For the small and medium-sized urban agglomerations in China, such as Guanzhong-Tianshui urban agglomerations, the middle reaches of the Yangtze River, and Sichuan-Chongqing urban agglomerations, the expansion trend of core cities will be further enhanced due to the gradient transfer of industries, whereas the spatio-temporal convergence speed tends to be slower than that of large urban agglomerations. As the change in urban spatial characteristics, regional spatio-temporal convergence can properly balance the internal contradictions of cities and develop urban economy. Due to their different economic scales, production factors, and their different flow characteristics, different industries lead to different industrial concentration locations. The structure of the industrial chain does not allow these industries to be too far apart in space, and these industries are linked to each other. Smaller cities will appear around the core city. The industries in these smaller cities are the organic components of the whole regional industrial chain, and the whole region has formed the demand for the mobility of production factors (the connection of upstream and downstream industries, the sharing of knowledge, the sharing of social capital, the sharing of the common market, etc.), which is also the result of measuring the transfer costs and benefits of production factors. In the future, the efficiency of sustainable development can be explored based on the convergence of economic growth.

5.1.2.2 Convergence Trend of Urban Sustainable Development Efficiency in Beijing–Tianjin–Hebei Region

Urban sustainable development is an important factor in the process of urban development, and the convergence trend of urban sustainable development efficiency is also one of the important elements of urban development convergence in development geography. The efficiency of urban sustainable development is the main representation and basic component of a country's sustainable development ability. In the process of urban development, improving the living environment, relieving the pressure of population growth, and ensuring the coordinated development of the economy, society, and environment is necessary to adopt the values of sustainable development and ensure the economic benefits and quality of life of the city. It also minimizes the consumption and pollution of energy and other resources to meet the needs of current and future urban development. For cities at all levels in the Beijing-Tianjin-Hebei region, β and α tests were used to judge the convergence and divergence state and the development trend of the absolute gap, respectively, and to verify the convergence trend of their sustainable development efficiency. The above two tests are used to judge whether cities with low sustainable development efficiency can catch up with those with high sustainable development efficiency in the future and whether the absolute gap in development efficiency will increase or decrease. First, it is necessary to establish an evaluation index system. The design of the system includes input and output systems, in which the input system is designed for the input of production factors, resource consumption, and ecological environment loss, and the output system is designed for the development of individual residents' well-being. The input of production factors includes the total social water supply, total electricity consumption, investment in fixed assets and the number of employees at the end of the year. The consumption of resources includes the total amount of liquefied petroleum gas provided, the total amount of natural gas, the discharge of industrial wastewater, and the construction land in the municipal district. Ecological environment loss includes the industrial dust emission and the production of CO_2 and SO_2. The development and well-being of individual residents are the number of university teachers, the number of doctors, the amount of public expenditure, and the GDP per capita of the city. The data are from the China Urban Statistical Yearbook (2009–2016). Second, the conditional index of the choice is determined. Based on the reality of urban development in the Beijing-Tianjin-Hebei region and the accuracy of the data, this study selected the following condition index: Per capita urban GDP (Yuan)(X1), per capita fixed asset investment (yuan)(X2), actual use of foreign capital (US$10,000)(X3), industrial profit rate above designated size (%)(X4), per capita urban public financial 0expenditure (Yuan)(X5), per capita electricity consumption (KWH/person)(X6), per capita industrial wastewater discharge (tons/person)(X7), and industrial SO_2 production per capita (tons/person)(X8). Finally, its convergence is tested. Two sets of data were set, one containing all 13 cities in the Beijing-Tianjin-Hebei region, and the other containing 12 cities excluding Beijing, for the σ convergence test (Table 5.3).

The test results show that the sustainable development efficiency of the Beijing-Tianjin-Hebei region converges (σ 2013 < σ 2008) from 2008 to 2013, but diverges

Table 5.3 Test on the σ convergence trend of urban sustainable development efficiency in the Beijing–Tianjin–Hebei region from 2008 to 2015

Year	2008	2009	2010	2011	2013	2014	2015
Beijing–Tianjin–Hebei region	2.236	1.457	1.741	1.498	1.828	2.962	4.22
Tianjin–Hebei region	1.553	0.813	0.554	0.697	0.577	0.569	1.184

in 2014 (σ 2008 < σ 2014 < σ 2015) from 2014 to 2015. Since 2014, the efficiency gap has also been widening annually. From 2008 to 2015, sustainable development efficiency is convergent in cities in the Tianjin-Hebei region excluding Beijing (σ 2008 < σ 2015). The gap in sustainable development efficiency between cities in the Tianjin-Hebei region narrowed in 2014 and 2015 but widened between Beijing and the Tianjin-Hebei region. Table 5.4 shows the results of the β convergence test.

Table 5.4 shows that the absolute convergence of β does not exist from 2008 to 2015 (β regression coefficient is less than 0), indicating that in the Beijing-Tianjin-Hebei city cluster many years later, the efficiency of the cities with low development efficiency is still less than that of cities with high development efficiency. The β regression coefficient of the equation 2 > 0 from 2008 to 2013 has absolute convergence, indicating that after many years, the efficiency of cities with low development efficiency may catch up with that of cities with high development efficiency in the Beijing-Tianjin-Hebei city cluster. The two results are consistent, which proves the accuracy of the experimental results. Therefore, it can be concluded that ignoring other influencing factors, the sustainable development efficiency of the Beijing–Tianjin–Hebei region converges from 2008 to 2013 and diverges from 2014 to 2015. Generally, there is a high probability future cities with low sustainable development efficiency in the Beijing–Tianjin–Hebei region will not catch up with cities with high sustainable development efficiency.

Table 5.4 Absolute convergence of urban sustainable development efficiency in the Beijing-Tianjin-Hebei region from 2008 to 2015

Parameter	Regression equation 1 (2008–2015)	Regression equation 2 (20,082,013)
β	0.116^{**} (3.209)	-0.974^{**} (−5.157)
Constant term C	-0.099^{***} (3.209)	2.050^{***} (5.367)
Adjust R^2	0.666	0
F statistic	6.978^{**}	26.591^{***}
AR (n)	AR(3)*(−1.902)	/
D-W	2.675	1.629
Estimation method	OLS	WLS

Note *, **, and *** represent significance levels of 1%, 5%, and 10%, respectively, and values in brackets are t statistics. OLS: ordinary least squares; WLS: Weighted Least Squares

5.1.3 Driving Mechanism of Urban Spatial Expansion and Development in Developed Areas

The driving force of urban spatial expansion in developed areas mainly comes from the flow needs of production factors under the restriction of a specific industrial structure. Among them, the marginal production efficiency of land resources is an important factor and an important constraint in the development process. Research on the driving mechanism of urban spatial expansion in developed areas can also provide an important reference for future urban spatial development planning in less developed areas.

5.1.3.1 Driving Mechanism of Urban Spatial Expansion and Development in Developed Areas

The development of urban spatial expansion in developed areas is affected by many factors. China has made corresponding progress in the development of urban spatial expansion. Domestic scholars believe that the driving factors of urban spatial expansion and development mainly include national policies, population growth, transportation, and regional economic growth. In terms of urban strategic planning, scholars mainly study the human, environmental, and industrial factors that affect the change in spatial structure and form.

The spatial change in production factors will affect the development of cities, and in this process, the decision of manufacturers will affect the demand for factor flow, which will lead to the rise of corresponding cities or urban agglomerations. In other words, when manufacturers choose cities with less demand, they can achieve higher marginal productivity. However, it is still necessary for manufacturers and labor to gain access to the economy outside the region, and the convergence of regional urban expansion and development can balance this requirement. For example, experienced experts and engineers are scarce, and it is most advantageous for them to be located in the core of urban agglomerations, whose geographical radius of services will increase as urban spatial expansion converges. For example, the fixed cost of large medical facilities is very high, and if there are too few people in need of such services, they cannot share the cost, meaning that they can only be concentrated in core cities with large populations, and residents of small cities can enjoy these facilities through regional urban spatial development convergence. When the spatial expansion of regional cities converges, the external economy of labor and capital factors will increase, making the benefits of some production factors concentrated in smaller cities significant. The government is also an important influencing factor in urban spatial expansion and development.

Currently, urban spatial expansion is mainly restricted by urban development planning, and urban future development planning is dominated by the government. Regional governments can plan and determine the spatial distribution of production factors, including infrastructure construction, improvement of public service system, and land supply. These production factors will guide the direction and form of urban space expansion. In addition, the investment promotion policies, tax, and fiscal preferential policies of regional governments will further promote regional economic development. After the improvement of urban space accessibility, the level of social services, government service efficiency, and the urban environment will gradually attract the attention of the government, thus promoting the further expansion of urban space and regional convergence.

The national government determines the development strategy of national urban agglomerations, affects the construction of national infrastructure such as high-speed railways and expressways, and further affects the convergence of regional urban spatial expansion. Therefore, whether it is urban spatial expansion or urban agglomeration spatial expansion, national policies and government adjustment are important determinants. Transportation network accessibility is also an important factor affecting urban spatial expansion and development. Although there are different disputes between traffic development and economic development, there is a consensus on the impact of traffic network accessibility on the convergence of regional urban spatial expansion. The accessibility of the transportation network, particularly the accessibility of high-speed rail, transforms the regional spatial structure, promotes the re-differentiation and positioning of regional urban system functions, and promotes the transfer and transformation of the regional industrial structure. With the coordinated development of Beijing-Tianjin-Hebei rising to the national strategy, the national government first focused on the construction and improvement of high-speed railways and expressways in the Beijing-Tianjin-Hebei region. With the continuous enhancement of transportation network accessibility, Beijing's industries continue to transfer to Tianjin, Hebei, and Tianjin, and cities in Hebei actively undertake the "same-city effect" brought by economic integration and corresponding transportation convenience so that the Beijing-Tianjin-Hebei region's separate administrative regions truly realize coordinated development and become an economic development community.

5.1.3.2 Driving Mechanism of Urban Spatial Expansion in Zhejiang Province

Urban land use change is mainly affected by natural factors and economic factors. According to relevant studies, in which, GDP(X1), total social investment in fixed assets (X2), industrial output value (X3), total retail consumption (X4), per capita disposable income of urban residents (X5), structure ratio of secondary industry (X6), structure ratio of tertiary industry (X7), urban population (X8), urbanization rate (X9), and per capita road area (X10) are selected. These ten indicators from the data since 2005, the data after differential processing, and principal component analysis are used in this study. The data are obtained from the Zhejiang Statistical Yearbook (2005–2021). The statistical analysis results (Tables 5.5 and 5.6) indicate that the cumulative contribution rate of the first to the fourth principal components reaches 86.29%, and the extraction of these three factors can explain the urban land expansion. As shown in Table 5.5, the variables highly correlated with the first principal component include X2, X3, X4, X5, X6, and X7; the variables highly correlated with the second principal component include X1, X3, X4, X5, and X9; the variables highly correlated with the third principal component include X8; the variables most correlated with the fourth principal component are X10. In summary, the first principal component expresses economic development factors, the second principal component expresses economic and urbanization factors, the third principal component expresses population factors, and the fourth principal component expresses infrastructure factors.

Analysis of six main factors of the first principal component revealed that the main component is the economic factor. China's total investment in fixed assets

Table 5.5 Component matrix of driving mechanism of urban spatial expansion in Zhejiang Province

Factor	First principal component	Second principal component	Third principal component	Fourth principal component
X1	0.054	0.932	0.001	−0.122
X2	0.706	0.061	−0.431	0.358
X3	−0.604	0.735	−0.106	−0.009
X4	0.569	0.534	−0.106	0.034
X5	0.624	0.580	−0.250	0.096
X6	−0.917	0.047	−0.328	0.106
X7	0.897	−0.109	0.294	−0.233
X8	0.008	0.028	0.970	−0.009
X9	−0.241	0.725	0.459	0.026
X10	−0.018	−0.026	0.307	0.91

Table 5.6 Eigenvalues, contribution rates, and cumulative contribution rates of the driving mechanism matrix of urban spatial expansion in Zhejiang Province (%)

Principal component	Eigenvalue	Contribution rate (%)	Cumulative contribution rate (%)
1	3.282	32.823	32.823
2	2.575	25.751	58.575
3	1.71	17.103	75.678
4	1.062	10.615	86.293
5	0.689	6.891	93.184
6	0.381	3.807	96.991
7	0.187	1.869	98.861
8	0.074	0.741	99.601
9	0.034	0.342	99.943
10	0.006	0.057	100.000

reached 669.625 trillion yuan in 2005 and 55.288.4 trillion yuan in 2021. Industrial output value increased from 716.615 billion yuan in 2005 to 9496.99 billion yuan in 2021. Total retail sales of social consumption increased from 464.585 billion yuan in 2005 to 250.79 billion yuan in 2018. The per capita disposable income of urban residents will increase from 16,294 yuan in 2005 to 68,486.8 yuan in 2021. The ratio of secondary industry structure decreased from 53.4% in 2005 to 41.8% in 2018. The tertiary industry structure increased from 39.9% in 2005 to 54.7% in 2018, indicating that the industrial structure of Zhejiang Province has undergone fundamental adjustment, the pillar industry has changed from the primary to the tertiary industry, and Zhejiang Province has realized economic transformation. The second principal component mainly includes GDP and urbanization rate. China's GDP increased from 1341.768 billion yuan in 2005 to 7404.078 billion yuan in 2021, with an average annual growth rate of 81.87%. The urbanization rate rose from 56% in 2005 to 72.7% in 2021. The third component is the urban population, which will increase from 25.77 million in 2005 to 46.6 million in 2020. The fourth principal component is road area per capita. Per capita road area increased from 16.03 m^2 in 2005 to 18.03 m^2 in 2021.

In summary, factors affecting urban spatial expansion in Zhejiang Province mainly include economic development, urbanization, population, and infrastructure. Among them, economic development has the greatest impact, followed by urbanization, population, and finally, infrastructure.

5.2 Industrial Agglomeration Process in Developed Areas

5.2.1 Industrial Agglomeration Process

Industrial agglomeration is one of the important characteristics of the regional development process, and it is also an important component of development geography. Industrial agglomeration is a kind of industrial organization mode and a world economic phenomenon. Its main connotation is that the concentration of the same kind of industry in a certain region makes its capital elements gather in space. The research on industrial agglomeration began at the end of the nineteenth century. Marshall began to pay attention to the economic phenomenon of industrial agglomeration in 1890 and proposed two important concepts, namely "internal economy" and "external economy." When agglomeration evolves into a highly geographic concentration in a region, the concept of industrial agglomeration is combined with geographical characteristics, so geographical proximity becomes one of the basic characteristics defining agglomeration. With the development of industrial agglomeration theory in geography, a series of industrial agglomeration theories such as location agglomeration theory, agglomeration optimal scale theory, and the diamond model have been formed.

5.2.1.1 Formation Mode and Classification of Industrial Agglomeration

There are two forms of industrial agglomeration: market creation mode and capital transfer mode. Market creation mode refers to the emergence of a specialized market with characteristics within a certain range, which enables the region to have good information exchange and trading conditions, and thus leads to the upstream and downstream enterprises of the industrial chain to gather near the region. Industrial agglomeration in the traditional sense is guided by the market mechanism (Zhejiang Province is a typical case of the market creation mode) and finally forms an industrial cluster with a complete industrial chain. Capital transfer mode means that to get close to the market and reduce costs, large enterprises re-choose the position that is conducive to operation, resulting in upstream and downstream enterprises gathering around the enterprise, and generally need to move in a certain amount of capital from the outside. Many industrial clusters have been formed in China under the mode of capital migration, among which the migration capital that plays a driving role is mainly foreign investment. There are two types of industrial agglomeration: directional agglomeration and economic connection agglomeration. To benefit from a certain region, enterprises spontaneously conglomerate in the region and form industrial groups with economic characteristics. This method has been well developed in many cities in China. In particular, the future urban development planning of government departments has played a significant role in promoting directed agglomeration. The agglomeration of economic relations is through the addition of neighboring enterprises' economic links between the two countries enhancing their external development conditions. An agglomeration can be further divided into clusters caused by vertical and horizontal economic linkages. Vertical economic connection refers to the input–output relationship between firms. Horizontal economic connection refers to the relationship between industrial clusters formed around regional leading industries and departments.

Industrial agglomeration is a new high-efficiency industrialization road with Chinese characteristics. It can enhance productivity by strengthening enterprises' cooperation and division of labor. It can also enable enterprises to gain competitive advantages in cooperative competition and strengthen communication among enterprises and between enterprises and customers, thus enhancing enterprise innovation, generating economies of scale and competitive and innovation benefits, and contributing to social and economic development. China's current industrial agglomeration has problems such as insufficient innovation ability, imperfect infrastructure, lack of systematic planning, unclear positioning, imperfect industrial chain system, and low management quality. In the future, development can be studied from the perspective of the cultivation of talents, the cultivation of enterprises, the optimization of structure, the optimization of markets, the attention of government, and the upgrading of industrial clusters.

5.2.1.2 Industrial Agglomeration Measuring Method

Research on industrial agglomeration should measure the degree of industrial agglomeration first and provide guidance for the development of the regional economy by judging the development of the regional industry. Currently, the commonly used ways to measure the degree of industrial agglomeration are: industry concentration (Bain, 1959), the Herfindal-Hirschman index (Herfindahl, 1950), the EG index (Ellison and Glaeser, 1997), spatial Gini coefficient (Krugman, 1991), industrial agglomeration index CIP (Competitive Industrial Performance), location quotient index (Haggett et al., 1977), etc.

Industry Concentration Ratio

Industry concentration, also known as industry concentration rate or market concentration, is measured by the share of the entire market in the largest regions or related values of an industry. Its calculation formula is as follows:

$$CR_n = \frac{\sum_{i=1}^{n} X_i}{\sum_{i=1}^{N} X_i} \tag{5.3}$$

where CR_n is the industrial agglomeration degree of X; X_i is the value related to the i place; n is the number of regions with the largest industrial scale; N indicates all regions. It is generally believed that the higher the CR_n value is, the more agglomeration is likely to occur, and the higher the CR_n value is, the greater the agglomeration degree is. This method is simple and easy to use for calculating the industrial agglomeration degree, and can reflect the industrial agglomeration level in an image. However, this method is susceptible to the influence of n value selection, ignores the scale distribution of other regions outside the largest region, and fails to reflect the differences in industrial structure and distribution within the largest region.

Hirschman-Herfindahl Index

The Herfindahl Index was initially used to measure the interaction between monopoly and competition in a market, and has since been widely used for other purposes. Its calculation formula is as follows:

$$H = \sum_{i=1}^{N} Z_i^2 = \sum_{i=1}^{N} \left(\frac{X_i}{X}\right)^2 \tag{5.4}$$

where X represents the total market size; X_i represents the scale of enterprise i; Z_i represents the market share of enterprise i; N represents the number of firms in the industry. In practical analysis, the reciprocal of the H index is often used to measure industrial diversification. This method is easy to calculate and also takes into account the total number and size of enterprises, which can explain the competition and monopoly among enterprises and the internal merger and decomposition.

However, while this result can measure absolute concentration, it cannot measure relative concentration, cannot explain the inter-regional relationship, and the level of its value cannot judge whether the industry is clustered.

Space Gini Coefficient

The spatial Gini coefficient was proposed by Krugman to measure the degree of industrial agglomeration. The formula is:

$$G = \sum (S_i - x_i)^2 \tag{5.5}$$

where G is the spatial Gini coefficient; S_i is the proportion of employment in a certain industry in region i to the total employment in that industry in the whole country; x_i is the proportion of employment in the region to the total employment in the country. When $G = 0$, the spatial distribution of industry is uniform, and the greater the G value (the maximum value is 1), the higher the degree of regional industry agglomeration. Using the spatial Gini coefficient to calculate the degree of industrial agglomeration can transform it into an intuitive graph, which is relatively intuitive and easy to understand; however, it ignores the differences in scale among enterprises and industries and among regions. Therefore, a Gini coefficient greater than zero does not mean the existence of the agglomeration phenomenon, and the measurement of industrial agglomeration degree often contains false components. It only applies to a reasonable range of tests of the central-periphery pattern of an economy.

EG Index

Before using the EG index, it is assumed that there are N enterprises in industry A in a certain region, which is divided into M small areas, and N enterprises happen to be in M small areas. Its calculation formula is as follows:

$$\gamma = \frac{G - \left(1 - \sum_i x_i^2\right)H}{\left(1 - \sum_i x_i^2\right)(1 - H)} = \frac{\sum_{i=1}^{M} (s_i - x_i)^2 - \left(1 - \sum_{i=1}^{M} x_i^2\right) \sum_{j=1}^{N} Z_j^2}{\left(1 - \sum_{i=1}^{M} x_i^2\right)\left(1 - \sum_{j=1}^{N} Z_j^2\right)} \tag{5.6}$$

where S_i represents the proportion of employment in an industry in region i to the total employment in the industry; x_i is the proportion of employment in the region to the total employment in the country. The Herfindahl Index H represents the distribution of firms in the industry as measured by the number of employees. In general, $\gamma > 0.05$ indicates a highly concentrated industry, and γ indicates no geographical concentration of an industry.

The EG index has a perfect mathematical form, and considering the scale and regional differences in enterprises and industries, it can be compared across industries, across time, and even across countries, and is one of the more widely used industrial agglomeration measurement methods. However, the EG index retains the shortcomings of the Gini index and fails to keep complete industrial agglomeration information, resulting in incomplete statistics. The industrial agglomeration degree calculated using the EG index is often smaller than the actual agglomeration degree.

CIP

The CIP industrial agglomeration index can measure the uneven degree of industrial distribution in the region. The formula is:

$$CIP_i = \sqrt[2]{\frac{1}{c}\sum_i \left(\frac{x_{ij}}{\sum_j x_{ij}} - \frac{\sum_i x_{ij}}{\sum_i \sum_j x_{ij}}\right)} \times 100 \tag{5.7}$$

Among them, CIPi is the industrial regional agglomeration index of industry i; c is the total number of analyzed regions. i is the industry label, j is the region label, and x_{ij} is the output of industry i in region j. The lower the value, the more evenly distributed the industry. The industrial agglomeration index reduces the impact of the difference in economic scale between regions, but its calculation is relatively complex.

Location Quotient

Location quotient is an index proposed by P'Haggett for calculating concentration. The formula is:

$$LQ_{ij} = \frac{q_{ij}/q_j}{i} \tag{5.8}$$

Among them, LQ_{ij} is the location quotient of i industry in j region; q_{ij} is the relevant index of i industry in region j (such as output value, employment number, etc.); q_j is the relevant index of all industries in j region; q_i refers to the relevant indicators of *i* industry in the whole country; q is the relevant index of all industries in the country.

The higher the location quotient, the greater the level of regional industrial agglomeration. This value can be used to reflect the specialization rate of the industry, the spatial distribution of the factors, the role of the leading sector, the role of a region in the overall situation, and other aspects. We can find out the advantageous industries in a region with a certain status in the whole country and measure the specialization rate according to the size of the location quotient. In the following section, we take the Beijing-Tianjin-Hebei region as an example and calculate the industrial agglomeration degree of the region by calculating the location quotient.

5.2.1.3 Analysis of Industrial Agglomeration in Beijing-Tianjin-Hebei Region Based on Location Quotient Index

The Beijing-Tianjin-Hebei metropolitan area originated in the Capital Economic Circle, including Beijing, Tianjin, and prefecture-level cities in Hebei Province. In the 12th Five-Year Plan, the Beijing-Tianjin-Hebei coordinated development has been promoted as a national strategy. The coordinated development of the Beijing-Tianjin-Hebei region aims to focus on regional integration, carry out industrial division of labor and cooperation for its own advantage, improve regional public services, promote transport infrastructure construction, achieve coordinated regional development, realize inter-regional connectivity, and optimize and upgrade the industrial structure. To improve the sustainable carrying capacity of the region, the coordinated development of the Beijing-Tianjin-Hebei region will help remove the non-core functions of the capital, optimize the structure of the city cluster, and enhance regional competitiveness. To achieve this goal, it is necessary to create an industrial layout for the coordinated development of the Beijing-Tianjin-Hebei region, develop their respective advantageous industries according to local conditions, and strengthen the complementarity among regional industries according to their respective urban function positioning and resource advantages to avoid disorderly competition in regional industries. By accurately measuring the degree of industrial agglomeration in the Beijing-Tianjin-Hebei region and rationally arranging the industrial structure, the spatial economic ties within the Beijing-Tianjin-Hebei region can be strengthened, the huge siphon effect of Beijing's policy orientation on surrounding cities over the years can be reversed, the competitiveness of the Beijing-Tianjin-Hebei region in the country and even the world can be improved, and the model window for regional economic development in China can be provided. Therefore, it is necessary to study the industrial agglomeration in the Beijing-Tianjin-Hebei region. Using the statistical yearbook of the Beijing-Tianjin-Hebei region and China, the industrial location quotient of the Beijing-Tianjin-Hebei region was calculated based on the industrial data of three major provinces (Beijing, Tianjin, and Hebei) and the entire country from 2002 to 2016. By calculating the location quotient of each industry in the Beijing-Tianjin-Hebei region, the degree of industrial agglomeration in the region is analyzed.

Among the estimated results of the three major industrial locational quotients in Beijing shown in Table 5.7, the locational quotient of the tertiary industry is far greater than 1, which indicates that Beijing has the highest degree of agglomeration in the tertiary industry. The tertiary industry is the dominant industry in Beijing, the degree of specialization has been at a high level, and the advantages of the tertiary industry are far more than in Tianjin and Hebei. From 2002 to 2011, the location quotient of the tertiary industry in Beijing generally showed an upward trend, and from 2012 to 2016, the location quotient of the tertiary industry in Beijing generally showed a downward trend. From 2002 to 2016, although the location quotient of the primary industry fluctuated in some years, the location quotient of the primary industry showed an overall downward trend, and the location quotient of the primary industry was far less than 1, which indicates that the agglomeration of the primary industry was not obvious. From 2002 to 2016, the location quotient of the secondary industry generally showed a downward trend and was higher than that of the primary industry, indicating that the concentration degree of the secondary industry in Beijing was higher than that of the primary industry but far lower than that of the tertiary industry.

Table 5.7 Changes in location quotient of three major industries in Beijing from 2002 to 2016

Year	Primary industry	Secondary industry	Tertiary industry	Year	Primary industry	Secondary industry	Tertiary industry
2002	0.143	0.652	1.636	2010	0.094	0.517	1.704
2003	0.138	0.651	1.632	2011	0.095	0.504	1.732
2004	0.108	0.671	1.646	2012	0.085	0.501	1.689
2005	0.112	0.619	1.684	2013	0.086	0.493	1.66
2006	0.104	0.568	1.72	2014	0.077	0.494	1.628
2007	0.097	0.544	1.715	2015	0.086	0.481	1.587
2008	0.098	0.503	1.761	2016	0.058	0.484	1.556
2009	0.102	0.512	1.703				

According to Table 5.8, from 2002 to 2016, the location quotient of the secondary industry in Tianjin is greater than 1, which indicates that Tianjin has the highest degree of agglomeration in the secondary industry. The location quotient of the tertiary industry in Tianjin is generally greater than 1, but some of them are less than 1 in some years. Compared with Beijing, the degree of tertiary industry agglomeration in Tianjin is lower than that in Beijing. Tianjin is concentrated in the secondary and tertiary industries. From 2002 to 2016, the location quotient of the primary industry showed an overall downward trend, and the location quotient of the primary industry was less than 1, which indicates that there is no agglomeration of the primary industry in Tianjin.

Table 5.9 shows that the location quotients of the primary and secondary industries in Hebei Province from 2002 to 2016 are both greater than 1, which indicates the agglomeration of the primary and secondary industries in Hebei Province, with the primary industry having the highest degree of agglomeration in the Beijing–Tianjin–Hebei region. From 2002 to 2016, the location quotient of the secondary industry in Hebei Province showed an overall upward trend, and the degree of agglomeration of the secondary industry gradually surpassed Tianjin. However, the tertiary industry in

Table 5.8 Changes in location quotient of three major industries in Tianjin from 2002 to 2016

Year	Primary industry	Secondary industry	Tertiary industry	Year	Primary industry	Secondary industry	Tertiary industry
2002	0.293	1.118	1.098	2010	0.168	1.132	1.044
2003	0.283	1.138	1.061	2011	0.148	1.131	1.046
2004	0.263	1.181	1.032	2012	0.138	1.142	1.037
2005	0.258	1.153	1.028	2013	0.14	1.145	1.034
2006	0.254	1.205	0.961	2014	0.144	1.142	1.037
2007	0.214	1.231	0.945	2015	0.136	1.139	0.8
2008	0.185	1.281	0.885	2016	0.14	1.061	1.096
2009	0.174	1.155	1.022				

Hebei Province lags, and there is no agglomeration. From 2002 to 2016, the regional quotient of the tertiary industry in Hebei Province was lower than 1, declining in some years.

According to Table 5.10, among the primary industries represented by agriculture, forestry, animal husbandry, and fishery, the location quotient of Hebei Province is greater than 1 from 2002 to 2016, whereas those of Beijing and Tianjin are both less than 1, indicating that Hebei Province has a higher degree of agglomeration in the primary industry than Beijing and Tianjin. In the secondary industry represented by industry and construction, the degree of industrial agglomeration in Hebei Province has gradually surpassed that of Beijing and Tianjin. This is mainly related to the industrial structure of the three places. Beijing and Tianjin have entered the stage of post-industrialization, and the tertiary industry plays a dominant role in the industrial structure, while Hebei Province is still in the middle stage of industrialization, with the industrial structure dominated by heavy industry and the secondary industry playing a dominant role in economic development. Although the location quotient of the tertiary industry in Beijing is higher than that in Tianjin, in the subdivision of the tertiary industry, the concentration of the wholesale and retail industry in Tianjin is higher than that in Beijing and Hebei Province; however, in the real estate, finance, and catering industries, the industrial concentration in Beijing is higher than that in Tianjin and Hebei Province. Among the subdivisions of tertiary industry in Hebei Province, only the transportation industry has a higher location quotient than in Beijing and Tianjin. This is mainly related to the acceleration of infrastructure construction in Hebei Province, particularly the establishment of the Xiongan New Area and the holding of the 2022 Winter Olympics, which have boosted the development of the transport industry in Hebei Province.

Table 5.9 Changes in location quotient of three major industries in Hebei Province from 2002 to 2016

Year	Primary industry	Secondary industry	Tertiary industry	Year	Primary industry	Secondary industry	Tertiary industry
2002	1.195	1.089	0.845	2010	1.322	1.132	0.792
2003	1.247	1.083	0.838	2011	1.272	1.166	0.783
2004	1.216	1.105	0.813	2012	1.274	1.164	0.779
2005	1.203	1.121	0.808	2013	1.28	1.182	0.773
2006	1.205	1.121	0.813	2014	1.291	1.183	0.780
2007	1.293	1.129	0.789	2015	1.302	1.180	0.800
2008	1.239	1.157	0.771	2016	1.273	1.194	0.805
2009	1.308	1.133	0.794				

Table 5.10 Changes in industrial location quotient in Beijing-Tianjin-Hebei region from 2004 to 2016

Year	Area	Agriculture, forestry, animal husbandry, fishery	Industry	Building	Wholesale and retail	Transportation	Catering	Finance	Real estate
2004	Beijing	0.11	0.634	0.919	1.266	1.028	1.195	2.907	1.631
	Tianjin	0.256	1.226	0.813	1.555	1.246	0.801	1.082	0.77
	Hebei	1.222	1.106	1.071	0.826	1.201	0.552	0.606	0.615
2008	Beijing	0.096	0.465	0.756	1.566	0.876	1.192	2.385	1.647
	Tianjin	0.173	1.234	0.736	1.303	1.268	0.791	0.956	0.735
	Hebei	1.205	1.196	0.859	0.754	1.631	0.672	0.459	0.696
2012	Beijing	0.087	0.477	0.627	1.352	1.038	1.182	2.179	1.203
	Tianjin	0.137	1.228	0.614	1.413	1.206	0.976	1.193	0.603
	Hebei	1.237	1.218	0.822	0.826	1.894	0.829	0.528	0.639
2016	Beijing	0.058	0.471	0.598	0.964	0.93	0.866	2.024	1.006
	Tianjin	0.14	1.141	0.658	1.316	0.912	0.909	1.22	0.695
	Hebei	1.281	1.252	0.88	0.825	1.662	0.764	0.657	0.716

5.2.2 *Industrial Diffusion*

The process of industrial expansion is an important stage in regional industrial development, and it is also an important concern of development geography. Industrial diffusion refers to the change of industrial distribution in space, which makes the industry expand or transfer to the outside. Its root cause is the heterogeneity of industrial development in space, which makes economic activities spread or transfer in space through investment capital, technology transmission, or labor migration. Industrial diffusion is a result of industrial agglomeration. In developed areas with economic growth poles, after a certain level of industrial development, due to the continuous expansion of the industrial scale and the increasing number of enterprises, the industrial demand for labor, land, water resources, and other production factors will inevitably lead to the rise in the cost of these production factors, thus intensifying competition among enterprises. It is difficult to improve production efficiency in the short term; therefore, reducing product prices is the most direct means to enhance product market competitiveness. Therefore, to ensure the appropriate profit margin for the product, the industry has to move to the surrounding areas with lower labor costs and more abundant land resources, and the industrial diffusion effect is generated.

For example, some regions of a country have a self-reinforcing advantage in manufacturing. Due to the improvement of trade liberalization, manufacturing will be concentrated in these regions, and with the strengthening of this geographical advantage, the average wage in these regions will be slightly higher. Similarly, the increase in demand for manufacturing has strengthened the concentration of manufacturing and thus increased the wage level of the agglomeration. Therefore, regions with strong economic activities have a high wage level, but with time, the wage gap between regions increases, and some enterprises will not be able to pay high wages. Therefore, these enterprises are likely to move to regions for plant investment, resulting in industrial diffusion. These enterprises will also begin to accumulate and form strong advantages in new regions, leading to higher wages in the new regions After a while, third and fourth regions are created.

5.2.2.1 Industrial Expansion Theory and Research Progress

Industrial diffusion theory was developed from traditional industrial diffusion theory based on new economic geography. Traditional industrial diffusion theory is based on a series of comparative advantage theories proposed by Smith (1776), David (1817), Ohlin (1993), and others. It is believed that the differences in industrial spatial distribution depend on the differences in exogenous technology and factor endowments. According to industrial diffusion theory based on new economic geography, even in the two regions with symmetrical technical conditions and factor endowments, differences in market size, price differences, and crowding effects may lead to industrial diffusion. According to this theory, trade freedom affects the market proximity effect (that is, enterprises tend to choose regions with a large market size), the cost-of-living effect (that is, the cost of enterprises in agglomeration areas is less), and the market congestion effect (that is, the dispersion force caused by the continuous accumulation of industries and the reduction of enterprises' competitiveness), thus becoming an important variable of industrial diffusion.

Many recent studies on industrial diffusion have been conducted in China. Scholars have not only fully studied foreign industrial diffusion, but also conducted related studies on domestic industrial diffusion. However, some scholars directly apply international industrial diffusion theory to domestic research, ignoring the differences in the degree of market development, labor flow mechanism, and industry position in the industrial chain between the two; Therefore, the conclusions obtained need to be contextualized. In fact, industrial diffusion is neither a simple diffusion to the surrounding area nor a separate transfer of an enterprise, but a large number of enterprises move out from a concentrated area, and the scope of diffusion is not limited to the surrounding area. Therefore, future studies on industrial diffusion should consider the current situation of the country.

5.2.2.2 The Process of Financial Industry Diffusion in China

The financial industry is an important force for national economic development, and the development of the financial industry can bring both opportunities and risks to the country. Therefore, it is necessary to study the financial industry, part of which is the diffusion of the financial industry. Local and international scholars have conducted studies on the diffusion effect of the financial industry. For example, Palmberg (2012) studied the financial industry of New York, Shanghai, Tokyo, and other cities and found that the free flow of capital production factors and labor force among financial centers was the main reason for the diffusion of the financial industry. Guo et al. (2016), investigated the impact of technology on the diffusion of the financial industry and concluded that the Internet has accelerated the external diffusion of the financial industry. Relevant research predicted the diffusion of the financial industry in Tangshan through mathematical methods, while some took the Beijing-Tianjin-Hebei region as the research object and made recommendations from the perspective of the integration of Internet technology and the financial industry.

Many studies have been conducted on industrial diffusion, particularly the influencing factors of industrial diffusion. Extensive studies have been conducted on the methods of identifying the influencing factors of industrial diffusion. Here, we use the panel data model to identify the factors affecting the diffusion of China's financial industry. First, the total volume of import and export trade (to measure the degree of opening-up), the number of employees with a bachelor's degree or above (to measure the human capital), the consumption level of residents (to measure the consumption level), the proportion of the tertiary industry (to measure the industrial structure), the proportion of the urban population to the permanent population (to measure the urbanization rate), and the research and development (R&D) expenditure at the national level from 2006 to 2017 are collected data to measure technological innovation. The following panel data model is then used for regression:

$$gm = a + \beta_1 lngdp + \beta_2 lncz + \beta_3 lnopen + \beta_4 lnhr + \beta_5 lnunemp \\ + \beta_6 lncost + \beta_7 lnad + \beta_8 lntec + \beta_9 lnurban + \varepsilon \tag{5.9}$$

where *gm* is the size of the financial industry and is the explained variable. *gdp* is the level of economic development, *cz* is the government control, *open* is the opening up to the outside world, *hr* is the human capital, *unemp* is the unemployment rate, *cost* is the consumption level, *ad* is the industrial structure, *tec* is the technical level, *urban* is the urbanization level, and these variables are explanatory variables.

Through the logarithmic processing of the data and multicollinearity test of the model using the variance inflation factor, the values are all below 10, indicating that there is no multicollinearity. The F test and Hausman test are used to find that the financial industry diffusion model is suitable for regression. The results in Table 5.11 show that the diffusion of the national financial industry is proportional to the level of economic development and the marketization index, indicating that the higher the level of economic development, the greater the degree of financial industry diffusion, and the reduction of government intervention in the financial industry will also promote industrial diffusion. The spread of the financial industry is also positively correlated with human capital and opening up to the outside world, but negatively correlated with the unemployment rate, indicating that the increase in unemployment will inhibit the spread of the industry. The industrial structure, technical level, consumption level, and urbanization level all promote the diffusion of the financial industry, but the effect is not obvious.

Based on this, it is believed that in the future development of the financial industry, China should first adjust the direction of economic development and increase the training of talents and technologies. Second, the state should strengthen the opening up to the outside world, establish good cooperative relations, improve the financial system, and reduce government intervention to accelerate the development of the financial industry.

Table 5.11 Regression results of China's financial industry diffusion model

Model Variable	Ingdp	Incd	Inopen	Inhr	Inunemp	Incost	Inad	Intec
The whole country	0.577	0.016	0.429	0.071	0.018	0.031	0.086	0.355
Model variable	Inurban	Incons	H test	R^2	F	Observed value		
The whole country	0.009	0.016	0.005	0.89	132.37	450		

5.2.3 *Industrial Agglomeration Mechanism*

The development mechanism of industrial agglomeration in developed areas is important research content of economic development in development geography and an important reference for industrial development in underdeveloped areas. The development mechanism of industrial agglomeration in developed areas is influenced by many factors. Research shows that the formation of industrial agglomeration behavior is related to the market mechanism, demand, government policies, talent flow, and other factors, and the complex connections among these factors affect enterprises. The development of industrial agglomeration will also lead to industrial diffusion, which is a process of opposition and unity and an important cause of spatial change in industrial development. Moreover, industrial agglomeration and diffusion are complex, accompanied by diffusion in the process of industrial agglomeration, and industrial diffusion cannot be separated from industrial agglomeration.

5.2.3.1 Industrial Agglomeration Mechanism

The formation mechanism of industrial agglomeration includes internal and external mechanisms. The internal mechanism of industrial agglomeration is cooperation, and the reduction of transaction and transportation cost is one of the key factors affecting industrial cooperation and industrial agglomeration. The regional agglomeration of industries has resulted in the reduction of transaction and transportation costs, and it has also enabled the rapid diffusion of information on the demand, production, and industrial chain of different industries. When the transaction cost is lower than the management cost generated by the internal division of labor, the internal division of labor of the enterprise will be replaced by a more efficient inter-enterprise division of labor, so that the total and average costs of the enterprise will be reduced, the production efficiency will be increased, and more enterprises will enter the industry and expand the scale of the industry. The upstream and downstream enterprises, driven by the benefits and costs, also moved into this region and eventually formed various specialized markets and industrial groups of supporting services in the region. The external mechanism of industrial agglomeration is the national policy. Since the 1990s, China has strengthened the development and construction of industrial parks and science and technology parks in its urban development planning, and this policy has played an important role in the beginning of China's industrial agglomeration. In the early stage of the formation of industrial clusters, there is a lack of corresponding policy restrictions and constraints, which also makes it possible for industries to gather in spatial locations.

The mechanism of industrial diffusion development includes promoting and pulling mechanisms. The promotion mechanism of industrial diffusion development is mainly caused by the unreasonable division of labor within industrial clusters or excessive industrial clusters. Excessive industrial agglomeration will lead to a series of social problems such as environmental pollution, excessive consumption of resources, traffic jams, and regional vicious competition. These problems often lead to industrial diffusion after industrial agglomeration. Industrial diffusion is also a specific stage of industrial development. Proper industrial diffusion can not only ensure the appropriate scale of industrial clusters but also ensure the rational optimization of industrial structure. The driving mechanism of industrial diffusion is mainly caused by the guidance of national policy, the pulling of the traffic network, and the spontaneous pulling of the market. The guidance of the national policy first led to the spread of the secondary industry to the peripheral areas, and finally, the population and commerce also spread outward. The distance and scale of industrial diffusion are directly related to the strength of national policies and the improvement of service industries such as transportation, education, and medical care. For the spontaneous pull of the market, the labor market, the land market, and the competitive product market are the key factors. Labor costs, land prices, and product competitiveness will promote the diffusion of industries to ensure product profit margins.

From the perspective of world cities in developed regions, industrial agglomeration and diffusion tend to coexist (Fig. 5.3). Overall, producer services are concentrated in the core, whereas manufacturing services are scattered in the periphery. From the perspective of the industry, the agglomeration of some industries in the manufacturing sector, such as high-end manufacturing and big city industries, is influenced by path dependence and then forms the advantageous manufacturing industry in the differentiation of cities around the world. Under the conditions of developed electronic information technology and sound infrastructure, the production and consumption process of the manufacturing industry can be separated, which leads to the proliferation of the manufacturing industry. The production, sales, and other departments spread to sub-central or marginal areas. The branch of producer services and the financial industry continue to expand, whereas the producer services (precisely because of the development of electronic information technology, communication, infrastructure, and so on), are increasingly concentrated in urban centers. Through the cumulative circular effect, large metropolitan industries and producer services are concentrated in the city center, which improves production efficiency and reduces costs, thus expanding productivity. From the perspective of the agglomeration and dispersion mechanism of the manufacturing industry, the overall trend of advantage agglomeration and overall diffusion appears, and it is believed that the separated manufacturing value chain leads to the spatial diffusion of the industry. From the perspective of the agglomeration and decentralization mechanism of the producer service industry, the subordinate institutions of the producer service industry show a trend of decentralization, the headquarters show a trend of agglomeration, and the business service industry show a trend of agglomeration.

5.2.3.2 Industrial Agglomeration and Diffusion Mechanism of Creative Industries: A Case Study

Creative industry refers to the industry that has the potential to create wealth and employment through individual creativity and the use of knowledge. In the process of

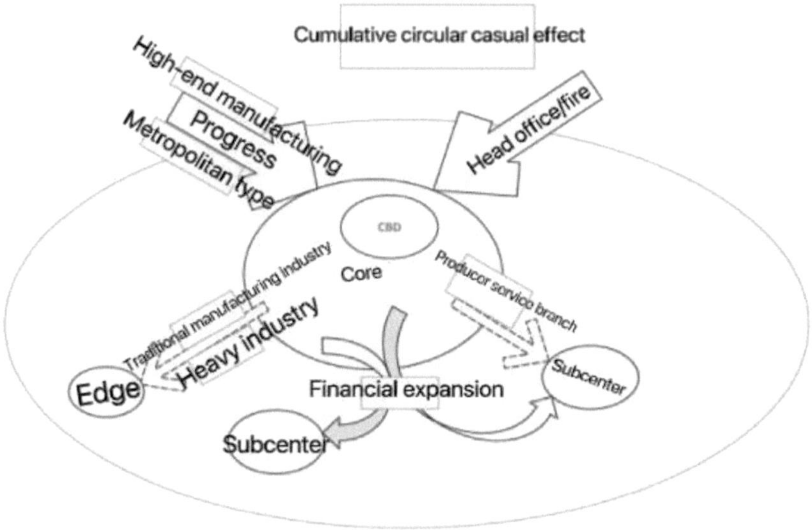

Fig. 5.3 Mechanism of industrial agglomeration in world cities

their development, creative industries are often accompanied by industrial agglomeration and diffusion. According to the literature, the agglomeration and diffusion of industries have certain formation and evolutionary mechanisms. From the perspective of industrial agglomeration, the formation of a creative industry first refers to the aggregation of each creative unit within the region. Under the economic conditions and social recognition of the creative industry, creative units (such as individuals, organizations, and enterprises) gather to form an agglomeration of creative knowledge, talents, and technologies. Second, the creative units form connections, influence each other, cooperate, promote the development of creative industries, and form a stable capital chain. Finally, after gathering, pairwise interactions between units form a rich network structure, so that funds, technologies, and talents can flow more frequently in this network system and form an industrial chain (Fig. 5.4).

In terms of industrial diffusion, creative industries will begin to spread after they have gathered to a certain extent. Creativity is the core of the creative industry; therefore, the diffusion of the creative industry is essentially the diffusion of creative value. In terms of creative value, the diffusion of the creative industry can be divided into the diffusion of the creative source, the diffusion of the creative scheme, and the diffusion of the creative industry. Its diffusion path is for people to develop creative ideas in various environments and spread them within the group to form creative solutions. The creative solutions become creative products through production. Finally, through consumers purchasing goods, the marketization of creative products and the diffusion of creative industries are completed. This shows that creative products can have economic benefits through marketization diffusion. In addition, the diffusion of creative industries will also radiate to the surrounding industries to accelerate their development and generate certain economic benefits.

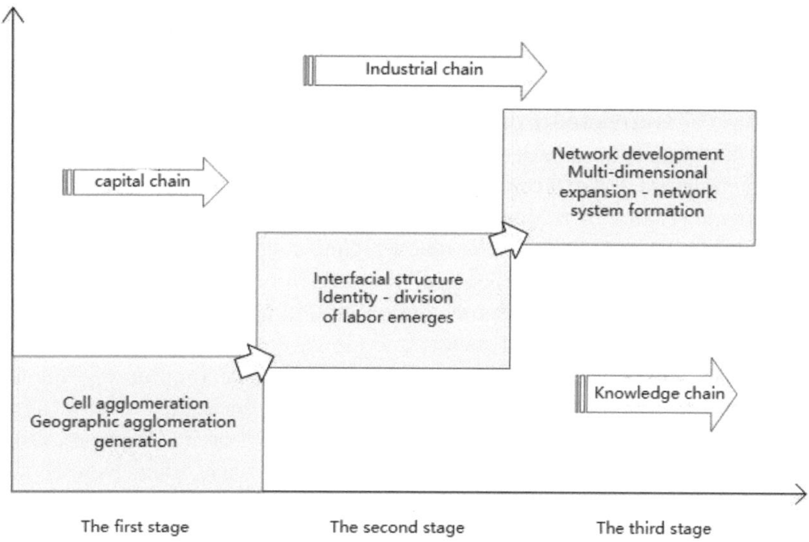

Fig. 5.4 Evolutionary mechanism of creative industry agglomeration

5.3 Agglomeration and Convergence of Economic Development

5.3.1 Agglomeration Process of Economic Development

Economic development is an important aspect of the development of developed areas, and it is also one of the important contents of development geography. Economic agglomeration refers to the phenomenon that various economic factors affect each other and have economic effects on a specific region. The spatial distribution of economic activities is often not uniform, but has the characteristics of agglomeration distribution, that is, different enterprises form a certain scale of local concentration in space, which creates conditions for economic agglomeration. Throughout history, the spatial agglomeration of the economy has always been a universal phenomenon. The agglomeration of economic development is evident in the process of industrialization, and the economic development course of industrialized countries is basically the process of economic agglomeration and change. Therefore, the rapid development of world industrialization makes the agglomeration economy an important reason for the formation and expansion of cities. The agglomeration of enterprise economic activities, the agglomeration of social economic activities, and the rise of cities promote the development of the agglomeration economy. When the economic activities of enterprises reach a certain state of concentration, compared with the state of decentralization, the production efficiency will also be improved.

5.3.1.1 Theoretical Research on the Agglomeration Process of Economic Development

Richardson (1978) proposed to divide economic agglomeration into industrial, social economic, and family economic agglomeration. Industrial economic agglomeration refers to the internal and external economy related to the scale of enterprises, which has the characteristics of regional agglomeration, detailed division of specialization, etc., and its competitiveness is much higher than that of "large and comprehensive" and "small and comprehensive" production modes or regional production modes. Currently, industrial economic agglomeration is the most widely used form, which affects the development of society and family economy, and the two forms of economy will, in turn, affect the development of industrial economic agglomeration. The two forms of economy (society and family) not only serve the industrial agglomeration economy but also have a certain restricting effect on it. Therefore, different types of economic agglomeration interact and restrict each other. New economic geography theory provides an important method for quantitative research on spatial economic agglomeration.

New economic geography theorist Krugman (1991) proposed a model that uses the equilibrium method to solve the spatial economic distribution process. Therefore, the spatial agglomeration characteristics of economic activities and the formation mechanism of economic agglomeration at different geographical scales have received increasing attention. Then, based on Krugman's model, several extension and expansion models appeared successively, including D-S (Dixit-Stiglitz Model) model theory (1977) and OTT model theory (2002). The D-S model proposes an economic agglomeration centripetal force including the central market effect and the price index effect, whereas the OTT model proposes an economic agglomeration centrifugalism including the market crowding effect. These theories hold that economic agglomeration is caused by the transfer of labor and capital under the influence of centripetal and centrifugal forces of economic agglomeration. All these models provide important technical methods for analyzing economic spatial agglomeration characteristics.

Urban economic agglomeration makes full use of and absorbs various resource elements and positive factors of the city itself, surrounding areas, and locally and internationally, which enhances the economic strength and the city's development potential. The neoclassical location theory in the agglomeration theory of urban economic development divides economic activities into three categories according to region: urban economy, local economy, and industrial center economy. In contemporary urban economic theory, the research on economic agglomeration is mainly conducted in two directions: The first is the research on the dynamic changes of the causes of agglomeration in the urban economy, which holds that the agglomeration of the urban economy is composed of the agglomeration of enterprises and population, whereas external economic agglomeration is constantly changing with the progress of technology. The second is the research on the mechanism of agglomeration and dispersion equilibrium to promote urban economic development. This type of research holds that the agglomeration and dispersion equilibrium of the urban economy depends on the mutual transformation mechanism between agglomeration and diffusion in the urban economy.

There are two kinds of economic agglomeration impacts on regional economic development. First, economic agglomeration promotes economic development, which is reflected in the improvement of labor productivity. For example, Shanghai's economy is highly concentrated, and its labor productivity and per capita disposable income are much higher than those in other regions. Second, industrial agglomeration inhibits economic development, leading to unbalanced economic development and gradually increasing the income difference. For example, the development of China's central and western regions lags behind that of the eastern region because the eastern coastal areas absorb a large amount of labor and capital.

5.3.1.2 Economic Agglomeration in Yangtze River Delta Region

The Yangtze River Delta region in this case only includes 25 prefecture-level and higher cities in Shanghai, Jiangsu, and Zhejiang provinces. The Yangtze River Delta region is the most economically concentrated region in China, and compared with the Beijing-Tianjin-Hebei region and the Pearl River Delta region, it has the smallest area and the highest economic output value and has agglomeration advantages. Therefore, taking the Yangtze River Delta region as an example, the Gini coefficient is used to measure and analyze the economic agglomeration phenomenon in this region.

The Gini coefficient is a common index used internationally to measure the income gap of residents in a country or region. The closer the Gini coefficient is to zero, the more equal the distribution of income becomes. International practice considers a Gini coefficient of < 0.2 as the absolute average income, 0.2 ~ 0.3 as the average income comparison, 0.3–0.4 as a relatively reasonable income, 0.4–0.5 as a relatively large income gap, and > 0.5 as a very large income disparity. According to the statistical yearbook of each province and city, the Gini coefficient of the urban agglomeration in the Yangtze River Delta region during 1998–2018 was calculated (Fig. 5.5). The results show that the Gini index of the urban agglomeration in the Yangtze River Delta fluctuates between 0.414 and 0.497. The Gini index continued to rise from 1998 to 2014 and reached its highest value of 0.497 in 2014. After that, the Gini index showed a downward trend. In general, the Gini index of urban agglomerations in the Yangtze River Delta region is in the range of 0.4–0.5, indicating a relatively large income gap in urban agglomerations in the Yangtze River Delta region.

From 1998 to 2014, the Gini index of urban agglomerations in the Yangtze River Delta region increased, indicating that the economic agglomeration benefits of large cities continued to increase, and the economic gap between cities continued to widen. According to the statistical data, this is mainly because Shanghai has driven the rapid economic development of Suzhou, Wuxi, Hangzhou, Nanjing, and other surrounding cities, whereas the cities far away from Shanghai, such as Yancheng, Chuzhou, Taizhou, and Jinhua, have not been clearly reflected in the radiation driving effect; thus, the gap between cities has widened and the Gini index has risen. After

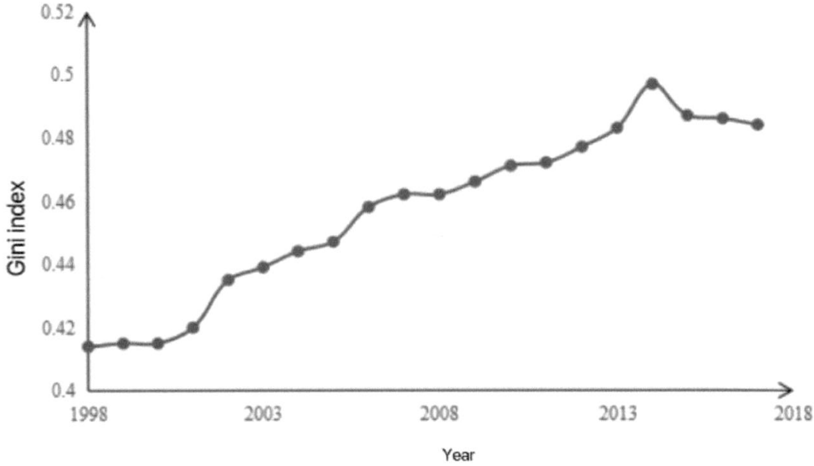

Fig. 5.5 Gini index of urban agglomerations in the Yangtze River Delta region, 1998–2018

2014, the Gini index of urban agglomeration in the Yangtze River Delta region has declined, mainly because of the gradual improvement of facilities in the north–south comprehensive corridor along the Yangtze River, TongSujia, Hexuanhang, Hujianyong, TaixiChangyi, and other emerging corridors, a high degree of transportation network, and more convenient transportation in the Hefei and Nanjing metropolitan areas, which promotes the balanced economic development of urban agglomeration in the whole Yangtze River Delta region. As a result, the Gini Index gradually declines.

5.3.2 Convergence Process of Economic Growth

Convergence of economic growth means that under closed economic conditions, there is a negative correlation between the initial static indicators (per capita output, per capita income) and the economic growth rate of different economic units in an effective economic scope (countries, regions, and even households), that is, the less developed regions have a higher economic growth rate than the developed regions. The process led to the gradual disappearance of the difference between the static indicators of each economic unit at the beginning of the period. Since the 1990s, the convergence of economic growth has gradually become a topic of great interest in development economics and international economics.

5.3.2.1 Convergent Development Theory and Progress of Economic Growth

Economic convergence is the concern of economic growth theory and economic development theory. In neoclassical economic theory, the study of convergence includes α and β convergence, conditional and absolute convergence, and conditional and convergence clubs. New economic growth theory converges from α convergence, β convergence, and club convergence (Dobrescu & Dobre, 2014). The traditional theory of economic growth holds that differences in per capita income or capital between countries lead to differences in economic growth rates, but these differences may gradually disappear over time. Therefore, countries or regions with low per capita income have higher convergence of the economic growth rate. New economic growth theory holds that differences in economic growth rates between countries are not related to differences in per capita income, but to differences in technological progress, and these differences do not necessarily disappear over time.

Currently, there are two kinds of research on the convergence of regional economic growth. One is the convergence test, which does not consider the spatial factor in the traditional sense. Most of the early foreign studies were based on the conditional convergence model of cross-section or panel data, which was used to explore the relationship between economic growth rate and initial output level and determine whether there is convergence in the economic growth of a country or region and empirically analyze the convergence speed. Inspired by this kind of research, the convergence of China's regional economic growth has also aroused the attention of domestic scholars. This type of research basically aims to judge whether China's regional growth has local convergence or conditional convergence from the perspective of experience. Another type of research considers convergence tests for spatial factors. Such studies use various models, such as spatial error, hysteresis, and the Dubin model, to analyze the convergence of regional economic growth under the premise of considering spatial dependence and heterogeneity. However, due to different proxy variables, data processing, and data sources, the existing literature offers different views on the convergence of China's regional economic growth. Relevant studies found that China's regional economic growth has a global convergence trend; however, the convergence speed tends to decline over time. Some researchers suggest that China's regional economic growth is in the process of divergence, and the divergence phenomenon is more obvious in the economic growth of relatively less developed regions.

In China, most scholars have conducted empirical research on China's economic convergence based on GDP data. The tertiary industry is an important growth point of future economic development. Through the analysis of the tertiary industry, we examine the differences in inter-provincial economic growth in China and the economic convergence of China's inter-regional development. Therefore, it has an important guiding role in predicting the future development trend of the entire Chinese economy.

5.3.2.2 Convergence Analysis of National Economic Growth

To develop China's economy more effectively, in-depth research on the country's economic growth is necessary. An α test can directly describe the change in the economic gap in different regions. Therefore, α is used to study the convergence of China's economic growth to better understand the state of China's economic growth. First, the national data is collected, and then the coefficient of variation (CV), Theil index (TEL), and locational Gini coefficient (LGC) are used to measure how much the data is offset. If the index shows a downward trend over time, it indicates that there is α convergence in the country's economic growth.

Using the above method, the difference change in China's economic growth from 1980 to 2015 is measured. Figure 5.6 shows the change in the economic growth imbalance indicator. Evidently, since 1980, the CV, TEL, and LGC values of the national economic gap have tended to decline, among which the CV index is the most obvious. If the values of the three indexes all tend to follow the same downward trend, it can be judged that the economic growth has convergence. To determine the degree of correlation between indicators, Pearson's correlation coefficient between indicators is used. The value of Pearson's correlation coefficient is between − 1 and + 1, and the greater the absolute value, the greater the correlation. Positive numbers represent positive correlation and negative numbers represent negative correlation. As shown in Table 5.12, the coefficients among indicators are all greater than 0.8, indicating that the indicators are positively correlated, and their changing trends are basically the same. Figure 5.6 also shows that generally, the change follows a consistently downward trend. Therefore, α convergence is observed in China's economic growth, and the gap between the economic growth of provinces is gradually decreasing.

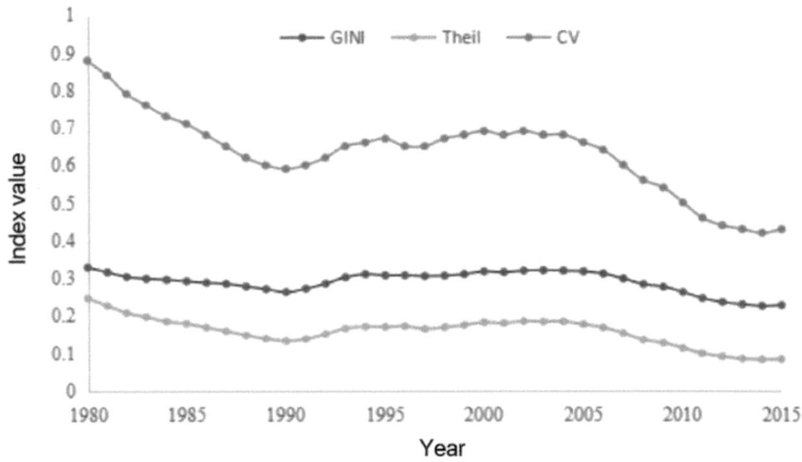

Fig. 5.6 Inter-annual change in the index value of economic development imbalance in China from 1980 to 2015

Table 5.12 Pearson's correlation coefficients among different indicators

	Gini	Theil	CV
Gini	1.000	0.918	0.866
Theil	0.918	1.000	0.990
CV	0.866	0.990	1.000

5.3.3 *Agglomeration and Convergence Mechanism of Economic Development*

Economic agglomeration in developed areas is often accompanied by economic diffusion, and the two intercross and intermerge. Every economic agglomeration produces new requirements for economic diffusion, and every economic diffusion will bring higher-quality economic agglomeration. The urban economy in developed areas develops in the continuous transformation of economic agglomeration and economic diffusion. The spread of the industrial economy in developed areas to less developed areas will promote the balanced development of regional economic growth and help realize the convergence of regional economic growth. Therefore, economic agglomeration and economic convergence in developed areas are also mutually integrated.

5.3.3.1 Economic Agglomeration and Convergence Development Mechanism

The economic agglomeration mechanism can be divided into three types according to the causes of economic agglomeration. The first is the agglomeration mechanism of economies of scale within the same place and firm. The extent of this economic agglomeration is affected by the scale of production of the firm at that location. The second is the localization economic agglomeration mechanism of multiple enterprises in the same industry in the same place. This degree of economic agglomeration is primarily influenced by the industry's nationwide sales volume and the overall level of the industry's output in that location. The third is the urban economic agglomeration mechanism of multiple industries in the same place, and the degree of economic agglomeration is affected by the correlation degree of different industries in the city and the total output level of all enterprises. These economic agglomeration mechanisms may lead to the phenomenon of spatial polarization in cities, trigger the Matthew effect in the development relationship between developed and less developed areas, urban and rural areas, etc., form developed areas or economic centers, and further aggravate the spatial differences and imbalances in economic development.

The driving factors of the formation of the economic agglomeration mechanism include the location orientation of economic activities, the internal relations of economic activities, and the pursuit of economic activities for an agglomeration economy. The location orientation of economic activities is mainly guided by national or regional government policies. Future economic development planning, future urban development planning, improvement of the transportation network, and various economic incentive policies will exert regional guidance on national or regional economic agglomeration. The internal connection of economic activities mainly refers to the division of labor among economic industries, that is, whether it can break the boundaries of departments or regions, organically combine the division of labor among industries and regions, and then promote the further development of economic agglomeration. Economic agglomeration has the properties of accumulation and self-strengthening. The pursuit of economic agglomeration by economic activities will often strengthen the self-strengthening and accumulation of the agglomeration economy by countries or regions and strive to realize the expectation of becoming the center of economic agglomeration, which also promotes the development of economic agglomeration.

There are two main types of economic convergence mechanisms, namely neoclassical convergence theory due to diminishing marginal rate of return on investment and new growth theory due to technology diffusion. Neoclassical convergence theory holds that in the long run, less developed countries can catch up with developed countries in per capita income level by virtue of their higher growth rate. According to neoclassical economic growth theory, the only reason for the difference in per capita income and growth rate between regions is the technological difference. If the technical level is similar among regions, the per capita income and economic growth rate of different regions will have a strong convergence. It is found that both neoclassical growth and new growth theories have convergence mechanisms in terms of China's regional economic convergence, but for different regions, different convergence mechanisms may play a role in different periods. Before the 1990s, China's capital accumulation had a significant effect on convergence; however, at the beginning of this century, technology diffusion had a significant effect on convergence. Relevant studies have also used Jiangsu Province as an example and found that the capital convergence mechanism replaced the technology convergence mechanism as the main mechanism influencing county economic convergence in Jiangsu Province in recent years, and the policy of coordinating regional development in Jiangsu Province played its due role.

The driving factors of the economic convergence mechanism include the economic diffusion effect, technological progress, human capital, and international trade. The economic diffusion effect of developed areas can promote the transfer of some industries to less developed areas to promote the economic development of those areas. Therefore, development at the scientific and technological levels reflects development at the economic level. The influence of science and technology diffusion in developed areas on less developed areas is also an important factor for the less developed areas to catch up with developed areas and achieve national economic convergence. The role of human resources in economic development has become increasingly prominent. In fact, an important driving force for long-term economic growth can effectively control regional development gaps and achieve economic convergence. With the acceleration of the process of global economic integration, international trade can eliminate the economic gap between regions and become an important factor affecting economic convergence. Developing the driving forces of national economic convergence and realizing national economic convergence are conducive to the overall planning of the government's macro policies and the coordinated and rapid development of the inter-regional economy, which has special significance for the country's long-term sustained, stable, and healthy development.

5.3.3.2 China's Economic Agglomeration Mechanism and Economic Growth Convergence Mechanism

The current situation of China's economic development indicates that in the past, China's market competition was not fierce, the central advantage was obvious, and the central market effect and the price index effect were the dominant factors. Now, the market competition in the country is fierce, the central advantage has decreased, and the role of the market crowding effect is gradually increasing. China's economic development shows a trend of declining center advantages. For example, Shanghai is the economic center of the Yangtze River Delta, and the Yangtze River Delta is the economic center of the eastern coastal region; however, its status is constantly decreasing under the influence of Jiangsu, Zhejiang Province, and the Beijing-Tianjin-Hebei region respectively. The mechanism of economic agglomeration in China is different from that of foreign countries, in which labor is the leading factor. The leading factor of economic agglomeration in China is capital flow, including capital stock and total investment.

Based on the current situation of China's economic growth, the gap between regions in the country was large in the past, but in recent years, under the guidance of national policies, the gap between regions has gradually decreased, and the inter-regional economic growth has shown a convergence trend. Therefore, we analyze the national economic growth data from 1993 to 2012 to examine the convergence mechanism of China's economic growth based on new growth theory and neoclassical theory.

First, using the C-D function commonly used in economics, the provincial production function of our country is estimated, and then the capital stock margin and total factor productivity (TFP) from 1993 to 2012 are calculated. Then, using 2003 as the time node, the spatial Durbin model (Eqs. 5.10 and 5.11) is used to analyze the labor capital and total factor productivity (TFP) data from 1993 to 2003, 2004 to 2012, and 1993 to 2012, respectively.

$$ln\left(\frac{K_{i,t}}{L_{i,t}}\bigg/\frac{K_{i,t-1}}{L_{i,t-1}}\right) = \alpha_i + \beta ln\frac{K_{i,t}}{L_{i,t}} + \rho Wln\left(\frac{K_{i,t}}{L_{i,t}}\bigg/\frac{K_{i,t-1}}{L_{i,t-1}}\right) + \delta Wln\frac{K_{i,t}}{L_{i,t}} + \mu_i + \eta_t + \varepsilon_{i,t} \tag{5.10}$$

$$\ln\left(\frac{TFP_{i,t}}{TFP_{i,t-1}}\right) = \alpha_i + \beta lnTFP_{i,t-1} + \rho Wln\left(\frac{TFP_{i,t}}{TFP_{i,t-1}}\right)$$

$$+ \delta WlnTFP_{i,t-1} + \mu_i + \eta_t + \varepsilon_{i,t} \tag{5.11}$$

Here, $K_{i,t}$ and $K_{i,t-1}$ refer to the area i in t and $t-1$ years of material capital stock; $L_{i,t}$ and $L_{i,t-1}$ refer to region i in t and t − 1 years of labor employment; $K_{i,t}/L_{i,t}$ and $K_{i,t-1}/L_{i,t-1}$ in t and $t-1$ years of labor capital in area i; $TFP_{i,t}$ and $TFP_{i,t-1}$ is the total factor productivity of region i in t and $t-1 years$; α_i is the top of the intercept; ρ is the coefficient of the spatial trailing term; λ is the coefficient of the spatial error term; δ is the spatial lag coefficient of the explanatory variable; W is the spatial weight matrix of economic distance; and β is the elasticity of regional capital growth rate to other initial capital levels. It is a measure of regional capital convergence. μ_i is a fixed effect in space, and η_t is a fixed effect in time. Tables 5.13 and 5.14 show the final results.

From the perspective of the neo-classical convergence mechanism (Table 5.13), β is negative during 1993–2012, indicating the convergence of capital stock, and ρ and δ both pass the significance test, indicating that during the process of regional capital convergence, the spatial spillover effect of Chinese capital is negatively transmitted through the initial level of labor capital and the growth rate of average labor capital in the neighboring region. During the entire period of 1993–2012 and the sub-periods of 1993–2003 and 2004–2012, the capital stock showed convergence. From the perspective of the growth convergence mechanism (Table 5.14), TFP showed convergence during 1993–2012. In addition, TFP showed divergence from 1993 to 2003, whereas ρ was significantly negative from 2004 to 2012, indicating that this period was a stage of technological convergence, and the spatial spillover effect was negatively conducted through the initial technical level of neighbors. The δ value, as the evidence, passed the significance test from 1993 to 2003 but did not pass

Table 5.13 Estimates of labor capital convergence in China from 1993 to 2012

Year	1993–2003	2004–2012	1993–2012
Regression coefficient			
ln(Ki, t−1/Li, t−1)	−0.918(0.000)	−0.108(0.034)	−0.389(0.000)
P	−0.434(0.000)	−0.827(0.000)	−0.674(0.000)
Wln(Ki, t−1/Li, t−1)	−0.532(0.228)	−0.149(0.348)	−0.660(0.000)
Model test			
R^2	0.496	0.304	0.246
LIK	125.160	164.060	197.120
Sigma. sq	0.025	0.013	0.026
SDM utility test			
Wald_spatial_lag	6.021(0.000)	2.568(0.000)	5.332(0.000)
Wald_spatial_error	5.512(0.002)	2.357(0.000)	5.425(0.001)
LR_spatial_lag	5.925(0.001)	3.894(0.000)	5.669(0.000)
LR_spatial_error	5.835(0.003)	1.569(0.000)	5.698(0.000)

Table 5.14 Estimates of the convergence of total factor productivity in China from 1993 to 2012

Year	1993–2003	2004–2012	1993–2012
Regression coefficient			
ln(Ki,t−1/Li,t−1)	0.442(0.000)	−0.068(0.043)	−0.144(0.000)
ρ	−0.324(0.003)	−0.529(0.000)	−0.421(0.000)
Wln(Ki,t−1/Li,t−1)	−0.202(0.031)	−0.245(0.114)	−0.015(0.915)
Model test			
R^2	0.310	0.401	0.172
LIK	446.400	500.660	866.090
Sigma. sq	0.003	0.001	0.003
SDM utility test			
Wald_spatial_lag	9.301(0.000)	10.021(0.000)	6.321(0.000)
Wald_spatial_error	7.502(0.000)	9.502(0.000)	8.012(0.002)
LR_spatial_lag	8.025(0.000)	7.325(0.001)	8.325(0.001)
LR_spatial_error	6.326(0.000)	7.335(0.000)	8.335(0.000)

the significance test from 2004 to 2012, indicating that the TFP level of the neighboring region has a positive effect on the TFP growth rate of the region; however, this positive effect is no longer obvious after 2003.

In fact, in the convergence mechanism of China's economic growth, the two convergence mechanisms do not operate separately but affect and cooperate with each other. Therefore, when judging the convergence of economic growth, we should not consider only one mechanism but make comprehensive considerations to reflect the results of China's regional economic development more accurately.

5.4 Agglomeration and Convergence of Innovative Talents

5.4.1 The Gathering Process of Innovative Talents

Innovative talents are an important driving factor for sustainable development in developed areas and an important component in the development of geographical talents. Talent gathering refers to a certain time range, according to a certain connection, a large number of the same type or related talents in an area (physical environment), gathered together in one industry (virtual space), the scale of which is an important feature. The agglomeration effect stems from a talent agglomeration scale, and the formation of the talent agglomeration scale is a dynamic accumulation process. Influenced by various factors, talent resources in one region may flow to other regions, making some regions become talent agglomeration highlands. On the one hand, talent agglomeration acts on sustained economic growth; on the other hand, economic growth in turn acts on talent agglomeration, thus forming a virtuous cycle. Therefore, many countries and regions take the talent gathering project as the development focus.

5.4.1.1 Innovation Talent Gathering and the Development of Industrial Cluster

The agglomeration of innovative talents is a result of innovative talents pursuing a kind of "consensus domain," that is, the synthesis of the induced conditions of the benefit maximization and the economically developed areas. There are both similarities and differences between innovative talents and ordinary talents in their flow. Although the flow purpose is the same, the talents basically aim to obtain higher economic returns than those obtained in the original place. However, their expectations of future income are low because of their own physical quality and ability to restrict their flow. As a special group, innovative talents with higher quality and ability have greater spatial mobility, and they have higher expectations for future returns. Therefore, innovative talents are easily attracted by economically developed regions from areas with low efficiency or where their own capabilities are difficult to give full play.

The gathering of innovative talents is synchronized with the development of industrial clusters. In the era of knowledge economy, disciplines have the characteristics of high differentiation and synthesis. With the socialization of production services, innovation activities are replaced by unity and cooperation. The concentration of innovative talents can generate more energy by changing the spatial structure, thus reducing the risk of innovation. The development of regional industrial clusters needs talents, especially the high concentration of innovative talents. The competitiveness of regional industrial clusters is enhanced by the agglomeration of innovative talents. The social quality resources in the cluster area are also highly integrated. On the contrary, the development of regional industrial clusters has produced significant demonstration and imitation effects, resulting in the simultaneous formation of the "economic growth pole" and the "innovative talent growth pole," which has a strong temptation for innovative talents in economically underdeveloped areas.

The flow of innovative talent into a region will bring vitality and improve the overall quality of the region. With the continuous inflow of innovative talents, different groups interact and expand at many levels. As the innovative talents "do as the Romans do" and become indigenous, the cultural environment and ideology of the inflow collide. The long-term gathering of innovative talents in the inflow places makes foreign and local cultures increasingly integrated and friendly, and the overall quality of the region is subtly changed. This is a kind of invisible, potentially long-term penetration and interaction.

5.4.1.2 The Changing Process of Talent Gathering Competitiveness in Shanghai

Currently, Shanghai's export growth is weak, the infrastructure investment industry is difficult to grow significantly, and the binding industry of resource factors is becoming increasingly prominent. The first is the increasing role of land constraints. Shanghai's per capita land area is only one-third of Beijing and Tianjin; however, the construction land area is far higher than that of other global cities, and if the current model continue to develop, there will be no land to use in the future. Second, the capital constraints are becoming increasingly prominent, and the introduction of foreign capital in Shanghai is facing severe challenges. Third, the shortage of cheap labor resources requires the development of high-end industries to fill the growth space, improve the overall quality and energy level of the industry, and ensure rapid economic growth. Therefore, through the analytic hierarchy process, this study analyzes the competitiveness of talent gathering in Shanghai and finds out the factors that have a decisive impact on talent gathering.

First, economic development level, urban construction level, living conditions, business environment, international and domestic exchanges, higher education, vocational training, public education funds, talent structure, talent quality, and talent flow are selected as evaluation indicators. The annual data of Shanghai since 2000 are used. Second, each index is dimensionless. Then, the annual effect weighting score is calculated according to the weight of each indicator and the processed indicator data. Finally, the study analyzes the process of talent gathering in Shanghai. The comprehensive index model of talent agglomeration adopted in the analysis is as follows:

$$a_t = \sum_{k-1}^{s} w_k \sum_{j-1}^{m} w_j \sum_{i-1}^{n} w_i \frac{P_{it} - P_{it-1}}{P_{it-1}} \tag{5.12}$$

where a_t is the comprehensive index of talent agglomeration in t years; w_k, w_j, andw_i are the weights of factors, index categories, and the upper level of indicators; P is the index value; n is the number of indicators of a certain index category; m is the number of index categories of a certain factor; and s is the number of factors affecting talent agglomeration (Tables 5.15 and 5.16).

Therefore, Shanghai strengthens the carrier of talents gathering in the region, sets up various formal and informal knowledge-based teams, integrates outstanding talents into an organic whole through lectures, seminars, and research projects and cultivates diverse talent echelons in various fields. The entire echelon system is innovated with the driving effect of outstanding figures in the echelon to realize the best combination of talent value and cost and promote the benign growth of regional talent capital. At the same time, external talents as well as external knowledge and technology can be introduced through the implementation of appropriate policies, such as the open door and international development of talents.

Table 5.15 Index system and data of talent agglomeration competitiveness in Shanghai

Primary criterion layer	Second criterion layer	Index level	2000	2006	2007	2008	209
Talent environment	Level of economic development	Per capita GDP (Yuan/person)	30,047.00	57,695.00	66,367.00	75,109.00	78,989.00
		Import and export per capita (US $)	3401.00	12,533.00	15,229.00	17,058.00	14,455.00
		Foreign trade utilization per capita (US $/person)	196.00	392.00	426.00	534.00	548.00
		Per capita fixed assets investment (Yuan/person)	11,623.00	21,625.00	23,996.00	25,573.00	27,446.00
	Level of urban construction	Actual road length at year end (km)	6641.00	14,619.00	15,458.00	15,844.00	16,071.00
		Green coverage rate of built-up area (%)	22.20	37.30	37.60	38.00	38.10
		Urban per capita public green space (m^2)	4.60	11.50	12.01	12.51	12.80
		Total collection of public libraries (10,000 volumes)	5500.00	6062.40	6253.90	6394.10	6593.40
		Average annual real wage level of employees (Yuan)	15,420.00	29,569.00	34,707.00	39,502.00	42,789.00
		Tap water production capacity (10,000 m^3/day)	1048.00	1138.00	1050.00	1069.00	1096.00
	Living conditions	Total electricity consumption (KWH)	559.50	990.20	1072.40	1138.20	1153.40
		Sewage treatment capacity (tons/day)	63.10	426.70	418.90	485.20	470.20

(continued)

Table 5.15 (continued)

Primary criterion layer	Second criterion layer	Index level	2000	2006	2007	2008	209
		Number of beds in medical and health institutions (10,000)	7.30	9.40	9.60	7.80	8.10
	Career environment	Output value of new products in science and technology activities as a percentage of GDP (%)	29.80	24.40	25.60	32.30	30.30
		Number of scientific and technological activities (number)	891.00	1162.00	1108.00	1314.00	1356.00
		Proportion of science and technology expenditure in local government expenditure (%)	1.60	5.20	5.10	4.60	7.20
		Foreign trade enterprise book	1730.00	1592.00	1670.00	1854.00	1789.00
		Number of domestic enterprises (number)	3865.00	2051.00	2075.00	2711.00	2567.00
		Number of private enterprises	452.00	1422.00	1500.00	2126.00	2039.00
	International and domestic exchanges	International and Domestic Academic Exchange Conference (Session)	1085.00	2068.00	2101.00	905.00	2189.00
		Number of international and domestic exhibitions (times)	270.00	309.00	423.00	445.00	460.00

(continued)

Table 5.15 (continued)

Primary criterion layer	Second criterion layer	Index level	2000	2006	2007	2008	209
		Number of foreign students (persons)	13,000.00	14,100.00	13,121.00	14,273.00	15,447.00
Talent development	Higher education	Number of colleges and universities (number)	37.00	60.00	60.00	61.00	66.00
		Number of college students (persons)	172.00	341.00	352.00	266.00	267.00
Talent structure	Vocational training	Number of participants in vocational skills training (10,000)	31.00	34.20	36.50	43.50	40.40
	Public education expenditure	Education as a proportion of government expenditure (%)	15.10	13.10	12.90	12.50	11.60
	Age	Talent under 45 years old (10,000)	47.10	52.50	57.80	58.30	60.00
		Management talents (10,000)	17.90	20.10	22.90	23.10	34.60
		Science and technology activity personnel (10,000)	113.00	143.90	151.20	177.90	185.00
	Specialty distribution	Number of teachers (persons per 10,000)	79.80	117.60	113.20	68.90	74.40
		Number of lawyers (persons per 10,000)	3.10	4.40	4.70	5.20	5.60
		Number of doctors (persons per 10,000)	31.00	25.00	26.00	27.00	27.00

(continued)

Table 5.15 (continued)

Primary criterion layer	Second criterion layer	Index level	2000	2006	2007	2008	209
Talent quality	Educational background	Number of college graduates (10,000)	4.10	11.10	11.90	12.20	12.70
		Number of master's and doctorate degrees (persons)	5853.00	19,729.00	23,583.00	25,217.00	28,283.00
	Ability	Number of personnel with senior titles (persons)	6.40	7.60	7.70	7.90	8.50
		Number of personnel with intermediate titles (persons)	25.40	26.10	26.10	25.90	27.00
		Number of staff with junior titles (persons)	36.10	28.30	28.20	27.60	26.00
		Scientific and technological achievements (examples)	1102.00	1953.00	2396.00	1866.00	2166.00
Talent flow	International flow	Number of talents studying abroad who choose to stay in their home city (persons)	975.00	4317.00	4000.00	3422.00	4200.00
	Domestic flow	Number of domestic talents residing in the city (persons)	12,898.00	2640.00	2563.00	5245.00	6000.00

Source Based on the data from the Shanghai Statistical Yearbook in 2001, 2007, 2008, 2009, and 2010

Note The data of talents aged below 45 years are adopted from the statistical yearbook. The data of high, middle, and junior titles are obtained from professionals in state-owned enterprises and institutions. The number of domestic talents who settled in the city in the past year and the number of international and domestic academic exchange conferences of the Association for Science and Technology system are obtained from the Statistical Yearbook

Table 5.16 Correlation between talent agglomeration and economic development in Shanghai from 2000 to 2009

Index	2000	2006	2007	2008	2009
Comprehensive index of talent concentration	0.554	0.839	0.893	0.926	0.973
Annual GDP growth rate (%)	11.000	12.700	15.200	9.700	8.200
Correlation degree	0.199	0.151	0.170	0.105	0.084

5.4.2 *Regional Convergence of Innovative Talents*

The convergence effect of innovative talents is an important component of the convergence mechanism in development geography. The aggregation effect of innovative talents refers to the effect of the relative concentration of relevant innovative talents according to certain interconnections in a harmonious environment within a certain time and space, which exceeds their independent roles. The term "convergence effect" first appeared in the study of economics. The convergence effect in economics describes how, between economies at different levels of development, the laggard closes the gap faster. Particularly in developed countries, after reaching a certain economic level, regional balanced development is an important factor restricting national economic development. Later, the convergence effect becomes an important element of development geography research. Therefore, developed regions or developed countries emphasize the balanced development of regional economy and talents; thus, the various level differences among economies may gradually reduce.

5.4.2.1 Research on Regional Convergence of Innovative Talents

The convergence effect of talents is an important factor restricting the balanced development of a regional economy. The convergence effect of talents refers to the phenomenon of talents converging to the same place, influenced by different economic environmental and other factors, that is, the balanced development of talents. This effect can be obtained through an analogy of the properties of continuous points in mathematical analysis. Talents can be analogized as continuous points in the convergence effect, giving talents the properties they have, and then abstracting the convergence effect function.

The convergence of innovative talents can prevent a series of problems caused by the gathering of innovative talents. Although the agglomeration of innovative talents in the early stage of development in developed regions has had several positive effects, mainly embodied in the innovation effect, knowledge spillover effect, information sharing effect, scale effect, and regional effect, there are also certain problems, mainly including talent personality conflict, interest conflict, goal conflict, and talent value conflict. Convergence of innovative talents can prevent these negative effects.

Several studies have been conducted on the agglomeration of innovative talents in developed regions, but relatively few studies have been conducted on the convergence and development of innovative talents. This is mainly because the concentration of innovative talents is a natural talent distribution situation with the development of a social economy in developed areas. Currently, the convergence and development of innovative talents rely more on national policies to encourage innovative talents to support the development of less developed regions. Although the convergence of innovative talents is also developing, it has not yet formed the mainstream compared with talent agglomeration.

5.4.2.2 Case Study on Convergence Effect of Innovative Scientific and Technological Core Talents

The convergence effect of core talents in science and technology can be divided into the internal effect and external convergence effect. The core talents of science and technology are compared to continuous points, and their properties are assigned to the core talents of science and technology, then, the convergence effect function is introduced, and the corresponding convergence function diagram is created (Fig. 5.7). The convergence effect of the innovative core talents of science and technology is analyzed through the function, and the relationship between the core talents of science and technology and the agglomeration environment system is obtained (Fig. 5.8). The convergence function of regional innovative scientific and technological core talents is as follows:

$$E(h, A) = ah + A \tag{5.13}$$

where a represents the influence factor coefficient of regional innovative scientific and technological core talents, and $0 \leq \alpha \leq 1$; h is an unknown parameter of the influence factor of regional innovative science and technology core talents, the value range is $0 \leq \mathrm{h} \leq 100$, that is, the definition domain of function $\mathrm{E} = \alpha\mathrm{h} + \mathrm{A}$; and A is the clustering effect of regional scientific and technological talents.

The internal convergence effect of the core talents of science and technology includes three types. The first is the interaction effect of knowledge and information. The clustering of core talents in a region increases the sources of information and improves the quality of information. Due to the Matthew effect, this clustering of talents makes it easier to complete a breakthrough in a certain field, thus forming internal convergence (Rossoni & Rosa, 2024). The second is the competitive learning effect of internal talents. In a group with abundant talents, the status and treatment of core talents motivate the internal staff to keep striving, and the general staff will take the core staff as an example to make progress in competition and cooperation, forming a competitive learning effect. Third is the knowledge cross effect of core talents. Most of the core talents are senior talents or experts in a certain field, and they have the right to speak in their own fields. With the development of science and technology, various disciplines need to cross-reference, which makes the crossover of core talents and disciplines increasingly important, resulting in the knowledge crossover effect.

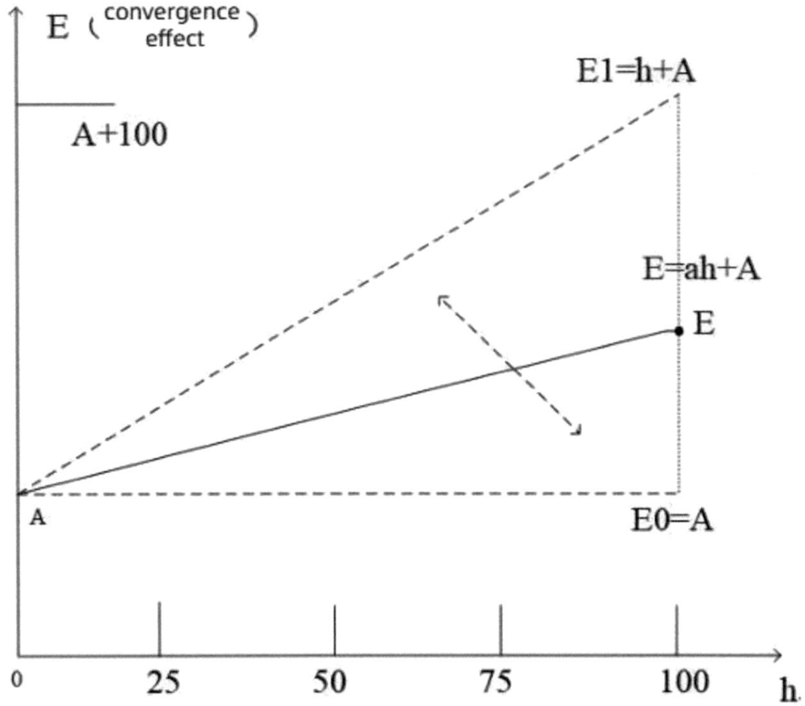

Fig. 5.7 Illustrates the convergence effect function

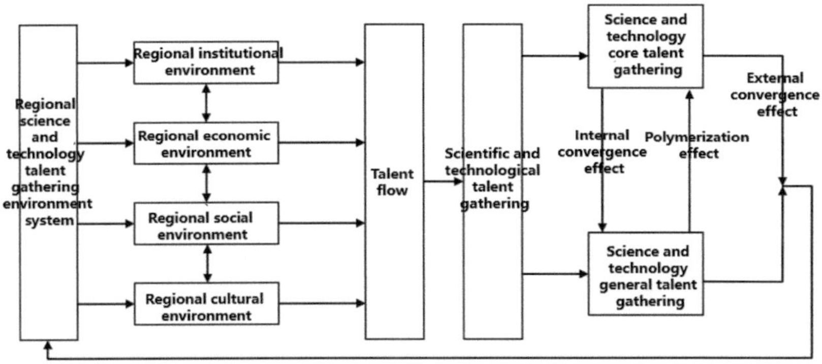

Fig. 5.8 Relationship between core talents in science and technology and their regional convergence environment

The external convergence effect is divided into two categories: the strong gravitational effect and the benchmarking effect. In terms of the strong gravitational effect, the core talents of science and technology show the ability of academic and scientific research in the aspect of explicit or implicit knowledge. Tacit knowledge can only be acquired through face-to-face understanding. Due to personal differences, tacit knowledge is more difficult to acquire than explicit knowledge. One of its important characteristics is scarcity, which has a gravitational effect on the outside world, gathering talents who want to acquire this ability to the core talent gathering area. In terms of the benchmarking effect, the contribution of talents to social development can be investigated from three angles: The first one is the investigation of the amount and conversion rate of scientific and technological achievements; the second is the calculation of the development amount of new scientific and technological products and their earnings; and the third is the contribution rate of scientific and technological progress to economic growth. Generally, groups with core talents have a greater contribution to the economy and greater influence in the industry, which can lead to the progress of a certain industry and produce an external benchmarking effect.

5.4.3 Agglomeration and Convergence Mechanism of Innovative Talents

Under the background of the new economic form, China's economic development has gradually shifted from element-driven to innovation-driven, and the fundamental driver of innovation is innovative talents. Innovative talents bring innovative technology, and then promote the development of an innovative economy. Therefore, both internationally and in China, the developed regions are due to good medical and educational conditions and more development opportunities, resulting in the continuous transfer and agglomeration of innovative talents to the developed regions, and the driving force of regional agglomeration of innovative talents is increasing. This study shows that the driving factors and mechanism of the agglomeration and convergence of innovative talents in developed regions are mainly reflected in the following four aspects:

Relatively efficient system design. The increasing reward mechanism and self-reinforcing mechanism of institutional change will be strengthened along its direction in future development. For example, Shanghai has a certain driving effect on the Yangtze River Delta, and the economic integration of the Yangtze River Delta is more a conscious act of collaborative innovation than the product of top-level design. Due to the correctness of the initial path selection, the marginal benefit increases and enters a virtuous circle, influenced by the self-reinforcing mechanism. Good institutional environment promotes talent gathering.

A relatively satisfactory marginal return. The flow of innovative talents is the behavior of pursuing individual benefit maximization. Similar to other economic decisions, the flow of innovative talent is subjected to a benefit–cost analysis. Income, environment, and opportunities vary among different regions. Generally, developed regions have obvious advantages in terms of location and economic development. From the perspective of personal income, individuals can obtain a higher marginal output rate of human capital through their own mobility, and thus gain more income, improved working and living conditions, improved welfare, and more development opportunities. Such marginal benefits outweigh the opportunity costs and direct costs incurred in the flow process.

A relatively rich vehicle for innovation. The flow of innovative talents should not only consider the maximization of economic benefits but also the development opportunities of the career, the effectiveness of the innovation platform, and carrier. The "three carriers" of the introduction of innovative talents in Jiangsu Province include high-tech development zones, colleges and universities, and "three-capital enterprises," particularly the IT industry. This is a convergence point for innovative talent. For example, the agglomeration effect of Wuxi Science and Technology Park on high-level talents was initially formed. The main body of the park has absorbed high-level talents mainly in large enterprises and their R&D institutions, and its agglomeration effect on senior talents is particularly obvious. Institutions of higher learning, such as universities in the Jiangsu province, recruit talents and even carry out global recruitment to seek academic experts. The IT industry in Suzhou, especially in Kunshan City, has exerted a gathering effect on many industry elites and promoted Kunshan's economy to become the first of the top 100 counties in the country.

Relatively high talent density. Because of the agglomeration effect of knowledge production, the flow of talents leads to the agglomeration phenomenon. Economically developed areas have advanced science and technology, high knowledge level, fast information flow, and high academic cooperation, which provide a good environment for their knowledge production. Increasing marginal income is an important manifestation of knowledge production, which produces the "scale economy" effect and positive external economy. Knowledge production has the characteristics of a group that needs to achieve a certain density of talents and a working environment with collaborative innovation among talents. This kind of environment is conducive to study and research activities. Therefore, the "Matthew effect" of the gathering of innovative talents is gradually emerging. In this respect, state-owned enterprises have unparalleled talent-gathering potential and need to be revitalized.

Agglomeration and diffusion mechanism of innovative talents in Silicon Valley. Silicon Valley is a gathering place for high-end talents and top technologies in the world. The composition of talents in Silicon Valley shows the following characteristics: a large number of talents with a high education level, a high degree of internationalization of talents, a wide source of talents with a high inflow rate, hard-working talents, talents with innovative and entrepreneurial spirit, and strong mobility of talents. The talent gathering in Silicon Valley, from the analysis of its talent gathering mechanism, mainly includes the following reasons.

(i) Unique Silicon Valley culture. Silicon Valley culture first advocates and encourages individual struggle, advocates the talents gathered here to create wealth through their own wisdom and ability, and promotes the progress of human science and technology, which provides the deepest level of motivation for talents in Silicon Valley.
(ii) The spirit of Silicon Valley. The spirit of Silicon Valley is a series of excellent spiritual cultures formed in the process of continuous development and growth of Silicon Valley, including a series of spirits such as the spirit of innovation and adventure, the spirit of freedom, and the spirit of inclusiveness. The Silicon Valley spirit continues to evolve and expand today, maintaining its vitality and attracting and supporting innovative talent.
(iii) Talent gathering platform. Silicon Valley is the gathering place of tens of thousands of high-tech companies. The enterprises in Silicon Valley will adopt various recruitment methods, adopt various incentive measures, create a fair and competitive working environment, and pay attention to the training of employees, which provides an innovation platform for innovative talents and forms a scale effect of talent agglomeration.
(iv) Talent development environment. The US government has enacted many laws and policies to encourage the Silicon Valley system.

Exercise

The phenomenon of agglomeration and convergence in the process of regional development in developed areas is a problem worthy of in-depth discussion. Here are some possible questions to consider. These questions aim to guide thinkers to explore the agglomeration and convergence of developed regions in the process of regional development from multiple perspectives, so as to better understand this complex problem and propose corresponding countermeasures and suggestions.

1. Please briefly describe the meanings, connections, and differences between agglomeration and convergence in the process of development in developed regions.
2. Why is there agglomeration in the development process of development regions? What is the impact of this agglomeration on the development of developed regions?
3. Why do developed regions converge in the process of regional development? How does this convergence affect the development of developed regions?
4. What is the relationship between agglomeration and convergence in the process of development in developed regions? How can the balanced development of agglomeration and convergence be achieved?
5. In the process of regional development, how should China's developed regions cope with the challenges brought by agglomeration and convergence and achieve sustainable development?
6. What are the implications of the agglomeration and convergence of developed countries in the process of regional development in China? What experience

can our country use for reference to promote the balanced development of the developed areas?

7. Please analyze the actual situation of agglomeration and convergence of developed regions in the process of regional development and discuss their development strategies based on specific cases.
8. In view of the agglomeration and convergence of developed regions, what measures should the government and enterprises take to optimize the regional development pattern and promote the all-round development of economy and society?
9. In the context of globalization and informatization, how can developed regions take advantage of agglomeration and convergence to enhance regional competitiveness?
10. Looking into the future, how do you think the agglomeration and convergence of developed regions will evolve in the process of regional development? What trends are worth watching?

Extended Reading

The following are some suggestions for the extended reading of "Agglomeration and Convergence of regional development process in Developed Regions,". Through these extended readings, we can have a more comprehensive understanding of the agglomeration and convergence phenomenon in the process of regional development in developed regions and provide a rich theoretical basis and practical cases for in-depth research and discussion of this issue.

1. Regional development theory: We can start with classical economic geography theories, such as Weber's industrial location theory and Hoover's agglomeration economic theory, to understand the theoretical basis of agglomeration and convergence of developed regions.
2. Agglomeration economy: The research on agglomeration economy related literature, such as Marshall's industrial cluster theory, and Krugman's new economic geography, is helpful in deeply understanding the formation mechanism of agglomeration phenomenon in developed regions.
3. Regional integration: Read the literature on regional integration, such as the cases of the European Union and the North American Free Trade Area, to understand the performance of agglomeration and convergence in developed regions in the process of regional integration.
4. Globalization and global cities: Study the literature related to globalization and global cities, such as the theory of the world city system, and global city network, to understand the new characteristics of agglomeration and convergence of developed regions in the context of globalization.
5. Policy regulation: Read the literature on the impact of government policies on regional development, such as regional development strategy, industrial policy, and urban planning, and analyze how policies guide developed regions to achieve a balance between agglomeration and convergence.

6. Sustainable development: Study the theory of sustainable development to understand how to achieve the sustainable development of agglomeration and convergence in developed areas to avoid excessive agglomeration of resources and environmental problems.
7. Empirical research: Read the empirical research literature on development in developed regions, such as the case studies of Silicon Valley in the United States, the Yangtze River Delta in China, and the Pearl River Delta, to understand the actual situation and development strategy of agglomeration and convergence in these regions.
8. Interdisciplinary research: Read literature involving economics, geography, sociology, environmental science, and other disciplines to comprehensively understand the multiple dimensions of agglomeration and convergence in developed regions.

References

Adam, S. (1776). *An inquiry into the nature and causes of the wealth of nations.* China Financial Publishing House.

Akihiro, O. (2023). Impacts of enhancing regional network economies on regional productivity and productive efficiency in Japan: Evaluation from stochastic frontier analysis. *Asia-Pacific Journal of Regional Science, 8*(1), 25–43.

Bain, J. S. (1959). *Industry organization.* Harvard University Press.

Burgess, E. W. (2015). *The growth of the city: An introduction to a research project.* Urban Ecology.

Couch, C., & Karecha, J. (2006). Controlling urban sprawl: Some experiences from Liverpool. *Cities, 23*(5), 353–363.

David, R. (1817). *On the principles of political economy and taxation.* Huaxia Publishing House.

Clarke, C.G., Dickenson, D.J.P., Dickenson, J.P., Gould, W.T.S., Mather, S., Mather, S., Prothero, P.R.M., Prothero, R.M., Siddle, D.J., Smith, C.T., Smith, M.C.T., & Thomas-Hope, E. (1996). A Geography of the Third World (2nd ed.). Routledge.

Dobrescu, E. M., & Dobre, E. M. (2014). Theories regarding the role of the growth poles in the economic integration. *Procedia Economics and Finance, 8*, 262–267.

Ellison, G., & Glaeser, E. L. (1997). Geographic concentration in U.S. manufacturing industries: A dartboard approach. *Journal of Political Economy, 105*(5), 889–927.

Ficcadenti, V., Cerqueti, R., & Varde'I, C. H. (2022). A rank-size approach to analyse soccer competitions and teams: The case of the Italian football league Serie A. (1):85–113

Guo, F., Kong, S. T., & Wang, J. (2016). General patterns and regional disparity of internet finance development in China: Evidence from the Peking university internet finance development index. *China Economic Journal, 9*(3), 253–271.

Haggett, P., Cliff, A. D., & Frey, A. (1977). *Locational analysis in human geography.* Edward Arnold.

Harris, C. D., & Ullman, E. L. (1945). The nature of cities. *The Annals of the American Academy of Political and Social Science, 242*(1), 7–17.

Herfindahl, O. (1950). *Concentration in the steel industry.* Columbia University.

Hoyt, H. (1939). *The structure and growth of residential neighbourhoods in American cities.* Government Printing Office.

Krugman, P. (1991). Increasing returns and economic geography. *The Journal of Political Economy, 99*(3), 483–499.

Mueller, D. P. (1981). The current status of urban-rural differences in psychiatric disorder. An emerging trend for depression. *Journal of Nervous and Mental Disease, 169*(1), 18.
Ohlin, H. (1993). *Interregional and international trade.* Harvard University Press.
Palmberg, J. (2012). Spatial concentration in the financial industry. *Ratio Working Papers, 16*(7), 849–854.
Richardson, H. W. (1978). The state of regional economics. *International Regional Science Review,* (3), 1–48.
Rossoni, L., & Rosa, R. A. (2024). Reducing the Matthew effect on journal citations through an inclusive indexing logic: The Brazilian spell (Scientific Periodicals Electronic Library) experience. *Publications, 12*(1), 5.

Chapter 6
Coordinated Regional Development in China

Abstract Regional development factors, quality, model, and system differences contribute to regional development imbalance and development differences. In the process of development, both efficiency and fairness should be taken into account. China's development and core proposition of development geography research in the new era are aimed at ensuring quality and coordinated development among regions. The comprehensive analysis method of factors in geographical research is an integrated analysis of the differences in regional development factors, which makes it possible to achieve coordinated regional development in eastern, central, and western China under the guidance of the government's macro-policy and the market regulation mechanism, and finally realize the convergence of China's economic, social, and ecological environmental development. The evolution of regional development is the concrete reflection of regional economic, political, social, and cultural evolution, and also the further interpretation of a regional development mechanism. The planning and optimization of territorial space in the new era is the basis of the national macro-control and the embodiment of the national development concept, which not only requires the balance of inter-regional development but also emphasizes the resilience of the regional system, that is, to achieve green and sustainable development and take the path of ecological civilization development. The objectives of this chapter are as follows: (1) To explore the evolution of regional development as a reflection of economic, political, social, and cultural dynamics and elucidate regional development mechanisms. (2) To underscore the significance of spatial planning and optimization in contemporary territorial governance, which are essential for national macro-control and embody sustainable development principles.

6.1 Coordinated Regional Development

The evolution of China's development status at different stages shows that the human understanding of development is also constantly advancing. Through the in-depth interpretation of China's coordinated regional development thought and further analysis of its development strategy, we can have a glimpse of the exploration process of

X. Deng and M. Song, *Development Geography*,
https://doi.org/10.1007/978-981-96-6917-2_6

China's construction of socialist development theory with Chinese characteristics, and also provide an effective way for future design of coordinated regional development path. The development ideas at different stages determine the country's resource allocation, capital and technology investment, fiscal and tax policies, and opening up policies. Based on an analysis of the spatial differences of regional development factors, development geography actively explores the effective path of regional development convergence.

6.1.1 Coordinated Regional Development Since the 1950s

In the early days of the founding of New China, China is in the primary stage of socialism with a relatively low level of comprehensive development. The industrial foundation is relatively weak, the level of scientific and technological development is low, and the regional economic development is uncoordinated. To improve the comprehensive development level of the country, the Party Central Committee has studied and grasped the basic connotation of socialist production distribution theory, which originates in the Soviet Union and takes "coordinated distribution theory" and "determinism of production relations" as the core. The Party Central Committee proposed to gradually shift the economic center of gravity, vigorously promote the economic development of the central and western regions, and developed a regional economic development strategy of "strategy and strategy westward." Specifically, to balance the distribution of industrial development across the country, while encouraging industrial development in coastal areas, it is also necessary to comprehensively promote industrial development in the mainland. To break the inherent layout of productive forces and speed up China's national defense construction, taking into account the international political and domestic economic pattern, the state has implemented a "big push" strategy for the relatively less developed southwest and northwest regions of our country, increasing investment and promoting preferential policies at the same time to speed up the pace of internal economic construction, especially industrial construction.

The coordinated regional development spans the period from the "First Five-Year Plan" to the "Fifth Five-Year Plan." Under the guidance of the idea of coordinated regional economic development, the state began to implement the strategy of coordinated development. The main aim of the First Five-Year Plan is to vigorously promote industrial construction with the help of the Soviet Union and encourage the development of the handicraft industry and the establishment of agricultural production cooperatives in various regions to promote the progress of local agriculture and the socialist transformation of agriculture, handicraft industry, and individual industry and commerce. After that, the state began to implement the "Second Five-Year Plan," still focusing on industrial construction, with heavy industry as the center, based on the existing collective ownership by all people, the main development of industry, agriculture, and handicrafts, and further develop the corresponding transportation and commerce. In the 60 years since the founding of New China, due to the rapid development of economy and society and the constant changes of the national situation, the basic framework of China's economic division should be further explored. China's economic regions are basically divided into eastern, central, western, and northeastern regions. During the "Third Five-Year Plan, the state made a comprehensive adjustment to the three-tier layout of the country, classifying the eastern coastal areas with relatively developed industrial technology and the Sino-Soviet border areas as first- tier regions and some central and western areas with relatively low economic development as third- tier regions. For example, Guizhou, Shaanxi and Gansu, Western Shanxi, Western Henan, and Western Hunan, and the rest of the country were classified as second-tier regions. The mid-1960s entered the Third Five-Year Plan; Sino-Soviet relations deteriorated; the surrounding situation changed significantly; and the Party Central Committee began to adjust the development tiers, actively develop agriculture, and solve the problem of food and clothing of the people first. At the same time the development of national defense and other basic industries was strengthened to ensure that the national economic construction has a focus, ensure proportional concentration in the third-tier areas, promote the economic development of the third-tier areas, thus forming spatial strategy transfer. In the 1970s, with the gradual improvement of foreign relations and the "Fourth Five-Year Plan" to ""Fifth Five-Year Plan" period, the investment intensity of the big three tiers slowed down while ensuring the inland strategy and the construction of the rear, increasing investment in coastal provinces, and the introduction of new projects mainly in the eastern and central regions. At this time, China's regional development strategy began to focus on coastal areas, the economic center of gravity began to shift, and regional economy began to gradually shift to non-equilibrium.

The "Western promotion" strategy adopted in the period of coordinated regional development made great achievements under the national conditions at that time, which solved the need for the vigorous development of industry and agriculture, further narrowed the economic gap between the eastern and western regions, alleviated the extremely uncoordinated development of regional economy in the early days of the founding of China, and laid a stable foundation for the future development of the central and western regions. However, during the period of coordinated regional development, attention was paid to the synchronous and stable development of all regions, and the differences in basic conditions and local development advantages were not fully taken into account, while the internal economic relations between the two and the region were ignored. Due to their weak conditions the return on investment of the central and western regions is lower than that of the eastern coastal areas. Although the egalitarian development route has narrowed the regional development differences on the surface, it ultimately makes the eastern coastal areas miss the development opportunity. For example, in 1955, Shanghai and Tianjin received significantly less investment than before, only 76% and 108% of the depreciation cost in the same year, which delayed the development of the local economy, failed to improve the economic and industrial level timeously, and missed the development opportunity. During the period of coordinated regional development, the development process of China's national economy is slow and the overall economic efficiency is low, which eventually makes the western region gradually lose the impetus for further development.

6.1.2 Uncoordinated Regional Development Since the 1970s

At the end of the 1970s, in the context of the international environment in which China's geopolitical and military situation relatively slowed down, the economic pattern of neighboring countries was upgraded and the economic growth rate was significantly increased, China's disadvantage in the international market competition became increasingly prominent, the industrial structural adjustment became increasingly urgent, and the regional economic development strategy also urgently needed to change. The second-generation leaders of the Party creatively devised the idea of uncoordinated regional development, and China officially began to implement the uncoordinated regional development strategy, which accelerated the speed of the country's economic development. The strategy was to develop rich areas faster to promote the development of slower areas and accelerate the speed of the country's economic development while promoting industrial development. This may be the first shortcut to achieving common prosperity in China. During the period of uncoordinated development, China's eastern region has set up five special economic zones, 14 economic zones, and coastal port central cities, as well as five development zones and coastal economic and technological opening up pilot zones throughout the country, forming a relatively open coastal zone. The country's strategy is also to increase the amount of investment in the eastern region and provide policy support, break the equalitarianism in the stage of coordinated development, and promote the rapid development of the eastern region first, thus driving the development of the entire national economy. The period of uncoordinated regional development mainly includes two stages, namely the "Sixth Five-Year Plan" and the "Seventh Five-Year Plan" periods. The "Sixth Five-Year Plan" stipulates that China should make full use of the location advantages of the coastal areas to promote the rapid economic development of those areas, but also make full use of its own technical conditions, management advantages, and scientific research capabilities; actively develop international trade through the energy advantages of the central and western regions to slow down the transportation and energy problems in the eastern region; and prioritize the rapid development of the coastal areas, thus driving the development of the mainland. During this period, the special economic zone has played a positive role in promoting the development of the eastern region of China, which is reflected in the fact that it is far ahead in technology, management, knowledge, and foreign policy, which have had a profound impact on the formulation of China's uncoordinated regional development strategy and promoted the continuous improvement of the uncoordinated strategy. In 1986, China officially began to implement the "Seventh Five-Year Plan," which still focused on the economic development of the coastal areas, hoping that it would play a leading role in the economic development of the mainland, and for the first time, China was uniformly divided into eastern, central, and western regions based on the differences in the development level of the regions. Meanwhile, it was proposed that the development relationship among the three economic belts should be properly handled. In other words, the eastern region should give priority to and focus on development, and the central region, which is rich in energy and raw materials, should develop its advantageous industries. The harmonious and stable development of the eastern and central regions cannot be separated from the strategic support of the western region and should follow the development priority sequence of the eastern, central, and western regions, that is, from the high level of development to the low level of development. As shown in Table 6.1, the proportion of investment in infrastructure construction in the eastern part of China has gradually increased.

The strategy of uncoordinated regional development has made remarkable achievements. By shifting the focus of development to the eastern region, the speed of China's economic and industrial development has been greatly accelerated. The first achievement is the rapid economic development in the eastern region, the continuous implementation of the reform and opening up strategy, which have achieved remarkable results, effectively promoting the establishment of the socialist market economic system. Second, with the economic development of the eastern region, many people in the central and western regions actively work in the east, which indirectly solves the employment problem in some areas, fundamentally promotes the overall economic growth of China, and effectively improves economic efficiency. Since 1978, the proportion of GRP (Gross Regional Product) in the eastern region of China (GRP share) has been increasing annually, particularly after 1990, increasing to 55.7% in 2010, an increase of 8.3%, and the population share is lower than the overall GRP share (Fig. 6.1). Third, the establishment of special economic zones and the opening up of cities in the eastern region have effectively promoted economic development, boosted the region's economic cooperation with the central and western regions, provided valuable experience, played a leading and exemplary role, and thus promoted the development of the central and western regions. Fourth, it has greatly alleviated the imbalance in China's industrial structure, made local economies develop flexibly, comprehensively promoted the diversified development of market players, and established a unique market system.

Table 6.1 Proportion of investment in infrastructure construction in eastern, Central, and western China (1978–1995)

Year	Eastern region	Central region	Western region	Ratio
1978	40.1	30.6	20.4	1.97:1.50:1
1980	44.5	29.5	20.3	2.19:1.45:1
1985	48.4	28.2	16.8	2.88:1.68:1
1990	50.9	23.2	16.8	3.03:1.38:1
1995	54.9	22.7	15.2	3.61:1.49:1

Data sources China fixed asset investment statistical yearbook 1950–1995; compilation of statistical data of new China in the past 60 years

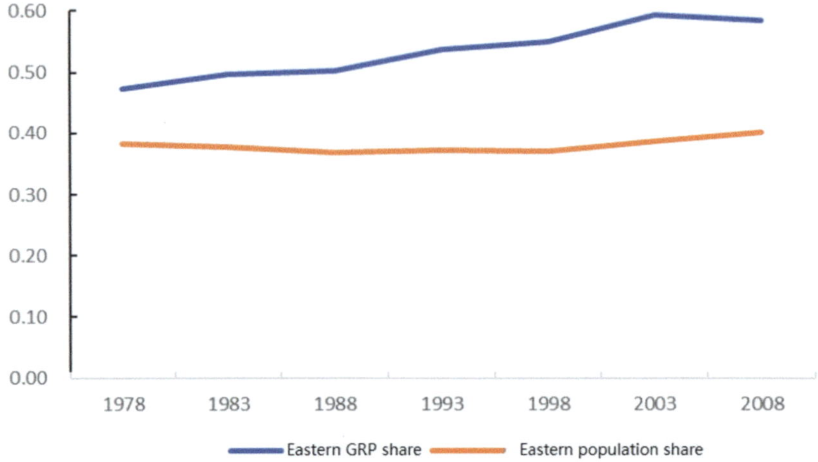

Fig. 6.1 Evolution of GRP in China from 1978 to 2010. *Data source* China Statistical Yearbook (2010)

The uncoordinated regional development strategy puts more emphasis on efficiency. For the developing countries in the primary stage of economic development, implementing the uncoordinated development strategy is more appropriate and has practical guiding significance for their future development. By implementing this strategy, China's economy has gradually developed and has formed a multilevel, focused, and all-round uncoordinated development pattern. However, with the prosperity and development of the eastern region, regional imbalance is gradually reflected, and the bottleneck of coordinated regional economic development is prominent.

It took approximately seven years (1979–1986) from the formation of the first decision to the initial construction of the Shenzhen Special Zone, the development of which was difficult. In the late 1970s, many developed countries began to save labor, land, and other aspects of investment, focusing on improving their own scientific and technological level. After experiencing 10 years of "Cultural Revolution," China urgently needs to change its backward face.

Since the establishment of the Shenzhen Special Economic Zone, it has focused on the construction of infrastructure and carried out many reforms to promote rapid economic development. However, there are conflicts with the original economic system, and the construction of the special economic zone is faced with severe challenges. In the early stage of reform, the Shenzhen Special Economic Zone focused on market development and took it as a breakthrough. First, when the SAR (special administrative region) was first established, the Government focused on the reform of the infrastructure system, and relevant departments worked together to actively invite tenders and cooperate with contractors. Second, the Government reformed the price system. At the beginning of the establishment of the special zone, the implementation of the allocation plan encountered serious obstacles, and the construction of the special zone was relatively slow. In 1982, the SAR began to reform the price system and fully liberalized commodity prices. Two years after the implementation of the reform, considerable results were achieved. Third, the country reformed the wage system. The "iron rice bowl" was completely abolished, and enterprises and public institutions began to widely implement the wage system. The implementation of this system mobilized the enthusiasm of the staff and played a model role in the wage system in other regions of the country, and enterprises in other regions gradually began to adopt the system. Fourth, China reformed the employment system. A large number of entrepreneurs set up enterprises here, but at that time there was a contradiction between the labor and employment system implemented in Shenzhen and foreign businesspeople, and the reform of the labor and employment system was imperative. Shenzhen began to abandon the "iron rice bowl" system, and gradually implemented a two-way choice mechanism between enterprises and employees, gradually improving the employment system by making the employment problem market-oriented and ensuring smooth reform. Since then, the employment system of the Shenzhen Special Zone has further implemented the labor contract system, which has met the employment needs of most enterprises with diverse ownership. The Shenzhen Special Zone focuses on basic construction, the investment environment has been preliminarily formed, and the city has begun to take shape.

In 1985–1986, the Shenzhen Municipal Party Committee proposed to comprehensively start the reform. First, many state-owned enterprises in Shenzhen began to implement shareholding and a new state-owned assets management system. Second, the Shenzhen Special Economic Zone began to reform its financial system and further promote the multi-level and open development of its financial market. Third, the SAR proposed paid land use for the first time, and the relevant departments began to develop a management system, gradually implement it, and auction land publicly. Fourth, the SAR began to promote the civil service system. Fifth, the housing system of the special zone would be comprehensively improved and the commercialization of housing gradually realized. Sixth, the development of private enterprises would be vigorously promoted.

Since 1992, the economy of the Shenzhen Special Zone has developed rapidly and has the legislative power, the speed of urban construction has been greatly accelerated, and the full implementation of the reform and opening up has made export-oriented industries become the "leader" of development.

6.1.3 A Period of Coordinated Regional Development Since the 1990s

Since the 1990s, China's regional development has taken on new characteristics. First, all regions in the country still emphasize the principle of efficiency and focus on both efficiency and equity in policy formulation. In the 1990s, China began to implement the "Eighth Five-Year Plan" and entered a new period of regional development. With the guiding principles of "overall planning, rational division of labor, complementary advantages, coordinated development, taking into account interests and common prosperity," China began to adjust the economic strategy of prioritizing the development of the eastern region in the 1980s, and put the economic development of the central and western regions on its national agenda. Second, China endeavored to open its doors wider to the outside world to actively attract foreign investment and expand investment areas. The eastern coastal areas were to remain open to the outside world, and efforts were intensified to promote the economic development of the coastal areas through the establishment of economic development zones while driving the economic development of the central and western regions. In the western region, the country aimed to focus on the implementation of the "opening up along the border" strategy, actively carry out foreign trade and economic cooperation with neighboring countries, make full use of its own regional advantages to promote the economic development of the western region, and further promote the coordinated development of the entire country. Third, China aimed to focus on the economic development of the central and western regions by increasing the construction of basic industries and infrastructure. Fourth, the country aimed to encourage the development of township enterprises, accelerate the "87 Poverty Alleviation Plan," promote economic development in poor areas, and comprehensively promote poverty alleviation.

Uncoordinated regional development drives China's economic development to a certain extent; however, its defects are gradually revealed, and the goal orientation of both efficiency and fairness is gradually formed in the process of economic and social development. In 1999, China began to implement the western development strategy and actively promote the economic development of the central and western regions. The Party Central Committee pointed out that it should focus on supporting the economic development of 12 regions such as Chongqing and Sichuan. Jiang proposed that the development of the central and western regions should be promoted holistically, and these regions should also actively cater to the needs of the market economy vigorously develop the advantageous industries in each region, turn the resource conditions into an economic driving force, and continue to promote reform and opening up. Jiang further pointed out that the central and western regions should pay attention to infrastructure construction. While making full use of the local rich natural resources, the local government should also provide strong support and speed up the establishment of a more standardized financial transfer payment system to promote the economic development of the western region. Since China began to implement the western development strategy, the speed of economic development in the western region has accelerated significantly, and the strategy has made great progress. In 2000, the ratio of the present value of state-owned industrial fixed assets to GDP in the western region of China was higher than that in the eastern and central regions (Fig. 6.2). Local infrastructure construction has been carried out holistically and accelerated, and the local economy and people's living standards have been greatly improved. With the completion of projects such as the Qinghai-Tibet Railway and the West–East gas pipeline, the economy of the western region will develop at a faster pace. While promoting economic development, the western region has not ignored environmental protection. The state has stepped up efforts in ecological restoration and environmental protection, and the government has issued corresponding policies to support and achieve significant results. For example, the project of returning farmland to forest and grassland in the western region has played a good role in protecting the ecological environment. Notable achievements have also been made in projects such as the protection of natural forests and the comprehensive management of sandstorms.

After the implementation of the "Ninth Five-Year Plan," the traffic arteries and central cities of various provinces and regions in China have actively led the coordinated and stable development of the inter-regional economy and formed seven inter-provincial economic regions. During the implementation of the "Tenth Five-Year Plan," the government has actively introduced many preferential policies, particularly in the areas of opening up, taxation, land, and talents, increasing financial transfer payments to the western region and investment in urban construction, and the local government has provided corresponding policy support. After the implementation of the Tenth Five-Year Plan, it was further emphasized that the current economic development of the eastern, central, and western regions should be fully considered, and local conditions should be adapted. The eastern region should continue to open up to the outside world to promote rapid economic development, and areas with advanced economic development should realize modernization ahead of schedule and play a leading role. The central region makes full use of its geographical advantages and rich comprehensive resources to promote rapid economic development. In 2003, the regional GDP of Jiangxi Province reached 283.046 billion yuan (Table 6.2). The

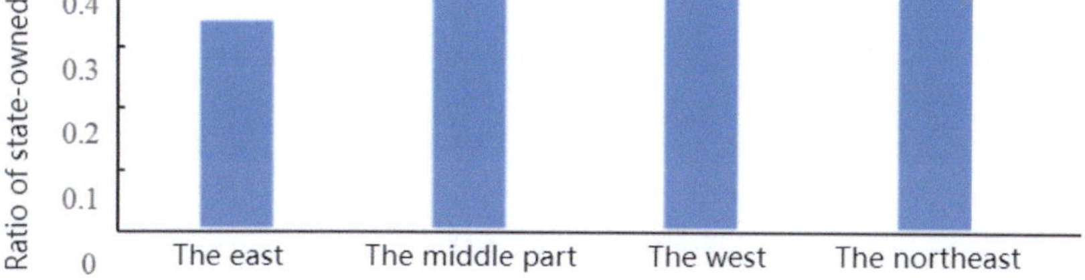

Fig. 6.2 Ratio of present value of fixed assets of China's regional state-owned holding industry to GDP in 2000. *Data source* China Statistical Yearbook (2001)

western region should proceed from its own reality, make overall planning and scientific development, improve the investment environment, actively attract the introduction of projects, and participate in the development and construction of the western region.

Kunming actively participates in the development of the western region, accelerates its economic development, and strives to build a modern ecological city. In addition, as a national comprehensive pilot city to improve transportation and environmental protection, Kunming firmly grasps the development opportunity, develops infrastructure, pays attention to environmental protection, and actively promotes sustainable development. The Government aims to aims to vigorously improve its level of science and technology, accelerate the implementation of the policy of opening to the outside world, and actively introduce external investment, accelerate the development of science and education and the urbanization process, and strongly promote the rapid economic development of Kunming city. Simultaneously, the Government aims to focus on accelerating infrastructure to lay the foundation for Kunming's economic development, pay attention to environmental protection, cultivate the awareness of ecological environmental protection of all people, accelerate the construction of environmentally friendly pilot cities, and accelerate the construction of related environmental protection projects, including the Kunming Miyuhe urban drinking water supply project and the Dianchi Lake ecological protection construction project. It also aims to accelerate the adjustment of industrial structure and promote the rapid development of Kunming's economy, develop science and technology education, and raise the level of science and technology holistically. Kunming takes advantage of the development opportunity of the western development, vigorously develops the economy, and actively builds the pilot city of environmental protection. In recent years, Kunming has continued to implement the strategy of sustainable development and strived to build the city into a forest ecological city, which not only protects the ecological environment but also promotes the city's economic development, cultivates the public's concept of environmental protection and ecological awareness, improves the basic literacy of the public, promotes the rational development and utilization of resources, and promotes the coordinated

Table 6.2 Main indicators of economic and social development of the five central provinces in 2003

Index	Jiangxi	Hunan	Hubei	Henan	Anhui
Total population at year end (10,000)	4254	6663	6002	9667	6410
Gross regional product of five central provinces (100 million yuan)	2830.46	4638.73	5401.71	7048.59	3972.38
GDP per capita (Yuan)	6678	7554	9011	7570	6455
Per capita fiscal revenue (Yuan)	395.31	403.19	432.80	349.70	344.38
Government expenditure (100 million yuan)	382.0981	573.7453	540.4356	716.5978	507.4398

Data source China Statistical Yearbook (2004)

development of humanity and nature. Kunming has not only achieved its own development goals and vigorously promoted ecological environmental protection but also played a leading role in establishing a good ecological city image. The scientific and reasonable development of Yilinze agriculture and Yilinze forest has resulted in Kunming forming a unique ecological greening system.

Based on the above, Kunming has formulated the main tasks and development goals during the "Tenth Five-Year Plan" period, speeding up the infrastructure construction of water, roads, trees, and gas and building a characteristic ecological greening system. To speed up the adjustment of industrial structure, the relevant government departments should identify some key projects and accelerate the pace of implementation and actively promote the implementation of the western regional development strategy. A project bank of the Tenth Five-Year Plan will be established to speed up the economic development of Kunming (while paying attention to environmental protection) and strive to build it into a national garden city within three years.

The coordinated regional development strategy has achieved remarkable results, which has effectively enhanced China's comprehensive national strength and the overall economic strength of the country. The western region's development strategy has also made gratifying achievements. While the economic and technological level of the western region has greatly improved, it has not affected the economic development of the eastern region, which is still in a good state of economic development and maintains a positive momentum.

6.1.4 Regional Overall Development Since the Twenty-First Century

As a cross-century project, the western region's development strategy has achieved outstanding results, which are reflected in the speed of economic development and the improvement of the ecological environment. However, due to the relatively poor natural conditions in the western region, low economic level, and weak infrastructure, the conversion rate of funds invested in the region is low, medium and long-term projects cannot produce benefits in the short term, which is bound to conflict with the relative scarcity of national funds. In addition, the status of northeast and central regions in the overall pattern of the country has also declined, and the development strategy of the western region needs to be adjusted. In 2003, the Third Plenary Session of the 16th Central Committee proposed five overall plans for the future development direction of China, namely the overall urban and rural development, overall regional development, overall economic and social development, overall harmonious development of man and nature, overall domestic development, and economic development strategy of opening up. On this basis, the Central Committee of the Party pointed out that under the guidance of the scientific outlook on development, the overall development of the regional economy should be fully realized, and proposed that in the future, China should fully implement the strategy of developing the western region, pay attention to the region's economic development and infrastructure construction, and promote the economic growth of the region. It is also necessary to actively promote the comprehensive revitalization and orderly development of the old industrial base in Northeast China, continue to promote the rise of the central region while continuing to prioritize the development of the eastern region, and comprehensively promote the overall regional development. Regional coordination and interaction mechanisms should be established and improved to comprehensively promote coordinated regional development. From 2003 to 2010, the economic concentration index of the central region of Jilin Province generally showed a downward trend, and the inter-county economy tended to be coordinated (Table 6.3). The population concentration index increases every year; however, its value is low, indicating that the population presents an insignificant inter-county flow phenomenon. Regional overall development further deepens coordinated regional development, which not only refers to its basic content but also refers to its basic connotation.

According to the level of regional economic development, that is, geographical location, China's economic area is basically divided into four major zones. In the outline of the 11th Five-Year Plan, according to this division, the macro-strategic deployment of China's future regional economic development is carried out. It is pointed out that in the future, the eastern region will still be encouraged to take the lead in economic development, promote further economic development in the eastern region while making improvements in independent innovation, establishing and improving market mechanisms, and driving the economic development of the central and western regions while developing itself. The western region should speed up the pace of reform and opening to the outside world, pay more attention to infrastructure construction and natural environment protection, and actively develop local advantageous industries to effectively improve the local development capacity and economic performance. Northeast China should pay attention to the development of modern agriculture and promote the adjustment of industrial structure. There are many state-owned enterprises in Northeast China, which occupy the main market. It

Table 6.3 Regional equilibrium index analysis of the central urban agglomeration of Jilin Province from 2003 to 2009

Year	Economic concentration index	Population concentration index	Degree of concentration and dispersion
2003	42.8	25.5	60.2
2004	42.4	25.6	58.2
2005	40.8	26.2	58.2
2006	40.1	26.5	58.3
2007	39	26.7	58.3
2008	39.4	26.7	58.3
2009	38.7	26.7	58.7

Data source Li et al.

is necessary to actively promote the transformation of existing enterprises, make full use of local natural resources, develop advantageous industries, strive to promote economic transformation and upgrading, comprehensively improve the economic level of Northeast China, and realize the revitalization of old industrial bases. To improve its economic level, the central region should promote grain production, strive to build major grain-producing areas, utilize its own energy and manufacturing advantages, develop local advantageous industries, further increase local infrastructure construction, increase market development efforts, and build a modern market system to play the role of connecting the east with the west and promote economic development at the same time.

6.1.4.1 Northeast Revitalization: Resource-Based Cities Get Rid of Difficulties, and Deep-Seated Problems Have Not Been Fundamentally Solved

In its 16th National Congress report, the Party Central Committee stressed the necessity of promoting the adjustment and basic transformation of the industrial structure in the northeast region and encourage the development of alternative industries in resource-based cities. In 2003, the State Council proposed a series of measures for revitalizing Northeast China, made corresponding strategic arrangements, and achieved significant results. The economic structure of Northeast China has been optimized accordingly, which not only improves the independent innovation ability of local enterprises but also provides employment opportunities for shanty towns and some laid-off workers by promoting the development of enterprises, effectively increases the employment rate, thus further solving the problems of people's livelihood and improving the level of opening up. However, compared with the eastern region, the level of economic development in Northeast China is still low. In the future, it is necessary to further enhance its innovation ability, optimize the market environment, promote the development of enterprises, optimize the industrial structure, and attract resident talents to effectively solve the problem of brain drain.

6.1.4.2 The Rise of the Central Region: Fully Taping the Development Potential of the Central Region and Vigorously Developing the Economy

In 2004, Premier Wen Jiabao proposed to fully stimulate the development potential of the central region and vigorously promote the economic development of the region. The State Council has proposed measures for the rise of central China and made corresponding strategic deployment. In 2006, The State Council pointed out that in the future, we should promote the economic growth of the central region, give full play to its advantages in natural resources, build and develop the central region into a key strategic grain production and processing base, a major import and export base of clean energy and raw materials, further promote the rapid development of a modern equipment manufacturing industry, and focus on the development of high-tech industrial bases. It is also necessary to pay attention to the development of transportation industry and build large-scale comprehensive transportation hubs. Specifically, the CPC Central Committee has made detailed arrangements for improving the central region, and it is necessary to learn from the relevant experience of the revitalization of the northeast and the development of the west. Twenty-six cities are learning from the successful experience of the revitalization of the northeast while referring to the implementation of relevant policies, and 243 counties (cities and districts) are learning from the successful experience of the development of the west while referring to the implementation of relevant policies. The economic level of the central region has greatly improved, and the strategy for improving the central region has achieved remarkable results. The specific performance is as follows: First, the central provinces began to actively introduce investment and the scale of investment continued to increase, resulting in the rapid economic development of the central provinces, and the economic growth rate is higher than that of other regions. Since then, the state has paid more attention to the economic development of the central region, and the local government has actively introduced new investments, improved and promoted the investment environment, and gradually begun to use foreign capital, and the scale has been expanding. Second, the rapid economic growth of the central region has further shortened the gap with the eastern region. The central region continues to introduce new investment to increase investment, so that its economic growth rate even exceeds that of the eastern region.

The strategy of the eastern region taking the lead in development has achieved remarkable results, not only the rapid economic growth of the eastern region but also effectively driving the economic growth of other regions. Since the period of uncoordinated regional development, the rapid economic development of the eastern region, the last year of the implementation of the "Eighth Five-Year Plan," the eastern region accounted for a relatively high GDP in China, reaching 62%, and during the implementation of the entire "Eighth Five-Year Plan," its average growth rate has been at the domestic leading level, reaching 16%. Since the implementation of the strategy of taking the lead, the economic growth rate of the eastern region has maintained a high level, thus driving the economic development of other regions, resulting in a period of rapid development in the entire Chinese economy, effectively improving the level of economic development in China, and accelerating the pace of reform and opening up. The eastern region has developed rapidly, and its economic and industrial structure has been further optimized, which has effectively promoted the transformation and upgrading of the industry. In the future, the eastern region should achieve rapid economic growth, promote the rational use of resources and environmentally friendly development, create a green and open economic system, and promote economic, social, ecological, and government transformation while promoting economic development.

The eastern region itself has developed at a relatively fast pace and has built an open economic system in an all-round way, further accelerating economic development. The region also actively participates in international cooperation, fully integrates with the "Belt and Road" project, and enhances its international influence while developing the economy. To quickly and effectively improve the innovation capability of the eastern region, the government has introduced investment and gradually increased the investment scale of industrial research and development (R&D). The innovation-driven development momentum of the eastern region is gradually becoming prominent, and innovation will evolve into a new advantage and a new engine for the region's development, and its economic development speed is still relatively fast. Since then, China has entered a period of coordinated development of the four major sectors, and China's per capita GDP began to increase annually from 2000 to 2010 (Table 6.4).

Using the strategy of coordinated regional development, a pattern of coordinated development in the eastern, northeastern, central, and western regions has gradually taken shape, and China's economy and society have developed steadily and in an orderly manner.

As one of the six rising provinces in central China, Jiangxi has developed rapidly. During the Tenth Five-Year Plan period in Jiangxi, relevant departments actively promoted local economic development, vigorously implemented the strategy of revitalizing Jiangxi through science and education and strengthening the province with talents, introduced a large number of scientific and technical talents, and effectively promoted scientific and technological innovation and progress. Scientific and technological progress has promoted the rapid economic development of Jiangxi Province, and its contribution rate has reached 45.6%, which has exceeded the national average, and has made great progress compared with that of 2000, about 5.8%. In 2002, Jiangxi Province actively introduced foreign capital to promote economic development, the actual utilization of foreign capital in the province reached 540 million US dollars,

Table 6.4 China's regional economic development (2000–2010)

Year	National per capita GDP (yuan/person)		Western per capita GDP (yuan/person)	
	The highest	Minimum maximum error rate (times)	The highest	Minimum maximum error rate (times)
2000	34,547	266,212.98	7470	26,622.81
2005	51,474	505,210.19	16,331	50,523.23
2010	73,297	132,215.54	47,174	132,213.57
Year	Per capita GDP (yuan/person in Central China)		Northeast per capita GDP (yuan/person)	
	The highest	Minimum maximum error rate (times)	The highest	Minimum maximum error rate (times)
2000	7188	48,511.48	11,226	85,621.31
2005	12,495	86,751.44	18,983	133,481.42
2010	27,615	206,111.34	41,782	267,151.56

a large increase compared with that of 2001, and the total amount of the actual utilization of foreign capital in the first half of 2003 jumped to the first place in the central region. During this period, the embryonic form of industrial agglomeration began to emerge, various industries in Jiangxi Province developed rapidly, and the economy grew rapidly; for example, Xinjiang-Baosteel Industrial Park and Xinghuo Silicone Industrial Park. The emergence of a large number of industrial clusters will comprehensively and effectively promote the economic development of Jiangxi Province, which is a key factor in its rising strategy in central China. Jiangxi Province's economy is developing rapidly; however, its economic development level is still at a low level in the central part of the country, and there are still some shortcomings in economic development, and the scientific and technological innovation ability needs to be further improved.

The strategic focus and breakthrough of promoting scientific and technological progress in Jiangxi lies in implementing the science and technology for all, public relations with great achievements, training leading talents, and opening up strategies. Specifically, it is necessary to speed up the reform of the scientific and technological system and provide institutional guarantees for scientific and technological progress. China should also actively introduce scientific and technological talents, promote the industrialization of scientific research achievements, accelerate scientific and technological progress and innovation, vigorously promote the optimal allocation of resources, rational use of local resources, the development of characteristic industries, and the construction of five industries for industry-university-research strategic alliance. With the rise of central China, Jiangxi Province vigorously promotes scientific and technological innovation and strives to make breakthroughs in key areas and related advantages of Jiangxi Province. Based on existing science and technology, the relevant incentive mechanism of science and technology innovation should be fully established to promote science and technology progress and innovation in Jiangxi Province. It is also necessary to pay attention to improving the incentive mechanism of scientific and technological innovation, increase the enthusiasm of enterprises to participate, comprehensively promote scientific and technological progress, comprehensively create a new atmosphere of scientific and technological innovation, stimulate the power of scientific and technological innovation in the province, and fully recognize the role of scientific and technological talents. In addition, China should actively promote the innovation and development mechanism integrating digestion and absorption re-innovation, integrated innovation, and original innovation and fully tap Jiangxi's advantageous industries to provide the source of Jiangxi's scientific and technological progress, aim to achieve the international leading level. Original innovation, integrated innovation, and the progress of science and technology should be the focus.

6.1.5 Coordinated Regional Development Since 2020

The diversified strategy of promoting coordinated regional economic development proposed by the 20th National Congress covers many aspects, such as building a high-quality regional economic layout with complementary advantages, improving the inter-regional flow mechanism of production factors, deepening inter-regional cooperation to promote the orderly flow of factors, and implementing the "multi-point and multi-polar" regional economic development strategy. These strategies aim to promote the coordinated development of China's regional economy through various means and measures and solve the long-term formation of economic development differences, uncoordinated social development, cultural development imbalance, and other problems. China should build a high-quality regional economic layout with complementary advantages; break through the constraints of geographical nature through transportation, population-industry agglomeration, and informatization; crack the problem of "Hu Huanyong line," and establish a modern regional system. This involves breaking the phenomenon of "administrative region economy," constructing the spatial layout structure of the "central city—urban agglomeration—economic belt (economic zone)," and adjusting the exercise method of industrial policy. China's regional economic development is in good condition from 2011 to 2022, and the national per capita GDP will reach the highest of 85,310 yuan/person in 2022 (Table 6.5).

Table 6.5 China's regional economic development from 2011 to 2016

Year	National per capita GDP (yuan/person)		Western per capita GDP (yuan/person)	
	The highest	Minimum maximum difference	The highest	Minimum maximum difference
2011	86,246	70,222	38,276	22,252
2013	100,569	78,480	46,320	24,231
2016	123,391	95,995	58,327	30,931
2019	161,776	127,069	74,337	39,630
2022	189,988	145,342	97,433	52,787
Year	Per capita GDP (yuan/person in Central China)		Northeast per capita GDP (yuan/person)	
	The highest	Minimum maximum difference	The highest	Minimum maximum difference
2011	34,719	8791	37,353	11,260
2013	43,835	11,883	43,956	11,888
2016	56,844	22,872	47,069	13,044
2019	76,712	35,556	58,019	16,863
2022	90,358	39,485	68,422	17,549

Data source China Statistical Yearbook (2011–2023)

It is necessary to fully utilize regional advantages, develop the characteristic industries of each region, implement the coordinated and interconnected economic development mode of each region, and establish a coordinated green and sustainable development system for each region. Thus, while fully implementing the coordinated regional development strategy, a new pattern of regional development will be formed and the spatial layout of the modern economic system will be comprehensively promoted. The modern regional development system expected and established by the Party Central Committee coordinates the east, west, north, and south, promotes a new pattern of coordinated regional development across the country, and promotes the rapid economic development of various regions across the country. It is also necessary to promote the rapid flow of economic factors and optimize rational allocation, accelerate the urbanization process of various regions, adapt to the gradual transformation from small towns to metropolitan areas, gradually realize the modernization of the entire country, form a new development pattern with urban agglomeration, development axis, and economic zone as the main body, and comprehensively realize the coordinated development of regions.

China's urban agglomerations are large in number and scale; however, many aspects need to be further improved. It is necessary to improve the structure and function of urban agglomeration and give full play to its driving role to promote the economic development of small towns. China should rationally promote the internal integration of urban agglomerations, actively improve the internal transportation facilities of urban agglomerations, and accelerate the realization of a network to achieve national modernization and improve the overall competitiveness of the country. China should pay attention to the development of cities with a large population, ensure timely de-population, avoid excessive pressure caused by overpopulation in the central urban area, rationally de-stress its non-core functions, adjust the main industry and service chain, and form a metropolitan area with the surrounding cities.

Many three-dimensional cities and towns of different sizes have been formed between major cities and urban agglomerations, often referred to as economic belts. China should pay attention to the construction of major transportation arteries between urban agglomerations, realize the construction of a transportation network, accelerate the construction of an economic belt that has already taken shape, rationally arrange the development axis, fully improve the urban infrastructure, comprehensively gather economic factors, build an economic uplift belt that addresses the overall situation, and coordinate the development of the axis region while developing the economy. In the future, China should give full play to the driving role of the economic belt with a good economic foundation and build more new economic belts, accelerate urban construction, promote the coordinated development of the economic belt, and build it into a development axis that coordinates the east, the west, the south, and the north.

In the "13th Five-Year Plan," to better adapt to and lead the new normal of economic development and comprehensively promote the rapid development of China's economy, the government put forward the Beijing-Tianjin-Hebei coordinated development strategy. To reasonably and effectively undertake the non-capital functions of Beijing, China should take the lead in making breakthroughs in the transportation and environmental protection industries. To promote the integrated development of Beijing-Tianjin-Hebei and relieve the pressure on the capital, relevant departments should further plan the spatial layout of regional development and the core functions of the capital and ensure the industrial transfer and industrial development of the undertaking area. Therefore, the functional positioning of Tianjin should be clarified and non-capital functions should be removed in a rational and orderly manner.

The function of Tianjin is to effectively relieve the non-capital functions of Beijing, utilize its location and transportation advantages, undertake relevant work, and ensure industrial transfer. Therefore, it is strategically important for Tianjin to relieve the non-capital functions of Beijing and the pressure of the capital. Tianjin has a good industrial base and transportation conditions. In recent years, Tianjin has undertaken several investment projects, including Beijing projects (about 2100), and the total investment reached 146 billion yuan. Tianjin vigorously promoted scientific and technological innovation, built the Binhai-Zhongguancun Science and Technology Park, and actively introduced a large number of enterprises. The Science and Technology Park began to be gradually utilized, and the registered capital reached 5.82 billion yuan, promoting the collaborative innovation and development of Beijing and Tianjin. In addition, Tianjin has undertaken several projects, including Baidu Innovation Center, Wumart, and other key projects, which have effectively promoted Tianjin's economic development, and some key projects and Beijing-Tianjin strategic cooperation demonstration zones are actively promoted. In recent years, the scale of industrial development in Tianjin has been growing, and the registered investment has gradually increased, which has effectively relieved the non-capital functions of Beijing, promoted the economic development of Tianjin, and eased the pressure on the capital. The Outline of the Beijing-Tianjin-Hebei Collaborative Plan points out that Tianjin's industrial development should be comprehensively promoted while effectively alleviating the pressure on the capital. In recent years, Tianjin has focused on the development of emerging and high-tech industries, vigorously promoted economic development, effectively inherited the non-capital functions of Beijing, and actively promoted industrial transfer.

Unlike non-economic zones, economic zones can be divided into multiple levels. General economic zones are city-centered, whereas economic zones at the national level are usually larger in scope and the economic level of each region within the economic zone is the same. Therefore, urban agglomerations are mostly centered to further promote the development of surrounding areas, give full play to their radiating driving role, narrow the development gap among regions, and form more economic zones with higher economic levels. China will form a trend of economic development in which economic regions,, such as the pan-northeast economic zone, pan-Bohai economic zone, and pan-Yangtze River Delta economic zone, draw on each other's strengths. The economic zones can complement each other's advantages and cooperate with each other. By relying on urban agglomerations with higher economic levels, it can radiate the surrounding regions, promote their economic

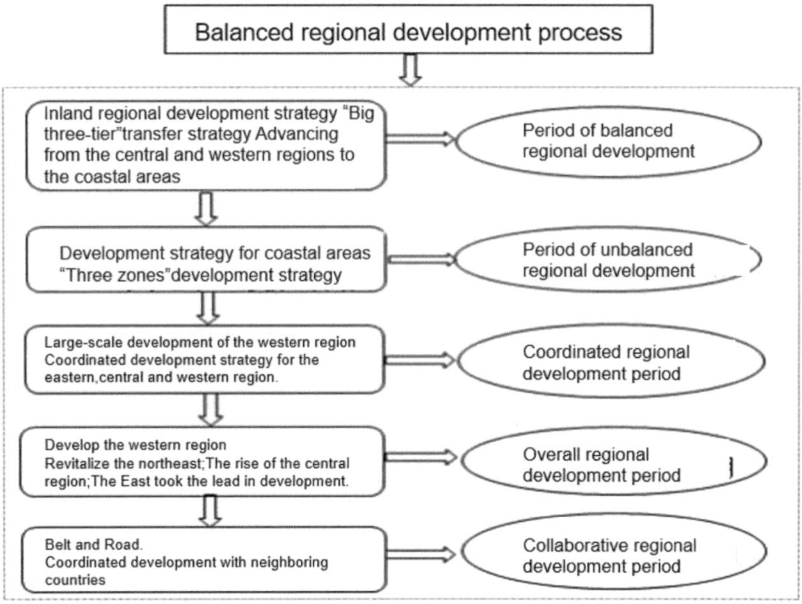

Fig. 6.3 Evolution of coordinated regional development in China

development, improve their economic growth rate, and also support urban agglomerations. Through the economic driving role of urban agglomerations, the active cooperation between them and the surrounding regions can be promoted, thus reducing the cooperation obstacles and promoting the complementary advantages and strong cooperation among regions. Further, urban agglomerations help promote international and regional cooperation, establish strong economic ties with neighboring countries, and promote the rapid economic development of border areas to further realize the strategic goal of coordinated regional development (Fig. 6.3).

6.2 Spatial Planning System

As an important means of regulating and controlling coordinated regional development, China's spatial planning system has evolved, and its development at each stage also reflects the phased needs of China's regional development. Spatial planning is the basis for the government's regulation and control of coordinated regional development and high-quality development. Considering the differences between regional natural and human factors, spatial planning considers the coordination of regional functions to realize the overall development of production, life, and ecological space. The development and change characteristics and driving factors of spatial planning are also the realistic embodiment of regional convergence analysis in development geography. Currently, there is no unified definition of spatial planning. Based on the existing relevant international research and relevant domestic practice, spatial planning refers to some strategies, plans, or policies established in accordance with the corresponding regulations, including the rational layout of territorial space and the overall development and utilization direction of the country, considering China's national economy and the direction of social development and goal requirements. One of the most obvious features is a reasonable and precise spatial scope. Usually, spatial planning includes the comprehensive utilization of national resources, rational distribution of productive forces, overall land management, and comprehensive environmental protection. A spatial planning system is an objective requirement for promoting the construction of ecological civilization. It should not only consider the corresponding restriction and coordination of inter-governmental interests and powers but also effectively intervene in our market economy, considering social development.

6.2.1 Evolution

In 2018, China established the Ministry of Natural Resources to comprehensively supervise the entire process of the construction of China's spatial planning system. The establishment of the new ministry reflects the country's reconstruction of the spatial planning system, a major and fundamental reform from the level of institutional construction, and it is also a positive response to the problems previously existing in China's spatial planning system. Previously, the lack of connections between different levels of planning led to repetitions (and even differences) in some aspects, and the approval process and cycle of superior departments were long. Since the reform and opening up more than 40 years ago, China has made continuous progress in economic, political, and social development and achieved great results. The spatial planning system has also experienced a process from scratch along with the national development, and has been improved and adjusted through continuous exploration and reference. The main planning includes urban and rural planning and main functional zone planning (Table 6.6). During this period, the spatial planning system went through four stages.

The first development stage of the spatial planning system is the period from the First Five-Year Plan to the Fifth Five-Year Plan, which can be called the period of rapid development and innovation of spatial planning. Before the reform and opening up, China was in the planned economy era, when the economic level was low, the market showed a shortage economy, the overall goal of the five-year plan formulated at that time was only economic development, and the main market players were only affiliated state-owned enterprises and local state-owned small and medium-sized enterprises. The state of spatial development presents a low level of relative balance. In the early days of the founding of the People's Republic of China, although the superior departments required that relevant departments should actively cooperate after the corresponding urban planning was issued, that is, the industrial layout

Table 6.6 Core components of spatial planning in China

Planning name	Index control	Directory control	Zoning control
Urban and rural planning	Urban population size, construction land scale, plot ratio, building density, green space rate, etc.	List of recent construction projects	Three districts and four tiers; City or town overall land use in planning; detailed planning
Main functional area planning	Intensity of land development	Optimize development zones and focus on developing key ecological functional areas, an agricultural products zone, a restricted development zone, a prohibited opening list of major producing areas, and urbanized areas	
Woodland protection and utilization planning	Forest stock; expropriation and occupation of forest land quota index	List of key forestry projects	Public welfare forest and commercial forest; forest geology volume level management
Water function zoning	• Two-level zoning		
Marine functional zoning	• Classification and division		

should actively cater to urban planning and the production organizations of various departments should actively cooperate, the spatial development was mostly viewed as the result of inter-regional distribution, limited by the low level of economic development at that time. Instead of the rational layout of territorial space and the overall development and utilization of the country, there is no reasonable planning for the population, resources, environment, and economy. Therefore, in the preliminary construction and exploration stage of the spatial planning system, China has not yet determined the specific content of the system, nor has it formed a complete system; however, the space can effectively support the five-year plan, whose implementation plan needs to rely on the corresponding space, and can further affect the regional economy in terms of the spatial layout. In this preliminary exploration stage, the Party Central Committee comprehensively studied land improvement in several countries, particularly some Western European countries, and then made preparations for China's macro land and regional planning, so that China's existing national economic development plans could be displayed in space, and spatial planning could play a major role in the comprehensive development pattern of national land and guide urban construction. In this period, China's spatial planning system showed top-down characteristics, from the country to the region and then to the local area, but some concepts and methods need to be clarified. During the preparation of the spatial planning system, the definition of the legal relationship between the corresponding legal protection and other plans was not synchronized, and the planning system at that time did not effectively play a role in restraint and control. In 1987, the state began to formally formulate effective spatial planning, laying the foundation for the formation of a unified planning system, and adopted the "Measures for the

Preparation of Land Planning," which specified the main tasks of land planning. The promulgation of the Measures for the Preparation of Land Planning has realized the transformation from single urban planning decision-making to dual urban planning and land use planning.

The second development stage of the spatial planning system spans the Sixth Five-Year Plan to the "Tenth Five-Year Plan," which can be called the adjustment and reform period of the development of the system. In this period, China was in the parallel period of planning and market, and the state began to gradually decentralize power, delegate power to lower departments, and implement decentralized reform. The overall goal of the Five-Year Plan is not only economic development but also coordinated economic and social development. In addition to several state-owned enterprises, there are also many private enterprises and foreign-invested enterprises. The establishment of special policy areas of different types and geographical scales, such as special economic zones, open areas, and economic development zones, has accelerated regional economic development. However, the spatial development status of different regions varies according to the differences in geographical location and economic levels and the prominent urban–rural dual structure, and some environmental pollution and other problems have gradually become prominent. In 2000, the National Development and Reform Commission gave its opinion on the reform of the planning constitution and proposed the concept of spatial coordination and balance. This shows that the relevant departments should consider industrial distribution when planning, guide the formulation of relevant plans at all levels, and coordinate the interrelationships among elements such as space, people, resources, and the environment. The spatial planning of this stage also changed with the change in the market economy structure, that is, from planning to guiding. The initial stage of spatial planning focuses on the exploitation and utilization of resources, and the adjustment and reform period realizes the unified transformation to exploitation, utilization, and protection; from the initial goal of economic development to gradually realizing the common development of the economy, population, resources, and environment; and from industrial development as the core in the primary stage of spatial planning to the coordination of regional spatial layout in the later stage. In addition, in 1990, the Urban Planning Law established the legal status of urban planning (Table 6.7). This means that the constraining and controlling role of the planning system gradually comes into play.

Since the twenty-first century, with the vigorous implementation of the scientific outlook on development and other policies, the Party and the state have increasingly paid attention to the issue of coordinated regional development. With the victory of the fifth Plenary Session of the 16th Central Committee of the Party, the Party Central Committee determined in the "Eleventh Five-Year Plan" to promote the development of old industrial bases and proposed the revitalization and development strategy of the northeast and other bases, and the state began to vigorously promote the comprehensive transformation of old industrial bases.

The construction of regional spatial structure in Northeast China is the key to the revitalization of the region. First, to promote the revitalization of Northeast China, efforts should be made to build a coastal economic belt. Through the development

Table 6.7 History of China's land development rights control system

Year	Legal basis	Control content and characteristics	Administrative authority
1984	Town planning clause case	The construction land permit and construction permit system have been clarified	Endemic
1986	Land administration law	Unified classification quota approval	Endemic
1990	Urban planning law	One book and two certificates	Endemic
1998	Land Management as amended Dharma	Land use plan management, pre-examination of construction land	Central and provincial
2007	Urban and rural planning law	Planning permits for rural construction will be increased	Endemic

of coastal cities to drive regional revitalization, a "five-point" strategy will promote economic development:

(i) Focus on building key industrial bases such as Dalian Changxing Island and Dandong Port, vigorously promote the development of growth industries, tap the development of advantageous industries, and form a modern industrial cluster.
(ii) Rationally use local characteristic resources, vigorously promote the structural adjustment of the light industry, timely grasp development opportunities, promote the development of local industries, and form a unique light industry system.
(iii) Give full play to Dalian's resource advantages, utilize the resources of the existing development zones, and firmly grasp the opportunity of building a Northeast Asia shipping center, carry out reasonable development, promote the economic development of Dalian, further develop it into a demonstration zone for adjustment and transformation and a development zone for emerging industries, and strive to build it into a state-level comprehensive supporting reform pilot zone.
(iv) The northeast should actively develop new industries while not ignoring the development of petrochemical, metallurgy, shipbuilding, heavy equipment, and other industries, use its own unique natural resources rationally, and pay attention to the development of the above industries in the coastal areas.
(v) With the over-exploitation and utilization of resources, Northeast China should timely adjust its industrial structure, vigorously promote industrial transfer, and build a labor-intensive industrial transfer base.
(vi) Dalian is a key area for the revitalization of Northeast China, and its characteristic industries should be fully developed and turned into an important service industry development center. Second, to promote the revitalization of Northeast China, we should optimize the economic belt of Harbin and Daqing and promote the local economic development. It is necessary to promote the

rapid economic development of the northeast region, optimize and improve the spatial management layout of the four economic belts, and guide the rational and orderly economic development of the four regions. China should also learn from the relevant economic development experiences of other countries, promote the orderly development of various advantageous industries, and pay attention to the vertical and horizontal division of labor. Third, promoting the revitalization of Northeast China should focus on the continuous construction of six secondary axes. In other words, it is necessary to pay attention to regional economic development, build six routes represented by Dandong–Huolinhe, Suifenhe–Manzhouli, etc., promote the development of the above areas along the "point," utilize the resource advantages of the regions along the routes, form industrial clusters, and develop characteristic industries on this basis, and promote the comprehensive construction of the secondary axis. In addition, the construction of urban economic zones (belts) should be promoted, thereby driving the economic development of the abovementioned areas while promoting the rapid development of Dandong, Tongliao, Heihe, and Yanbian. Fourth, to promote the revitalization of Northeast China, the state should pay attention to agricultural development and build a unique modern agricultural base. This means comprehensively promoting the construction of agricultural bases such as Songliao, Songnen, and Sanjiang Plain. To promote the revitalization of Northeast China, the state should fully the region's own resource advantages and emphasize agricultural development, particularly grain production. Modern science and technology, modern management, and modern industry should be combined.

Dalian Metropolitan Economic Zone, Central Liaoning region, Changji-ji Economic Zone, and Hada-Qi region are listed as key areas in the revitalization plan of Northeast China. Dalian should strengthen the "window" and leading role, highlight the development of a modern service industry, and increase the development of port industries, to provide space for the development of the entire northeast region. The central region of Liaoning has better economic conditions and comprehensive advantages. It is necessary to utilize regional advantages and strengthen cooperation and linkages. The central region of Liaoning should play a leading role in prioritizing development and become the pilot region for the revitalization of Northeast China,. In addition, it is necessary to focus on building an international manufacturing base and establish a relevant mechanism conducive to economic development with healthy, rapid, coordinated, and sustainable development. The Changji-ji economic zone should utilize local resources and develop machinery manufacturing and agricultural processing bases to further demonstrate its core position while developing new growth regions and centers. With rich natural resources, the region should be developed into a new industrialization pilot zone, a key green food processing base in the country, and a base for economic and technological cooperation in Northeast Asia.

The third development stage of the spatial planning system is from the Eleventh Five-Year Plan to 2017, which can be called the transformation period of the development of the spatial planning system. Since this period, China's market economy has been further developed, market-oriented reform has been gradually deepened, and the process of globalization has been continuously deepened and accelerated. The overall goal of the Five-Year Plan has been further extended to the comprehensive, sustainable, and coordinated development of the economy, society, and the ecological environment. Major market players have also changed, mainly state-owned enterprises, private enterprises, and multinational corporations. In this period, the uncoordinated spatial development status gradually became prominent, which was more serious in the inter-regional development. The ecological environment gradually deteriorated due to the excessive pursuit of economic development in the early stage, and the different levels of economic development between regions widened the urban–rural income gap. In 2006, the Eleventh Five-Year Plan was introduced and, at the same time, the connotation of the main functional area was introduced. The functional zoning of China's land space will be carried out according to the environmental carrying capacity, land development density, and development potential. The main functional zoning is to further combine the actual situation of China's regional population distribution, economic layout, land utilization, and urbanization development pattern in the region and obtain four types of main development functional zones, namely the optimized development zone, key development zone, restricted development zone, and prohibited development zone. Based on this, China's relevant departments will formulate regional main development policies, reasonable development and performance evaluation of the main functional areas, and reasonable maintenance of spatial development order to build a scientific spatial development structure. In 2010, the National Functional Zoning Plan (2011–2020) was released, which began to divide four categories of space at the national level according to regional resource endowments and centered on industrialization and urbanization construction as an important means of space control. Thus, future development strategies for different regions were proposed and further instructions were given. In 2013, the Outline of the National Land Plan (2014–2030) (Draft) pointed out the main tasks of centralized development, classified protection, unified governance, development of related security systems, and supporting policy guarantees of the future land and fully highlighted the comprehensive and strategic characteristics. However, it is precisely because spatial planning has played an important role in strengthening macro-control in our country that it has reached its due height, while other regional spatial planning has also improved greatly. According to the division of functions, different departments customize their "own" spatial planning so that there are cross-dislocations, levels, technical specifications, and standards among different plans, and there are conflicts. Later, with the slight enhancement of regional development, the problems of contradictions and incoordination in the planning of various sectors became increasingly prominent, resulting in the inability to implement the planning and the difficulties in implementation and management. Generally, China's spatial planning system in this period had serious problems of multi-regulation conflicts in terms of a hierarchical system, division of powers, and preparation of a technical system. Since the reform and opening up, China's space policy has gradually evolved, and the economic system has realized the transformation from a planned economy to a market economy (Table 6.8).

Table 6.8 Evolution of space policy in China's Five-Year Plan

Economic system environment	Era of a planned economy	Parallel planning and marketing	Further development of a market economy
Main market entities	Affiliated state-owned enterprises and local state-owned small enterprises	State-owned enterprises, collective enterprises, private enterprises, foreign-invested enterprises	State-owned enterprises, private enterprises, multinational corporations
Space elements in the five-year plan	Coastal and inland, third-tier, overall project layout, national unity	Special policy zone, Three major zones (east, west, central), and urban–rural development	Overall regional development strategy, main functional areas, and urbanization strategic pattern
Space development status	Low-level relative equilibrium	Regional differences are evident, urban–rural dualism is emerging, and environmental issues are becoming more prominent	Coordination with the Ministry of Regional Development has intensified, the ecological environment has deteriorated, and the income gap between urban and rural areas has widened

The fourth development phase of the spatial planning system is from 2018 to the present. The year 2018 marks a turning point for China in reshaping its national planning system. In March 2018, after merging the spatial planning functions of several departments, in accordance with the relevant reform programs of national institutions and with the former Ministry of Land and Resources as the main body, the National Ministry of Natural Resources was established to supervise the unified planning and control of all the space and land in China. In 2019, the State Council stressed the need to establish a territorial spatial planning system, integrating functional zoning planning, land use planning, urban and rural planning, and other spatial plans into a unified territorial spatial planning to integrate multiple plans. Starting from 2018, China aimed to gradually realize the turning point of the spatial planning system, focusing on resource and environmental protection and further management, rearranging the powers of governments at all levels, and developing the system into a key means to regulate local development in the future. In recent years, the constraints of resources and the environment are increasing daily, and the contradiction between humanity and land is gradually becoming prominent, which promotes the construction of ecological civilization in China. China's newly released territorial spatial planning can also help promote the country to rationally plan the territorial space, promote ecological environmental protection, scientifically delineate the planning boundary of the control area, scientifically and reasonably plan the spatial structure and layout of the permanent use of the territorial land, and protect the barrier of the

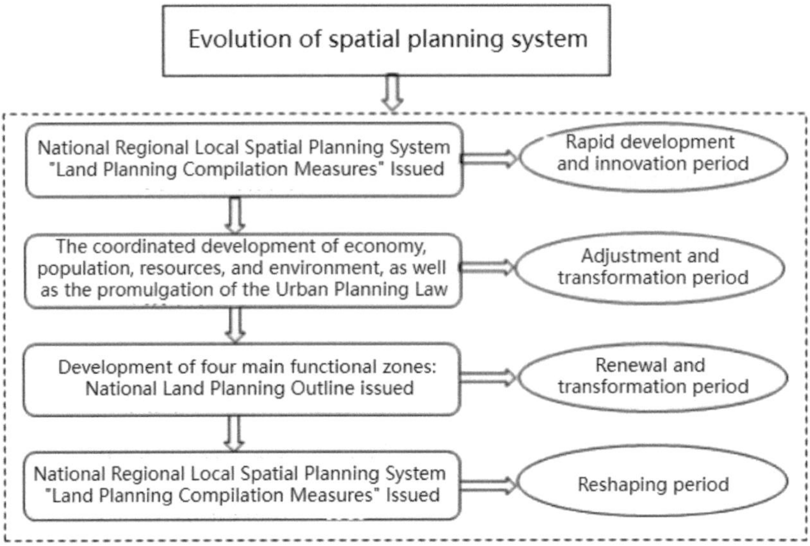

Fig. 6.4 Evolution of China's spatial planning system

territorial ecology based on the environmental carrying capacity and the evaluation of the spatial suitability of the territorial ecological resources.

China's spatial planning system has evolved from a period of rapid development and innovation to that of remodeling (Fig. 6.4).

The coordinated development of the Beijing-Tianjin-Hebei region is progressing in a reasonable and orderly manner, and the "integration of multiple planning" in the Xiongan New Area is also proceeding well. Xiongan New Area is under the jurisdiction of three counties, namely Xiongan, Rongcheng, and Anxin, and their adjacent individual areas. In coordinating the planning of the Xiongan New Area on a map, the map boundaries can be reasonably delineated to achieve "one new area, one plan, one blueprint." The orderly progress of "multi-regulation integration" is a key way to realize the intensive, efficient, and sustainable use of "three-life space" (ecological, production, and living space), and can effectively prevent the existing multi-regulation conflict, space control failure, and other obvious drawbacks.

To ensure that the Xiongan New Area is established, "a blueprint is drawn to the end." Since the beginning of the construction of the Xiongan New Area in 2017, the Party Central Committee has proposed to focus on drawing a blueprint, clarify the main function of the relevant land plots, further delineate the non-construction boundaries of various regions, which must be approved by relevant departments, provide reasonable legal protection, and try to protect the end and rigid constraints to the end.

The objective is to facilitate collaboration between the two distinct categories of land use within the Xiongan New Area. Since the commencement of construction of the Xiongan New Area in 2017, the Party Central Committee has proposed the integration of above-ground and underground resources and space into a unified to facilitate the unified planning, coordination, and utilization of relevant spaces and resources. It is essential to emphasize the three-dimensional linkage and to unify and rationally utilize the two types of space above and below ground. Before this, it is essential to the sequence of utilization for the relevant heterogeneous resources, guarantee the proper utilization of resources, view the intensive development and utilization of above-ground and underground space as a fundamental objective, and ensure the comprehensive prevention and control of above-ground and underground disasters in the Xiongan New Area.

The objective is to establish a seamless connection between the three life spaces in the Xiongan New Area. Following the successful convening of the 19th National Congress, the Party Central Committee has stressed the need to promote harmonious coexistence between humans and the natural environment. This will entail improving relevant content, further achieving the scientific goals of production, development, a rich life, and a conducive ecology. The three-life space of the Xiongan New Area will then be clearly delineated, and suitable construction areas, restricted construction areas, and prohibited construction areas will be properly planned. Finally, natural resources will be fully utilized in the new area, and ecological capital will be employed. It is necessary to gather the production capital of the Xiongan New Area and then increase the living capital. By delineating the "Sansheng" space in Xiongan New Area and gathering the capital invested in it, it is necessary to initially determine a reasonable spatial development order, identify the areas that require protection and those that can be developed, and then implement a spatial plan that balances these two objectives.

6.2.2 Problems and International Reference

Currently, China's spatial planning system is divided into horizontal and vertical, and there are different levels of local planning. Generally, all levels of planning are from different departments, and a scientific and reasonable pattern has not yet been formed. This section summarizes the problems in China's current planning and combines the historical experience of the international space planning system to help China's space planning.

(i) **Foreign spatial planning system construction for reference**

To further improve China's spatial planning system, China can learn from the experiences of developed countries. The spatial planning system of developed countries has the following common points:

Spatial planning hierarchy system

The spatial planning systems of some developed countries show obvious integrity and are relatively unitary, with only one spatial planning at each level, and departments at each level make respectively relevant spatial development and land use plans for different regions. The hierarchical system of spatial planning is more common in developed countries. For example, the United Kingdom has a single planning system. After the issuance of the Housing and Urban Planning Act, it gradually improved the urban–rural integrated planning system, with national, regional, county, and town as the main levels, and it is still in place today. Germany also has a single planning system, in which all levels of departments, namely federal, state, and regional, formulate respectively relevant spatial planning. The federal government is primarily responsible for setting principles and patterns of land use and developing spatial plans for the country. States are expected to refine the principles and models established by the federal government and further lead their regional spatial planning. Regional (municipal) governments are responsible for the specific planning of infrastructure such as buildings. In addition, Japan represents parallel spatial systems in developed countries (Fig. 6.5), which are usually reflected in the simultaneous existence and respective roles of territorial space planning and territorial utilization planning, and the two are parallel and coordinated.

The basic content focuses on cohesion and clear division of labor

The spatial planning of some developed countries started earlier, with a rich experience, formed a more complete spatial planning system, and paid more attention to the connection between different levels of planning. The superior department is responsible for formulating the dominant national planning, developing the country's spatial development policy, playing a leading role in the overall situation, and effectively guiding the lower departments to formulate lower-level planning. According to the instructions of the superior and the current situation of the region, under the guidance

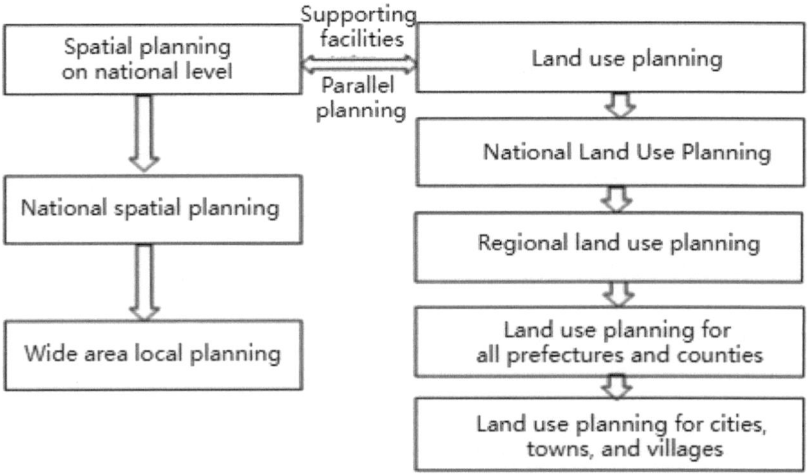

Fig. 6.5 Spatial planning system of Japan. *Data source* According to Xiamen Urban Planning and Design Institute (Xiamen Spatial Planning System Research 2015)

of the superior planning, the lower departments should begin to draw up a land planning scheme that can be implemented in the region. For example, the United Kingdom formulated national and regional policy guidelines in its spatial planning, with a clear division of labor between the two sides, and the German federal government issued federal and federal state planning when formulating spatial planning, both of which fall under national strategic planning (Fig. 6.6). However, the sub-planning in the United Kingdom and the regional planning in Germany only represent the transformation from the national spatial policy to land use and have no direct effect on the specific implementation of the sub-plots. The local planning issued by the United Kingdom and local construction planning issued by Germany are the direct basis for relevant institutions to carry out development.

Perfection of supporting laws

Many countries with relatively perfect spatial planning systems pay close attention to the improvement of supporting laws. The main methods can be divided into two categories: First, the state establishes corresponding laws and regulations at various levels, namely administrative levels, to form an interconnected and complete system. For example, Germany has successively promulgated the Federal Spatial Order Planning Law, State Territorial Spatial Planning Law, Construction Utilization Regulations, and State Construction Regulations at the national, state, and local levels.

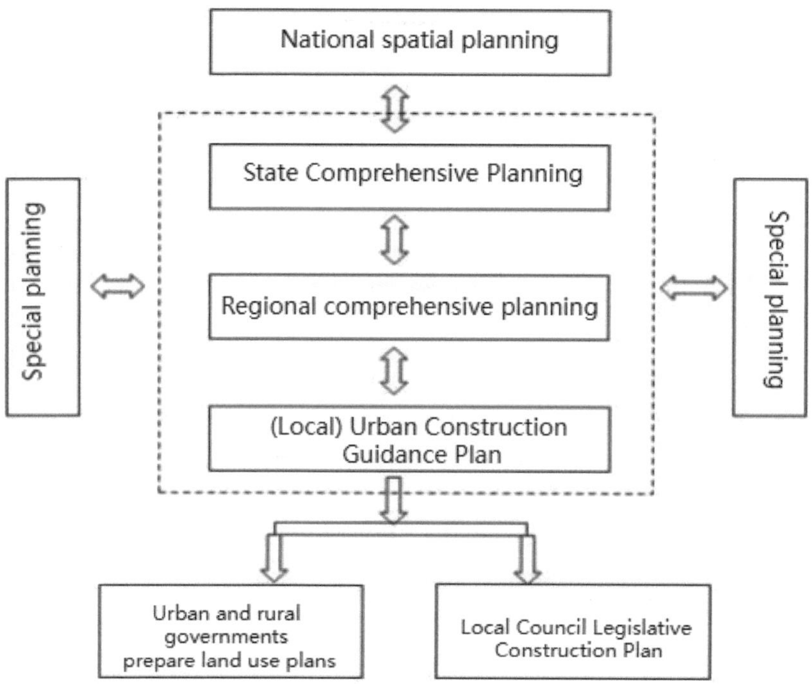

Fig. 6.6 German spatial planning system

Second, after spatial planning is introduced, it immediately becomes a bill, and if it is necessary to appeal and amend the bill, it must go through legal procedures. For example, in the United States, states and below can formulate corresponding spatial planning; however, its implementation must be approved by the legislative body, and the approval of planning at all levels is a bill, and it has the corresponding legal effect; to modify it, one must go through legal procedures (Fig. 6.7).

(ii) **Problems with China's spatial planning system**

There are more than 80 legal and many more non-legal types of planning in China. Too many relevant plans have been formulated by various departments without the relevant legal basis. To date, China has not established a reasonable and orderly

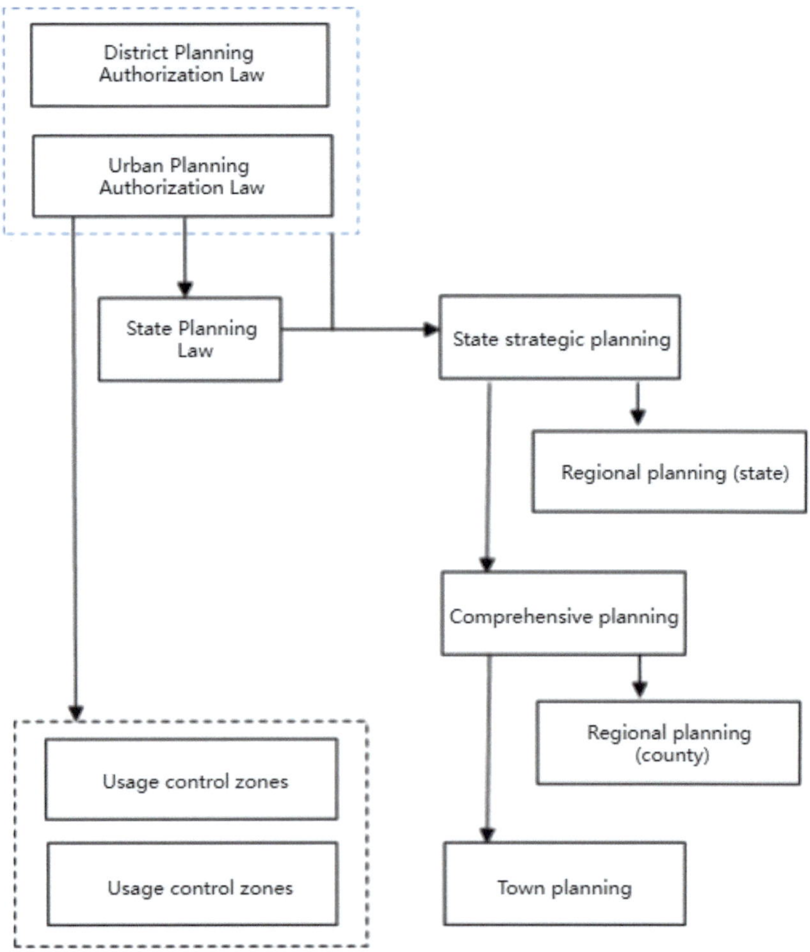

Fig. 6.7 American spatial planning system. *Data source* According to Cai (Construction and Enlightenment of the American Spatial Planning System, 2017)

spatial planning system, and the existing system is uncoordinated, unsystematic, complex, and imperfect, and the management, preparation, approval, implementation, and supervision processes differ (Table 6.9). This causes problems such as lack of coordination and conflicts among departments in planning, which are manifested in three aspects.

This causes problems such as a lack of coordination and conflicts among departments in planning, which are manifested in the following aspects.

Lack of a top-level spatial planning design

Spatial planning is comprehensive and helps with overall planning. The formulation of spatial planning requires the cooperation of various departments, and if necessary, transcends departmental responsibilities. It is necessary to strategically consider various space needs. In promoting holistic economic development, environmental protection should be considered and strengthened. Currently, spatial planning at different levels in China is divided based on different departments at different levels according to their respective division of responsibilities and different characteristics of various industries and their unique professional advantages, and all departments plan together from different levels and angles (Fig. 6.8). However, China's current spatial planning lacks a comprehensive coordination department that plays a regulating role, cannot rationally plan the overall territorial space, and lacks a top-level design and leading planning to plan the overall situation. On the one hand, the planning content of parallel departments overlapped, the boundaries of powers were not clear, and the order of space development was disturbed, which made the management efficiency and social efficiency generally low. On the other hand, more and more levels are responsible for planning, their functions are unclear, and their contents are similar. These factors affect the unity and effectiveness of space policy in different aspects.

There is no unified standard for formulating the basic contents of departmental planning

Currently, basic data are collected and processed differently in the planning of various departments in China, and many technical aspects such as different land classification standards and space control zoning standards are slightly different, which makes the planning of various departments difficult to unify. For example, in terms of data collection, different departments use different coordinate scales and platform systems for geographic information. No shared data information platform has been established yet, leading to different planning base periods. The urban and rural planning departments at lower levels formulate plans based on the requirements at higher levels and follow the Urban Land Classification and Planning and Construction land Standard (GB50137-2011) to formulate relevant plans within their respective levels, whereas the relevant land planning departments at higher levels formulate relevant plans in accordance with the Land Use Classification Standard (GB/T21010-2007). However, the two standards followed by the upper and lower levels are very different, which leads to disunity in the planning of the two departments. In terms

Table 6.9 Comparison of major space plans

Type		Project			
		National economic and social development plan	Urban and rural planning	Overall land use planning	Ecological environment protection plan
Administration	Planning by a competent department	Comprehensive development and reform departments	Comprehensive nature of urban and rural planning departments	Special nature of the Ministry of Land and Resources	Special nature of environmental protection departments
Approval	Feature compilation method	Independent	Independent	From top to bottom, unified	Independent
	Approval authority	Local People's Congress	Superior government	The State Council and higher-level governments	County-level People's Government
	Implementation intensity	Guiding	Constraints	Mandatory	Constraints
Implementation	Implementation methods	Annual Government Work Report	"One Book, Three Certificates" Management System	Land use control; Annual land use plan; Construction project management and control; Agricultural conversion system; Emphasize "three lines" and two boundary control	Environmental Impact Assessment
Supervision	Planning period	5 years	10 –20 years	10 years	5 years
	Supervisory agencies	Local People's Congress	Superior government and Local People's Congress	The State Council and higher-level governments	Local People's Congress
	Implementation evaluation	Government annual report and mid-term evaluation	Planning and editing	Law enforcement supervision	Report annually to the Local People's Congress

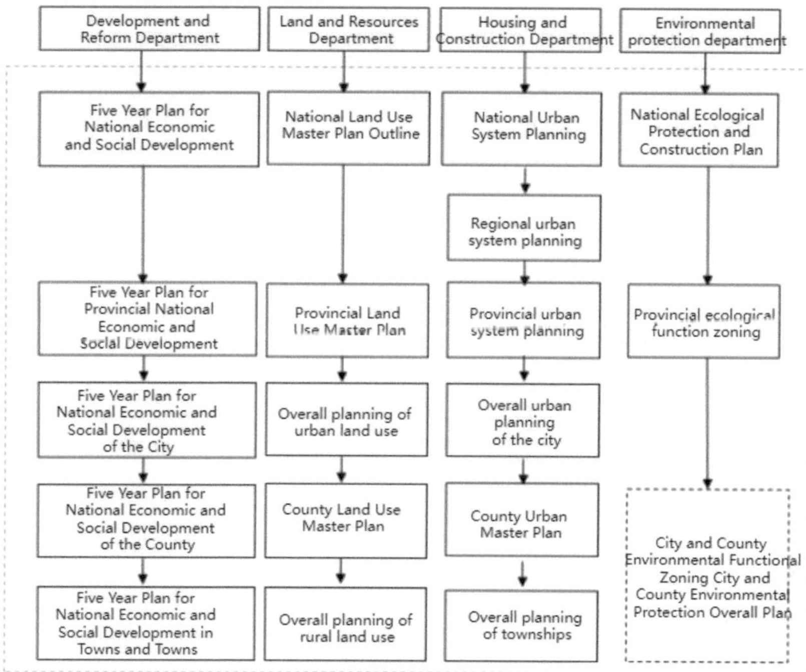

Fig. 6.8 Framework of China's existing planning system

of different spatial control zones, for example, the lower-level urban and rural planning departments have delimited "prohibited construction areas, limited construction areas," whereas the higher-level land planning departments have delimited "prohibited construction areas, restricted construction areas, conditional construction areas" and so on.

A sound security system has not yet been established

In addition to the functions of control and utilization, spatial planning also has the function of guarantee. However, the existing spatial planning in China has different perspectives, and no programmatic planning plays a leading role in coordinating the planning at all levels, resulting in insufficient spatial planning guarantee ability. For example, the preparation and formulation of urban, rural, and national planning refer to the Urban and Rural Planning Law and the Land Management Law, which are statutory. However, China's main and ecological functional zone planning is done in accordance with the relevant administrative rules and regulations of the State Council, and their legal status is lower than that of statutory planning. However, due to the increasingly prominent functional status of the main and ecological functional zones, their actual guiding role is equal to or even higher than the statutory planning at the same level. The existence of multiple upper-level plans at the same time leads to the lack of specific guidance for the lower level and the planning, that is, multiple upper-level plans guiding the preparation of the lower-level plans at the same time. Due to the lack of coordination among the leading plans of legal significance, the lower planning planners and planning executives lose when the content is contradictory. As a result, regional coordination mechanisms that should be considered in planning are not fully utilized.

Addressing the contradiction and conflict between land use index planning and space optimization layout

As the spatial carrier of multiple factors, the scientific overall planning of land use index is one of the important issues in the construction of a spatial planning system in China. Currently, China has adopted the spatial planning method of "supply according to demand." After multi-level decomposition, the higher-level government cannot completely and timely collect important land use information in the lower-level government area, as well as the actual land use needs of various departments in different places. Moreover, the performance evaluation of land resource efficiency has not been fully valued. As a result, the distribution of indicators tends to be average, causes an imbalance between the determination of indicators and the distribution of benefits, and lacks a quantitative scientific basis. In addition, the scale structure and layout of the "three life" space is also the core issue of spatial planning. With China's new urbanization, food security, and ecological civilization construction, the competition between the "three life" land is becoming increasingly fierce. To recognize and alleviate the land use contradictions between urbanization development (life), grain production (production), forest and grassland resources protection, and development and utilization (ecology) in China is a scientific problem and a realistic proposition that should be solved urgently to realize the division and sustainable utilization of the three-life space. The goal of economic and social development planning should be connected with the spatial layout, and the performance of regional land use should be analyzed with reference to the goal of national economic and social development planning. Based on the systematic idea that mountains, rivers, forests, fields, lakes, and grasses are a community of life, the model of "demand determines supply" should be reasonably proposed. Space carrier has a typical uniqueness. A series of space development rights related to production, ecology, and life should be maintained. In this process, a reasonable and effective bottom line should be drawn and rules of tolerance and cooperation should be formulated for each other; otherwise, the existing sustainable space development order cannot be guaranteed.

6.2.3 Development

China lacks a top-level spatial planning design, and the existing spatial planning system lacks leading planning qualities. Therefore, it is necessary to establish a

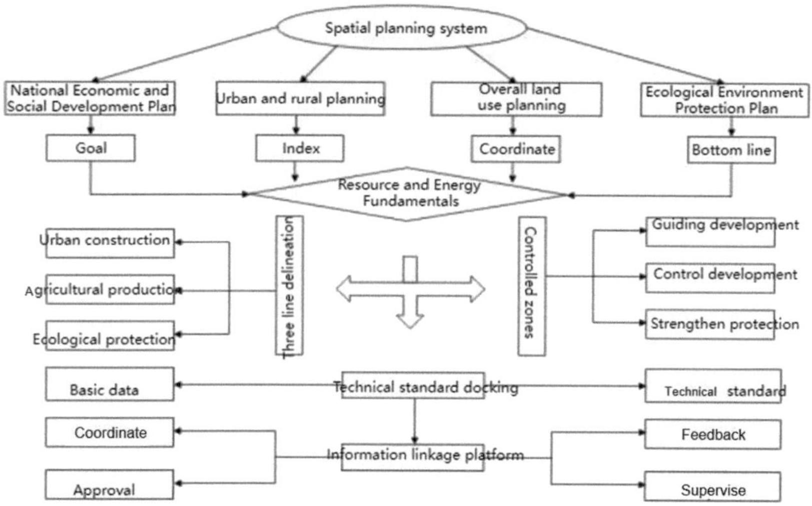

Fig. 6.9 Multi-plan integration logical framework

spatial planning system with clear responsibilities at all levels in terms of institutional reform, hierarchical relationship, legal construction, and public participation (Fig. 6.9).

(i) Accelerate the reform of the national planning system

Realize "multi-plan integration"

Since the establishment of the Ministry of Natural Resources in China, the "Several Opinions on the Establishment and Supervision of the Implementation of the Territorial Space System" has been issued one after another, and the spatial layout has undergone corresponding changes and entered a milestone reform stage. Compared with the original development of main functional zones, land use, urban and rural planning, and other types of spatial planning, China's new spatial planning has been greatly improved. First, it is conducive to realizing "multi-planning into one," its formulation integrates the above-mentioned, a new territorial spatial planning system is formulated, and the basic information platform of territorial space is formed. At this point, the new spatial planning system of our country has been further improved, forming the national "one picture" land space planning, and realizing "multi-planning integration" (Fig. 6.9). In addition, the process of developing and implementing a plan needs to go through a series of steps, and each step is implemented and supervised by the relevant department. After developing the plan, it passes through approval preparation, supervision, and implementation stages, in that order the development and operation of the plan are guided by a regulation and policy system and a technical standard system, respectively. The new territorial spatial planning is not a simple change of name, nor is it a patchwork of forms, shows the complete essence and demand of the structural reform of the spatial supply side in the new era, fully reflects the strategy, makes the new spatial planning more scientific, so that the upper-level planning plays a better leading role in the lower-level planning, and improves the integrity of the entire spatial system. The formulation of China's new spatial planning system should fundamentally realize the development and protection of national land space and work out a suitable national land space plan timeously to ensure high quality, high efficiency, fairness, and sustainability.

Geographical science underpins "three-line demarcation"

To effectively improve the binding force and guidance of China's spatial planning and layout, reasonably protect ecological security, force development and transformation, strengthen the bottom-line thinking, support the formulation of the spatial planning system with the "three-line delineation," rationally maintain its main function positioning, and further combine China's existing ecological civilization, industrial development, town and village layout, infrastructure construction in various regions of the country, and other factors. It is important to reasonably delineate the "three-life control line" to further achieve the scientific goal of protecting national ecological security, production safety, and life safety, that is, the permanent basic farmland red line, ecological protection red line, and the urban development boundary, through "three-line" support to formulate all kinds of spatial planning in China and use it as the boundary to delineate different spatial control measures. Demarcating the "three lines" can help slow down urban sprawl, improve land use, and enhance urban development efficiency. It will help ensure that the country's actual cultivated land and basic farmland area remain unchanged and accelerate the steady implementation of the strictest cultivated land protection system. In addition, it will help strictly adhere to the bottom line of ecological security; ensure that the total amount of national ecological land use "only increases, not decreases"; effectively promote the formation of national functional zones such as ecological environmental protection space, agricultural production and development space, and urbanization and industrialization construction space; further plan a consistent and reasonable pattern of territorial space protection, development, and utilization; and thus accelerate the promotion of functional zone planning. Define the level and effectiveness of each program in different spatial dimensions.

Space control

Spatial regulation from the perspective of complex geographic systems emphasizes the harmonious symbiosis between nature and human systems. When formulating the relevant plans of China's traditional national economic and social development, the horizontal integration of space regulation may have put too much emphasis on macro economy, industrial economy, social development and people's lives, resulting in many plans with strong orientation and objectivity, and the "emphasis on indicators, light on space" idea is more prominent. China's development and reform department is responsible for formulating the relevant "main function zone planning," to promote the reasonable division of national land space and introduce relevant policies for the differentiated development of national land space, and recover some existing defects from a certain point of view. In terms of vertical spatial control, the development and reform departments at the city, county, and township levels in China should fully consider and implement the relevant elements of the main functional zoning at the next level when formulating and introducing the relevant local development and performance appraisal policies. In the implementation of spatial control in urban and rural planning and overall land use planning in China, we need to pay more attention to the formulation and implementation of the principle and mandatory spatial control, and strengthen the connection with lower-level planning. China has no relevant and effective functional zoning planning at the city, county, and township levels. Relevant development and reform departments should actively cooperate with corresponding departments of urban and rural construction planning and land use planning and work out reasonable and effective policies and specific implementation measures according to the relevant components of the national and provincial functional zoning planning at the two main levels to play a leading role. When formulating and realizing the relevant components of the national economic and social development plan, the urban planning department and the land department actively cooperate to formulate reasonable and fine technical guidelines for the coordination of space control in various departments.

(ii) **Clarify the relationship between different levels of planning**

The different levels and types of existing spatial planning in China country are contradictory and not smoothly connected with each other. The reform divides the territorial spatial planning into five levels, namely national, provincial, municipal, county, and township levels. The "Several Opinions" clarified the focus of the overall planning of territorial space at all levels, the hierarchical relationship of planning at all levels, and the direction of future spatial planning. The National Territorial Space Plan is programmatic and strategic planning, which is led by relevant departments of the Ministry of Natural Resources and plays a leading role in lower-level planning. It is the policy and general outline for the protection, development, utilization, and restoration of the national territorial space. Provincial territorial spatial planning is led by the relevant departments of the provincial government, focusing on coordination, which further determines the focus of the preparation of the lower level of territorial spatial planning based on the implementation of the higher level of territorial spatial planning and improves the cohesion and coordination of the upper and lower levels of planning. City, county, and township territorial spatial planning is organized by the municipal government, focusing on implementation. Its main responsibility is to formulate the relevant measures of local territorial spatial planning under the adequate guidance of the superior planning. The formulation and implementation of the territorial spatial planning of other cities, counties, and towns should be based on the actual situation, and the provincial government should prepare, guide, and supervise the implementation (Fig. 6.10). The formulation of city, county, and township planning is relatively loose, and the planning of a city, county, and township or several townships can be combined according to the situation.

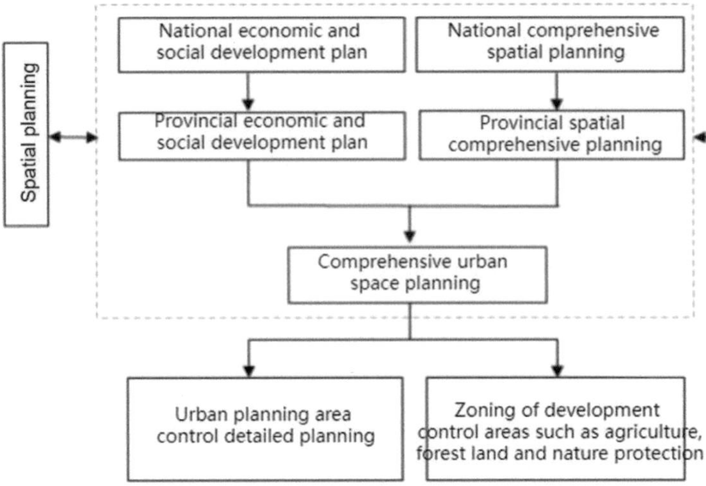

Fig. 6.10 Conception of China's spatial planning system

(iii) **Improve relevant laws and regulations**

To improve the laws and regulations of spatial planning, China's spatial planning system should have a leading plan, and the specific content of the spatial planning of each administrative region formulated by various levels of departments should be given legal planning status. China's new territorial space planning system should be given greater authority. Departments at all levels should ensure that all levels of planning are statutory, with legal status, all planning is approved and implemented, responsible agencies at all levels have no right to change the system, and previous arbitrary adjustments, changes, and other issues are improved. Second, departments should clarify the implementation and supervision of strict planning and strengthen the further management of control boundaries and binding indicators, and relevant departments should strengthen supervision. Finally, it is necessary to strengthen the implementation of relevant laws and regulations and strictly investigate and supervise the behaviors that violate the planning to make the implementation of supervision more stringent. Therefore, it is necessary to further promote the formulation and implementation of the Spatial Planning Law and related laws and regulations, promote the improvement of local administrative regulations, supervise the effective implementation of relevant departmental rules and approval regulations, and fully guarantee the reasonable implementation of spatial planning regulations.

(iv) **Strengthen publicity and popularization of the plan and encourage public participation**

Territorial planning is not only a national plan but also a plan for the well-being of the people. Therefore, the planning supporting system is expressed through administrative leadership and expert opinions and also encourages the public to actively participate in the whole process to ensure that the spatial planning formulated by institutions at all levels of our country under the market economy system plays an active role in promoting and comprehensively balancing the interests of multiple parties. To encourage the active participation of the public, it is necessary to rationally use the media, increase the importance of publicizing the participation in the formulation of spatial planning, promote the planning awareness and participation enthusiasm of people from all walks of life in China, and comprehensively promote the reform of the integrated management system of natural resources in China. Encouraging the public to actively participate in territorial spatial planning is a democratic process, and it is also a process of planning implementation and publicity. In the process of planning preparation, only by addressing the doubts and contradictions of the public can the planning process be more scientific and the possibility of landing greater.

6.3 Industrial Transformation and Development

Industrial transformation and development are an important support for regional coordinated development. They refer to the change of development mode and development concept and are the way to achieve high-quality regional development. The development of industrial and regional political, economic, social, cultural, and environmental factors complement each other. Industrial development has led to population flow, but it is also accompanied by some social and environmental challenges. Industrial development is not only the main element of spatial planning but also the core of high-quality regional development. Industrial transformation refers to the process in which a country or region changes the existing industrial structure according to its own economic development situation to improve its coordination ability between the economy and the environment. To promote industrial transformation, relevant government departments usually formulate or introduce a series of policies related to industrial adjustment, actively introduce new industries, fully develop local characteristic industries, expand or reduce the scale of existing industries and industrial organizations, and actively introduce advanced science and technology equipment to improve performance. The effective implementation of industrial transformation requires cooperation among various departments, which is not only a fundamental requirement to promote China's new industrialization but also a major measure for coordinating the development of China's economic region and coordinating the development of the social economy and environment.

6.3.1 Industrial Transformation Lessons from Other Countries

Industrial transformation is a long-term process based on economic development, which requires cooperation between the government and enterprises to formulate long-term strategic planning for the country or region. The report of the 19th National Congress of the CPC Central Committee pointed out that all regions should promote supply-side structural reform and economic transformation. According to the current development status of the three major industries, the development of the primary industry is relatively slow, the output value created is relatively low, the output value created by the secondary industry is still the highest, whereas the tertiary industry has developed rapidly in recent years, and the economic development speed of the eastern region is much faster than that of other regions.

(i) **Phase and question of industry transition in China**

The fundamental purpose of industrial transformation is to develop alternative industries and build a new system of regional industrial development. The internal mechanism of industrial transformation indicates that technological progress and institutional innovation can effectively promote industrial development, promote the evolution of the industrial level from low to high, and lay a solid foundation for industrial transformation. The improvement of the industrial level is a process of quantitative change, and waiting for the qualitative change of quantitative change agglomeration will further create a new industrial development system with high technology, high management level, and high efficiency. From the current industrial development situation of China, the lack of innovation, imperfect market mechanism, and imperfect government function seriously restrict the further development of the country's industrial transformation.

The industrial technology innovation system is not sound

The lack of scientific and technological innovation is the main obstacle and bottleneck in industrial transformation. Currently, China still lacks the key core technology to promote industrial transformation, the industrial division of labor is unclear, the lack of advanced technology and the key technology content and added value are low, the innovation system is imperfect, and it is still in the state of "manufacturing—processing—assembly." This is because of the serious lack of rules and mechanisms that can provide enterprises and relevant scientific and technological personnel with good profits, and the government has not given corresponding policy support. From the current situation in China, although enterprises can independently declare high-tech enterprises, there are still many difficulties, the high conditions for identification lead to many outstanding enterprises cannot declare success or enjoy preferential policies, some basic technologies are not popular enough, and R&D investment is low.

Measures to encourage the development of the real economy are inadequate

The real economy is the basis for the orderly development of China's industry. In the new round of industrial competition in the future, China can promote industrial transformation through the development of its real economy, which is an important means to occupy the commanding heights. If the virtual economy is vigorously developed and unchecked, it will squeeze the development space of the real economy. If the problem is serious, the orderly and healthy development of the real economy will be affected, the operating cost will increase, and the industrial hollowing and asset bubble problems will be serious. Currently, the imbalance of the inter-industry profit rate seriously restricts the sustainable development of the real economy, particularly industrial development. First, the slowdown in industrial development has reduced the prices of industrial products and the benefits of enterprises. Second, industrial costs remain high for a long time, reducing the working capital of enterprises, which will affect the normal production and operation of enterprises and further development. Third, the level of enterprise profitability and overall income of enterprises is low, and the direction of enterprise investment is limited.

Industry access standards are not perfect

Currently, China's existing industry access standards are not perfect. Relevant departments need to establish open and fair industry access standards to promote reasonable industrial competition and increase the vitality of competition. Currently, some industries such as those with a high degree of state-owned capital monopoly, still have high access standards, which need to be further improved, and a large number of private capital still cannot enjoy relevant policies due to ownership problems. In addition, the impact is greater for the non-monopoly industries such as automobiles and petrochemicals, more attention is paid to the relevant projects before they apply for construction, and the relevant departments should approve or record the aspects of capital and energy conservation, whereas the benefits obtained after the project has been implemented and completed are rarely paid attention to.

The industry exit mechanism is not sound and funds are insufficient

The structural excess of production capacity is the key problem that must be addressed in China's industrial transformation. The excess capacity structure is affected by various factors, which is not only closely related to economic cycle fluctuations and the inherent laws of the market but also affected by the mechanism of production capacity withdrawal, thus affecting the pace of industrial transformation. Currently, China's industrial exit mechanism is still not perfect, the market supply and demand cannot be adjusted in a timely manner because of price changes, some enterprises with limited development cannot retreat in a timely manner, resulting in a large number of resources that cannot be released timeously, such as human capital, land, and environmental capacity, limiting the transformation of advantageous enterprises. China's resource-based cities are rich in a large number of natural resources such as minerals; however, their prices are low, the output price in resource-based cities is low, and the input price is high, resulting in a serious imbalance between input and output, a high loss of benefits of resource-based cities, and a lack of industrial transformation capital, and most resource-based cities in Northeast China cannot make ends meet (Table 6.10).

The industrial structure of Northeast China is dominated by resource-based industries; however, the excessive consumption of natural resources abruptly increases the downward pressure of the economy, and the growth rate of key industrial economic indicators begins to slow down. The over-exploitation of resource-based cities has caused vibration and stagnation; therefore, it is necessary to further promote industrial transformation.

The overexploitation of natural resources in Northeast China has led to the contradiction between local resources, environment, and development, and the further development of industrial transformation has inevitable setbacks. First, the economic growth momentum during the industrial transformation is insufficient. Northeast China has many resource-based industries. After the reform and opening up, its economic development level has declined, and its economic proportion has been declining and even stagnated. In recent years, the over-exploitation of resources has attracted the attention of the government. After the implementation of industrial transformation, the level of economic development has begun to improve; however, the follow-up capacity of industrial development is insufficient, and the funds cannot be introduced in timeously in the later stage, resulting in the slowing down of the development of related industries. The increasing downward pressure on the economy has led to the increasing difficulty of industrial transformation. Second, the similarity of leading industries leads to the delay in industrial transformation. The

Table 6.10 Fiscal revenue and expenditure of some resource-based cities in Northeast China in 2021

City	Local fiscal revenue (10,000 yuan)	Local government expenditure (10,000 yuan)	Difference value (10,000 yuan)
Fushun	769,500	1,641,300	– 871,800
Fuxin	473,300	1,425,700	– 952,400
Panjin	1,585,600	1,964,300	– 378,700
Lioayuan	189,200	1,020,300	– 831,100
Baishan	255,200	1,461,700	– 1,206,500
Songyuan	513,800	2,415,000	– 1,901,200
Jixi	396,300	1,805,500	– 1,409,200
Hegang	282,500	1,212,900	– 930,400
Shuang Yashan	317,300	1,502,800	– 1,185,500
Daqing	1,847,900	3,368,500	– 1,520,600
Qi Taihe	239,900	848,700	– 608,800

Data source Jilin Province 2022 National Economic and Social Development Statistical Bulletin, Liaoning Province 2022 National Economic and Social Development Statistical Bulletin, and Heilongjiang Province 2022 National Economic and Social Development Statistical Bulletin industries in Northeast China are relatively similar, and the local economic structure is relatively simple, manifesting as a more obvious heavy weight characteristic. Since the beginning of the implementation of industrial transformation in Northeast China, it has relied on the local industries that had been well developed previously. Although some advanced technologies have been introduced and the industrial chain has been extended, this form of "industrial transformation" does not meet the real sense of "industrial transformation" and does not achieve its fundamental purpose. Third, there is lack of independent R&D and core technical support, and backward equipment. Northeast China not only contains rich natural resources but also has a good development of its equipment manufacturing industry, which used to be the leading level in the country. However, in the later period, natural resources were overexploited and the industrial level improved slightly. The development of the main industry still comes from its own resource advantages. It still relies on technology imports or can only be responsible for basic tasks such as parts processing, and a large number of core technologies and important equipment still rely on external support.

The industrial transformation in Northeast China is advancing steadily, which is a gradual process. The implementation of relevant strategies and industrial transformation should be actively promoted. In addition, regional marketization should be further promoted, the decisive role of the market in the allocation of resources should be fully demonstrated, the utilization efficiency of resources should be improved, the rational use of resources should be promoted, and resources should be guided to actively flow into high-tech industries. Second, efforts should be made to build a foreign investment platform, further open up the field of foreign investment, and improve the utilization level of foreign investment. In recent years, Northeast China has introduced foreign investment; however, its utilization rate needs to be further improved. For example, petrochemical, natural gas, and other industries can invest part of their capital to further explore the value of the industrial chain and also use part of the capital, technology, and other resources to vigorously develop agriculture, manufacturing, and high-tech industries and fully explore and show the follow-up development capacity of intensive industries in Northeast China. Third, the government should give full play to its functions, actively train and introduce high-tech talents, and provide corresponding policy support. The industrial transformation in Northeast China needs to rely on technological innovation, and the realization of technological innovation needs to rely on a large number of professionals. Industrial transformation can not only be achieved through the support of enterprises and the market but also needs the corresponding support of the government to create a reasonable and fair market competition environment, increase the existing industry types, and create a diversified industrial transformation financing system. Currently, many enterprises have the problem of insufficient funds in the process of transformation. The government should consider establishing a special fund to provide special support and prioritize subsidizing the transformation of enterprises with broad transformation prospects.

In short, there are more inevitable setbacks in the industrial transformation of Northeast China, and further improvement is needed. Since the implementation of industrial transformation, some historical problems such as those related to structural and institutional systems have not been effectively solved, and there have been problems of "three-phase superposition" and the repair of independent economic cycles. Therefore, the steady implementation of industrial transformation and the harmony and stability of Northeast China should be fully based on the sustained and stable development of its economic system and the degree of marketization in all aspects, which requires the cooperation and joint efforts of the local government and related enterprises.

(ii) **Foreign industrial transformation for reference**

Although the measures and countermeasures adopted by various countries in terms of industrial transformation are different, their successful experience can be used as a reference for industrial transformation in China. Based on the basic information of each country's own economic development status and industrial structure, we analyzed the specific practice of industrial transformation in each country, summed up the experience and lessons of industrial transformation in each country, and put forward the following experiences for reference.

Adopt the government-led industrial transformation model

Recognizing the leading role of the government and establishing a unified leading organization is the core of industrial transformation. Some developed countries such as Germany had an earlier industrial transformation and a more mature development system. The establishment of special leading institutions to carry out unified planning, guidance, and implementation of industrial transformation is important for urban agglomerations or individual cities, and the establishment of relevant research, cultivation, and management institutions can effectively promote industrial transformation. Some countries with successful industrial transformation have management institutions and training institutions with clear responsibilities, which can effectively promote coordination and cooperation among various departments and the implementation of policies and regulations (Table 6.11). It is an important task to accelerate the industrial transformation of local areas in our country. Industrial transformation is a huge and complex project requiring the cooperation of various departments. Therefore, special leading institutions should be set up to coordinate the smooth progress of transformation.

Prioritize improving the investment environment

Actively introducing external investment and advanced management technology is an important means to implement industrial transformation, and developing alternative industries is a shortcut to increasing external investment. The economic development level of each region should be fully considered to find suitable alternative industries. The government should actively introduce foreign investment, advanced technology, and concepts and actively introduce corresponding policies on talent introduction to introduce a large number of high-tech talents to comprehensively promote industrial transformation. For example, Japan implemented the coal industry transformation (Fig. 6.11). The successful transformation of the industry was affected by many

Table 6.11 Foreign organizations related to industrial transformation

Mechanism type	Examples
Governing body	The European Union "Steel and Coal Organization," Germany "Ruhr Coal Management Association," France "Ministry of Land Improvement and Production Transformation," Japan "Regional Revitalization and Rectification Corporation," and so on
Breeding institution	France "Enterprise Garden," the EU "European Enterprise Innovation Center," and so on
Intermediary agency	Japan's "Public Employment Stabilization Office," the UK's "Re-employment Resettlement Office," and so on
Financing institution	The European Regional Development Fund, the European Steel Industry Development Fund, the European Structural Foundation, the French Foundation, and the Financial Company for the Promotion of Industrialization of Mining Areas
Training institution	"Cross-industry Training Center" in Ruhr, Germany

factors, among which the quality of the soft environment was an important factor for the introduction of foreign investment, which was inseparable from the strong support of the government and the harmonious business atmosphere.

Change the ownership structure of cities in industrial transformation

The proportion of state-owned economy in China's resource-based cities is relatively high. With the national exploitation of resources, the industrial transformation of resource-based cities is imperative. The ownership structure of resource-based cities is single, and there are many state-owned enterprises, making it difficult to transform the industry and the industrial structure. Some developed countries have rich experience in industrial transformation. China can learn from relevant experience and actively promote small and medium-sized enterprises to establish an ownership structure with various economic components and diversified interests. Germany's Ruhr region vigorously develops emerging industries to promote industrial transformation while innovating traditional industries (Fig. 6.12). Focusing on the diversified

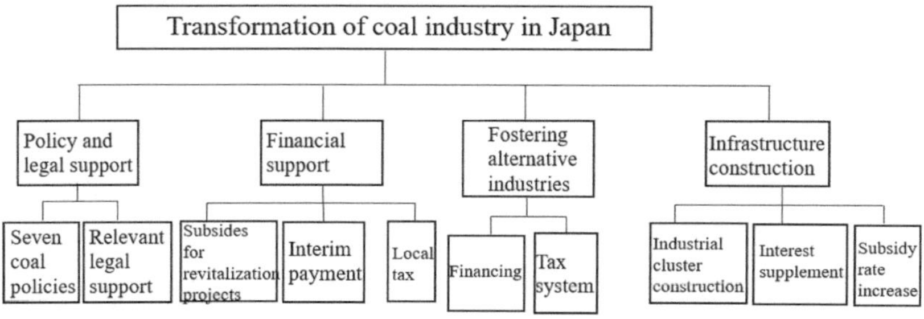

Fig. 6.11 Industrial transformation in Japan

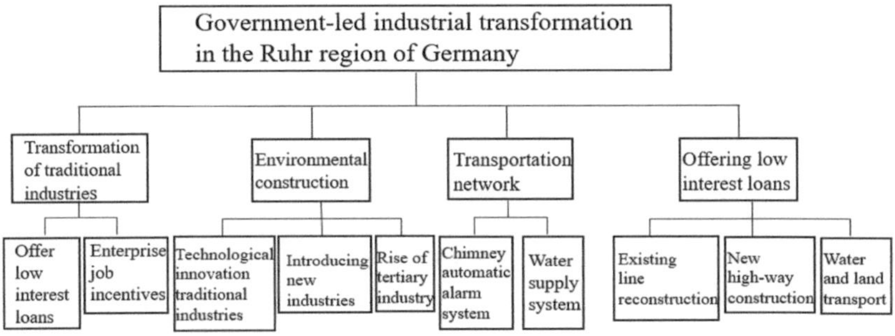

Fig. 6.12 Industrial transformation in the Ruhr region, Germany

development of small and medium-sized enterprises and accelerating the pace of implementation can effectively promote the optimal allocation of various production factors. Accelerating the industrial transformation of resource-based cities can not only promote the economic development of resource-based cities but also effectively alleviate the pressure of urban employment.

Actively provide professional training and encourage individuals to start businesses

Industrial transformation is a long and lasting process, and its biggest problem is the "transformation" of personnel. In particular, some personnel in resource-based industries are only responsible for simple manual labor, which makes re-employment and changing jobs difficult and makes it necessary to continue to learn to adapt to the new job. To promote industrial transformation, while fully considering the needs of urban industrial development combined with the personal wishes of personnel, the French Lorraine region actively provides professional training and vocational skills training, has introduced corresponding preferential policies for entrepreneurship, and provides relevant employment information and corresponding technical support. The government should encourage individual entrepreneurship and provide capital, technology, and other support to entrepreneurs to create a harmonious entrepreneurial environment and atmosphere, solve the employment problem, and attract and retain talents (Fig. 6.13).

The government should provide policy support

To vigorously promote the transformation of urban industry, the government should issue corresponding policies, particularly in resource-based cities. The government should vigorously introduce foreign capital, set up special funds to provide financial support for industrial transformation, and actively seek and develop alternative industries. The government should provide capital, technology, talents, and other resources to promote urban industrial transformation. The government should actively use macro-control measures to promote the transformation of regional industrial structure and economic structure, actively promote the transfer of advantageous industries

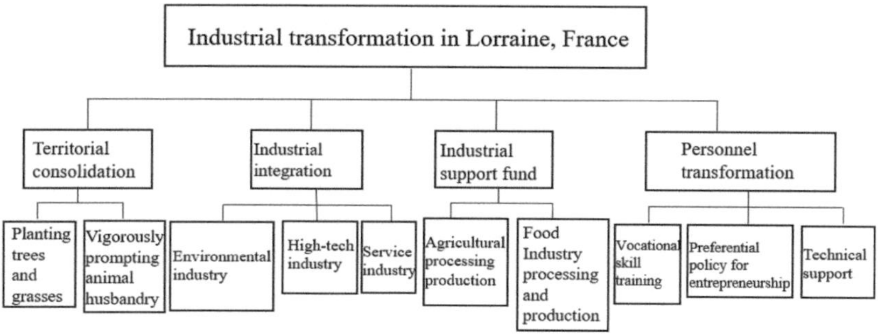

Fig. 6.13 Industrial transformation in Lorraine, France

from economically developed areas to resource-based cities, and provide adequate technical support.

6.3.2 Industry Transformation Development Train of Thought

Currently, China's industrial transformation is facing a bottleneck period. To actively promote the contact and interaction between the government and enterprises, it is necessary to carry out comprehensive sustained strategic planning (Fig. 6.14), that is, pay attention to the realization of the strategic objectives of industrial transformation and fully consider the local reality, expand or shrink the existing industrial scale and industrial organization, and actively build a new industrial system. China should also improve coordination among industries so that industrial transformation in the country can smoothly realize return on investment, product market, and high-quality industry development.

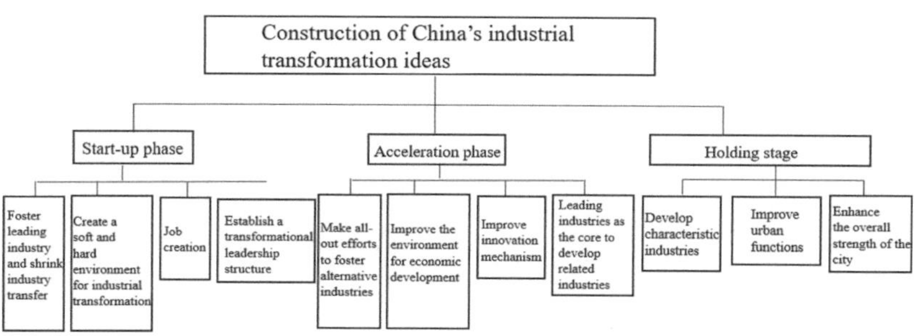

Fig. 6.14 Construction of China's industrial transformation ideas

(i) Improve the mechanism for innovative development

The root of industrial competition is that of innovation ability. China should vigorously promote the ability of scientific and technological innovation, take the promotion of scientific and technological innovation as an important step in China's industrial transformation, and drive the vigorous development of the industry through innovation. Enterprises should be regarded as the main body of innovation and their leading role should be recognized. The government and relevant departments should actively explore leading industries, further promote the development of relevant key enterprises, rationally handle the upstream and downstream relationship of the industrial chain, enhance the independent innovation ability of enterprises, and actively create industrial innovation strategic alliances. Industrial transformation should also pay attention to expanding the talent team, focus on the development of key core technologies and basic common technologies, and actively and effectively promote the transformation of government functions from R&D management to innovative services.

(ii) Accelerate industrial restructuring

Industrial structure affects the quality and development path of regional development. High-quality industrial development is a green, low-carbon, intelligent, and innovative development model. Exploring the development path of industrial transformation, achieving industrial development, and achieving high-quality development are topics of interest in development geography. The optimization of industrial structure is still the key task of industrial development, but it is a long process. In the process of industrial transformation and development, it is necessary to always implement the concept of quality and efficiency as the center, actively promote the development of strategic emerging industries, pay attention to the development of traditional industries and reasonably optimize them, timely eliminate low-end backward production capacity, and further promote the leap of the industrial value chain from the low end to the middle and high end.

(iii) Vigorously promote the development of the real economy

Constructing a new system for China's industrial development and actively developing the real economy are the basis for the orderly development of the industry. To promote China's industrial transformation, the government should create an environment conducive to the development of the real economy. In other words, the government should introduce relevant policies and measures; increase fiscal, tax, and financial support; create a better development environment for the real economy; and promote the further integration and coordinated development of "Internet + manufacturing," laying a solid foundation for industrial transfer. The government should give full play to its leading role, actively implement fiscal and taxation policies, promote the development of the real economy, establish a sound market environment, build and improve a modern financial system that promotes the development of the real economy, further improve the system and mechanism for the integration of manufacturing and information technology, and promote high-level two-way opening-up to the maximum extent. to the government should also make efforts to reduce the difficulties faced in production, such as labor, land, resources, and environmental challenges.

(iv) Comprehensively promote the green and safe development of enterprises

To fully realize industrial transformation and build a new system of industrial development, it is necessary to vigorously promote the green and safe development of enterprises. To seize the development opportunities timeously and seize the first opportunity in industrial competition, China must promote the green and safe development of enterprises, increase the restrictions on some hard indicators such as energy conservation, environmental protection, and safety, and force the formation of a green and sustainable production mode with high scientific and technological levels, low energy consumption, safety, and reliability. to the government should also make efforts to reduce pollutant emissions, promote green and low-carbon production of enterprises, and form a safe and reliable waste treatment system. While actively improving the industrial environment access mechanism, establishing the market incentive and restraint mechanism, and improving the environmental supervision mechanism, it is necessary to further strengthen the main responsibility of enterprise safety production and promote the green and safe development of enterprises.

(v) Establish a fair and equitable market competition environment

A reasonable market competition environment is an important link to building a new industrial system. It is necessary to utilize the synergistic role of the government and the market and build an "effective market" and "promising government" to effectively promote the reasonable free flow of various resources and give full play to their role. To build a new industrial system, it is necessary to actively implement industrial policies and clarify their basic requirements, further improve the mode of industrial support services, optimize enterprise-related services, and innovate the assessment and public opinion guidance mechanism.

The labor-intensive industry plays an important role in China's economic development. Guangxi's light industrial labor-intensive industry involves all aspects of "clothing, food, housing, and transportation." "Clothing" is mainly concentrated in textile and apparel, leather products, and other related industries. "Food" is mainly concentrated in agricultural and sideline food processing, food and beverage manufacturing, and other related industries. "Housing" is mainly concentrated in wood processing, wood, bamboo, rattan, palm, and grass products, furniture manufacturing, and other related industries. "Line" is relatively sparse, mainly reflected in plastic products and other related manufacturing industries.

For a long time, Guangxi has paid attention to the development of labor-intensive industries, which has played an irreplaceable role in the local economic development. Guangxi can meet the local market demand, provide several employment opportunities for the local people, expand import and export trade, and increase export foreign exchange earnings. More importantly, it can provide sufficient development funds for the local area and promote local economic development. According to incomplete statistics, in 2009, the number of labor-intensive enterprises in Guangxi reached 1895, solving the employment problem for nearly 360,000 people, and the total income of enterprises exceeded 150 billion yuan. The increase in labor-intensive enterprises has provided many employment opportunities in Guangxi. The local industrial added value reached 51.117 billion yuan, effectively promoting the development of the local economy. Guangxi has made rational use of its natural resources and vigorously developed advantageous industries, forming a unique development system of agricultural and sideline food processing industry with sugar production, meat and poultry processing, etc., providing sufficient materials for all parts of the country. After 2000, the agricultural and sideline food processing industry in Guangxi developed rapidly, and the number of enterprises gradually increased, exceeding 500 in 2009, providing many jobs and solving the employment problem for nearly 120,000 people. In addition, Guangxi has also formed the food and beverage manufacturing industry represented by canned food and mineral water, and the number of enterprises has gradually increased, exceeding 300 in 2009. The food and beverage manufacturing industry is developing at a rapid pace and has made certain achievements; however, there is still a large gap compared with developed regions, and its development level needs further improvement. The vigorous development of the garment industry in Guangxi benefited from the opportunity of "moving mulberry from the east to the west," which promoted the development of the local cocoon and silk industry.

Although the light industry, textile, food, and other industries in Guangxi have developed rapidly, there is still room for improvement. There are still some defects in the development of the above industries, and problems such as small scale and unreasonable structure need further improvement. First, to promote the development of intensive industries in Guangxi, the government should increase support. This can be done by providing relevant policy support and providing enterprises with advanced technology, talents, and financial support. The government should establish national, autonomous regional, and other levels of special funds to support industrial transformation, create a healthy financing environment, and introduce foreign investment. In addition, the government should promote the technological progress of relevant industrial transformation enterprises and the construction of an enterprise base and independent brand construction to assist enterprises to transform and upgrade, simultaneously expand domestic and foreign markets, and train outstanding talents. Second, the construction of a raw material base and public service system should be accelerated. It is necessary to utilize local resource advantages, promote industrial development, improve industrial development conditions, create a healthy market environment, and prioritize the construction of mulberry cocoon, aquatic products, and wood processing raw material forest bases to promote the development of intensive enterprises in Guangxi and guarantee the provision of adequate raw material. Third, the government should vigorously promote industrial transfer and change the industrial development model. It is also necessary to utilize local resource advantages, policies, and location advantages, provide funds, advanced technology, and outstanding talents to enterprises that actively promote industrial transformation, introduce advanced development concepts and management models, and promote the industrial transformation of Guangxi. Based on the existing level of industrial development, the construction of a characteristic industrial gathering base should continue to improve the level of industry specialization and product quality to achieve the modernization of enterprises and economic development.

6.4 Coordinated Regional Development Modes

A country's development goal determines its development path and affects its basic principle of dealing with the contradiction of elements in the development process. The sustainable development of ecological civilization should strengthen the top-level design, formulation of standards, and legislative protection. In 2015, the United Nations adopted the 2030 Agenda for Sustainable Development, which further formulated the SDGs in the economic, environmental, and social aspects. The Sustainable Development Summit finally established 17 development goals with a total of 169 specific indicators. The conference was convened and the agenda was adopted to achieve the Sustainable Development Goals (SDGs) of green economic growth, social peace and inclusion, and ecological environment by 2030. In March 2016, the Chinese government actively implemented the relevant contents of the agenda, incorporated the agenda into the "13th Five-Year Plan," and integrated it into China's future development planning, achieving integration with China's future planning. In 2016, the United Nations Sustainable Development Agenda released China's National Plan for the Implementation of the 2030 Agenda for Sustainable Development (the China Plan). The China Plan sets out five major points that summarize China's experience in achieving the Millennium Development Goals, opportunities, and challenges in promoting the 2030 Agenda for Sustainable Development, guiding ideology and overall principles, the overall path, and implementation plans for the 17 SDGs.

To implement the goals of the Sustainable Development Agenda, China has formulated measures in three aspects: economic, social, and ecological environment. In terms of sustainable economic development, including specific development goals for sustainable economic growth, sustainable industry, sustainable agriculture, animal husbandry and fisheries, sustainable modern resources, and sustainable urbanization. Social sustainable development mainly includes five development goals in the areas of poverty eradication, healthy lives, quality education, gender equality, and social inclusion. Environmental and ecological sustainable development can be roughly summarized as the specific development goals for coping with climate change, sustainable use of terrestrial ecosystems, protection of oceans and marine resources, protection of wildlife, and prevention of air, soil, and water pollution. According to the national regional coordinated development process, combined with China's measures to implement the goals of the sustainable development agenda, the paths of regional coordinated and sustainable development are divided into ecological agriculture development, ecological immigration, eco-economic-social coordinated development, and ecological civilization development.

6.4.1 Agricultural Modernization Development

Agriculture is the foundation of the survival and development of human society, and its development status directly affects the quality of regional development. The report of the 18th National Congress pointed out that in the future, it is necessary to gradually realize the requirements of the "four modernizations," comprehensively promote the rapid development of modern agriculture, and lay a solid foundation for the realization of the "four modernizations." In 2014, the Party Central Committee issued the "Several Opinions on Comprehensively Deepening Rural Reform and Accelerating Agricultural Modernization" to comprehensively promote the development of modern agriculture. The Committee proposed to promote the rapid development of eco-friendly agriculture, protect the ecological environment, actively promote the smooth construction and operation of the pilot project of agricultural resources rest and growth, and further strengthen the ecological protection construction. To further realize the basic requirements of the "synchronization of the four modernizations" and realize the sustainable development of agriculture, the Party Central Committee proposed to establish a long-term development mechanism. Since 2015, China has provided a series of policy support for the development of ecological agriculture, which aims to promote comprehensive ecological transformation in terms of development objectives, guiding ideology, development mode, implementation means, and management system. After that, China's ecological agriculture concept has been widely recognized by all sectors of society, and a green list plan for the construction of an ecological agriculture system has been proposed (Table 6.12).

Until now, agricultural development has experienced primitive slash-and-burn farming and traditional small-scale agriculture to achieve industrial development. In the past, environmental awareness was relatively weak. Although industrialization was realized, the irrational use of resources caused serious damage to the environment, the greenhouse effect was serious, and the safety of agricultural products was greatly challenged. To integrate and match agriculture with resources and environment and adapt to the ecological environment in advance, the ecological transformation of agricultural direction has emerged. Social and economic benefits could not meet the agricultural goals until later when a new agricultural goal (i.e., ecological environmental benefits and social and cultural benefits) was put forward.

From the 1980s, domestic scholars have referred to the current situation of agricultural development in China and the global trend of science and technology, integrated the advantages of traditional agriculture and modern science and technology, and gradually proposed a new concept and scientific practice of ecological agriculture. This means that after China's agriculture began to implement ecological transformation, its agricultural practice that fully considers ecology is "ecological agriculture." In the past 30 years, China has fully considered its own agricultural development

Table 6.12 Green inventory scheme for China's ecological agriculture system construction

Green list type	Concrete measures	Government programs and grants
Biodiversity utilization	Pest control, rational crop rotation, beneficial microbial utilization, natural vegetation, and germplasm resources protection in agricultural areas	Agricultural invasive biological control project, fertile soil project, soil organic matter enhancement project, germplasm resource protection project
Material recycling	Reusing waste straw, comprehensive utilization of livestock and poultry manure, recycling the waste agricultural film, and reusing organic waste	Straw raising and livestock joint household demonstration project, straw energy utilization subsidy project, circular economy development special project, agricultural clean production demonstration project
Rational landscape layout	Farmland forest network construction, horizontal hedgerows, crop Mosaic layout, rural landscape renovation, windbreak and sand fixation belt construction	Coastal shelterbelt project, plain area forestry project, beautiful countryside construction project, three North Shelterbelt project
Resource saving, substitution, and value-added technology	Fertilizer saving, water saving, energy saving, medicine saving technology, soil fertilizer technology, fishery value-added technology	Dry farming water-saving project, soil testing and fertilizer formula project, rural energy comprehensive construction project
Agricultural pollution and treatment technology	Ecological interception ditch, plant buffer zone along the river, polluted land restoration and replanting, construction of wetland	National water pollution control project and pilot project to repair heavy metal pollution in cultivated land

situation and proposed a set of ecological agriculture theory and technology system suitable for the country's agricultural development. Meanwhile, due to the different natural conditions in different parts of the country and the difference in the level of social and economic development in the same period, China began to implement an ecological agriculture pilot. Initially, the aim of economic benefits, now the core of ecological agriculture, is not only to pursue economic benefits and improve the economic level but also to further reflect its economic, social, and ecological benefits. The existing ecological agriculture in China is a new comprehensive agricultural development system that combines agricultural production, the economic development level of most rural areas, and the overall goal of protecting the environment and reducing pollution. Currently, with the strong support of the government and the active participation of the people, China's agricultural ecological transformation is being actively and orderly promoted. The following observations have been made:

(i) China has been an agricultural country since ancient times and has accumulated great experience in traditional agricultural production. In addition, China has carried out in-depth discussions on the methods and mechanisms of ecological agriculture in the late period and has accumulated much experience during this process. Most importantly, ecological agriculture and agricultural ecological transformation have received strong support from the government.

(ii) China's per capita resource endowment is insufficient; while realizing the transformation of ecological agriculture, the ecological environment bottleneck has appeared. The basic ecological environment has been destroyed, and the awareness of farmers and agricultural enterprises to protect the agricultural ecological environment is not strong; therefore, the enthusiasm for environmental protection should be mobilized. Various measures are required to promote the transformation of agricultural ecology in China. For example, the government should strengthen the pilot construction of different types of ecological agriculture, promote scientists to continue to conduct high-level scientific research, improve innovation capacity, increase the relevant research on agricultural ecological environment, increase environmental protection measures, carry out a comprehensive comparison of agricultural production methods, and preserve scientific and technological achievements. The field application test of the existing results is carried out to confirm the technical achievements and make corresponding adaptive improvements. It is important to reach out to grassroots farmers, discover and summarize the practical experience of the masses, use all kinds of publicity media, widely publicize and carry out the promotion and matching of agro-ecological transformation knowledge, actively enhance the ecological awareness of farmers, and teach ecological knowledge at no cost. All regions should formulate effective relevant systems with strong operability and delineate red lines for the rule of law in the agricultural ecological environment (Luo 2017).

Leisure agriculture emerged in some European countries in the 1930s and 1940s, then, other developed countries adopted it. The Jiangxi Provincial Department of Agriculture established Jiangxi Modern Agricultural Ecological Demonstration Park, drawing on the local and international experience. The establishment of the ecological park is mainly based on the establishment of the Jiangxi Sericulture and Tea Research Institute, which utilizes its rich ecological resources and existing industrial base, relying on its advanced scientific and technological strength and unique geographical advantages. The establishment of the ecological park not only protects the ecological environment but also builds a comprehensive platform integrating demonstration, promotion, sightseeing, leisure, entertainment, experience, and science popularization. The establishment of the ecological park has promoted the rational utilization and in-depth development of agricultural ecological resources in Jiangxi Province, which helps effectively promote the adjustment of agricultural structure, reduce environmental pollution, and improve the ecological environment, and is a comprehensive base platform for improving agricultural science and technology.

The Jiangxi Modern Ecological Agriculture Demonstration Park is a comprehensive platform based on industry, with agriculture and tourism as the model, ecology as the theme, science and technology as the driving force, innovation as the life, and culture as the soul. It now contains nine major industries such as silkworm, tea, seedlings, and fruit, and has developed in an orderly way, gradually evolving into a development pattern of "eight districts, three museums, one station, and one base." Led by the ecological park, ecological agriculture in Jiangxi Province began to develop gradually. In 2012, the annual income of the Jiangxi Modern Ecological Agricultural Demonstration Park reached 50 million yuan, and that year, more than 600,000 local and foreign tourists, 30,000 local primary and secondary school students, 16 foreign experts and scholars, and more than 220 groups (more than 10,000 people) of national and foreign delegations at all levels, generating certain economic benefits. In recent years, the park has won many honors such as national AAAA tourist attractions.

6.4.2 Resettlement of Population in Ecologically Fragile Areas

The differences in natural factors and resource endowments among regions determine the ecological and environmental carrying capacity of regions. The resettlement of a population in ecologically fragile areas is an effective way to realize a well-off society simultaneously. The report of the 17th National Congress of the Communist Party of China stipulates the requirements of a "harmonious development of man and nature," "good ecological environment" and "ecological civilization." The 18th National Congress officially included ecological civilization construction in the overall layout, stressing the need to establish the idea of respecting, adapting to, and protecting nature, fully integrating ecological civilization construction with economic, political, and cultural construction, and all departments actively participating in building a beautiful China and promoting sustainable development. Effective measures to deal with the relationship between environmental protection and human development are the key to the implementation of sustainable development strategies around the world. In 2003, China began to implement the "Regulations on Returning Farmland to Forests," which for the first time proposed to encourage the people to return farmland to forests, implement ecological immigration, and provide corresponding subsidies in accordance with set standards. After that, Northwest China began to promote the ecological immigration project. In 2010, China began to gradually establish an ecological compensation mechanism and formulate corresponding compensation standards, and it took ecological immigration as one of the regional policies to restrict and prohibit development zones. So far, ecological migration has effectively protected the ecological environment in the western region and has become a key measure for local economic poverty alleviation and development.

The central and southern regions of Ningxia are dry and rainless; therefore, there is a need to improve the ecological environment. To respect nature, adapt to nature, protect the ecological environment, improve the level of local economic development, and solve the problem of poverty, Ningxia began to implement the ecological immigration policy. Intercounty ecological immigration is the main immigration strategy in the Twelfth Five-Year Plan period. The stability and sustainable development of the intercounty immigration area is the key to the success of the entire immigration project. As the capital city of Ningxia, Yinchuan has actively participated in the ecological immigration policy, and the three districts of Xingqing, Jinfeng, and Xixia alone have resettled 5800 immigrants, totaling 1387 families, accounting for 15.4% of the total number of immigrants in Yinchuan. Yinchuan City plays a leading role and actively participates in ecological immigration. The relevant departments organized the implementation of ecological immigration in Jingyuan County, and Lingwu City docking with Jingyuan County, settling 10,850 ecological immigrants in 2279 households.

In 2012 and 2013, Jingling New Village carried out four ecological migrations in seven batches. The implementation of an ecological immigration policy aims to effectively protect the ecological environment of Jingyuan County, further improve the living standards of local people, and solve the problem of poverty. Twenty-seven administrative villages in seven townships of Jingyuan County participated in these four migrations, and Jingling New Village received and provided living accommodation for approximately 7000 residents from a total of 1686 households. Among them, Jingling South Village accepted more than 4000 ecological immigrants from a total of 1128 households, and Jingling North Village accepted nearly 3000 ecological immigrants. The labor force (most of whom are manual workers) accounts for less than one-third of the total number of immigrants, and the cultural level is low; 70% have junior-high-school culture. To solve the problem of migrant life, the Jingling new village built a long-date base and breeding park, implemented the circulation contract system, a land circulation fee, and an enclosure circulation contract fee according to the amount of land charges (350 yuan per mu per year and 100 yuan per year, respectively). Considering the living conditions of immigrants, the Jingling new village has a farmer's market for the convenience of immigrants. It also has clinics, village-level activity places, primary schools, kindergartens, etc. to ensure the normal life of immigrants. Local units and enterprises provide support in all aspects, and the government provides assistance through land transfer and policy inclines, so as to constantly change the concept of immigrants and help them adapt to the new life while alleviating poverty.

The concept of "ecological immigration" was proposed in 1900 by an American scholar, who first proposed that "ecological immigration" is a type of immigration carried out for the purpose of environmental protection. With the rapid development of China's economy, the economic conditions have improved; however, the ecological environmental problems are prominent and the poverty problem is still unresolved. The country began to gradually implement the ecological immigration project. In theory, ecological migration is a big project that benefits the country and the people, and its related research can enrich the theory of sustainable development, protect the ecological environment, and further deepen the theory of ecological civilization. Since the ecological migration project began in China, the ecological environment has greatly improved, and poverty problems have been further reduced, people's living standards have greatly improved, and the coordinated economic development of China has been promoted timeously and effectively. The types of ecological migration in China can be divided into several categories according to the purpose, that is, protecting the ecological environment, mainly the source of big rivers; reducing natural disasters and preventing and controlling floods to safeguard and maintain the grassland ecological environment; building water conservancy and hydropower projects to maintain and safeguard the ecosystem of nature reserves or scenic spots. In the 1980s, China began to pay attention to the theoretical research on the relocation of residents in ecologically fragile areas, and gradually carried out the implementation. Currently, China's immigration areas are implemented in the northwest and southwest, and most of these areas have fragile ecological environments and urgently need protection. The project of relocating residents in ecologically fragile areas in China has achieved some results. For example, the degradation of grassland in the "Sanjiangyuan" area has effectively stopped, the ecological environment in Inner Mongolia and other places has been further protected and improved, and the living conditions of the relocated people have greatly improved. Ecological immigration policies effectively protect the ecological environment, improve the living standards of local people, and promote social and economic development to a certain extent; however, the social and cultural adaptation brought by the immigration process is also worth noting. In China, ecological migration has a history of only approximately 20 years, and Ningxia Autonomous Region has implemented five major poverty-alleviation relocation projects in inhospitable areas (Table 6.13). There are still problems that need to be addressed in the current resettlement policy of residents in ecologically fragile areas in China, and there is a lack of comprehensive scientific planning and laws and regulations. All departments should actively provide support to establish and improve the policies and regulations of ecological migration. People are reluctant to leave their hometowns where their ancestors lived for generations, and the acceptance of ecological immigrants is low. Some immigrants are once again struggling to find jobs due to problems such as insufficient protection of minorities. The resettlement of residents in ecologically fragile areas is still a challenge. In the future, the government should give full play to the advantages of ecological migration projects in establishing and improving ecological migration management mechanisms, ecological migration publicity, and education, improving relevant laws and regulations, and developing migrant assistance to promote the sustainable development of ecological civilization.

6.4.3 Economic and Social Transformation and Development

Although China has provided great help in promoting the realization of the United Nations Millennium Development Goals, with the improvement of the economic

Table 6.13 Five poverty-alleviation relocation projects in inhospitable areas of Ningxia

Project	Year	Relocation population (thousand)	Per capita input (Yuan)
A surname	1983 ~ 2000	198	931
Poverty alleviation and Yanghuang irrigation project	1998 ~ 2015	308	9059
Poverty alleviation project for relocation from inhospitable areas	2001 ~ 2007	153	9036
"Eleventh Five-Year Plan" central arid zone county ecological immigration	2008 ~ 2012	153.6	18,000
"Twelfth Five-Year Plan" central and southern region ecological immigration	2011 ~ 2015	346	28,800

Data source Poverty Alleviation Bureau of Ningxia Autonomous Region

level, environmental and ecological problems are increasingly aggravated, and the poor population is still more than 10 million, China still faces many problems and challenges on the road to sustainable development in the future. With the three major strategies of "One Belt and One Road," Beijing-Tianjin-Hebei coordinated development, and Yangtze River Economic Belt development, regional ecology-economy-society coordinated sustainable development has become a new trend of China's regional development strategy.

In 2016, to promote the coordinated healthy and sustainable development of the global economy, ecology, and society, the Global Action Goal Network for Sustainable Development actively joined the Bertelsmann Foundation in the United States and jointly developed a set of indicators, one for each of the 17 Global Sustainable Development Goals (SDGs). A set of specific analytical methods and indicators are designed to help countries around the world identify the issues and challenges they face in achieving the SDGs. In this standard system, three colors (green, yellow, and red), are used for coding, and the degree of realization of the goal is presented based on color differences. Green means that the goal has been achieved. Yellow means that the goal has begun to be implemented but needs improvement, and reddish-brown means that the goal is facing severe challenges. Some scholars have ranked China's relevant indexes according to this index system, and the 2016 SDG index ranked 76, with a score of 59.1, which is in the middle of the world level. Nine items are red, five are yellow, and three are green. The nine red goals are SDG2 (ending hunger), SDG3 (good health and well-being), SDG9 (industry, innovation, and infrastructure), SDG10 (reducing the inequality gap), SDG11 (sustainable cities and communities), SDG13 (climate action), SDG14 (underwater life), SDG15 (terrestrial life) and SDG16 (peace, justice, and strong institutions). In 2018, China's SDG index ranked 54th, and the overall ranking was in the upper reaches. The evaluation results show that China has been facing major challenges in economic and social equity, agricultural water resources pollution, environmental protection, and other aspects. It also points out the guiding ideology, coping strategies, and specific policy implementation measures and programs. In the future, in the process of further realizing the SDGs, we should promote the process of sustainable development by improving the working mechanism, attaching importance to the construction of ecological civilization, strengthening the coordinated development of ecology, economy, and society, and raising the awareness of public responsibility.

6.4.4 Promoting Ecological Progress

Entering the new era of the development of socialism with Chinese characteristics, the principal contradiction in Chinese society has undergone a fundamental transformation, and there are still contradictions between the demand and supply of beautiful environments and high-quality ecological products in the people's ever-growing pursuit of a better life. This is also the major strategic judgment of the Party Central Committee on the construction of ecological civilization in the "three-phase superposition." The starting point of the Party's governance is people-oriented. In the conflict between development and environmental protection, it is emphasized that economic development "cannot be at the expense of the ecological environment." Specifically, the critical period refers to when the industrial structure is biased, the energy structure is biased toward fossil energy, the industrial layout is more human-led, the ecological environment protection between urban and rural areas is uncoordinated, the phenomenon of polluting enterprises "going up the mountain to the countryside" is more prominent, the ecological degradation is still serious, and the pollution brought by new business forms cannot be ignored. Presently, we are in the decisive period of building a moderately prosperous society in all respects. If the ecological and environmental issues are not properly managed, it will be difficult to achieve high-quality development and build a moderately prosperous society recognized by the people. In the difficult period, the current distance from the people's yearning for a better life is still long. For example, although the 2017 "atmosphere" was successfully concluded, air pollution prevention and control, including industrial structure adjustment, energy distribution adjustment, transportation structure optimization, and other deep-seated problems still need to be solved. China is now emphasizing the fight against pollution prevention and control, focusing on these aspects. During the past 40 or more years of reform and opening up, people have not only written a magnificent epic of national and national development but also accumulated rich material, financial, institutional, and technical conditions for solving ecological problems and pollution prevention and control. This shows that we have the foundation, ability, and confidence to achieve the goals of ecological conservation. The great significance of the "three phases of superposition" is that it explains the difficulty of the task of waiting for no time and falling back if we do not advance. It points out the extreme importance of accelerating the improvement of the weak links in ecological and environmental protection and the objective foundation of having the conditions and ability. Therefore, to promote the construction of ecological civilization, we must have an objective understanding, guard against arrogance and impetuousness, overcome difficulties, and strengthen efforts to unswervingly follow the path of ecological civilization development.

The concept of ecological civilization has been fully integrated into China's major regional development strategy in the new era. It includes ecological progress in the five-in-one overall plan for the cause of socialism with Chinese characteristics, clearly states that we should vigorously promote ecological progress, strive to build a beautiful China, coordinate the five-in-one overall plan for the sustainable development of the Chinese nation, and realize the comprehensive and coordinated development of China's material, political, spiritual, social, and ecological progress.

In accordance with the revised vision of development, we will closely monitor the evolution of the most pressing issues confronting Chinese society. In addition, we will adhere to the stipulations pertaining to high-quality development, pursue the overarching five-in-one plan and the four-pronged comprehensive strategy in a unified manner, and strive to capitalize on the distinctive advantages of diverse regions, thereby reducing regional disparities in development. To achieve equal access to basic public services, coordinated access to infrastructure, and a similar level of basic living security for the population, it is necessary to deepen reform and opening up. This will entail the resolute breaking down of interest and policy barriers among regions, as well as the acceleration of the formation of a new mechanism for coordinated regional development featuring strong overall planning, orderly competition, green coordination, and mutual benefit.

The promotion of coordinated regional development in the country will be a key objective. In line with the overarching strategies of the Belt and Road Initiative, the coordinated development of the Beijing-Tianjin-Hebei Region, the Yangtze River Economic Belt, and the Guangdong-Hong Kong-Macao Greater Bay Area, China aims to promote inter-regional integration and complementarity, with a particular focus on the western, northeastern, central, and eastern regions. The Belt and Road Initiative will facilitate the coordinated opening up of coastal, inland, and border regions, enhance major infrastructure connectivity with international economic cooperation corridors as the backbone, and establish a new regional development pattern that coordinates domestic and international development, encompassing the east, west, south, and north. The Beijing-Tianjin-Hebei region will be promoted in a coordinated manner, with adjustments made to both the regional economic and spatial structure. The development of the Xiongan New Area in Hebei Province and the subcenter of the city of Beijing will be facilitated, and an optimized development model will be explored to disperse the functions of densely populated and economically dense areas, including megacities, in an orderly way, and effectively controlling the phenomenon known as the "big city disease."

The geographical advantages of the Yangtze River Economic Belt, which spans the East, West, and South, will be fully exploited. The coordinated development of the upper, middle, and lower reaches of the Yangtze River, along with the high-quality development of the regions situated along the river, will be promoted. This will be achieved through joint efforts to protect the Yangtze River and to refrain from large-scale development. Furthermore, the ecological priority and green development principles will be adhered to.

A new model will be established, in which central cities will assume a leading role in the development of urban clusters, which in turn will drive regional development. This will facilitate integrated and interactive development among regional sectors. The Beijing-Tianjin-Hebei city cluster will be developed as a priority, with Beijing and Tianjin at its center, to facilitate the coordinated development of the Bohai Rim region. The city of Shanghai will serve as the focal point for the advancement of the Yangtze River Delta urban agglomeration, thereby propelling the growth of the Yangtze River Economic Belt. The Guangdong-Hong Kong-Macao Greater Bay Area will be led by Hong Kong, Macao, Guangzhou, and Shenzhen, with the objective of driving the innovative and green development of the Pearl River-Xijiang Economic Belt. With Chongqing, Chengdu, Wuhan, Zhengzhou, and Xi'an as the center, the objective is to lead the development of city clusters in Chengdu-Chongqing, the middle reaches of the Yangtze River, the Central Plains, and the Guanzhong Plain and drive the integrated development of relevant sectors.

We intend to enhance the coordination and synergy between the major strategies, including the Belt and Road Initiative, the coordinated development of the Beijing-Tianjin-Hebei Region, the Yangtze River Economic Belt, and the Guangdong-Hong Kong-Macao Greater Bay Area, with a view to promoting inter-regional cooperation. Hainan will pursue the comprehensive deepening of reform and opening up, with the objective of advancing the establishment of a pilot free trade zone and investigating the potential for developing a free trade port in accordance with Chinese characteristics.

6.5 Summary

This chapter explores the factors influencing regional development imbalances in China, including differences in elements, quality, models, and systems. It highlights the necessity of balancing efficiency with fairness, aiming for improved development quality and regional equity. The chapter uses a comprehensive analysis approach to examine regional development disparities and outlines strategies for achieving balanced growth across eastern, central, and western China, guided by government policies and market mechanisms. The chapter emphasizes that regional development evolution reflects broader economic, political, social, and cultural changes, and further clarifies regional development mechanisms. Additionally, the chapter underscores the importance of spatial planning and optimization in national macro-control, advocating, balanced regional development and resilience. It stresses the need for green, sustainable development and ecological civilization.

Exercise

1. What are the major challenges facing China's coordinated regional development?
2. How important is the matching of population and economic distribution in China's coordinated regional development?
3. Analyze the sustainability of the trend of coordinated regional economic development in China, particularly the necessity of industrial restructuring in the western region.
4. Describe the evolution of China's regional development thinking and suggest how policy and institutional safeguards can be improved in the future outlook.
5. Analyzes the impact of the five development concepts on the coordinated development of China's regional economy and its heterogeneity among different regions.
6. What is the impact of the western development strategy on the coordinated regional development in China? Please evaluate it from the aspects of social and economic output, people's living standards, and ecological environment.

Extended Reading

Chen, B., Lu, M., Timmins, C., & Xiang, K. (2019). *Spatial misallocation: Evaluating place-based policies using a natural experiment in China* (Working Paper 26148). National Bureau of Economic Research. https://doi.org/10.3386/w26148
Deng, X., Liang, L., Wu, F., Wang, Z., & He, S. (2022). A review of the balance of regional development in China from the perspective of development geography. *Journal of Geographical Sciences, 32*(1), 3–22. https://doi.org/10.1007/s11442-021-1930-0

References

Luo, S. M. (2017). Agroecology transition and suitable pathway for eco-agricultural development in China. *Chinese Journal of Eco-Agriculture, 25*(1), 1–7.
National Bureau of Statistics of China. (2001). China statistical yearbook 2001. Beijing, China: China Statistics Press.
National Bureau of Statistics of China. (2004). China statistical yearbook 2004. Beijing, China: China Statistics Press.

Chapter 7
Income Inequality and Poverty

Abstract Income disparity is a prevalent economic phenomenon and a key area of study in the field of income distribution. Understanding the growth of income distribution inequality in a nation or region, reducing the gap in income distribution, enhancing the income distribution system, and eradicating poverty all depend on the research on income inequality. The primary topics of discussion and analysis in this chapter are poverty and income inequality. Before proposing appropriate policy measures to address income disparity and poverty, the theoretical underpinnings of these issues are highlighted, followed by a measurement and characterization of income inequality and poverty. The objectives of this chapter are as follows: 1. To master the measurement method, representation, and theoretical basis of income inequality and poverty. 2. To understand policies to address income inequality and poverty.

7.1 Measurement of Income Inequality

7.1.1 Measurement Methods

Measuring income inequality is the fundamental tenet of income inequality research. Absolute and relative indicators can be used to categorize the traditional measurement markers of income inequality. Examples of absolute indicators include range, mean difference, and standard deviation. Relative indicators (Kovacevic, 2010), such as the coefficient of variation,Gini coefficient,generalized entropy (GE)index,Theil index,and Atkinson index,are frequently used to quantify income inequality as they exclude the impact of dimensions.

7.1.1.1 Absolute Index

(i) Average difference

The mean difference is the arithmetic mean of the deviation as shown in Eq. 7.1:

X. Deng and M. Song, *Development Geography*,
https://doi.org/10.1007/978-981-96-6917-2_7

$$D = \frac{\sum_{i=1}^{N} \left| Y_i - \overline{Y} \right|}{N} \tag{7.1}$$

where Y_i is the resident income of the i th sample, $\overline{Y}$ is the average income of all residents, and N is the number of samples.

(ii) Standard deviation

Standard deviation, also known as mean square deviation, is an index that reflects the degree of dispersion and is mostly used in statistics. Similar to the mean difference, standard deviation also reflects the deviation degree of each sample index value from its arithmetic mean value. The larger the value, the smaller the absolute difference.

$$S = \frac{\sum_{i=1}^{N} (Y_i - \overline{Y})^2}{N} \tag{7.2}$$

where Y_i is the resident income of the i th sample, $\overline{Y}$ is the average income of all residents, and N is the number of samples.

(iii) Extremely poor

Range, also known as full range, refers to the difference between the maximum and minimum value of a variable, which is used to show the range and magnitude of the variable, as shown in Eq. 7.3:

$$R = Y_{\max} - Y_{\min} \tag{7.3}$$

where Y_{max} is the highest income of residents in the sample and Y_{min} is the lowest income of residents in the sample.

7.1.1.2 Relative Index

(i) Coefficient of variation

The coefficient of variation is the ratio of the standard deviation of the sample value to its mean value, which reflects the relative degree of inequality in the deviation of residents' income from the national per capita resident income.

$$V = \frac{\sqrt{\frac{1}{N} \sum_i (Y_i - \overline{Y})^2}}{\overline{Y}} \tag{7.4}$$

where Y_i is the resident income of the sample i, $\overline{Y}$ is the average income of all residents, and N is the number of samples.

(ii) Gini coefficient

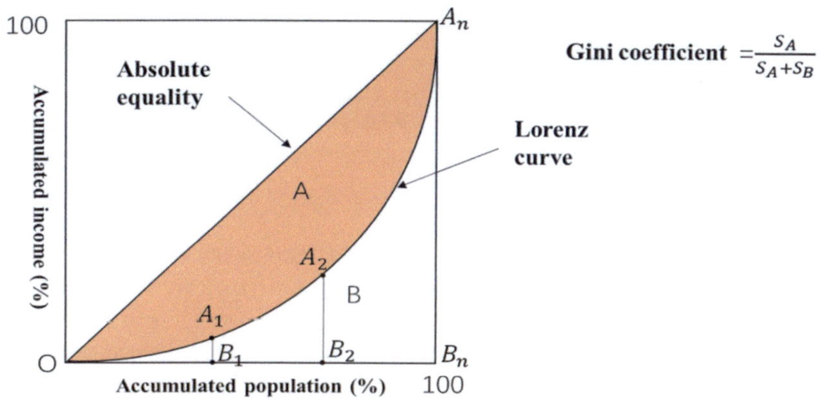

Fig. 7.1 Lorenz curve

The Gini coefficient is a commonly used indicator for measuring and comparing income inequality worldwide. It is computed using the Lorenz curve, which was developed by statistician Lorenz in 1907. The Lorenz curve has a 45-degree diagonal line known as the "line of absolute equality" (Fig. 7.1), which denotes that everyone has exactly the same income share. The horizontal axis of the curve displays the cumulative population in the order of income from lowest to highest. The vertical axis shows the cumulative income percentage. The Lorenz curve falls below the line of absolute equality whenever there is a deviation from equality; the bigger the divergence, the greater the inequality. The Gini coefficient and other scale-independent measures of inequality are based on the Lorenz curve.

We can assess the degree of income inequality among inhabitants by examining the Lorenz curve's curvature. The greater the curvature of the Lorenz curve, the more unequal the income distribution. A drawback of the Lorenz curve analysis is its inability to produce a precise measure of income inequality. This limitation was later resolved by Italian economist Gini, who initially distinguished between two types of areas, namely the "unequal area" and the "equal area" (denoted as areas A and B in Fig. 7.1, respectively), based on the Lorenz curve. The "completely unequal area" is the sum of the two areas. As shown in Fig. 7.1, the ratio of the area of inequality to the area of perfect inequality represents the Gini coefficient.

$$Gini = \frac{A}{A+B} = 2A = 2(0.5 - B) = 1 - 2B \tag{7.5}$$

Although the definition of the Gini coefficient is intuitive, the equation above is not applicable, and economists have developed a direct method to calculate the area of region B according to the definition. As shown in Fig. 7.1, divide the axis into n parts, each length is *1/n*, and the corresponding point is $B_1, B_2, \ldots, B_n$, and make the vertical line of the horizontal axis $A_1B_1, A_2B_2, \ldots, A_nB_n$. The vertical line represents the cumulative percentage $P_1, \ldots, P_n$ of income and the proportion of the total income of each class in the total income, respectively.

$$\begin{aligned} A_1B_1 &= P_1 \\ A_2B_2 &= P_1 + P_2 \\ &\cdots \\ A_nB_n &= P_1 + P_2 + \cdots + P_n = 100\% \end{aligned} \tag{7.6}$$

When the Gini is below 0.2, the distribution of income is perfectly equal. When Gini is between 0.2 and 0.3, income distribution is more balanced. When Gini is 0.3–0.4, it is relatively equitable. When Gini is 0.4–0.6, the income gap is substantial. When Gini is higher than 0.6, the income distribution is heavily skewed toward some people over others.

7.1.1.3 Calculation Method

(i) Simple measurement method

In the case of five income groups, it is assumed that the income proportions of the lowest, lower middle, middle, upper middle, and highest income group form an arithmetic series, and the tolerance is D; then, there is $P_2 = P_1 + D$, $P_3 = P_1 + 2D$, $P_4 = P_1 + 3D$, $P_5 = P_1 + 4D$. The direct formula for calculating the Gini coefficient is:

$$\begin{aligned} G &= \frac{1}{n}\sum_{i=2}^{n}\sum_{1=j<i}^{n}\left|P_i - P_j\right| \\ &= \frac{1}{5}[(P_2 - P_1) \\ &\quad + (P_3 - P_1) + (P_3 - P_2) \\ &\quad +(P_4 - P_1) + (P_4 - P_2) + (P_4 - P_3) \\ &\quad +(P_5 - P_1) + (P_5 - P_2) + (P_5 - P_3) + (P_5 - P_4)] \\ &= P_5 - P_1 \end{aligned} \tag{7.7}$$

This calculation shows that the Gini coefficient is approximately equal to the difference between the income percentage of the top and bottom income groups. The simple formula yields a Gini coefficient that is fairly close to the value provided by the World Bank. However, it is important to note that only the lowest income group, the lower middle-income group, and the middle-income group follow an arithmetic sequence, whereas the other groups do not.

(ii) Population equalization method

The population equalization method is also used to find the Gini coefficient using the Lorenz curve. It is a practical method that does not depend on the functional relationship of the Lorenz curve The disadvantage of this method is that it can underestimate the area A by approximating the arc with a line segment. However, the larger the number of groups, the smaller the error. Therefore, the size of the error can be controlled by changing the number of groups (n). The basic steps to calculate the Gini coefficient using the population equalization method are as follows. First, according to the statistical survey data, the residents (or families) are ranked from low to high according to income level, and then the residents (or households) are divided into different groups according to income and the same number of residents (or households). Second, the average income level of each income group mentioned above is calculated, which is represented by the average income level. Third, by comparing the average income level of each income group, we can obtain an intuitive measure of the overall income distribution of residents.

The approximate calculation formula of the Gini coefficient in the population equalization method is as follows:

$$G = 2A = \frac{2}{n}(P_1 + 2P_2 + \cdots + nP_n) - \frac{n+1}{n} \tag{7.8}$$

(iii) Grouping method

In the grouping method, the entire population (or households) is first arranged in increasing order according to income and divided into n groups. Let the population share and income share of group i be denoted by ω_i and P_i, respectively, and the cumulative income share be denoted by I_i then, the Gini coefficient is calculated as follows: $\omega_i P_i I_i$

$$G = \sum w_i P_i + 2\sum w_i(1 - I_i) - 1 \tag{7.9}$$

Calculating the Gini coefficient by grouping will underestimate it, and the courser the grouping, the greater the underestimation. Research by development economist Sundrum (1990) shows that if there are more than 10 groups, the impact on the Gini coefficient's accuracy is negligible.

7.1.1.4 Calculation of the National Gini Coefficient

In many developing countries, including China until 2012, urban and rural household income and expenditure surveys are conducted separately, which makes calculating the national Gini coefficient a statistical problem. Sundrum (1990) proposed a decomposition method for calculating the Gini coefficient that applies to groups of non-overlapping populations (rich and poor), as follows:

$$G = w_1^2 \frac{\mu_1}{\mu} G_1 + w_2^2 \frac{\mu_2}{\mu} G_2 + |w_1 w_2 \frac{\mu_1 - \mu_2}{\mu}| \tag{7.10}$$

where w_1 and w_2 are the proportions of the urban and rural populations in the total population, respectively. μ_1, μ_2, and μ are the urban, rural, and national per capita income, respectively. G, G_1, andG_2 are the national, urban, and rural Gini coefficients, respectively. This means that the national Gini coefficient is calculated from those of the urban and rural areas through weights and corrections. However, when calculating the weighted urban–rural Gini coefficient, it should be noted that the national income gap exists not only between urban and rural areas but also within these areas. Therefore, these income disparities should also be considered when calculating the national Gini coefficient (Sicular et al., 2007). In addition, the "grouping weighting method" assumes no overlap in the income distribution of urban and rural areas, which may not have been the case before the 2013 urban–rural integration survey. Araar (2006) argued that, without considering the overlapping factors, the national Gini coefficient calculated by the urban–rural weighting method would be underestimated.

Since China's population does not meet the condition of non-overlapping groups, it is necessary to modify the above method. Accordingly, The formula is provided for calculating the national Gini coefficient under overlapping urban and rural incomes as follows:

$$\begin{aligned} G = {} & w_1^2\frac{\mu_1}{\mu}G_1 + w_2^2\frac{\mu_2}{\mu}G_2 \\ & + w_1w_2\frac{\mu_1-\mu_2}{\mu}\left\{1+\frac{1}{2}P[(X-Y)<0]\right\} \end{aligned} \tag{7.11}$$

where $P[(X-Y)<0]$ is the revised weighting coefficient, and XY is the random variable of the income of urban residents and rural residents, respectively. Compared with Eq. 7.10, Eq. 7.11 adds a correction coefficient in the third term to correct the underestimation of the national Gini coefficient. This method is called the modified urban–rural weighting method.

(i) Generalized entropy index

The generalized entropy (GE) index is commonly used to study income inequality worldwide, as shown in Eq. 7.12.

$$GE = \begin{cases} \sum_{i=1}^{n} V_i\left[\left(\frac{Y_i}{\overline{Y}}\right)^c - 1\right], & c \neq 0, 1 \\ \sum_{i=1}^{n} V_i\left[\left(\frac{Y_i}{\overline{Y}}\right)^c - 1\right], & c = 1 \\ \sum_{i=1}^{n} V_i\left[\left(\frac{Y_i}{\overline{Y}}\right)^c - 1\right], & c = 0 \end{cases} \tag{7.12}$$

where Y_i is the resident income of the i th sample, $\overline{Y}$ is the average income of all residents, and n is the number of samples. V_i is the population weight of the i th sample, and the parameter c reflects sensitivity to income transfer. The GE index is concretized into the Theil index, which gives the same weight to part of the gap with different levels of development. At that time, it is concretized into the mean log deviation index (MLD index), which gives more weight to the low-level part of the gap. The larger the GE index, the larger the income gap, and the higher the income inequality (Aaberge et al., 2002); otherwise, the lower the income inequality.

The GE index can be decomposed according to within-group and between-group inequality, and the within-group and between-group inequality can be integrated into overall inequality. The Theil index can decompose the same income inequality into the within-group and between-group income inequality and also measure the changes in income inequality and reveal the direction and range of changes in between-group and within-group income inequality.

(ii) Atkinson index

The Atkinson index is a method proposed to measure inequality using the social welfare function based on the reference and examination of other inequality measurement methods. The index y_ε first computes an equivalent sensitive average income, ε is the inequality aversion function, and μ is the average income. The Atkinson index is expressed as follows:

$$Atkinson = 1 - \frac{y_\varepsilon}{\mu} \tag{7.13}$$

According to the formula of the index, the fairer the distribution of social income, the closer the equivalent sensitive average income is to the average income, and the smaller the value of the Atkinson index. For any distribution, the value of the Atkinson index is between 0 and 1, where 0 means that the distribution of social income is perfectly equal.

7.1.2 *Analysis of Chinese Residents' Income Inequality*

The measurement and analysis of Chinese residents' income inequality include the changes in differences in household income between urban and rural Chinese residents and the measurement and analysis of the Gini coefficient for national, urban and rural residents.

7.1.2.1 Changes in the Differences in Household Income Between Urban and Rural Residents

The income differences between urban and rural households ar reflected both between and within urban and rural areas (Chen & Zhou, 2005). Tables 7.1 and 7.2 show the per capita income levels of rural and urban households in different groups from

2002 to 2017. Based on the income of different groups in different years (Turnovsky, 1977), the mean, standard deviation, and variation coefficient of the per capita income levels of urban and rural households are also calculated to reflect the internal income differences between urban and rural areas.

Table 7.2 shows that, from 2002 to 2017, the income of Chinese rural households greatly improved for both low- and high-income households. The per capita nominal income level of low-income households increased from 857.1 yuan in 2002 to 3301.9 yuan in 2017, a nearly 3.85-fold nominal increase. The per capita nominal income of high-income households increased from 5895.6 yuan in 2002 to 31,299.3 yuan in 2017, a nearly 5.3-fold nominal increase. The overall per capita nominal income increased from 2699 yuan in 2002 to 14,374.3 yuan in 2017, a nearly 5.3-fold nominal increase. In addition, there is a large difference in the per capita income within China's rural households. In terms of intra-group differences, in the rural quintiles in 2002, the per capita net income of high-income households was 6.88 times that of low-income households, the variance of the income group was 1751.2, and the coefficient of variation was 0.649. In 2017, the per capita disposable income of high-income households was 9.48 times that of low-income households, increasing nearly 2.6 times. The variance of the income group was 9567.7, and the coefficient of variation was 0.666. This shows that the income disparity among rural residents in China is relatively large, the income level of low-income households is much lower than that of high-income households, and the income gap between the two is increasing.

Table 7.2 shows the per capita income levels of urban households in different groups from 2002 to 2017. From 2002 to 2012, the per capita income of urban households increased significantly. The per capita disposable income of the lowest-income households increased 3.4 times from 2409 yuan in 2002 to 8215 yuan in 2012, and the per capita disposable income of the highest income households increased 3.36 times from 18,999 yuan in 2002 to 63,824 yuan in 2012. The per capita disposable income of low-income households (the lowest-income households + low-income households) increased from 6058 yuan in 2002 to 13,723 yuan in 2017, a 2.27-fold increase. The per capita disposable income of high-income households (high-income households + highest-income households) increased from 30,772 yuan in 2002 to 77,097 yuan in 2017, a 2.5-fold increase. On average, the per capita disposable income of urban households increased from 8184 yuan in 2002 to 38,863 yuan in 2017, a 4.75-fold increase. In terms of the income gap between urban households, the per capita disposable income of the lowest-income households in 2002 was 2409 yuan, whereas that of the highest-income households was 11,773 yuan. The latter was 7.88 times that of the former. In 2017, the per capita disposable income of high-income urban households was 38,863 yuan, 5.6 times that of low-income urban households. Although the income level of urban residents has greatly improved, the income difference among urban residents has also gradually increased.

The income gap exists not only within but also between urban and rural areas. As shown in Tables 7.1 and 7.2, in 2002, the per capita net income level of the highest-income rural households was only between the income levels of the middle- and low-income urban households and the middle-income urban households. In 2017, the per capita disposable income of low-income urban households was 13,723 yuan,

Table 7.1 Per capita income levels of rural households in different groups from 2002 to 2017

Year	Low-income households	Low- and middle-income households	Middle-income households	Middle- and high-income households	High-income households	Mean	SD	CV
2002	857.1	1547.5	2164.1	3030.5	5895.6	2699.0	1751.2	0.649
2003	865.9	1606.5	2273.1	3206.8	6346.9	2859.8	1906.4	0.667
2004	1006.9	1842.0	2578.5	3607.7	6930.7	3193.2	2055.4	0.644
2005	1067.2	2018.3	2851.0	4003.3	7747.4	3537.4	2316.1	0.655
2006	1182.5	2222.0	3148.5	4446.6	8474.8	3894.9	2529.5	0.649
2007	1346.9	2581.7	3658.8	5129.8	9790.7	4501.6	2922.9	0.649
2008	1499.8	2935.0	4203.1	5928.6	11,290.2	5171.3	3389.2	0.655
2009	1549.3	3110.1	4502.1	6467.6	12,319.1	5589.6	3733.9	0.668
2010	1869.8	3621.2	5221.7	7440.6	14,049.7	6440.6	4224.3	0.656
2011	2000.5	4255.7	6207.7	8893.6	16,783.1	7628.1	5108.3	0.670
2012	2316.2	4807.5	7041.0	10,142.1	19,008.9	8663.1	5779.3	0.667
2013	2877.9	5965.6	8438.3	11,816.0	21,323.7	10,084.3	6339.2	0.629
2014	2768.1	6604.4	9503.9	13,449.2	23,947.4	11,254.6	7247.5	0.644
2015	3085.6	7220.9	10,310.6	14,537.3	26,013.9	12,233.7	7844.9	0.641
2016	3006.5	7827.7	11,159.1	15,727.4	28,448.0	13,233.7	8669.7	0.655
2017	3301.9	8348.6	11,978.0	16,943.6	31,299.3	14,374.3	9567.7	0.666
2018	3666.2	8508.5	12,530.2	18,051.5	34,042.6	15,359.8	10,468.8	0.682
2019	4262.6	9754.1	13,984.2	19,732.4	36,049.4	16,756.5	10,898.5	0.650

(continued)

Table 7.1 (continued)

Year	Low-income households	Low- and middle-income households	Middle-income households	Middle- and high-income households	High-income households	Mean	SD	CV
2020	4681.5	10,391.6	14,711.7	20,884.5	38,520.3	17,837.9	11,622.0	0.652
2021	4855.9	11,585.8	16,546.4	23,167.3	43,081.5	19,847.4	13,074.7	0.659

Note The period from 2002 to 2012 is the per capita net income of rural residents. Since 2013, the National Bureau of Statistics has conducted the Survey on Income and Expenditure and Living Conditions of Urban and Rural Integrated Households. The data from 2013 to 2017 are from this survey, which is the per capita disposable income of rural residents. The data different from the survey scope, survey method, and index caliber of urban and rural households before 2013. The data are from the China Statistical Yearbook 2003–2018 and the author's calculations. SD and CV refer to standard deviation and coefficient of variation, respectively

Table 7.2 Per capita income levels of urban households in different groups from 2002 to 2017

Year	Minimum	Low income	Medium lower income	Medium income	Medium upper income	High income	Highest income	Mean	SD	CV
2002	2409	3649	4932	6657	8870	11,773	18,999	8184	5307	0.648
2003	2590	3970	5377	7279	9763	13,123	21,837	9134	6150	0.673
2004	2862	4429	6024	8167	11,051	14,971	25,377	10,412	7194	0.691
2005	3135	4885	6711	9190	12,603	17,203	28,773	11,786	8225	0.698
2006	3569	5541	7554	10,270	14,049	19,069	31,967	13,146	9097	0.692
2007	4210	6505	8901	12,042	16,386	22,234	36,785	15,294	10,439	0.683
2008	4754	7363	10,196	13,984	19,254	26,250	43,614	17,916	12,486	0.697
2009	5253	8162	11,244	15,400	21,018	28,387	46,826	19,470	13,334	0.685
2010	5948	9285	12,702	17,224	23,189	31,044	51,432	21,546	14,524	0.674
2011	6876	10,672	14,498	19,545	26,420	35,579	58,842	24,633	16,614	0.674
2012	8215	12,489	16,761	22,419	29,814	39,605	63,824	27,590	17,767	0.644
2013	9896		17,628	24,173	32,614	57,762		28,415	16,469	0.580
2014	11,219		19,651	26,651	35,631	61,615		30,953	17,307	0.559
2015	12,231		21,446	29,105	38,572	65,082		33,287	18,110	0.544
2016	13,004		23,055	31,522	41,806	70,348		35,947	19,646	0.547
2017	13,723		24,550	33,781	45,163	77,097		38,863	21,745	0.560
2018	14,387		24,857	35,196	49,174	84,907		41,704	24,473	0.587
2019	15,549		26,784	37,876	52,907	91,683		44,960	26,426	0.588
2020	15,598		27,501	39,278	54,910	96,062		46,670	27,911	0.598
2021	16,746		30,133	42,498	59,005	102,596		50,195	29,679	0.591

Note The data on per capita disposable income of urban residents from 2002 to 2012 are from the sample survey of urban households. The survey scope, method, and index caliber of the urban household survey before 2013 are different from those of the urban household survey after 2013. The data are from the China Statistical Yearbook 2003–2018 and the author's calculations. SD and CV refer to standard deviation and coefficient of variation, respectively

4.2 times that of low-income rural households (3301.9 yuan), and the per capita disposable income of high-income urban households was 77,097 yuan, 2.5 times that of high-income rural households (31,299.3 yuan). According to the average annual income from 2002 to 2017, the per capita disposable income of urban residents was three times that of the per capita net income of rural residents in 2002, and 2.46 times that of rural residents in 2017. The above analysis shows that the income gap among Chinese residents exists both between and within urban and rural areas. The degree of income inequality is high, and income inequality is universal.

7.1.2.2 Gini Coefficient Measurement

To further investigate the degree of income inequality between urban and rural residents in China, the Gini coefficient, an indicator of this inequality, is used for further analysis. This study calculates the urban Gini coefficient based on the quintile data of per capita disposable income of urban residents from 1985 to 2017. Based on the rural Gini coefficient from 1985 to 2001, the Gini coefficient from 2002 to 2017 is calculated using quintile data of per capita disposable income (net income per capita from 2002 to 2012). The overall Gini coefficient from 1985 to 2017 is calculated by weighting the urban and rural Gini coefficients. The rural–urban Gini coefficient is calculated using the grouping method, and the overall Gini coefficient is calculated using the modified rural–urban weighting method proposed by Chen et al. (2010). All the calculation results are shown in Table 7.3 and Fig. 7.2.

Table 7.3 and Fig. 7.2 indicate that from 1985 to 2017, China's urban and rural Gini coefficients, as well as the overall national Gini coefficient, all exhibited a slow upward trend. From 1985 to 2017, the Gini coefficient of rural residents generally exhibited a high and slow increasing trend. The Gini coefficient of rural residents increased 48.9%, from 0.241 in 1985 to 0.359 in 2017. The Gini coefficient of rural residents rose from 0.241 in 1985 to 0.31 in 1994, then decreased to 0.269 in 1997, and then slowly increased to 0.359 in 2017.

From 1985 to 2017, the Gini coefficient of urban residents showed a slow upward trend. It increased by 177.9%, from 0.109 in 1985 to 0.303 in 2017. From 1985 to 2017, the Gini coefficient of Chinese residents showed a slow upward trend of rising first, then falling, then rising again. It increased by 66%, from 0.259 in 1985 to 0.43 in 2017. The Gini coefficient of national residents increased from 0.259 in 1985 to 0.382 in 1994, then to 0.427 in 2005, 0.38 in 2012, and finally to 0.43 in 2017.

The above analysis indicates a large income gap among urban, rural, and national residents, and this gap reflects an increasing degree of income inequality. China's overall Gini coefficient increased by 66% in the whole sample interval, and the Gini coefficients of rural and urban residents increased by 48.9% and 177.9%, respectively. This means that the income gap among urban residents increased much more than that among rural residents and the overall country from 1985 to 2017, and the income inequality among urban residents increased more compared to the others. Table 7.3 and Fig. 7.2 also show that the Gini coefficient of rural residents is greater than that of urban residents in the whole sample interval from 1985 to 2017, and the overall national Gini coefficient weighted by the two is greater than the Gini coefficient of the two. This indicates that the income inequality among rural residents is higher than that among urban residents. Due to the large income differences between urban and rural residents, the overall Gini coefficient calculated by both is greater than that among rural or urban residents.

Table 7.3 Gini coefficients for national, urban and rural residents

Year	National	Urban	Rural	National (National Bureau of Statistics)
1985	0.259	0.109	0.241	–
1990	0.302	0.114	0.256	–
1991	0.340	0.110	0.309	–
1992	0.356	0.126	0.304	–
1993	0.376	0.149	0.304	–
1994	0.382	0.146	0.310	–
1995	0.364	0.159	0.285	–
1996	0.358	0.162	0.291	–
1997	0.339	0.172	0.269	–
1998	0.352	0.181	0.272	–
1999	0.364	0.186	0.278	–
2000	0.380	0.195	0.290	–
2001	0.391	0.206	0.290	–
2002	0.415	0.257	0.298	–
2003	0.422	0.263	0.306	0.479
2004	0.416	0.273	0.296	0.473
2005	0.427	0.278	0.302	0.485
2006	0.423	0.276	0.300	0.487
2007	0.419	0.268	0.295	0.484
2008	0.413	0.272	0.300	0.491
2009	0.404	0.271	0.309	0.49
2010	0.396	0.265	0.298	0.481
2011	0.389	0.263	0.303	0.477
2012	0.380	0.251	0.302	0.474
2013	0.447	0.312	0.339	0.473
2014	0.447	0.302	0.350	0.469
2015	0.444	0.295	0.348	0.462
2016	0.427	0.297	0.355	0.465
2017	0.430	0.303	0.359	0.467
2018	–	0.317	0.366	0.468
2019	–	0.317	0.351	0.465
2020	–	0.323	0.351	0.468
2021	–	0.320	0.355	0.466

Data source China Statistical Yearbook and author's calculations.

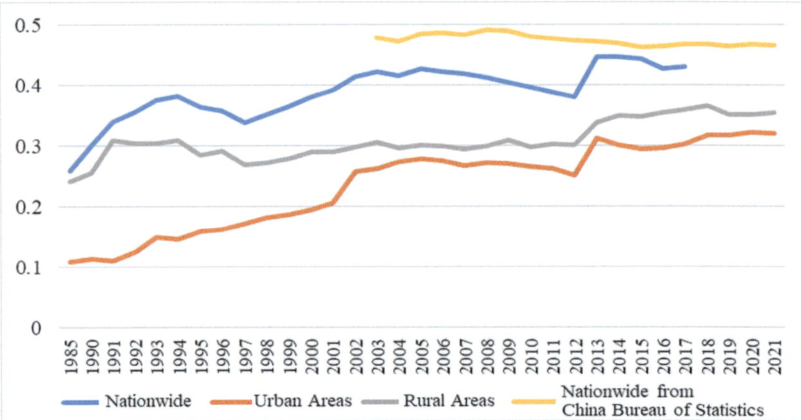

Fig. 7.2 Gini coefficients for urban and rural residents

7.2 Representation of Poverty

7.2.1 Poverty Criteria

The poverty standard is also called the poverty line. The standard definition of rural poverty in China refers to the basic cost of food and non-food (including services) that is necessary to maintain people's basic living needs under the conditions of a certain time, space, and social development stage (Wang et al., 2020; Zhou et al., 2023). The definition of China's poverty line is generally the same as that proposed by the World Bank. In the practice of poverty line measurement in China, the method recommended by the World Bank and the actual situation of rural areas in China are combined to calculate the poverty line. The basic logic and method of measuring the poverty line is to measure the food needs of the poor population through household surveys. By establishing a food demand model, the non-food demand line of the poor population is calculated, and then the poverty line is calculated based on both, that is, the sum of food and non-food demand.

With economic and social development, the poverty standard will be adjusted in two ways. First, it is adjusted according to changes in basic living needs. With the continuous improvement of economic development, people's living standards are constantly improving; therefore, it is necessary to adopt a higher standard to meet the basic living needs of the poor population. Second, it should be adjusted according to changes in the price index. By adjusting the price index of poverty standards in different years, the comparability of poverty standards between different years can be ensured.

Corresponding to the first type of adjustment, China has adopted three poverty standards with different living standards since the reform and opening up: the "1978 standard," "2008 standard," and "2010 standard." The "1978 standard" is based on the 1978 price of 100 yuan per person per year. Under this standard, the proportion of food expenditure is relatively high. The "2008 standard" has been in use since 2000. Under this standard, 865 yuan per person per year at the 2000 price is standard to meet basic food and clothing needs. The "2010 standard" is the current rural poverty standard in China. According to the 2010 price of 2300 yuan per person per year (Liu et al., 2018), this is the standard that provides basic stability in food and clothing for the poor population.

For the second type of adjustment, the poverty standard can be adjusted annually according to the "Consumer Price Index for the Rural Poor," to ensure comparability between years. Owing to the high proportion of food expenditure in the living consumption expenditure of the rural poor, when calculating the living consumption price index of the rural poor, the weight of food expenditure is increased according to the actual situation. According to the rural price level in the corresponding years, the current poverty standard is 2300 yuan per person per year in 2010, 2536 yuan in 2011, 2625 yuan in 2012, 2736 yuan in 2013, 2800 yuan in 2014, and 2855 yuan in 2015. The exact value of the same standard varies by year, but they all indicate that they are under the same standard of living and are comparable from year to year.

7.2.2 Poverty Measurement Methods

The poverty measurement index includes unidimensional and multidimensional poverty measurement indices. In the following sections, only a few commonly used and easily understood poverty measurement indices are introduced.

7.2.2.1 Single-Dimensional Poverty Measurement Methods

(i) Poverty incidence

This refers to the percentage of the total population that is poor. The specific expression is as follows:

$$H = \frac{q}{n} \tag{7.14}$$

where H is the incidence of poverty, n is the number of poor people, and q is the total population. The larger the index, the greater the number of poor people and the more severe the poverty level; otherwise, the lower the poverty level. The poverty incidence rate is a crude index, which contains little information and is not sensitive to the income distribution of the poor; thus, it cannot reflect the depth and intensity of poverty.

(ii) Average poverty gap rate of the poor population

To overcome the shortcomings of the poverty incidence index, Batchelder proposed the concept of poverty gaps in 1971. The poverty gap refers to the sum of the gaps between the income of the poor population and the poverty line. The expression is as follows:

$$g = \sum_{i=1}^{q} (z - y_i) \tag{7.15}$$

where g is the income gap, q is the number of poor people, z is the poverty line, y_i is the income of the i person used to measure the overall difference between the income of the poor people and the poverty line. The average poverty gap ratio of the poor people is obtained after standardized treatment, and its expression is as follows:

$$I = \sum_{i=1}^{q} \frac{z - y_i}{qz} \tag{7.16}$$

where I is the average poverty gap rate of the poor population. The income of the i person is used to measure the overall difference between the income of the poor population and the poverty line, which reflects the overall situation of the entire poor population, but cannot reflect the individual situation. Compared with the poverty incidence rate, the poverty gap rate is not sensitive to the number of poor people, but focuses on income distribution.

(iii) FGT index

Developed by economists Foster et al. (1984), the FGT index directly uses the poverty gap as a weight as shown in Eq. 7.17:

$$P_\alpha = \frac{1}{nz^\alpha} \sum_{i=1}^{q} g_i^\alpha \tag{7.17}$$

where α represents the sensitivity of poverty transfer, and the greater the value, the stronger the sensitivity of transfer. The values are usually 0, 1, and 2:

when $\alpha = 0$, $P_0 = \frac{q}{n}$ is the incidence of poverty;

when $\alpha = 1$, $P_1 = \frac{1}{nz} \sum_{i=1}^{q} g_i$ is the income gap ratio;

when $\alpha = 2$, $P_2 = \frac{1}{nz^2} \sum_{i=1}^{q} g_i^2$ is obviously higher than P_1, P_2 weight of poverty difference is added; thus, the transfer sensitivity principle is satisfied and the poverty difference within the poor population can be analyzed. The decomposable feature is the main advantage of the FGT index. This is because the FGT index can decompose poverty by both region and population group.

7.2.2.2 Multidimensional Poverty Measurement Method

Unidimensional poverty measures mainly define poverty based on the economic situation of individuals or families, which is income poverty and belongs in the category of absolute poverty. The root cause of family poverty is the "lack of ability," and income poverty is only the external manifestation of ability poverty. Since the ability poverty was proposed, the subsequent research on poverty has gradually been carried out to multidimensional poverty. The most commonly used index is the Multidimensional Poverty Index (MPI) jointly developed by the United Nations Development Program and the University of Oxford in 2010. Compared with the Human Development Index (HPI), the main dimensions of the MPI (i.e., education, health, and life) remain unchanged; however, the number of indicators used to measure each dimension have been increased to 10. The researches of Alkire & Foster (2011) and Yang et al. (2019) show that the A-F multidimensional poverty index is also commonly used to measure multidimensional poverty.

The A-F method assumes that there are n individuals whose welfare status is measured by d index, and the value of the individual on each index is x_{ij}. Generally, poverty can be identified by means of two critical values. z is used to determine whether an individual experiencing capacity deprivation on a specific indicator: g_{ij} is equal to the state of individual deprivation on a specific indicator. If $g_{ij} > z_j$, the individual suffers from capacity deprivation on a specific indicator; otherwise, it is considered not to be deprived. The importance of each indicator is reflected by setting weights w. The MPI can be obtained as follows:

$$MPI = \frac{1}{n}\sum_{j=1}^{d} c_j(k) \tag{7.18}$$

The MPI can be expressed as the product of multidimensional poverty incidence (H) and the average deprivation of the poor (A):

$$MPI = \frac{q}{n} \times \frac{1}{q}\sum_{j=1}^{q} c_j(k) = H \times A \tag{7.19}$$

where q represents the multidimensional poor population, and the average degree of deprivation of the poor population A is obtained by the ratio of the average number of deprived indicators of poor households to the total number of indicators.

7.2.3 Statistical Analysis of Poverty Status in China

The statistical analysis of poverty status in China in this part includes two aspects: China's overall poverty status and its multidimensional poverty status.

7.2.3.1 Overall Poverty in China

Table 7.4 and Fig. 7.3 show the rural poor population and poverty incidence in China from 1978 to 2017 under different poverty standards.

Table 7.4 and Fig. 7.3 show that, under the 1978 standard, the rural poor population in China in 1978 was 250 million, and the poverty incidence rate was 30.7%. Under the 2008 standard, the poverty incidence rate was 10.2% in 2000 and 2.8% in 2010. Under the 2010 standard, the number of poor people in rural China in 1978 was 770 million, with a poverty incidence rate of 97.5%. In 2017, the number of poor people in rural China dropped to 30.46 million, with a poverty incidence rate of 3.1%. Using China's poverty standard in 2010 as an example, China has lifted 740 million people out of poverty in 39 years, which shows that China has made remarkable achievements in poverty alleviation. According to the Millennium Development Goals Report published by the United Nations in 2015 (Haveman et al., 2015), China has contributed more than 70% to global poverty reduction.

7.2.3.2 Trends of Multidimensional Poverty in China

According to the MPI developed by Zhang (2015) with reference to the Poverty and Human Development Research Center of Oxford University, combined with the index of the China Health and Nutrition Survey data from 1991 to 2011, and according to China's specific national conditions, we selected 10 indicators from three dimensions of education, health, and living standard (Table 7.5) to construct China's multidimensional poverty index.

Determining the weight of dimensions and indicators has always been a core issue in multidimensional poverty measurement. Typically, poor individuals or families are identified by the weighted sum of their deprivation scores in each dimension. However, there is no unified method to determine the weights. Decancq and Lugo (2008) summarized three methods: the data mining, normative, and hybrid methods. However, each of these three methods has its own advantages and disadvantages. When the weight cannot be determined, the equal weight method is also commonly used in the literature (i.e., the same weight of each dimension index is used to identify poverty), which is used here.

Table 7.6 shows that from 1991 to 2011, the poverty incidence rate of school-age children in three dimensions and ten indicators showed an upward trend, whereas the other nine indicators showed a downward trend. Among them, the relative decline rate of poverty incidence of six indicators in the dimensions of health status and living standard was more than 48%. The relative decline of poverty incidence in health status was 89.21%, indicating that with the improvement in living standards and medical security, the health level of Chinese residents also improved. The poverty incidence of cooking fuel and sanitation facilities decreased from 85.4% and 82.87% in 1991 to 27.85% and 42.67% in 2011, respectively, showing a relative decrease of 67.39% and 48.5%. The incidence of poverty in durable goods dropped from 39.12% in 1991 to 1.45% in 2011, showing a relative decrease of 96.28%. The incidence of

Table 7.4 Rural poor population and poverty incidence in China from 1978 to 2017 under different poverty standards

	1978 Standard (100 yuan)		2008 Standard (865 yuan)		2010 Standard (2300 yuan)	
	Poor people (ten thousand)	Incidence of poverty	People living in poverty (ten thousand)	Incidence of poverty	People living in poverty (ten thousand)	Incidence of poverty
1978	25,000	30.7%			77,039	97.5%
1985	12,500	14.8%			66,101	78.3%
1986	13,100	15.5%				
1987	12,200	14.3%				
1988	9600	11.1%				
1989	10,200	11.6%				
1990	8500	9.4%			65,849	73.5%
1991	9400	10.4%				
1992	8500	10.0%				
1993	8000	8.8%				
1994	7000	7.7%				
1995	6540	7.1%			55,463	60.5%
1996	5400	6.3%				
1997	4962	5.4%				
1998	4210	4.6%				
1999	3412	3.7%				
2000	3209	3.4%	9422	10.2%	46,224	49.8%
2001	2927	3.2%	9029	10.0%		
2002	2820	3.0%	8645	9.2%		
2003	2900	3.1%	8517	9.1%		
2004	2610	2.8%	7587	8.2%		
2005	2365	2.5%	6432	6.8%	28,662	30.2%
2006	2148	2.3%	5698	6.0%		
2007	1479	1.6%	4320	4.6%		
2008			4007	4.2%		
2009			3597	3.8%		
2010			2688	2.8%	16,567	17.2%
2011					12,238	12.7%
2012					9899	10.2%
2013					8249	8.5%
2014					7017	7.2%
2015					5575	5.7%
2016					4335	4.5%

(continued)

Table 7.4 (continued)

	1978 Standard (100 yuan)		2008 Standard (865 yuan)		2010 Standard (2300 yuan)	
	Poor people (ten thousand)	Incidence of poverty	People living in poverty (ten thousand)	Incidence of poverty	People living in poverty (ten thousand)	Incidence of poverty
2017					3046	3.1%
2018					1660	1.7%
2019					551	0.6%

Data source China Rural Testing Report, China Statistics Press (2020)

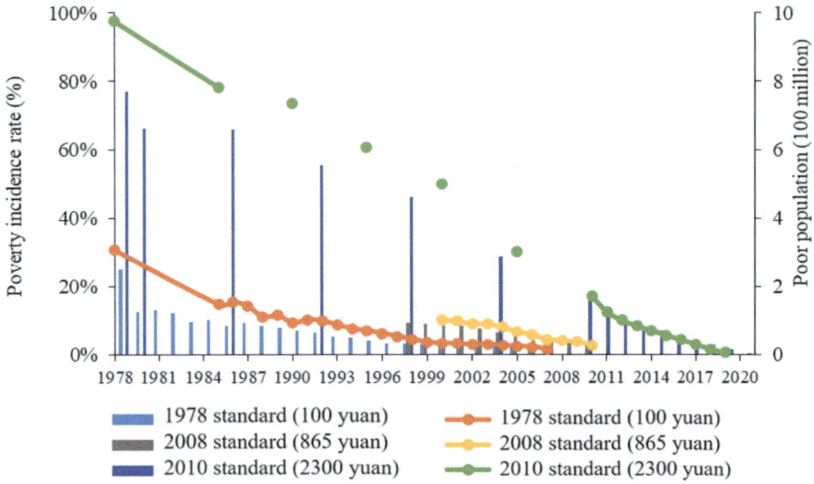

Fig. 7.3 Variation trend of rural poverty population and poverty incidence in China from 1978 to 2017 under different poverty standards

housing poverty dropped from 29.28% in 1991 to 4.8% in 2011, a relative decrease of 83.59%. In 2011, the incidence of poverty in terms of years of education, medical insurance, cooking fuel, and sanitation facilities were all above 10%.

Table 7.6 also shows that the poverty incidence of ten welfare indicators in the three dimensions is different from that in the income dimension. In 1991, the poverty incidence in the income dimension was 44.28%. Among the welfare indicators, the poverty incidence of cooking fuel, sanitation facilities, and medical insurance health status were all above 60%. In 2011, the poverty incidence rate of income dimension was 12.3%, and the poverty incidence rates of years of education, medical insurance, cooking fuel, and health care facilities were all above 20%. There were significant differences between the incidence of poverty in different dimensions of welfare indicators and that of income poverty, indicating that the incidence of poverty

Table 7.5 Dimensions, indicators, and deprivation thresholds of multidimensional poverty

Dimensions	Indicators	Deprivation threshold
Education (1/3)	Years of education (1/6) School-age children enrolled in school (1/6)	The highest level of education in the family is primary school or is over the age of 18 but has not completed 6 years of education, which takes the value of 1; at least one child aged 6 years and above in the household is out of school or not in school, taking the value of 1
Health (1/3)	Health status (1/6) Health insurance (1/6)	A current household member with a general or fairly serious illness takes the value of 1 If any family member does not have medical insurance, the value is 1
Standard of living (1/3)	Fuel for cooking (1/18) Sanitation facilities (1/18) Clean drinking water (1/18) Lighting (1/18) Housing (1/18) Durable goods (1/18)	Common cooking fuels are non-clean fuels (e.g., LPG, natural gas, electricity), which take the value 1 Toilets cannot use indoor flushing, indoor toilets (including no flushing) take the value of 1 Indoor or courtyard tap water or deep well water ($\geq$ 5m) cannot be used, which takes the value of 1 If no electricity is available at home, the value is 1 The value is 1 if the per capita living space is less than 12m^2 Having at most one of the following household appliances or means of transport takes the value 1: Color TV, washing machine, refrigerator, air conditioner, sewing machine, video recorder, electric fan, computer, microwave oven, rice cooker, telephone, mobile phone, camera, pressure cooker, VCD/DVD, electric assisted bicycle, motorcycle, car

Note In parentheses are the weights of the dimensions or indicators, where a value of 1 indicates that the individual is in poverty on this indicator
Data source Zhang (2015)

measured from a multi-dimensional perspective is more comprehensive than that measured using a single income indicator.

Table 7.7 shows the MPI, which integrates three dimensions and ten indicators into one indicator. The value of k is crucial when calculating the MPI. Here, we refer to the definition of multidimensional poverty in the United Nations' Multidimensional Poverty Measure, and set the value of k to 1/3, $k \geq 1/3$. The results show that the MPI decreased from 0.2229 in 1991 to 0.1575 in 2000, and then to 0.0265 in 2011, with a relative decrease of 88.11%. The MPI can be expressed as the product of the incidence of multidimensional poverty H and the average degree of deprivation A of multidimensional poor people. The reasons for the change in MPI can be determined by analyzing H and A. Table 7.7 shows that, from 1991 to 2011, China's multidimensional poverty incidence H decreased from 52.17 to 7.02%, with a relative decrease of 86.54%, which is the only factor for the decline of China's MPI. The average deprivation degree A of the multidimensional poor population decreased from 42.73% in 1991 to 37.73% in 2011, showing a relative decrease of 11.7%, and

Table 7.6 Poverty incidence rate by indicator in China from 1991 to 2011 Unit: %

Dimensions	Indicators	1991	2000	2011	1991–2011 rate of change
Education	Years of education	25.08	25.13	20.30	19.06
	School-age children enrolled in school	1.54	3.46	2.80	82.18
Health	Health status	87.55	73.48	9.45	89.21
	Health insurance	63.17	53.62	60.54	4.17
Standard of living	Cooking fuel	85.40	60.61	27.85	67.39
	Sanitation	82.87	63.09	42.67	48.50
	Clean drinking water	21.43	12.62	7.97	62.81
	Lighting	4.03	0.86	1.02	74.58
	Housing	29.28	11.24	4.80	83.59
	Durable goods	39.12	15.76	1.45	96.28
Income	Income	44.28	25.76	12.3	72.23

Note The income poverty line is based on the newly published 2300 yuan per person per year (2010 prices)
Data source Zhang (2015)

the relative change rate was much smaller than the multidimensional poverty incidence rate. Table 7.7 also shows that the decline of the MPI and multidimensional poverty incidence is much higher than the income poverty incidence. In 1991, the multidimensional poverty incidence rate was 7.89% higher than the income poverty incidence rate, whereas in 2011, the multidimensional poverty incidence rate was 5.28% lower. It also shows that the effect of poverty reduction from a multidimensional perspective is better than that from a single income dimension in China in recent years.

Table 7.7 China's multidimensional poverty index and its changes from 1991 to 2011

	1991	2000	2011	1991–2011 rate of change (%)
Multidimensional Poverty Index (MPI)	0.2229	0.1575	0.0265	– 88.11
Multidimensional poverty incidence (H)	52.17%	38.12%	7.02%	– 86.54
Average level of deprivation of the multidimensional poor (A)	42.73%	41.31%	37.73%	– 11.70

Data source Zhang (2015)

7.3 Theoretical Basis of Income Inequality and Poverty

Income distribution is an important link in economic activities, and the interest relationship and interest pattern in economic activities are finally realized through income distribution. Inequality in income distribution is a direct manifestation of social injustice and an inevitable phenomenon in the process of economic development. The theory of income inequality is based on the theory of income distribution. The common income distribution theories include functional income distribution theory, scale income distribution theory, inverted U-shaped theory, and regional income inequality theory.

7.3.1 *Functional Income Distribution Theory*

Functional income distribution, also known as factor income distribution, is an income distribution with production factors as the main body. It mainly examines the distribution of national income among land owners, labor owners, and capital owners. The income share of each factor in functional income distribution is income distribution at the macro level. Fang (2015) showed the representative theory of functional income distribution is the income distribution theory of the neoclassical school.

Neoclassical income distribution theory is based on the neoclassical micro distribution theory, which examines the change in and decision of the relative share in the functional income distribution. This macro distribution theory is based on the summation of the micro production function and the elasticity of substitution and establishes the aggregate production function based on the marginal productivity of production factors, which is used to describe the causal relationship between production and distribution. In the neoclassical framework, the relationship between factors of production and output is determined by a homogeneous production function. The typical aggregate production function was proposed by Cobb and Douglas (1928), and the specific form is $Y = AL^{\alpha}K^{\beta}(\alpha + \beta = 1)$. Y is the total output. L and K are the labor factor and capital factor of production input respectively. α and β are the labor input–output elasticity and capital input–output elasticity, respectively. The marginal products of labor and capital are $MP_L = \partial Y/\partial L = A\alpha L^{\alpha-1}K^{\beta}$ and $MP_k = \partial Y/\partial K = A\alpha L^{\alpha}K^{\beta-1}$, respectively. α and β denote the income shares of labor and capital in total output, respectively. Under the framework of neoclassical analysis, the contribution of the rate of return of each factor in the production process is determined; therefore, neoclassical distribution theory is also called the "functional" income distribution theory.

7.3.2 Scale Income Distribution Theory

Large-scale income distribution, also known as individual income distribution, mainly examines the size of the income gap among different individuals, aiming to reveal the characteristics of income distribution in microeconomic units and the measurement of income inequality related to it (Kohli, 2018). The research on the theory of individual income distribution generally has two aspects. The first examines how to improve individual internal efforts to increase income, which is the endogenous income pattern. These endogenous variables determine individual income and affect income inequality. The second examines the exogenous determinants of income, such as relevant systems and macro variables. The more typical one is the theory of inequality of opportunity, which mainly explores the influence of market segmentation or institutional discrimination on the unfair distribution of income; for example, the dual structure of urban and rural areas in China.

7.3.2.1 Income Distribution Theory Determined by Endogenous Model

The theory that a person's ability determines income differences is the oldest endogenous determination theory. It holds that the distribution of individual ability is normal; therefore, the difference in income is normal. However, this assumption is not supported by statistical data. Pareto empirical analysis shows that the distribution of income is not normal, but lognormal and skewed to the right. The most prominent theoretical development is human capital theory, which argues that income differences arise from differences in acquired human capital, rather than differences in income determined by innate ability, as in ability theory.

Modern human capital theory has two main branches. The first branch is represented by Schultz and Romer, who introduced human capital into the analysis of productivity and economic growth. Schultz argued that human capital is the knowledge, skills, and ability of workers, and the main factor that promotes economic growth. The second branch is represented by Becker and Minsel. Minsel first applied the theory of human capital to labor economics and took the lead in using the method of human capital investment to study income inequality. Becker introduced human capital into family analysis and studied the relationship between human capital and individual income distribution.

Life cycle theory, which is linked to human capital theory, not only analyzes the impact of human capital on income but also emphasizes the role of savings and personal assets throughout the life cycle. Personal income generally shows an inverted U-shaped trend with an increase in age. With increased age, personal income will gradually increase; however, after a certain stage, it will gradually decline with the arrival of the retirement age. Income distribution based on life cycle theory integrates the effects of labor income, income accumulation, and property income on income distribution into a theoretical analysis framework.

7.3.2.2 Income Distribution Theory Determined by Exogenous Model

The most common theory of income distribution determined by exogenous models is the theory of inequality of opportunity. Inequality of opportunity exists in both developed and developing countries. In developed countries, it mainly examines the discrimination of opportunities under the existing system. In developing countries, owing to serious market segmentation, dual economic structure between urban and rural areas, industrial monopoly, and other characteristics, as well as imperfect systems and laws, inequality of opportunity is more serious.

Inequality of opportunity in developing countries directly determines income inequality. The most typical example is the urban–rural dual economic structure, which refers to the coexistence of modern economic sectors (modern industrial sectors) and traditional economic sectors (traditional agricultural sectors) in the national economy. Traditional agricultural sectors are concentrated in rural areas, whereas modern industry cannot be concentrated in urban areas, thereby forming a situation of dual opposition between urban and rural areas. The urban–rural gap means that rural residents have much fewer educational opportunities than urban residents, resulting in labor market segmentation and unequal opportunities, which leads to income inequality.

7.3.3 Inverted U-Shaped Theory

In 1965, based on Kuznets' inverted-U hypothesis of income distribution, Williamson proposed the inverted-U theory in his paper "Regional Imbalance and National Development Process." In his speech at the American Economic Association in 1955, Kuznets proposed the inverted U-shaped hypothesis of income distribution, which holds that, in the process of economic development, the income gap has a trend of "first worsening and then improving" over time. Different from the previous income theory, Kuznets' inverted U-shaped theory is based on empirical research and introduces time series into the analysis of regional spatial structural changes. In the later analysis of regional economic development, Williamson further developed and applied Kuznets' inverted-U hypothesis of income distribution, and proposed the inverted-U theory of regional economic differences. Based on the empirical analysis of regional economic research, Williamson found that the inverted-U relationship is more obvious between the economic development stage and the regional development difference. This shows that the income distribution difference between regions will first increase and then decrease with a change in economic development stage (i.e., it will continue to expand in the initial stage of economic development).

7.3.4 Theory of Regional Income Inequality

The most common theory of regional income inequality is cumulative causality theory proposed by well-known economist Myrdal in 1957. Cumulative causality theory, also known as cyclic cumulative causality theory, was developed and concreted into a model by Kaldor, Dixon, and Thirlwall, after Muir Dahl. The theory holds, that in a dynamic social process, there is a cyclic and cumulative causal relationship between social and economic factors. The space in economic development is not generated and evenly spread at the same time. Once some regions develop ahead of other regions due to their initial advantages, they will continue to accumulate favorable factors based on the existing advantages and continue to develop, and repeat the cycle. This process leads to the dual economic structure of urban–rural division in economic geography and aggravates the imbalance in economic development among regions. This process also leads to the dual economic structure of urban and rural segmentation in economic geography and to the continuous expansion of regional gaps.

The cyclic cumulative causality will have two effects on regional economic development. One is the echo effect; as the factors of production such as labor, capital, and technology are affected by the difference in factor returns, they will flow from less developed regions to developed regions. The second is the diffusion effect. The echo effect is not infinitely circular; therefore, the expansion of the development gap between regions also has a certain limit. After the development of developed regions reaches a certain level, the production costs in developed regions will continue to rise, and the external economic benefits will gradually become smaller, thereby weakening economic growth (Aaberge et al., 2002). At this time, it will become uneconomical for developed regions to further expand the production scale, and capital, labor, and technology will continue to spread to the less developed regions. Muir Dahl called this process the "diffusion effect," and it contributes to the development of less developed areas.

After Myrdal, Kaldor developed the concept of efficiency wages and developed the theory of cyclic cumulative causality. Kaldor believed that each region's economic growth trend is determined by the efficiency wage (i.e., the ratio of monetary wage to productivity in each region). The lower the efficiency wage, the higher the economic growth rate. Conversely, the higher the efficiency wage, the lower the economic growth rate. In theory, efficiency wages should be the same across regions within a country. However, growth rates are higher in prosperous regions, where productivity is higher because of increasing returns to scale due to economic agglomeration, which reduces efficiency wages. An increase in the economic growth rate raises productivity, which lowers the efficiency wage, which in turn raises the economic growth rate. In this constant cycle, prosperous regions will become more prosperous, and the lagging regions will fall further behind.

7.3.5 Income Inequality and Poverty

Human understanding of poverty is a continuous evolution process that develops along with human society. The definition of poverty has also been deepened and enriched. Currently, some differences exist in how various disciplines define poverty. The economic definition holds that poverty occurs when people lack the means to meet their basic needs (too little property), emphasizing the importance of income. The sociological definition divides poverty into "missing out" and being "left out," emphasizing the social environment and adverse social factors affecting individuals or families. The development economics definition considers poverty in terms of "basic capabilities." Poverty is a type of multidimensional deprivation that people can experience. The political definition views poverty as a political problem, and the fundamental way to end poverty is to create more inclusive opportunities. Obviously, there are great differences in the dimensions and content of the definitions of poverty across disciplines.

Poverty has always been the focus of development economics, as well as sociology, political science, and demography. It is an eternal theme associated with the development of human society. Alleviating or even eliminating poverty is a common task in the process of human development and a long-term goal for developing countries. The study of poverty has always been at the core of scholarly attention. Development economist Makower (1953) proposed the famous argument that "a country is poor because it is poor." He believed that the root cause of poverty in developing countries is the lack of capital input, which leads to capital shortages on both sides of supply and demand, thereby resulting in low levels of per capita income. This is known as the "vicious cycle of poverty theory." In addition, other poverty theories, such as the Great Push theory (Rosenstein-Rodan, 1943), the low-level equilibrium trap theory (Nelson, 1956), and the unbalanced development theory (Hirshman, 1958), all comprehensively elaborate on the relationships among population growth, capital accumulation (or investment), and output growth in developing countries.

"Targeted poverty alleviation" and "two not to worry about, three security" are the latest developments in anti-poverty theory. The "six precision" fully considers the multidimensional and complex nature of poverty and notes that the causes of poverty among poor populations are unique and different. It is necessary to "take measures based on household and people, and help them to the point and the root" to achieve "no worries and no guarantees" by 2020. The "two guarantees" are to ensure that the rural poor have enough food and clothing, while the "three guarantees" refer to the rural poor having access to compulsory education, basic medical care, and housing security. The "three guarantees" consider not only the basic needs of living but also the needs of long-term development and scientifically outline the goal of stabilizing and eliminating multidimensional poverty. The "targeted poverty alleviation" and "no worries, no worries, and no guarantees" have charted the course for China's poverty standards and poverty monitoring work.

7.4 Policies to Address Income Inequality and Poverty

An important manifestation of unbalanced and inadequate income distribution is a large income gap and unreasonable income distribution structure. In the new era, problems remain in the field of income distribution in China, such as the large income gap and non-standard income distribution order, which directly lead to income inequality in China (Li, 2016). Based on China's income inequality and poverty, this chapter proposes corresponding measures from three aspects: deepening the reform of the income system and mechanism, implementing the strategy of regional coordination war, persisting in targeted poverty alleviation, targeted poverty alleviation, and winning the battle against poverty to improve China's income distribution policy, reduce urban and rural residents' income, reduce income inequality, and reduce poverty.

7.4.1 Improving the System and Mechanism for Distributing Income

Since the reform and opening up, China has continuously promoted the reform and improvement of the income distribution system, and gradually established an income distribution system in line with the basic national conditions and overall stage of development. With the deepening of economic restructuring, a distribution system in which distribution according to work is the main body and various forms of distribution coexist has been gradually established. The final formation of the income distribution pattern is the ultimate result of the interaction between the market and the government. Market mechanisms play a fundamental and decisive role in the final formation of income distribution patterns in the initial distribution. The government regulates the redistribution through taxation, transfer payment, social security, and other means. According to the operating mechanism of income distribution, the key to the growth of resident' income is to achieve the unification of the efficiency and fairness of primary distribution and redistribution. In the redistribution stage, more attention should be paid to fairness, and the role of the third distribution should also be considered. Therefore, to increase residents' income and reduce income inequality, several actions should be taken to deepen the reform of the income distribution system and mechanism, which are described in more detail below.

7.4.1.1 Improving the Policy System and Promoting Equal Opportunities for Income Distribution

Equality of opportunity is the basic component of the concept of social justice in a market economy and a prerequisite for the effective operation of the market economy. The income distribution system of equal opportunity defines the income distribution requirements of residents from the perspective of the balance of citizens' power. It is a type of equality with special provisions, and the market system directly endows people with this right. Aiyar & Ebeke (2020) reaearched that people having equal opportunities is the premise for the market mechanism to serve an effective regulating function, and a specific, fair, and deep-level institutional structure existing under the surface of market operation.

However, people's opportunities to obtain income are often unequal. This mainly comes from two aspects, namely innate factors and the inequality caused by other factors: the dual division of urban and rural areas, industrial monopoly, privilege, and unequal opportunities for citizens to obtain employment and promotion. To build a socialist system with Chinese characteristics for the new era and establish a fair, just, reasonable, and orderly income distribution system, it is necessary to eliminate institutional inequality in opportunities, and then realize the equality of opportunities in income distribution.

7.4.1.2 Developing and Popularizing Education to Equalize Educational Opportunities

Promoting equal opportunities in civic education is an important guarantee for equal employment and income opportunities for members of society, especially low-income groups. From the perspective of reducing the income gap among residents, the equality of basic education can gradually narrow the income gap caused by differences in human capital by improving the education level of low-income groups. Therefore, it is necessary to increase investment in basic education and provide special living subsidies to children from poor families to ensure that they can successfully complete their studies. It is necessary to reverse the trend of marketization and industrialization of higher education, increase financial investment in higher education, solve the shortage of higher education supply, and actively implement a policy of subsidizing higher education for children from poor families.

7.4.1.3 Expanding Employment Channels and Improving the Quality of Farmers' Employment

The report to the 19th National Congress of the Communist Party of China (CPC) noted the need to adhere to the strategy of prioritizing employment and the proactive employment policy to achieve higher quality and fuller employment. Employment is fundamental to people's livelihoods, and increasing employment is an effective way to raise the incomes of low-income people. China has a large population and abundant labor resources. Thus, whether China can fully improve quality employment and increase residents' income is closely related to reducing income inequality.

Under the new normal of the economy, poor people in China have low-income levels and lack employment opportunities. The main method for accelerating increases in income for farmers is to vigorously develop new types of agriculture, raise the education level of rural workers, remove barriers to urban and rural labor mobility, and closely integrate the process of urbanization with the process of citizenization of rural residents. Under the new normal of the economy, the power of farmers' traditional income-increasing channels is weakening. Therefore, implementing a rural revitalization plan is urgently needed to develop new channels, tap new potential, and cultivate a new driving force to build a stable, long-term, and sustainable mechanism to increase farmers' income.

It is also necessary to optimize the employment structure and improve employment quality. As a core component of the quality of economic development, employment quality is not only the main way to raise workers' wages and improve people's lives but also a necessary measure to improve the quality of economic development. For individuals, high-quality employment means sufficient employment opportunities, reasonable increases in labor remuneration, a fair and just employment environment, adequate social security, and decent social status. For the economy, higher quality employment includes more full employment, better employment structure, and higher labor productivity.

7.4.1.4 Expanding the Channels for People's Labor Income and Property Income

The report to the 19th National Congress of the Communist Party of China also noted the need to expand the channels for residents' labor and property income. Rural and urban households differ. Urban residents' investments in financial products bring more property income, whereas rural households rely more on real goods leasing and investment to obtain property income. For farmers, the key is to provide them with more property rights through deepening rural reform, so that they can receive more property income and ensure that their property rights are not lost in the urbanization process. We should deepen the reform of the rural collective property rights system and promote the reform of the rural collective assets and the stock cooperation system to allow farmers to use the land contract and management right and the ownership of collective assets to become shares in agricultural production cooperatives and rural economic cooperation organizations, develop various forms of stock cooperation, and let farmers become shareholders and participate in dividends in addition to their salary. The distribution of rents, salaries, bonuses, and shares in farmers' income should be increased.

7.4.1.5 Improving Fiscal and Tax Adjustment Mechanisms so that Residents Can Have a Share the Fruits of Reform and Development

(i) Dimensions of fiscal transfer payments

In fiscal adjustment, China should increase transfer payments to agriculture and rural areas. China's agriculture, rural areas, and farmers have made great sacrifices for the economic development of the country. Promoting the integration of urban and rural development and allowing rural farmers to share the fruits of reform and development is an important symbol of national modernization.

First, government subsidies to rural farmers should be increased. The most arduous task of building a moderately prosperous society in all respects lies in rural areas, particularly in poor rural areas. Therefore, it is necessary to break the dual division of urban and rural areas as soon as possible, promote the free flow and fair exchange of urban and rural elements, and implement the equal allocation of urban and rural public service resources to inject a new impetus into China's rural revitalization and development in the new era, allow many farmers to equally participate in the reform and development process and encourage farmers to find jobs and start businesses. The government should increase subsidies for agricultural production and operation, introduce preferential policies for agriculture, raise farmers' enthusiasm, inject impetus into agricultural modernization, increase farmers' incomes, and enable rural farmers to share in the fruits of reform and development.

Second, we must increase long-term investment in public services in rural areas. The report to the 19th National Congress of the CPC notes that a rural revitalization strategy should be implemented, and the Party should always make solving problems related to agriculture, rural areas, and farmers a top priority. Therefore, the government should establish and improve the system, mechanism, and policy support for the integrated development of urban and rural areas. By adjusting the investment pattern of public services, the government should implement the rural revitalization plan, increase the investment in rural areas and particularly less developed areas, and build modern agricultural industrial, production, and operation systems. The government should continuously improve the level of rural agricultural modernization to promote the ongoing equalization of residents' income between urban and rural areas and between regions.

(ii) Dimension of tax policy

China should speed up the reform of the tax system and improve the regulatory function of taxation. The government should raise the personal income tax threshold at an appropriate time, in line with the level of personal income growth, and change the situation in which ordinary wage-earners become the main taxpayers. Preferential policies and means should be used to encourage high-income people to pay more taxes and low-income people to pay less or no taxes. The government should account for the tax burden of taxpayers and reasonably determine the standard for deducting individual income tax. A sound tax environment should be established and good tax awareness among residents be cultivated, to give full play to the role of personal tax in regulating the income gap among individuals.

The government should improve tax policies and measures to support agricultural development. The basic status of agriculture and the characteristics of a weak industry require the government to further provide policy support to enhance the specialization and scale of agricultural production, vigorously develop agricultural industrial clusters, and improve the modern market system for agricultural products after abolishing agricultural taxes. Specifically, the government should adopt more preferential value-added tax rates for agricultural means of production, lower the prices of the means of production, and reduce the burden on farmers. Support for small and medium-sized enterprises should be increased, and preferential or reduced tax policies should be implemented. In addition, the tax support for the reemployment of laid-off urban workers should be increased to promote employment and reemployment.

7.4.1.6 Improving the Social Security System and Increasing the Level of Security for Low- and Middle-Income Groups

China should establish a multi-tiered social security system to improve the level of security for low- and middle-income groups. The construction of the social security system aims to narrow the income gap and more equitably distribute public service resources. Social security is crucial to the living standards and income of low-income groups. The 20th National Congress report noted that, to build a multi-tiered social security system in China, it is necessary to further deepen reforms in the areas of universal insurance, pension insurance, medical insurance, unemployment and work-related injury insurance, public social security services, and the urban and rural assistance system, as well as continuously enhance the important role of social security in increasing incomes.

It is also necessary to accelerate the equalization of basic public services and social security systems and steadily promote the urbanization process in rural areas. The government should regard institutional innovation as a breakthrough and reform the household registration system as an opportunity to improve the land transfer system such that the labor transfer can eliminate the shackle of land. We should also accelerate the equalization of basic public services and the social security system and gradually eliminate the gap between large, medium, and small cities and small towns and between urban and rural areas.

7.4.2 Implementing the Strategy of Coordinated Regional Development

This is a major strategic deployment made by the CPC Central Committee in response to the new features of coordinated regional development in the new era. The implementation of the coordinated regional development strategy is also greatly significant in narrowing the development gap between urban and rural areas and the gap in people's living standards, reducing income inequality, and generally realizing common prosperity for all.

7.4.2.1 Coordinating the Development of Both Developed and Less Developed Areas

For a long time, the economic development of China's central and western regions has continuously lagged behind that of the eastern region, and the less-developed regions have fallen far behind the developed regions. To implement a regional coordination strategy, it is crucial to coordinate the further development of developed and less-developed areas. This requires developed regions to continue implementing the regional integration strategy through reform and innovation, support the rapid development of less-developed regions and neighboring regions, and encourage the development and innovation platforms, such as national new areas and free trade pilot areas, to take root, germinate, and grow. China also needs to establish a long-term and inclusive support mechanism and a targeted and differentiated policy support mechanism to strengthen the weak links of underdeveloped regions in infrastructure, public services, and the ecological environment; resolutely win the battle against poverty; and ensure that underdeveloped regions achieve the goal of building a moderately prosperous society in all respects at the same time as the country as a whole. Simultaneously, we should vigorously support linkages among developed regions, support less developed regions in undertaking industrial transfer from developed regions, and facilitate the establishment of cross-regional industrial parks and other innovation platforms, to promote common development among regions by letting the group that gets rich first drive other groups to get rich.

In addition, we should vigorously support the development of old areas, ethnic minorities, border areas, and poor areas. Old and minority areas, border areas, and poor areas are special types of difficult areas in China. China needs to enact a series of measures to lift poverty-stricken communities out of poverty, support the development and construction of old revolutionary base areas, promote the sound development of ethnic minority areas, and advance the development and opening up of border areas. In the future, China should strengthen measures to promote the large-scale development of the western region and form new patterns. The country will give full play to the leading role of the Belt and Road Initiative, continue to open up cities in the western region, accelerate the development of internal and external corridors and regional hubs, improve interconnected transregional transportation networks, and increase the level of opening-up and export-oriented economic development in the western region.

7.4.2.2 Promoting the Free Flow of Factors of Production Between Urban and Rural Areas and Among Regions

China has an obvious urban–rural dual economic structure. The land system, household registration system, financial system, and other important aspects of the urban–rural dual structure have seriously restricted the development of modern agriculture and rural development. Under the background of the diminishing dividends of China's early reform and the increasingly acute contradictions between urban and rural areas, breaking down the barriers between urban and rural areas and stimulating the vitality of factors has become the primary problem to be solved in the urbanization process.

Therefore, it is necessary to implement a unified national negative list system for market access and eliminate discriminatory and exclusive restrictions on regional market access. China should actively break administrative monopolies, remove all regulations and practices that hinder market access and fair competition, create a favorable market environment, and stimulate market vitality. China should also deepen the reform of the household registration system, fully relax the requirements for settlement in cities, provide full coverage of basic public services for permanent urban residents, improve support facilities, and promote the free flow of human resources between urban and rural areas. The Chinese government should deepen the reform of the rural land system, explore ways to expropriate rural land, put rural collective construction land on the market, and reform the rural homestead system to effectively protect farmers' interests. Simultaneously, the effective allocation of scientific and technological resources should be guided in line with market demand and the free flow of innovation factors should be promoted.

7.4.2.3 Strengthening Cooperation and Mutual Assistance Among Regions

Although China's eastern region is relatively developed, its central and western regions are lagging. Thus, the coordinated development between the eastern, central, and western regions. China must integrate the efforts of all parties and promote the formation of a three-in-one poverty alleviation pattern, including special projects and industrial and social poverty alleviation. In addition, the government should strengthen industrial transfer from the central and western regions to the southeastern coastal areas and build cross-regional industrial parks. We will deepen overall and targeted pairing support, promote sustained and sound economic and social development of the three regions and three prefectures, promote exchanges and integration among regions and ethnic groups, and lay a solid foundation for long-term stability in local society. We should also actively guide social forces and various public welfare and non-profit organizations to extensively participate in poverty alleviation in poverty-stricken areas, to help those living in these areas increase their incomes and steadily lift themselves out of poverty.

7.4.2.4 Implementing Differentiated Regional Policies

China's regional development is uneven. When formulating and implementing policies, the country should consider the developmental characteristics of each region accurately and effectively design policies and measures related to finance, industry, land, and human resources; and adopt measures considering local conditions to foster and stimulate the unique development drivers of each region.

China should guard against and strengthen the awareness of ecological and environmental risks and strike a balance between economic development and ecological and environmental protection. In the process of industrial undertaking in the central and western regions, the country must strengthen environmental supervision, especially in areas with a fragile ecological environment. It must strictly implement ecological and environmental protections, enforce the law, and punish those who violate the law. The concept that clear waters and lush mountains are gold and silver mountains must be kept in mind and implemented. China should formulate policies and measures to promote the rise of the central region and the development of the western region in the context of the Belt and Road Initiative, and also apply preferential policies to the development of advantageous and suitable industries in underdeveloped regions.

7.4.3 *Consolidating and Expanding the Results of the War Against Poverty*

Consolidating and expanding the results of the war against poverty is an objective need for poverty-stricken households and areas to improve their ability to develop sustainably from a new starting point, and a need to further enhance the persuasive power to win the war against poverty comprehensively. Consolidating poverty eradication achievements needed to further strengthen the foundation of rural revitalization, and is inseparable from rural revitalization. In addition, consolidating and expanding poverty eradication achievements is also related to the construction of a new development pattern with the domestic macrocycle as the main part and the domestic and international double cycle promoting each other. Therefore, consolidating the achievements of poverty eradication means that comprehensive rural revitalization can be pushed forward from a higher starting point, thus promoting high-quality development of the national economy and achieving the goal of modernization. Consolidating and expanding the poverty eradication achievements require the following.

7.4.3.1 Strengthening Early Warning to Prevent the Return to Poverty and Eliminate Risks

The establishment of an information platform for preventing the return of poverty and for monitoring the dynamics of the low-income population is an important element in preventing the degradation of the achievements in the fight against poverty, be it rural population lifted out of poverty or low-income households, people facing specific hardships, and families on the edge of the low-income insurance all uniformly included in the platform so that, the relevant departments can detect such people and intervene promptly to help them. However, such a platform is not permanent. In the face of a rapidly changing society, we must, on the one hand, continue to optimize dynamic monitoring, which includes adjusting the scope of monitoring to prevent the return of poverty, and carrying out centralized inspections at the appropriate time while at the same time tightening and consolidating the responsibility for preventing the return of poverty to ensure that all those who should be included in the system are included. On the other hand, we need to promote the sharing of information between the monitoring of the prevention of a return to poverty and the dynamic monitoring of the low-income population, strengthen screening and early warning, and improve the timeliness of the monitoring to ensure that no household is left behind.

Detecting the risk of returning to poverty is not the main purpose; timely assistance is the ultimate need. To provide timeous help, the principle and requirements of the help must be accurate, and responsibilities of the contact person for help must be monitored. A single risk should be solved as soon as possible, and the risk of complexity and diversity should be stratified and categorized to implement comprehensive help measures. To further implement developmental support, based on the family members' ability to work and development needs for targeted assistance, not only food, clothing, housing, and transportation but also industrial and employment assistance projects should be provided to realize a sense of sustainability.

7.4.3.2 Precise Policy Measures to Help People According to Their Needs and Local Conditions

The reasons for possible degradation of poverty eradication gains can vary. At the individual level, degradation can be due to issues such as illness, disability, schooling, disaster, and so on. At the rural household level, unsustainable livelihoods due to a lack of human capital and assets are the main reasons rural households may experience a return to poverty. At the regional level, such as in remote areas and less developed regions, regional geographic, natural, and market conditions are important causes of large-scale poverty challenges. Therefore, there is a need to tailor the policy design to the local context and introduce different support measures for each situation.

The risk of returning to poverty that may be caused by the lack of education can be prevented by increasing education and training and enhancing human capital. For the risk that may be due to the lack of capital, a financial support policy should be developed. For the risk of returning to poverty that may be triggered by the lack of labor force due to illness and disability, sound social medical insurance and assistance for serious illnesses should be provided. For the risk that may be due to a lack of land resources, the income of specific populations and families should be increased through land transfer or the development of non-agricultural production. In summary, policy initiatives should be tailored to the different types of possible risks of returning to poverty and should be individualized and precisely designed to continuously improve the sustainable development capacity of the population that has been lifted out of poverty, so that the risk of returning to poverty can ultimately be eliminated and inequality reduced.

7.4.3.3 Sound Management System and Implementation of Precise Management

To further promote the consolidation and expansion of poverty alleviation efforts to achieve better results, we can also consider making efforts in terms of the management system, which can organically combine the supervision and inspection of the implementation and execution of some measures with assessment and evaluation. First, we can emphasize the responsibility for the work. During the transition period, we need to insist that secretaries at all five levels work together to consolidate and expand the results of poverty alleviation and comprehensively promote rural revitalization. Leaders at all levels should effectively take responsibility for prioritizing poverty eradication and ensuring that all policies and measures are implemented and effective.

Second, outdated and inefficient working methods should be adjusted, and successful experiences should be summarized based on inventorying and combing working methods and measures and initiatives to maintain effective and feasible working methods and abolish ineffective and inefficient ones. In addition, the institutional mechanism for poverty eradication should be considered, and multiple measures should be taken to consolidate the results.

Finally, it is necessary to strengthen supervision, inspection, and assessment and evaluation, establish a sound supervision, inspection, and assessment mechanism, and regularly supervise, inspect, and evaluate the work of poverty eradication. Identified problems should be rectified promptly and all policies and measures should be effectively implemented to achieve the desired results.

7.5 Summary

This chapter analyzes and discusses income inequality and poverty. It mainly includes the measurement, representation, measurement methods, theoretical basis, and relevant policy measures to address income inequality and poverty in China.

The results show that the income gap for Chinese residents exists both between and within urban and rural areas. The degree of income inequality is relatively high, and income inequality is universal. The national Gini coefficient increased by 66% over the sample range, whereas that of rural and urban residents increased by 48.9% and 177.9%, respectively. This means that, from 1985 to 2017, the income inequality of Chinese residents increased, and the increase in the income gap of urban residents was much higher than that of rural residents.

China has made world-renowned achievements in poverty alleviation. Using the 2010 poverty standard, China's rural poor population was 770 million in 1978, with a poverty incidence rate of 97.5%. The rural poor population dropped to 30.46 million in 2017. The incidence rate dropped to 3.1%, and in 39 years and 740 million people were lifted out of poverty. There is a significant difference between the incidence of poverty and the incidence of income poverty in different dimensions of welfare indicators, which means that the incidence of poverty measured from a multidimensional perspective is more comprehensive than that measured by a single income indicator.

To deal with income inequality and poverty, in this chapter, we propose a reform of income system and mechanism. Specifically, the mechanism includes improving the policy system, promoting equal opportunities for income distribution, improving fiscal and tax adjustment mechanisms, enabling residents to share the fruits of reform and development, improving the social security system, and improving the level of protection for low- and middle-income groups. We also suggest implementing a coordinated regional development strategy, including coordinating the development of developed and underdeveloped regions, promoting the free flow of factors between urban and rural areas, and strengthening inter-regional cooperation and mutual assistance, as well as implementing differentiated regional policies. We also propose adhering to targeted poverty alleviation to win the battle against poverty, including improving the identification mechanism of poverty alleviation targets, precisely targeting low-income populations, adapting measures to individual and local conditions, designing precise poverty alleviation policies and measures, and improving the poverty alleviation management system. These measures will help improve income distribution policies and reduce the income gap in urban and rural residents, thus reducing income inequality and poverty.

Exercise

1. Does poverty exacerbate income inequality? Is there a vicious circle in which poverty leads to inadequate resources for education, health care, etc., which in turn limits individuals' opportunities for development and exacerbates income inequality?
2. Several policies to address income inequality and poverty are discussed in this chapter. What challenges may these policies face in the course of implementation?
3. What are the global trends in income inequality and poverty? Are there certain regions or countries that have been successful in alleviating these problems? What lessons can other regions learn from these experiences?
4. China won the battle against poverty in 2021. How can we improve the endogenous development momentum of the poverty-stricken areas? How can we enhance the self-development capacity of poverty-stricken households through industrial cultivation and employment support?

Extended Reading

Duangkamon, C. (2008). *Modeling income distributions and Lorenz curves*. Springer.

Hong, D. V., Tuan, A. P., & Quan, T. (2023).The long-run effects of financial development on income inequality: Evidence from the Asia-Pacific countries. Heliyon.

Iradian, G. (2005). *Inequality, poverty, and growth: cross-country evidence*. Washington, D.C.
Milenko, F., & Fabrice, M. (2023).The geography of multi-dimensional poverty in France. *Social Indicators Research*.
Paolo, B., Ferreira, F. H. G., & Peragine, V. (2013). Inequality of opportunity, income inequality and economic mobility: Some international comparisons.
Shlomo, Y., & Edna, S. (2013). *The Gini methodology*. Springer
Servén, L., & Schmidt, H. K. (1999). Income inequality and aggregate saving: The cross-country evidence.
Thorsten, B., Asli, D. K., & Ross, L. (2004). Finance, inequality, and poverty: cross-country evidence.

References

Aaberge, R., Björklund, A., Jäntti, M., Palme, M., Pedersen, P. J., Smith, N., & Wennemo, T. (2002). Income inequality and income mobility in the Scandinavian countries compared to the United States. *Review of Income and Wealth, 48*(4), 443–469.

Aiyar, S., & Ebeke, C. (2020). Inequality of opportunity, inequality of income and economic growth. *World Development, 136*, 105115.

Alkire, S., & Foster, J. (2011). Counting and multidimensional poverty measurement. *Journal of Public Economics, 95*(7–8), 476–487.

Araar, A. (2006). On the decomposition of the Gini coefficient: An exact approach, with an illustration using Cameroonian data.

Chen, J., Dai, D., Pu, M., Hou, W., & Feng, Q. (2010). *The trend of the Gini coefficient of China*. Brooks World Poverty Institute Working Paper, (109).

Chen, Z., & Zhou, Y. (2005). Income distribution during system reform and economic development in China: The status and trend of income inequality of Chinese residents. Nova Publishers.

Cobb, C. W., & Douglas, P. H. (1928). A theory of production. *American Economic Review, 18*(1), 139–165.

Decancq, K., & Lugo, M. (2008). Settings weights in multidimensional indices of well-being and deprivation.

Fang, C. (2015). Important progress and future direction of studies on China's urban agglomerations. *Journal of Geographical Sciences, 25*, 1003–1024.

Foster, J. E., Greer, J., & Thorbecke, E. (1984). A class of decomposable poverty measures econometrics

Haveman, R., Blank, R., Moffitt, R., Smeeding, T., & Wallace, G. (2015). The war on poverty: Measurement, trends, and policy. *Journal of Policy Analysis and Management, 34*(3), 593–638.

Hirschman, A. O. (1958). *The strategy of economic development*. Yale University Press.

Kohli, V. (2018). Functional income distribution and effective demand in India: An applied post Keynesian model. *Journal of Post Keynesian Economics, 41*(3), 399–429.

Kovacevic, M. (2010). Measurement of inequality in human development—A review. *Measurement, 35*, 1–65.

Kuznets, S. (1955). Economic growth and income inequality. *American Economic Review, 45*(1), 1–28.

Li, S. (2016). Income inequality and economic growth in China in the last three decades. *The round Table, 105*(6), 641–665.

Liu, Y., Guo, Y., & Zhou, Y. (2018). Poverty alleviation in rural China: Policy changes, future challenges and policy implications. *China Agricultural Economic Review, 10*(2), 241–259.

Makower, H. (1953). Problems of capital formation in underdeveloped countries. *The Economic Journal, 63*(252), 897–899.
Myrdal, G. (1957). *Economic theory and underdeveloped regions*. Gerald Duckworth.
National Bureau of Statistics. China Rural Poverty Monitoring Report, 2010–2017.
National Bureau of Statistics of China. (2003). China statistical yearbook 2003. Beijing, China: China Statistics Press.
National Bureau of Statistics of China. (2020). China statistical yearbook 2020. Beijing, China: China Statistics Press.
Nelson, R. R. (1956). A theory of the low-level equilibrium trap in underdeveloped economies. *American Economic Review, 46*(5), 894–908.
Rosenstein-Rodan, P. N. (1943). Problems of industrialisation of Eastern and South-Eastern Europe. *The Economic Journal, 53*(210–211), 202–211.
Sicular, T., Ximing, Y., Gustafsson, B., & Shi, L. (2007). The urban–rural income gap and inequality in China. *Review of Income and Wealth, 53*(1), 93–126.
Sundrum, R. M. (1990). *Income distribution in less developed countries*. Routledge.
Turnovsky, S. J. (1977). *Macroeconomic analysis and stabilization policy*. CUP Archive.
Wang, H., Zhao, Q., Bai, Y., Zhang, L., & Yu, X. (2020). Poverty and subjective poverty in rural China. *Social Indicators Research, 150*, 219–242.
Yangyang, S., Alkire, S., & Peng, Z. (2019). Measurement and decomposition of multi-dimensional poverty in China. *China Economist, 14*(3), 12–28.
Zhang, Q. (2015). Dynamic change of multidimensional poverty in China: 1991–2011 (in Chinese). *Journal of Finance and Economics, 41*(4), 31–41.
Zhou, Y., Liu, Z., Wang, H., & Cheng, G. (2023). Targeted poverty alleviation narrowed China's urban-rural income gap: A theoretical and empirical analysis. *Applied Geography, 157*, 103000.

Chapter 8
Geographical Distribution of the Population

Abstract Population development is a prerequisite for sustainable economic development, and the level of sustainable regional development depends on the extent to which population size is harmonized with the economy, society, resources, and environment. An accurate understanding of the development history, structural distribution, spatial distribution, and future trends of a population is conducive to the formulation of national and regional development plans and population policies. This chapter provides an overview of China's fertility and migration policies and an understanding of the structural and spatial distribution of China's population by analyzing and summarizing the development of its population and predicting future trends in population change. The objectives of this chapter are as follows: 1. To master the measurement and characterization of population growth and structure. 2. To understand the differences in the spatial geographical distribution of a population. 3. To familiarize with population fertility and migration policies.

8.1 Measurement of Population Growth

8.1.1 Overview of Demographic Development in China

Plotted based on data from the China Statistical Yearbook, Fig. 8.1 shows trends in China's total population change over time. The figure illustrates that China's population increased from 963 million to 1.395 billion between 1978 and 2018, representing a net increase of 433 million people (45%) over a 40-year period. According to the graph of growth rate trends in China's total population over time, China's population growth rate between 1979 and 1987 continued to increase amidst fluctuations and despite the implementation of population-planning policies. However, the population growth rate over 27 consecutive years from 1988 to 2015 showed a continuous decline, with the population numbers showing a low and steady growth trend. With the introduction of the two-child policy on January 1, 2016, the population growth rate rose briefly in 2016, but only by 0.59%. Subsequently, the population growth rate was 0.53% in 2017 and 0.38% in 2018, setting a new 40-year low. According to

X. Deng and M. Song, *Development Geography*,
https://doi.org/10.1007/978-981-96-6917-2_8

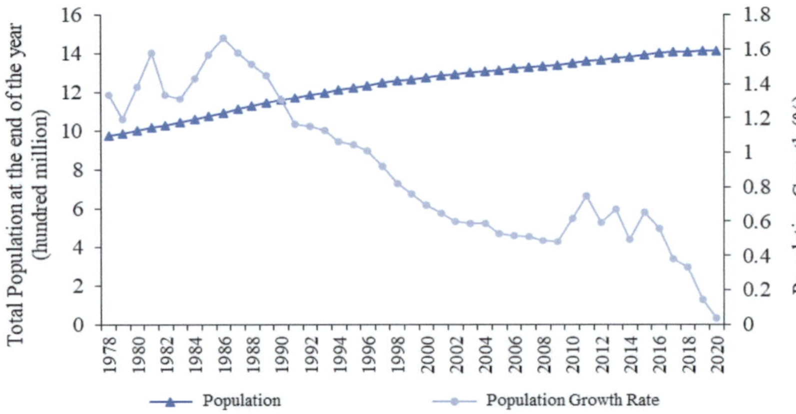

Fig. 8.1 Trends in China's total population over time

unpromising forecasts, in the near future, China will experience negative population growth, and the demographic dividend element of China's economic growth will disappear.

8.1.2 Population Projections

Population forecasting refers to making projections regarding future population data based on historical population data. By considering the various factors affecting population development, a scientific and reasonable modeling approach can be used to measure the population size and structure over a certain time in the future (Tayman et al., 2011; Wilson, 2013). This study used the most adapted statistical models, such as the auto-regressive and moving average (ARMA) model, linear regression, and grey forecasting, to predict China's population size and to obtain the most accurate results.

8.1.2.1 ARMA Model

Typically, a time series graph can approximately display the smoothness of the series. If the time series always fluctuates around a constant value and the fluctuation range is small, then the series is likely to be smooth. In contrast, if the series has an obvious trend or cycle, then it is usually not smooth.

As shown in Fig. 8.1, a clear growth trend in China's total population over time is observed in the first step; however, it does not have the characteristics of a smooth series. Additionally, the second step, which is the autocorrelation plot of China's

Table 8.1 Autocorrelogram of the total population series in China over time

Order	Autocorrelation	Partial correlation	Q-Stat	Prob
1	0.934	0.934	41.053	0.000
2	0.866	– 0.051	77.166	0.000
3	0.795	– 0.052	108.39	0.000
4	0.724	– 0.044	134.95	0.000
5	0.654	– 0.038	157.13	0.000
6	0.584	– 0.04	175.27	0.000
7	0.514	– 0.038	189.73	0.000
8	0.446	– 0.038	200.9	0.000
9	0.38	– 0.03	209.24	0.000
10	0.316	– 0.027	215.19	0.000
11	0.256	– 0.024	219.22	0.000
12	0.199	– 0.027	221.71	0.000
13	0.144	– 0.03	223.07	0.000
14	0.091	– 0.032	223.63	0.000
15	0.041	– 0.034	223.75	0.000
16	– 0.007	– 0.034	223.75	0.000
17	– 0.053	– 0.033	223.96	0.000
18	– 0.096	– 0.032	224.69	0.000
19	– 0.137	– 0.031	226.2	0.000
20	– 0.175	– 0.031	228.78	0.000

total population series over time, shows that the series has not been truncated, further indicating that the series is not smooth (Table 8.1).

In the third step, Table 8.2 shows the autocorrelation plot after the first-order differencing of China's total population series over time. The series is consistently non-truncated, again showing that the series is not smooth.

In the fourth step, Table 8.3 shows the autocorrelation plot after the second-order differencing of China's total population series over time. Although the series is smooth, it is also considered a white noise series, which is unsuitable for an ARMA model. Therefore, a linear regression model is suitable.

8.1.2.2 Linear Regression Prediction

First, the econometric model is as follows:

$$y = \beta_0 + \beta_1 x + \varepsilon \tag{8.1}$$

Table 8.2 Autocorrelogram after the first-order differencing of the total population series in China

Order	Autocorrelation	Partial correlation	Q-stat	Prob
1	0.878	0.878	35.504	0.000
2	0.775	0.02	63.865	0.000
3	0.694	0.044	87.193	0.000
4	0.603	– 0.079	105.26	0.000
5	0.552	0.119	120.76	0.000
6	0.502	– 0.012	133.92	0.000
7	0.424	– 0.128	143.56	0.000
8	0.375	0.057	151.32	0.000
9	0.304	– 0.118	156.57	0.000
10	0.269	0.133	160.8	0.000
11	0.222	– 0.129	163.77	0.000
12	0.149	– 0.099	165.16	0.000
13	0.079	– 0.09	165.56	0.000
14	0.026	0.028	165.6	0.000
15	– 0.027	– 0.023	165.65	0.000
16	– 0.083	– 0.146	166.15	0.000
17	– 0.127	0.043	167.35	0.000
18	– 0.171	– 0.072	169.63	0.000
19	– 0.219	– 0.02	173.5	0.000
20	– 0.26	– 0.089	179.17	0.000

where y is the explanatory variable representing China's total population and x is the explanatory variable representing the year. β_0 is the constant term, β_1 is the effect coefficient, and ε is the random error term, satisfying zero mean homoscedasticity. In addition, when using the linear regression model of Eq. 8.1 to predict the total population of China, the total sample is divided into four parts: 1978–2018, 1988–2018, 1998–2018, and 2008–2018. The results are shown in Table 8.4. The goodness of fit continues to improve from Model 1 to Model 4, and the prediction accuracy gradually improves.

The regression results in Table 8.4 verify the data for 1978–2018 and predict the total population data for 2019–2030. Figure 8.2 shows the results, with Models 3 and 4 having the highest prediction accuracy. The prediction results for Model 3 show a total population of 1.407 billion in 2020, 1.414 billion in 2021, 1.421 billion in 2022, 1.428 billion in 2023, 1.435 billion in 2024, 1.442 billion in 2025, 1.450 billion in 2026, 1.457 billion in 2027, 1.464 billion in 2028, 1.471 billion in 2029, and 1.478 billion in 2030.

Table 8.3 Autocorrelogram after the second-order differencing of the total population series in China

Order	Autocorrelation	Partial Correlation	Q-Stat	Prob
1	0.019	0.019	0.0168	0.897
2	0.039	0.039	0.0877	0.957
3	0.077	0.075	0.3666	0.947
4	– 0.004	– 0.009	0.3675	0.985
5	0.142	0.137	1.3738	0.927
6	– 0.122	– 0.136	2.1426	0.906
7	– 0.057	– 0.061	2.3155	0.94
8	– 0.034	– 0.046	2.3791	0.967
9	– 0.105	– 0.082	3.0013	0.964
10	– 0.171	– 0.186	4.6805	0.911
11	– 0.035	0.018	4.7549	0.942
12	– 0.015	0.009	4.7688	0.965
13	– 0.143	– 0.132	6.0665	0.944
14	– 0.042	– 0.026	6.1802	0.962
15	0.017	0.058	6.2004	0.976
16	– 0.041	– .087	6.3207	0.984

Table 8.4 Linear regression model fitted to the total population of China

Models	Number of samples	Linear regression model	*P*-value	Fitting optimality
1	Total	Y = 0.1094*X – 206.5203	0.0000	0.9691
2	1988–2018	Y = 0.0902*X – 167.8557	0.0000	0.9741
3	1998–2018	Y = 0.0717*X – 130.7684	0.0000	0.9972
4	2008–2018	Y = 0.0683*X – 123.8175	0.0000	0.9989

8.1.2.3 Grey Forecast

In addition to ARMA and linear regression model forecasting, the grey forecasting model, GM (1,1), which is commonly used in mathematics, can also be used for analysis (Guo et al., 2022). The principle of the grey forecasting GM (1,1) is as follows:

In the first step, assume that the original data series with n observations is expressed as $X^{(0)} = \{X^{(0)}(1), X^{(0)}(2), \ldots, X^{(0)}(n)\}$, then, it is summed over the cumulative amounts to obtain a new series $X^{(1)} = \{X^{(1)}(1), X^{(1)}(2), \ldots, X^{(1)}(n)\}$, which is used to fit the function.

In the second step, a fitting function is applied to find the predicted value $X_{(1)}$ of the new series:

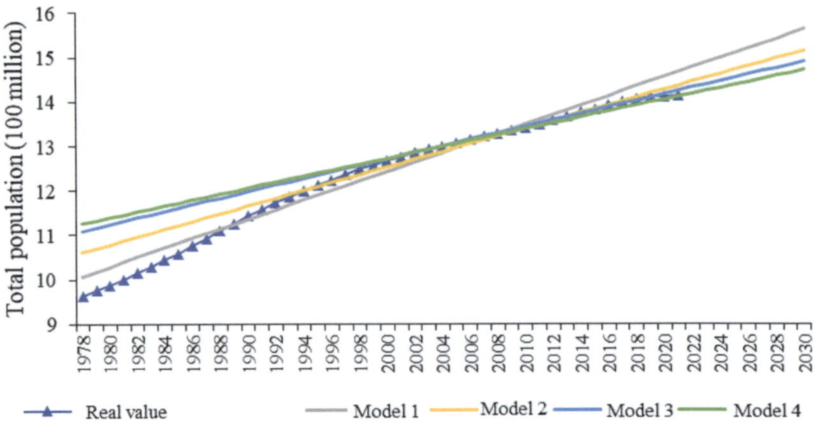

Fig. 8.2 Linear regression prediction results for the total population of China

In the third step, the cumulative subtraction operation using the Equation $X_{(0)}(\mathrm{k}) = X_{(1)}(\mathrm{k}) - X_{(1)}(\mathrm{k}-1)$ will generate a sequence of grey predicted values: $X_{(0)}(\mathrm{k}) = \{X_{(0)}(1), X_{(0)}(2), X_{(0)}(\mathrm{n+m})\}$ (i.e., predicting m periods backward for sample size n). According to the total Chinese population data from 1978 to 2018 shown in Fig. 8.1, using MATLAB to write the program and dividing the total sample into four parts (i.e., 1978–2018, 1988–2018, 1998–2018, 2008–2018), the grey prediction results shown in Fig. 8.1-6 are obtained.

As Fig. 8.3 shows, Models 3 and 4 have the highest prediction accuracy. The predictions from Model 3 show that the total population will be 1.413 billion in 2020, 1.420 billion in 2021, 1.428 billion in 2022, 1.436 billion in 2023, 1.443 billion in 2024, 1.451 billion in 2025, 1.459 billion in 2026, 1.466 billion in 2027, 1.474 billion in 2028, 1.482 billion in 2029, and 1.490 billion in 2030. This projection is similar to the linear regression results, and overall, it shows that China's population will be in a slow growth phase over the next decade.

8.2 Population Structure

As China's demographic development enters a transition period, the quality of the population's demographics becomes increasingly important. Therefore, further attention must be paid to both population structure and quantity. Menashe-Oren & Stecklov (2018) researched that population structure generally includes the gender, urban–rural, and age structure of the population.

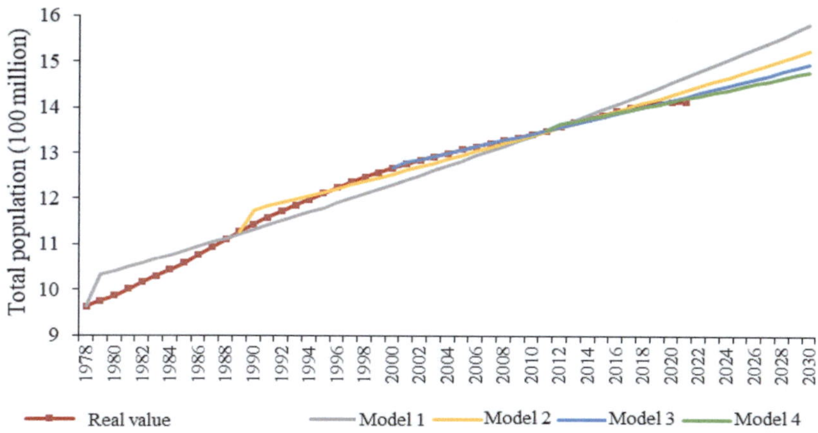

Fig. 8.3 Grey projection results for China's total population

8.2.1 Gender Structure

Generally, the normal range for the sex ratio of a country's total population is between 95 and 102 (Gautam et al., 2015). However, as shown in Fig. 8.4, the gender structure of China's population has a sex ratio that remains consistently around 105, which is higher than the critical value of 102. This means that the male population is much larger than the female population, which is mainly related to China's long-standing ideology of "preference for men over women." This causes the most immediate harm to marriage, as it prevents more men from marrying. It also leads to an increase in crimes such as marriage selling, child marriages, and women trafficking, lowering the ethical and moral standards in modern society and increasing the risk of family and social instability.

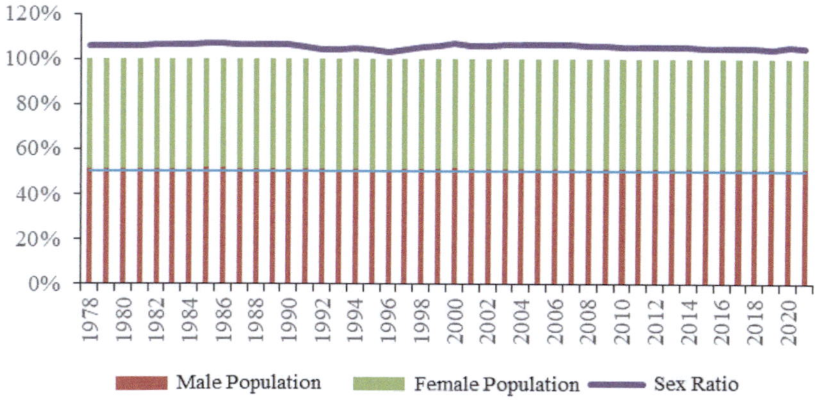

Fig. 8.4 The gender structure of China's population

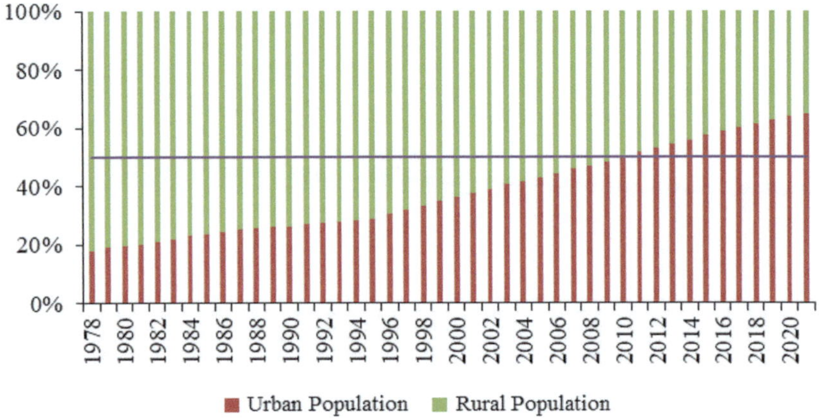

Fig. 8.5 Urban–rural structure of China's population

8.2.2 Urban and Rural Structure

As shown in Fig. 8.5, the urbanization rate of China's population increased from 17.92 to 59.58% between 1978 and 2018, representing a net increase of 41.66% over a 40-year period. According to the graph, the urbanization rate trend can be divided into two time periods: 1978–1995 and 1996–2018. The first is the early stage of reform and opening up in 1978, during which rural and urban reforms were implemented and the urbanization rate increased. The second stage occurred after 1996, when China's urbanization reforms were strengthened and the urbanization rate of the population began to increase at a faster pace and has maintained a high growth rate to date.

However, from a global perspective, although China's urbanization rate reached 59.58% in 2018, a large gap has remained compared to developed countries (Brass, 1974; Deng et al., 2008). For example (Population & Council, 1984), at the end of 2016, urbanization rates in other countries included 69.1% in Italy, 75.5% in Germany, 79.8% in France, 81.8% in the United States, and 82.8% in the United Kingdom.

8.2.3 Age Structure

Figure 8.6 shows the trend in the age structure of China's population from 1990 to 2018; three aspects can be observed. First, the proportion of the juvenile population aged 0–14 years shows a serious downward trend, falling from 17.69% in 1978 to 16.86% in 2018. Second, the share of the working-age population (15–64 years) has shown a trend of rising and falling, increasing from 66.74% in 1990 to 74.53%

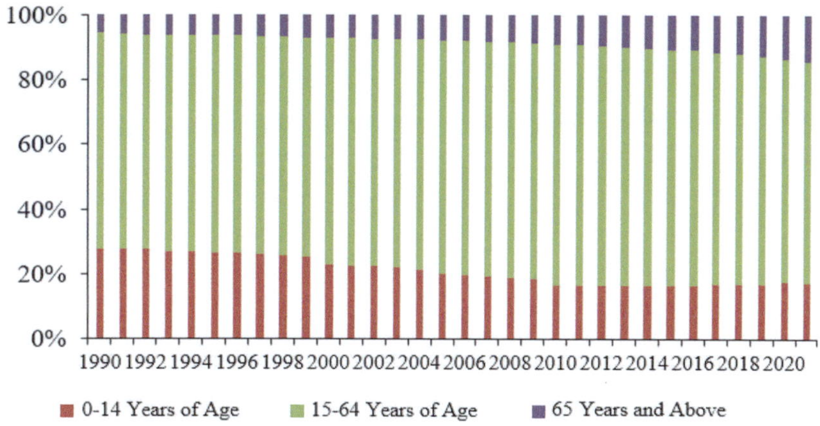

Fig. 8.6 Age structure of China's population

in 2010, and then falling to 71.20% in 2018. Third, the proportion of the working-age population in China has been continuously rising since 1990, increasing from 5.57% in 1990 to 11.94% in 2018, showing a serious trend of an aging working-age population. Forecasting future trends, with the declining share of the working-age population and rising share of the older adult population, future young adults in China will face greater pressure to raise young children and support older adults, and more comprehensive pension system protection and education systems will be required.

8.2.4 Population Dependency Ratio

The dependency ratio reflects the proportion of adults in society who bear the responsibility of providing support to children and older adults. It also largely reflects the relationship between them in terms of support and dependency (Bidisha et al., 2020). As Fig. 8.7 shows, the dependency ratio of China's juvenile and older adult populations has shown a trend of first decreasing and then increasing from 1990 to 2018, with the burden on young adults decreasing. In particular, the juvenile dependency ratio fell from 41.50% in 1990 to 22.10% in 2011, and then rose slightly to 23.70% in 2018. In contrast, the older adult dependency ratio rose from 8.3% in 1990 to 16.80% in 2018, a twofold increase in the population share. This shows that while the pressure on young adults to provide support is easing and the juvenile dependency ratio is falling significantly, the older adult dependency ratio is rising significantly, and the burden of supporting older adults will likely be a major social issue in the future.

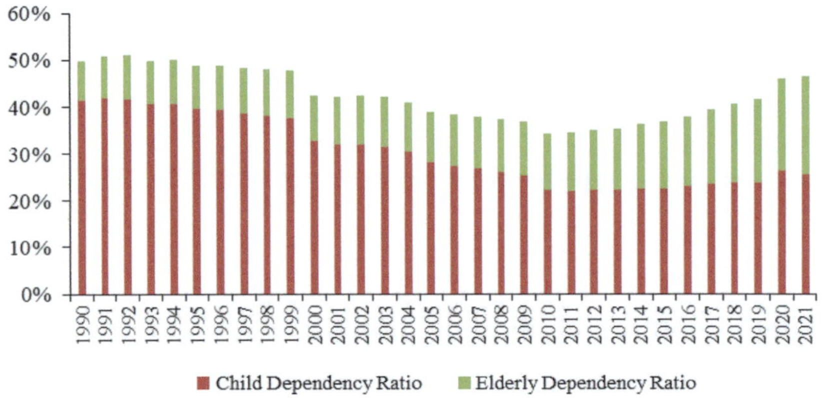

Fig. 8.7 Dependency ratio of China's population

8.3 Spatial Differences in China's Population

China is a vast country; thus, population issues should not only focus on the overall situation of population size and structure but also on differences in the population's spatial and geographical distribution, which can be divided into two indicators: absolute population size and population density.

8.3.1 Differences in the Geographical Distribution of Population Size

According to the national 2014 city size classification criteria, 2010 census data, and the latest decennial population census, 286 major cities in China are classified into five major classes. Cities with populations between 5 and 10 million are super-cities, cities with populations between 1 and 50 million are mega-cities (furthermore, cities with populations between 3 and 5 million are Class 1 megacities and those between 1 and 3 million are Class 2 megacities), cities with populations between 500,000 and 1 million are medium-sized cities, and the rest are small cities (Fig. 8.8).

Shanghai, Beijing, Chongqing, Tianjin, Guangzhou, and Shenzhen are super-cities. Ten cities, including Wuhan, are mega-cities, distributed in every direction of China, and are the core agglomerations of a region. Jinan and 120 other cities are megacities.

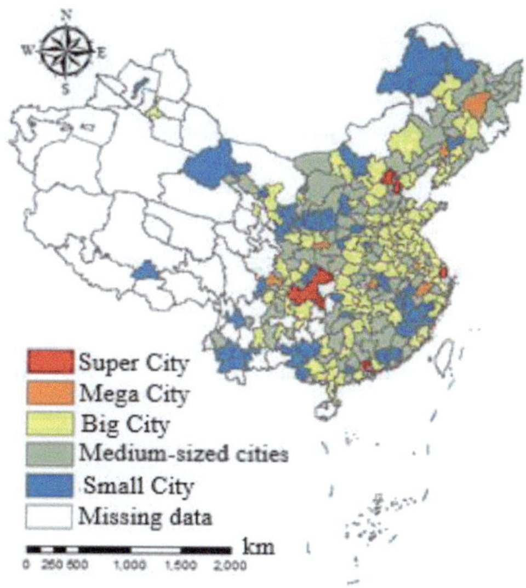

Fig. 8.8 Differences in the geographical distribution of population size in China

8.3.2 Differences in the Geographical Distribution of Population Density

The population and urban area data of 286 prefecture-level cities in China were obtained from the 2004–2017 China Urban Statistical Yearbook, and the results of the spatial geographical distribution of urban population density in China from 2003 to 2016 were calculated (Fig. 8.9). China's southeastern coastal areas and the Shandong and Hebei regions are high population density areas. Shantou had the highest population density in China with 2348 persons/km^2 in 2003. Jieyang ranked third with 1224 persons/km^2, and Shenzhen ranked fourth with 1168 persons/km^2. Until 2016, the population density of all prefecture-level cities in China showed an increasing trend.

Overall, serious geographical disparities exist in the geographical distribution of both population size and density in China, with the population being concentrated mainly in the eastern coastal regions, particularly in the regions of the North, Guangzhou, and Shenzhen and the four major municipalities.

8.4 Population Fertility and Migration Policy

Based on an analysis of the temporal trends of China's population size and demographic structure and the differences in the geographical distribution of its population size and density, the process of China's fertility policy and the current talent-attainment policy are further discussed from the perspective of policy changes.

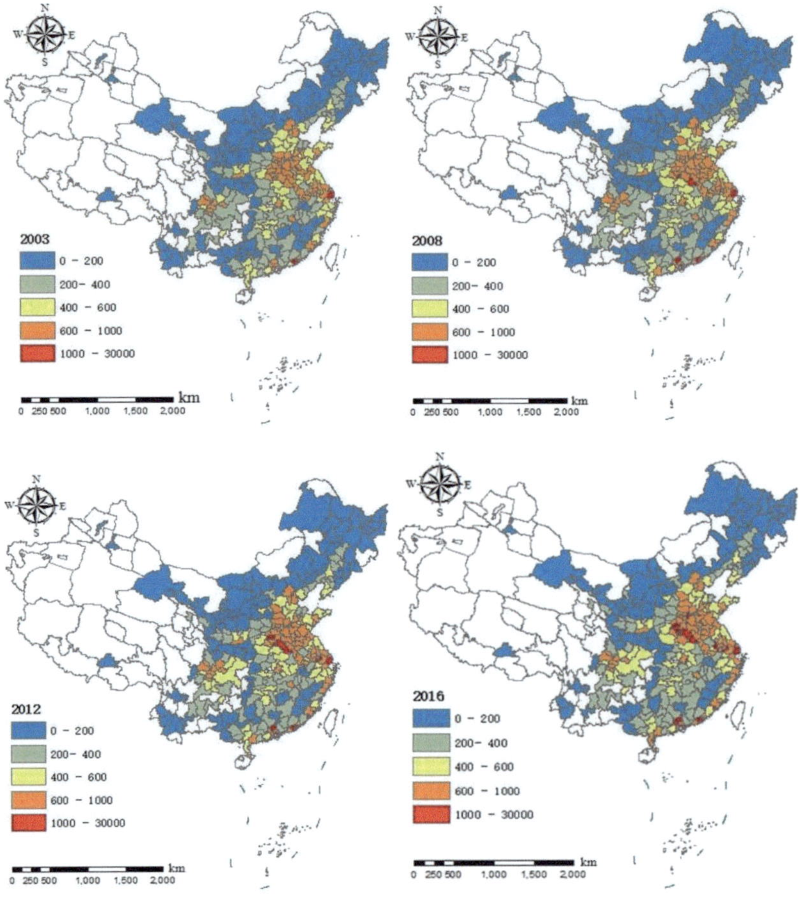

Fig. 8.9 Differences in the geographical distribution of the population density in China. *Note* See Tables 8.2 and 8.3 for the original data

Furthermore, additional analyses are presented to provide a better understanding of the antecedents and consequences of China's demographic changes and the way forward.

8.4.1 Population Fertility Policy in China

8.4.1.1 Exploring the Initial Stages of Family Planning

At the beginning of New China, the central government adopted a permissive birth policy to promote economic development, based on the principle that "there is strength in numbers." Simultaneously, the government encouraged families to have more children to ensure a sufficient labor force and gain the advantage of a demographic dividend. However, in 1953, the first population census found that China's population was growing too rapidly for society's capacity and, thus, constraining economic development. Thus, a "family planning" policy was proposed to conserve resources. In 1954, the government began to promote birth control and late marriage; however, at this time, the policy was still mainly based on propaganda and advocacy.

After the 1960s, the government first planned to implement "family planning" in rural areas, with subsequent leaders continuing it. In 1973, the national economic plan included population targets for the first time, and a family planning policy group was established to promote late marriage and childbirth. However, overall, the family planning policy at this stage was based on encouragement, promotion, and propaganda, with no real compulsory policy. Thus, family planning policies were still in the initial exploration stage.

8.4.1.2 Initial Formation of the "One Child" Policy

In 1978, as China's society stabilized and economic development took off, a high degree of importance was attached to family planning policy, which was established in the form of a constitution. In 1980, the Central Committee of the Communist Party of China stated at a national conference that it was clear that "a couple can only have one child." In 1981, the Family Planning Administration was established, and in 1982, family planning was implemented as a basic state policy, with the Constitution elevating family planning to a civic duty. This series of policies shows the importance that the central government imposes on family planning.

Between 1980 and 2009, China's family planning policy began to evolve from advocacy to enforcement to control the population, which resulted in a phenomenon of exogenous low fertility. In 1984, the central government made adjustments to the policy, piloting a two-child policy with rural families in need. In 1986, with the emergence of a gender imbalance in China's population, a new change in family planning policy was introduced, with families who had only one daughter being allowed to have two children after a certain number of years. The Regulations on the Administration of Technical Family Planning Services in 2001 and the Law of the People's Republic of China on Population and Family Planning in 2002 were important in improving the administrative and regulatory content of China's family planning policy and filling the gaps in the legal basis.

8.4.1.3 Implementation of the "Two Children Only" Policy

Since the turn of the century, the shortcomings of the family planning policy have gradually emerged, with the phenomena of "old age before wealth" "lost families" and "fewer children" becoming prevalent in society, and the fertility rate falling below 2.1. In 2004, eighteen population experts collectively suggested to the Family

Planning Commission that the "separate two-child" policy should be piloted first, then gradually extended to the whole country, before ultimately implementing the "full two-child" policy nationwide. However, it was not until 2009 that the Health and Family Planning Commission started exploring fertility policy reform, which was then applied nationwide.

At the Third Plenary Session of the 18th Central Committee in 2013, after several pilots and reforms, the "separate two-child" policy, which allows couples with one child to have two children, was implemented nationwide. This can be seen as an adjustment and improvement to the original family planning policy. The "separate two-child" policy has opened a new starting point for population policy.

8.4.1.4 Liberalization of the "Full Two-Child" Policy

In 2014, the Chinese Academy of Sciences noted in its Economic Blue Book that China's natural fertility rate was approaching the "ultra-low fertility trap" and that the "comprehensive two-child" policy must be launched immediately. By the end of 2016, only 1.39 million separate couples had submitted applications for a second child, and there was an urgent need to continue family planning policy reform. In 2016, China officially implemented the "comprehensive two-child" policy at the national level. After the implementation of the policy, the National Bureau of Statistics released the results of indicators, which showed that 17.86 million people were born in 2016, 1.31 million more than in 2015, and the natural population growth rate increased by 0.9 thousandths of a percentage point year-on-year, with the fertility rate reaching a peak.

By the end of 2016, family planning in China had undergone the entire process of "one-child" "separate two-child" and "comprehensive two-child" policies, achieving a complete transformation from "population control" to "population optimization."

8.4.2 Talent Settlement Policy

Currently, China's supply-side structural reform and high-quality economic development are inseparable from talent resources, and talent is the most critical and scarce resource. All major cities in China have successively introduced talent implementation policies to attract more high-end talent and obtain talent resources.

As shown in Table 8.5, which outlines the population settlement policies of major cities, all policies regarding nationwide talent focus on those who have received higher education in science and technology innovation, cultural and creative talent, sports talent, and talent for international interaction, according to their respective development positioning and characteristics. These policies aim to improve population quality rather than focusing on population size or population quantity alone.

Table 8.5 National population settlement policy for major cities

City	Population Implementation Policy
Nanjing	Persons with undergraduate and higher degrees and are within 40 years of age may settle directly and apply for a three-year housing subsidy Foreign students who have obtained a graduate certificate from a university in Jiangsu can apply for a residence permit for two to five years To achieve the development goal of a science and technology industry innovation center, a housing subsidy of no less than RMB 3 million is provided to exceptional talent
Zhengzhou	Those with specialized degrees and higher qualifications can settle in Zhengzhou after taking up residence in the city Housing subsidies for graduates of "double-class" universities and master's and doctoral students 100,000 RMB subsidy for the first home purchase for young talent with doctoral degrees
Chengdu	Direct settlement for people with a bachelor's degree or above up to the age of 45. They are also provided with a free young talent station
Hangzhou	Outsiders under the age of 45 with a vocational qualification of technician or above or those under the age of 35 with a vocational qualification of senior worker may obtain a household registration in Hangzhou, provided they have worked for the same employer in the city for at least three years and have a legal fixed residence in the city
Xi'an	Persons up to 45 years of age with a bachelor's degree may settle directly There is no age limit for persons with a Master's degree or above
Hefei	Foreigners can buy a house in Hefei after paying one month of social security Foreigners must have a college degree or above or be a senior skilled engineer to buy a second home in Hefei. They can buy after they have graduated in less than two years and settled down
Beijing	Five categories of highly-skilled talent with special allowances from the State Council Government can settle in the form of talent introduction, and their spouses and minor children can also enjoy direct settlement preference
Shanghai	For the time being, Peking University and Tsinghua University will be used as a pilot program to allow direct access to the green channel for direct settlement in Shanghai
Wuhan	Graduates from ordinary universities can settle in Wuhan directly within three years of graduation University graduates can enjoy preferential policies such as "20% off housing" and rental housing
Tianjin	Academic, qualified, skilled, entrepreneurial, and urgently needed talent, up to the age of 40 for bachelor's degrees, up to the age of 45 for master's degrees, and no age limit for doctorates, can settle in Tianjin directly
Guangzhou	The declaration of residence has been streamlined to only one-quarter of its original size
Shenzhen	No manual automatic settlement approval The introduction of "second approval" for university graduates, working talent, returning talent, post-doctoral students, and their spouses and children

8.5 Summary

This chapter examines the history of China's population development, future trends in population change, and differences in the structural distribution and geographic-spatial distribution of China's population, providing a brief overview of China's population fertility and migration policies. The main findings are discussed below.

First, between 1979 and 1987, despite the implementation of the planning policy, China's population growth rate generally maintained an increasing trend amid fluctuations. However, the population growth rate in the 27 consecutive years between 1988 and 2015 showed a continuous decline, with a low and steady growth trend. With the implementation of full liberalization of the two-child policy on January 1, 2016, the population growth rate rose briefly in 2016. Subsequently, the population growth rate was 0.53% in 2017 and 0.38% in 2018, setting a new 40-year low. Meanwhile, the forecasts presented in this chapter predict China's population over the next decade using three statistical models: the ARMA model, linear regression, and grey forecast. The obtained robust results indicate that China's total population will reach approximately 1.45 billion and 1.5 billion in 2025 and 2030, respectively, and that overall, China's population will be in a slow growth phase in the next decade.

Second, China's population faces certain problems regarding gender structure, urban–rural structure, age structure, and dependency ratio. China's population sex ratio has been around 105, which is much higher than the global maximum empirical value of 102. China has also experienced the rapid urbanization of its population over the past 40 years of reform and opening up; however, a large gap remains compared to developed countries. Further, while the overall pressure on the young and middle-aged populations and the juvenile dependency ratio in China are significantly decreasing, the dependency ratio of the older adult population is significantly increasing.

Third, serious geographical differences are observed in the distribution of China's population, both in terms of population size and density, with the population still concentrated mainly in the eastern coastal regions.

Fourth, both in terms of the evolution of the population's fertility policy and the current policy to obtain talent, China has entered a stage of "population optimization". In the future, major cities and regions will pay more attention to introducing high-end talent to improve population quality.

Exercise

1. What is the basic pattern of population distribution in China?
2. Analyze the main trends and characteristics of the evolution of China's population distribution pattern from 2000 to 2021.
3. Discuss how the natural geographic environment affects the stability of China's population distribution.
4. Discuss the influence of coastal urban agglomerations on the formation of population growth belts and its reasons.

5. Analyze the reasons for the uneven spatial distribution of China's population and its impact on regional development.
6. Analyze the unevenness of urban and rural population distribution and its causes.
7. Discuss the forecasts and challenges of the future geographic distribution of China's population.

Extended Reading

Feng, Z., Li, F., Yang, Y., & Li, P. (2017). The past, present, and future of population geography in China: Progress, challenges and opportunities. *Journal of Geographical Sciences, 27*, 925–942.

Henry, G. (1893). The geographical distribution of the population of the United States. *Publications of the American Statistical Association, 3*(24), 551–553.

He, B., & Ho, P. (1959). *Studies on the population of China, 1368–1953*. Harvard University Press.

Kan, H., Chen, R., & Tong, S. (2012). Ambient air pollution, climate change, and population health in China. *Environment International, 42*, 10–19.

Qi, W., Liu, S., Zhao, M., & Liu, Z. (2016). China's different spatial patterns of population growth based on the "Hu Line." *Journal of Geographical Sciences, 26*, 1611–1625.

Teng, F., Wang, Y., Wang, M., & Wang, L. (2022). Monitoring and analysis of population distribution in China from 2000 to 2020 based on remote sensing data. *Remote Sensing, 14*(23), 6019.

References

Bidisha, S. H., Abdullah, S. M., Siddiqua, S., & Islam, M. M. (2020). How does dependency ratio affect economic growth in the long run? Evidence from selected Asian Countries. *Journal of Developing Areas, 54*(2), 47–60. https://doi.org/10.1353/jda.2020.0015

Brass, W. (1974). Perspectives in population prediction: Illustrated by the statistics of England and Wales. *Journal of the Royal Statistical Society: Series A (General), 137*(4), 532–570.

Deng, X., Huang, J., Rozelle, S., & Uchida, E. (2008). Growth, population and industrialization, and urban land expansion of China. *Journal of Urban Economics, 63*(1), 96–115. https://doi.org/10.1016/j.jue.2006.12.006

Gautam, R. K., Jhariya, J., & Kumar, P. (2015). Globally declining population of women folk causing sex imbalance is a serious concern: An analysis of sex ratio around the globe. *Journal of Anthropology, 2015*(1), 431458. https://doi.org/10.1155/2015/431458

Guo, X., Zhang, R., Shen, H., & Yang, Y. (2022). An optimized damping grey population prediction model and its application on china's population structure analysis. *International Journal of Environmental Research and Public Health, 19*(20), 13478. https://doi.org/10.3390/ijerph192013478

Menashe-Oren, A., & Stecklov, G. (2018). Rural/urban population age and sex composition in sub-Saharan Africa 1980–2015. *Population and Development Review, 44*(1), 7–35. https://doi.org/10.1111/padr.12122

Population, C. O., & Council, N. R. (1984). *Rapid population change in China, 1952–1982*. National Academies Press.

Tayman, J., Smith, S. K., & Rayer, S. (2011). Evaluating population forecast accuracy: A regression approach using county data. *Population Research and Policy Review, 30*(2), 235–262. https://doi.org/10.1007/s11113-010-9187-9

Wilson, T. (2013). Quantifying the uncertainty of regional demographic forecasts. *Applied Geography, 42*, 108–115. https://doi.org/10.1016/j.apgeog.2013.05.00

Chapter 9
Geographical Differences in Culture

Abstract Culture is the sum of material and spiritual civilization and an important symbol of social development. Culture varies from region to region in five aspects: language, clothing, food, architecture, and transportation. This chapter discusses the geographical distribution differences of culture, specifically analyzing the role of culture and cultural industries in geographical development, the spatial distribution differences of culture and their causes, the spatial distribution differences of cultural industries and their causes, and the strategic spatial layout of China's culture and cultural industries to promote the development of China's economy through the development of culture and cultural industries. The objectives of this chapter are as follows: 1. To understand the role of culture and cultural industries in geographical development. 2. To understand the spatial distribution differences and reasons for different cultures. 3. To become familiar with the strategic spatial layout of Chinese culture and cultural industries.

9.1 Culture and Cultural Industries

The development of culture and cultural industries comprises an increasing proportion of the tertiary industry, which plays an important role in national economic development and improvement. It is also crucial in promoting the development of the geographical environment. The progress of culture and cultural industries is also a typical symbol of geographical development.

9.1.1 Passing on Human Civilization

Civilization is the sum of all social and natural behaviors of human beings born out of ignorance and barbarism; this specifically includes material and spiritual civilization (Bian, 2023). From a historical development perspective, human civilization has mainly progressed from primitive to agricultural, industrial, and now to ecological

X. Deng and M. Song, *Development Geography*,
https://doi.org/10.1007/978-981-96-6917-2_9

civilization. From the perspective of geographical regions, human civilization mainly originates in areas such as the Yellow River Basin, Nile River Basin, and India River Basin.

During the primitive civilization period, humans discovered the use of fire, learned to preserve it, and build simple bone and stone tools, which promoted human civilization to a certain extent. During the agricultural civilization period, humans began to use geology and natural resources to invent and create, with their most significant achievements including bronze ware, ironware, writing, and papermaking. During the Han Dynasty, humans learned how to make paper, which is a great treasure of Chinese civilization and even the world. The industrial civilization period is considered the age with the greatest transformation of the geographical environment. Humans use scientific and technological knowledge and natural resources to invent and create. Creations such as steam engines, trains, and factories have greatly improved the happiness of human life. At this stage, human beings are in a state of full, and even excessive, utilization of natural resources. The continuous exploitation of natural resources, industrial waste gas, and sewage discharge has damaged the environment. Humans have not properly reconciled the relationship between natural resources and human development and progress, which has led to green thoughts on the construction of ecological civilization. During the ecological civilization period, people have to consider more how balance can be achieved between industrial development and a clean environment. Many scientific development strategies have been proposed to promote the coordinated development of both.

The Yellow River Basin is considered the birthplace of Chinese culture. As the second longest river in China, the continuous development of both sides of the Yellow River created Chinese civilization. From the perspective of geographical environment, the Yellow River Basin consists of many mountains and is mostly temperate with sufficient sunshine and abundant rainfall due to the terrain, making it suitable for growing crops. Based on the blockage and interval of mountains, rivers, and lakes, and the enfeoffment of early lands, a system of superiority and inferiority based on land area and territory was also formed. Therefore, Chinese civilization has also been called "land civilization." This period echoed that of agricultural civilization, and mainly created bronzes, characters, and silk culture.

The Aegean culture in ancient Greece is often considered the origin of Western culture. The area around Mesopotamia and the Nile River Basin can be considered the distribution area of early ancient Greek civilization. The geographical environment of ancient Greek civilization did not include large areas of open land but instead had many small islands. The mountains on these islands had lush forests, where a culture gradually developed with a relatively unified ethnicity, set of customs, and language.

Drawing on both Chinese and Western civilizations, the development of Chinese culture mainly revolved around the development of land, whereas Western culture primarily developed as a maritime culture centered on the ocean. A comparison shows that Western civilization evolved utilizing the marine environment, with established sea traffic routes, flexible development, and developed commercial activities facilitating the development of mathematics to a certain extent. Simultaneously, the characteristics of the geographical environment of ocean routes also promoted the development of astronomy. Additionally, the capitalist civilization, which advocates equality, freedom, and technology in the West, has gradually begun to evolve. In contrast, Chinese culture developed mainly utilizing the land and evolved into a more agrarian civilization. The structure is relatively simple, the cultural development is stable but slow, there is little communication with other cultures, and the system of superiority and inferiority is more ideologically rigid than it is in the West.

9.1.2 *Regulating Human Behavior*

From the perspective of the development history of human civilization, the relationship between human beings and the geographical environment can be seen as a development process of respect, balance, destruction, and cherishment. In primitive societies, humans respected the natural environment. They often regarded natural disasters as the actions of gods and believed that the natural environment was a powerful force. At this time, the development and utilization of the natural environment was minimal, and human civilization progressed slowly. Over time, agricultural civilization began to develop, and humans began to use the materials of the geological environment to invent and create. During the industrial civilization period, to increase production speed and capacity, coal mines, oil, rare earth metals, and other substances were over-exploited, resulting in disordered ecosystems and extensive river pollution, which hindered the common development of human civilization and environmental advantages. Presently, we are in an era of ecological civilization construction. The continuous development and progress of today's era have resulted in continuous changes in people's ways of thinking and values, leading people to constantly consider how to approach the relationship between human beings and the environment and avoid the excessive development of resource-dependent cities, which leads to resource depletion.

People are the main body of all production and business activities. The above-mentioned problems encountered in the development process can be regarded as a conflict between humans and the environment. Improvement measures must be taken to change the unhealthy status quo of resource-based cities, starting with ourselves, to regulate in a timely manner, protect the environment, and realize the sustainable development of humanity and nature. Furthermore, more suitable development methods for such cities and ways to develop cultural industries for characteristic cities with rich cultural heritage must be sought. The resource-dependent development mode should be transformed into a market-dependent city, and urban transformation should be promoted, with cultural industry as the leading tertiary industry. The areas with ecological and geological damage should be considered as a whole to control development intensity, limit the land used for urban development, and build the ecological environment in a reasonable manner. These efforts have contributed significantly to accelerating the construction of an ecological civilization. The report of the 18th National Congress of the CPC proposes focusing on promoting the construction of ecological civilization, developing in a green and low-carbon direction, integrating ecological civilization and cultural construction, and establishing scientific civilization. We should strengthen the publicity of ecological civilization, highlighting the roles of government regulation and guidance and active public participation.

9.1.3 Cohesion of Social Forces

Culture is regarded as the sum of all material and spiritual civilization in the process of human development. A country's culture represents its unique characteristics and spirit. Without a rich culture as a support, a country will have difficulties achieving a higher development level to win the respect of all countries worldwide and ensure the happy lives of its people. With the gradual withdrawal of the primary and secondary industries from the stage of economic development, the cultural industry, as a part of the tertiary industry, quickly appeared. From its connotation, cultural industry can be analyzed as a material product that can be transformed, condensed, and evolved as spiritual civilization and service products.

The development of cultural industries is conducive to establishing a better ecological civilization and building better lives. The position of cultural industries in real industries is becoming increasingly important. The development of culture and cultural industries is important for environmental protection and plays a significant role in stabilizing society and realizing the healthy development of people and nature. Furthermore, the development of culture and cultural industries is conducive to changing the development model of cities that rely on geographical resources, improving the slow economic development caused by the lack of resources and enterprise competitiveness, and realizing industrial transformation and development to improve local economic levels. The development of the cultural industry has little destructive power over the environment and a low rate of natural resource consumption. It can help strengthen ecological civilization, enhance local cultural heritage, optimize industrial structure, and promote local economic development. Thus, the development of the cultural industry is conducive to improving people's respect for nature, compliance with nature, protection of nature, and improvement of cultural soft power.

From a national perspective, many channels are available to promote the development of the cultural industry, and China should work hard to support it. From the perspective of the development of the cultural industry, although its proportion in the tertiary industry has been increasing in recent years, much room is left for development. The development of various sectors of the cultural industry is also uneven. The products of the cultural industry involve all aspects, and the specific structure is extremely complicated. Therefore, scholars should conduct in-depth research on various regions and provide timely scientific feedback to China. China should adapt measures to different regions, achieve dynamic and balanced development for each region, adopt a hierarchical system for different cities, increase financial strength in weak regions, provide tax reduction and exemption and other support, strengthen the scientific and educational connotation of publicizing culture and rejuvenating China, and establish cultural policies to promote the development of the cultural industry under the premise of respecting the market.

From the perspective of enterprises, they comprise practical workers who respond to national policies. After China implements various policies for the development of the cultural industry in different localities, they should quickly adjust the industrial structure in light of their own development status, considering their own geographical environment, strengthening regional cultural propaganda, and gradually making the cultural industry more adaptable to market requirements and national development policies. For cities that were originally resource-exhausted and aging industrial bases, the geographical environment and ecological construction should be the focus of development while creating a civilized city that is suitable for the environment. In addition, the cultural and tourism industries can be integrated to form cultural tourism, allow reasonable planning and development for maintaining not only the economic development of local enterprises but also a good living environment, and realize the simultaneous development of nature and the economy.

From the perspective of the general public, each of us is responsible for the development of the cultural industry. The development and promotion of culture and the cultural industry are also greatly beneficial for environmental protection. Strengthening the construction of the cultural industry at this stage is beneficial to ourselves. Furthermore, improving cultural accomplishments and spiritual civilization is highly important. Therefore, from an individual perspective, we should start from ourselves to moderately develop minerals, oil, and trees; rationally classify garbage; promote the culture of one's place of residence; maintain the environment in which we live; and have a friendly attitude toward the environment, to selectively accept the opinions put forward by the state and enterprises for the development of the cultural industry. To cultivate one's own good cultural accomplishments, start from the small things around you and help in the development of the cultural industry to protect the beautiful environment.

At present, the problem of how industrial bases should be transformed is relatively prominent in regional transformation and development. The development and transformation of old industrial bases must first consider how to avoid overreliance on resource development and change the development mode from a conceptual perspective. The state should provide a helping hand in proposing a development strategy, enterprises should adapt to the market demand, and individuals should obtain more knowledge to adapt to this transformation along with the development of enterprises. Considering the above perspectives, the construction of ecological civilization, excellent development of the cultural industry, and protection of the environment in which we live all need to be based on the power of society. We should recognize where we are and do our jobs well, and then improve social cohesion, develop a healthy cultural industry by gathering social forces, strengthen ecological civilization, and build a beautiful and harmonious society.

9.1.4 Promote Economic Development

Cultural development is the basic symbol of social progress. From the past to the present, culture is a symbol of national development and prosperity. For example, China has exquisite silk and ceramics, and other countries have wonderful early architecture and different food cultures. These all reflect the characteristics of different cultures in various countries. The cultural development process can always drive rapid economic progress, political system reform, and improved ideological perception. The deeper the cultural development, the more advantages it has for regional economic development, and the faster the regional economic development, the more likely the reverse can be achieved. Therefore, the promotion of culture should be encouraged. The development of the cultural industry is an important factor affecting economic development. Areas with dense populations have profound cultural heritage and have highly developed culture since ancient times. These areas have always been relatively rich in China. The development of the cultural industry can also promote the development of the tourism industry, thereby driving economic development.

The development of the cultural industry plays a role in promoting economic development. For example, Hunan Province, as a major cultural province, has made a substantial contribution to economic development by relying on the development of cultural entertainment. In addition, Hunan, as a city of culture, has a relatively well-developed tourism industry. Local economic development is more favorable. The improvement of cultural soft power determines the growth of economic development. The development of cultural competitiveness is related to the development of national quality and plays an important role in industrial optimization and economic development (Yao et al., 2023). Differences in the cultural development process and between regions can be regarded as factors that influence the differences in economic development, and are also the qualitative constraint of various policies in the cultural development process. Regarding the current development situation in China, the imbalance of economic development in various regions is the status quo, and the geographical environment, as the main element of the human environment, is an important evaluation index for the development of humanity and nature. At this stage, geological environmental factors affect regional development intensity. Research shows that, in some regions with large fluctuations in geographical environment, the degree of economic development and utilization is relatively low, and substantial geographical restrictions are placed on promoting economic development. Moreover, the development of various industries in such areas will be affected differently. Therefore, we should adapt measures to local conditions; scientifically understand the causes of variations in economic development due to differences in population, environment, and terrain; use topography and landforms to scientifically plan industrial layout; develop better industrial chains; and promote economic development.

The development of the cultural industry cannot be ignored. We must pay attention to green development, maintain a friendly attitude toward the environment, develop cultural industries to improve environmental pressure, learn to adapt to nature, and form a harmonious relationship between humanity and nature. To establish a good ecologically civilized country, the cultural industry, as the support of national spiritual power, is an important guarantee for the strategy and tactics of China's development. We firmly believe that relying on the power of culture and technology can help develop a healthy national economy and that the cultural industry an important part of the national economy in achieving a non-negligible effect. Based on the above considerations, the development of culture and the cultural industry is helpful in improving our living environment and can alert us spiritually to take measures to protect it. In terms of the actual industry, it can replace resource-dependent industries and those with aging industrial equipment, develop emerging industries, promote the transformation and development of regional cities, and avoid excessive dependence. From a macro perspective, it can promote human civilization, constantly regulate people's behavior, develop in a direction that is beneficial to the environment, strengthen the cultural characteristics of different geographical environments, adapt to the development of the national economy, and continuously improve the overall comprehensive strength of China.

9.2 Differences in the Spatial Distribution of Cultures and Their Causes

9.2.1 Language and Culture

Language and culture share an inseparable relationship. When there was no common language, people in various regions had little communication with each other, and their development was relatively closed. With gradual development, communication gradually improved among people in various regions, promoting cultural exchange and prosperity. Different geographical environments and the barriers of mountains, rivers, and lakes imbue languages with unique local characteristics, forming dialects. In China, dialects are divided into ten mainly types: Mandarin, Wu, Cantonese, Fujian, Jin, Xiang, Hakka, Gan, Hui, and Pinghua dialects.

Mandarin broadly refers to a northern dialect, and is subdivided into Beijing Mandarin, Jilu Mandarin, Jiaoliao Mandarin, Northeast Mandarin, Central Plains Mandarin, Lanyin Mandarin, Jianghuai Mandarin, and Southwest Mandarin. This shows that Mandarin is distributed across a large area and terrain in China, is spoken by a large population, and is the dialect with the largest geographical distribution nationwide.

Wu dialect mainly refers to the dialects distributed in Jiangsu, Shanghai, southern Anhui, Fujian, and other places, and is also known by other names, such as Jiangdong and Jiangsu-Zhejiang dialects. Compared with Mandarin, which has many ancient sounds and has been passed down for more than 3000 years, Wu dialect was distributed in Wu and Yue in the early days and is closely related to the cultural background of both places. Owing to the dialect culture, there are operas with local characteristics, namely Wu Ballads and Yue Opera, which form the scenery of Jiangsu and Zhejiang and promote the development and progress of local culture and economy. Cantonese mainly refers to dialects distributed in regions such as Guangdong, Guangxi, and Hong Kong, and is also known as Cantonese vernacular. Hokkien dialect mainly refers to the dialects popular in Fujian, Guangdong, Taiwan, southern Zhejiang, and Jiangsu and includes Minnan, Minzhong, Mindong, Minbei, and Puxian dialects.

Jin dialect mainly refers to the dialects popular in areas such as Shanxi, Inner Mongolia, Shaanxi, Hebei, and Henan, and is considered the only local language other than Mandarin in northern China. The core language is Taiyuan dialect. The environment where Jin dialect was developed is in a closed area, resulting in the existence of a special language other than northern Mandarin. Xiangyu generally refers to the dialects popular in regions such as Changsha, Shaoyang, Xiangtan, and Yongzhou. It evolved from the Chu language in ancient times. Ancient texts indicate that it existed in the late Spring and Autumn period, and is now the main language in Hunan Province. Compared with other dialects, Hakka dialect is more closely related to Chinese people worldwide, resulting in a linkage effect between internal and external cultures, and the Hakka dialect accounts for a larger proportion in China.

Gan dialect is mainly distributed in cities such as Nanchang, Yichun, Xinyu, Pingxiang, Yingtan, Fuzhou, and Jiujiang. In ancient times, owing to the inconvenience of transportation and mainly hilly topography, Gan dialect itself had many dialect characteristics. Hui dialect mainly refers to the language popular in mountain areas such as the Huangshan Mountains, Qiantang River North, and Nanyuan Majinxi. Compared with Wu and Jiangxi dialects, it has the characteristics of Central Chinese. The development of this dialect has driven the development of scenic spots and architectural culture. Pingyu mainly refers to a dialect popular in the Guangxi area and is formed by mixing northern and southern languages. Words such as "you" commonly used in daily life belong to Pingyu. According to relevant information, the number of users of each dialect is shown in Figs. 9.1 and 9.2.

Figures 9.1 and 9.2 show that Mandarin is used by the largest number of people. This may be because, the geographical area where Mandarin is used is widely distributed, and the areas where Mandarin is popular have large populations, which is conducive to the popularity, spread, and future development of Mandarin. Compared with those using Mandarin, the number of people using Pinghua and Jiangxi dialects is small because these people are in geographically remote, hilly, or loess plateaus. Owing to inconvenient transportation in these areas in the early days, the development area was blocked, which was not conducive to the spread of dialects, leading to the formation of an independent dialect. Language and culture with local characteristics.

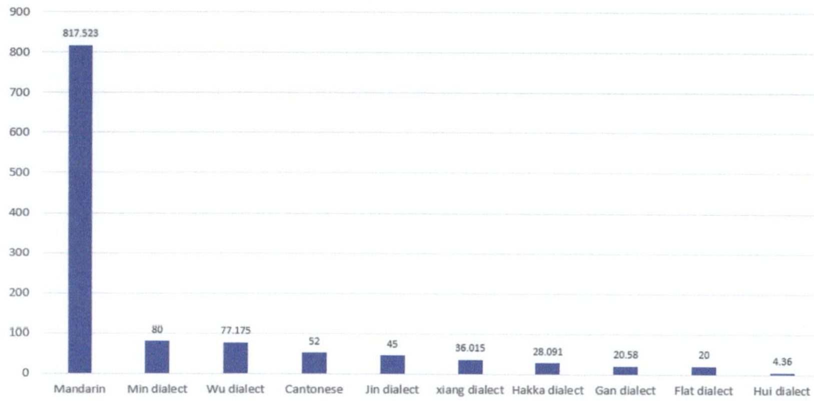

Fig. 9.1 Population of each dialect (million people)

Fig. 9.2 Proportions of dialect users

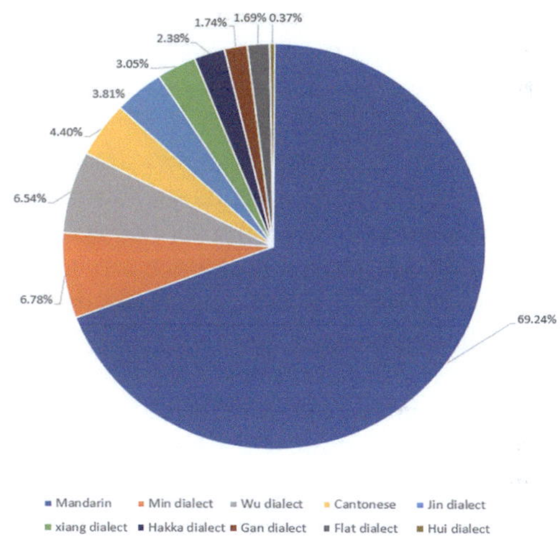

9.2.2 *Clothing Culture*

Clothing has been used as a symbol of human civilization since ancient times. Initially, human beings used leaves and animal fur to make simple clothes to resist the cold. However, with continuous development and progress, humans were no longer satisfied with the only function of clothing, that is, to avoid the cold, and gradually paid attention to beauty in their appearance and began to make exquisite decorations to match their clothing. During the Wenkou culture period, China developed silk, and the Maritime Silk Road emerged. In New China, Chinese tunic suits, men's suits, and women's cheongsams have become popular. Today, human-made materials have risen rapidly in popularity, and people use various materials to make beautiful clothing. From the perspective of historical development, the development of clothing over time reflects the continuous progress of human civilization.

Among the 56 ethnic groups in China, the Han ethnicity comprises the largest portion of the population and is the main ethnic group. It is distributed nationwide and in many other countries overseas. In the early days, the Han population mainly lived in the Yellow River Basin and Yangtze River Basin. The waist tie form, easy to wear Hanfu originated in the period of the Yellow Emperor, and then continued to evolve with the successive dynasties; however, there were only some changes the in details.

Tibetans reside on a high-altitude plateau with large temperature differences, including very low temperatures, and relatively long winters. Therefore, to resist the cold, the Tibetan people wear robes all year round. In winter, they combine some animal furs to keep warm, and in summer, relatively thin cotton robes are worn to prevent some mosquitoes and preserve heat at night. As for the Manchu costumes, since the Manchus were mostly in the Northeast region in the early days, they hunted, grazed, and engaged in farming all year round. Therefore, during the Qianlong period, it was argued that the Manchu costumes should be mainly for hunting convenience, and they made the "horseshoe sleeve" as characteristic clothing.

For some ethnic minorities in Xinjiang, such as Xinjiang Uyghurs and Kazakhs, because they are located in remote areas of China and border other countries, men like to wear embroidered shirts, and women wear embroidered skirts. The long skirts and embroidered vests can reflect the integrated characteristics of their nation with Russia and other countries. For the people living on the high slopes of the loess, due to the perennial wind and dust, the harsh geographical environment, and facing the loess plateau, most of them wear sheepskin jackets with white wraps around their heads in winter, and white sweaters in summer, which is a very local feature.

The costumes of the ethnic minorities in the southwestern region of China are colorful and beautiful. The local climate is suitable, with beautiful scenery, mountains, and rivers, and the cultural customs and landscapes echo each other. The men of the Dai people living there mainly wear sleeveless blouses and trousers, and a few wear long skirts, while women mainly wear skirts of various types, mostly with pleats and in the form of tubes.

In summary, we have introduced some characteristic ethnic costumes. China is civilized due to its multi-ethnic makeup, and multiple ethnic cultures adopt an incompatible model to adapt to the preservation of various ethnic costume cultures.

Table 9.1 shows the population sizes of different ethnic groups in China, using the 2010 national census to reflect the proportion of different clothing worn by each ethnic group. The Han ethnicity has the largest population, with 1,220,844,520 people, accounting for more than 91.64% of the total population. Since the Han population is spread across most of China, there are also various types of Hanfu developed. Differences in geographical altitude, climate, and temperature, have led to each ethic group developing their unique clothing culture, especially ethnic minorities with a small population at present (see Table 9.1), combined with the specific environment; thus, protective measures should be taken to preserve the precious traditional clothing culture.

Table 9.1 Population by ethnic group

Ethnicity	Frequency (person)	Proportion (%)	Ethnicity	Frequency (person)	Proportion (%)
Han	1,220,844,520	0.9164335	Tu	289,565	0.0002174
Zhuang	16,926,381	0.0127059	Mulao	216,257	0.0001623
Hui	10,586,087	0.0079465	Xibe	190,481	0.0001430
Manchu	10,387,958	0.0077978	Kyrgyz	186,708	0.0001402
Uighur	10,069,346	0.0075586	Jingpo	147,828	0.0001110
Hmong	9,426,007	0.0070757	Daur	131,992	0.0000991
Yi	8,714,393	0.0065415	Salar	130,607	0.0000980
Tujia	8,353,912	0.0062709	Brown	119,639	0.0000898
Tibetan	6,282,187	0.0047158	Maonan	101,192	0.0000760
Mongolian	5,981,840	0.0044903	Tajik	51,069	0.0000383
Dong	2,879,974	0.0021619	Pumi	42,861	0.0000322
Buyi	2,870,034	0.0021544	Achang	39,555	0.0000297
Yao	2,796,003	0.0020988	Nu	37,523	0.0000282
Bai	1,933,510	0.0014514	Ewenki	30,875	0.0000232
Korean	1,830,929	0.0013744	Jing	28,199	0.0000212
Hani	1,660,932	0.0012468	Jino	23,143	0.0000174
Li	1,463,064	0.0010983	De'ang	20,556	0.0000154
Kazakh	1,462,588	0.0010979	Bao'an	20,074	0.0000151
Dai	1,261,311	0.0009468	Russian	15,393	0.0000116
She	708,651	0.0005320	Yugur	14,378	0.0000108
Lisu	702,839	0.0005276	Uzbek	10,569	0.0000079
Dongxiang	621,500	0.0004665	Moinba	10,561	0.0000079
Gelao	550,746	0.0004134	Oroqen	8659	0.0000065
Lahu	485,966	0.0003648	Dulong	6930	0.0000052
Wa	429,709	0.0003226	Hezhen	5354	0.0000040
Shui	411,847	0.0003092	Gaoshan	4009	0.0000030
Naxi	326,295	0.0002449	Lhoba	3682	0.0000028
Qiang	309,576	0.0002324	Tatar	3556	0.0000027

9.2.3 *Food Culture*

Food culture has a long history in China. The different geographical environments in different regions have resulted in food cultures with local characteristics. Generally, food culture refers to the development of food technology, food production methods, and food preservation. China has the third-largest land area worldwide and a relatively large geographical area. Across regions, differences in latitude and longitude, climate, and terrain, as well as changes over time, have led to the formation of the diet of each region. Therefore, the geographical environment has a profound impact on human activities, and a rich food culture has also emerged from it.

An analysis of food culture indicates that the reasons for different food cultures may lie in differences in region, climate, and some religious beliefs. For example, in the Yangtze River Basin, the arid area in the center of the plain, mainly produces millet, whereas the middle and lower reaches mainly produce rice. In terms of the geographical environment, the middle and lower reaches have abundant rain, sunshine, and land nutrients, and are thus suitable for rice production. Regarding the diet in the northern and southern regions, the crops planted vary greatly by region, owing to the different geographical environments. The south has a suitable climate for rich products. Communities in the southeast of Guizhou mainly eat hot and sour food because the local area is suitable for growing of crops such as pepper. Rice is the staple food of the Yangtze River Basin. The climate in the north is dry, and the water may be less abundant than in the south. Taking the Guanzhong area as an example, most agricultural activities are focused on grain farming, and few vegetables are grown. Therefore, Biangbiang noodles (also known as "belt noodles") are the main food eaten in the area, and most dishes have a salty flavor.

In the Chinese traditional sense, owing to regional differences, either the "four major cuisines" or "eight major cuisines" are usually proposed. The eight major cuisines in China are Zhejiang cuisine, Fujian cuisine, Hunan cuisine, Anhui cuisine, Sichuan cuisine, Shandong cuisine, Cantonese cuisine, and Jiangsu vegetable cuisine. Geographically, Shandong cuisine is mainly popular in Shandong, Beijing, and Tianjin. Shandong has a mild climate and rich products, and Shandong cuisine is often famous for its use of seafood. Sichuan cuisine is popular in areas such as Sichuan and Chongqing because the environment in these areas is mostly humid, and the locals like spicy food; therefore, most of the dishes are spicy. Cantonese cuisine is mainly popular in Guangdong, Hong Kong, and Jiangxi, among other regions of China. Because it is located in a geographically subtropical region and is evergreen all year. Jiangsu cuisine is mainly popular in regions such as Jiangsu, Zhejiang, and Anhui. Jiangsu has always been known as the land of fish and rice, and the main ingredients in the local cuisine are aquatic products. Fujian cuisine is mainly popular in regions such as Fujian and Taiwan, and is based on Fuzhou cuisine. Fujian is near the sea and has a mild climate and vast area, and fish and shrimp are abundant throughout the year. The plains are rich in sugar, vegetables, and fruits. Anhui cuisine is mainly popular in Jiangsu, Zhejiang, Jiangxi, and Fujian. Several major cuisines have been combined over time to form today's Anhui cuisine, which is related to the unique geographical environment of ancient Huizhou. The biggest connection is that Anhui cuisine focuses on local products, dark colors, heavy oils, and strong flavors. Hunan cuisine is mainly popular in the Hunan area. Owing to the hot, humid climate, the local people like spicy food with strong flavors.

In summary, China's intricate geographical environment has resulted in the creation of several different food cultures. From a broad perspective, the main differences in food culture are between the north and the south. From a narrower point of view, the food culture also varies greatly across each city. The food culture of each city represents the characteristics of the regional culture. We should strive to carry forward and preserve these important food cultures.

9.2.4 Architectural Culture

Since the birth of human civilization, human activities have transformed the geographical environment and left historical traces, one of which is architecture. Architecture is produced by human beings to meet the needs of social life, and the different atmospheres of human activities have led to the creation of different architectural cultures. China's architectural culture has a long history. In the primitive period, humans did not have a fixed place to live and mainly relied on caves to avoid wind, rain, thunder, and lightning and keep warm. Therefore, humans did not have buildings then. Today, China's architectural culture mainly includes six major schools: the Fujian, Hui, Sichuan, Jin, Su, and Wan schools. The buildings of each school reflect the living habits of the people of China since ancient times and their respect for history and culture.

Hui-style architecture is mainly popular in Huangshan City, Jinhua, Yanzhou, and some ancient villages today. The local landforms are mountains, open plains, and complex geological structures. Huizhou architecture, with local cultural customs, was developed based on the large differences in terrain and shapes, with the support of the geographical environment. In the ancient villages built in the Hui style, the architecture is well-proportioned and the colors are light and elegant, which demonstrates integration with the natural environment. In addition, patios are a special feature of Hui-style architecture and often appear in ancient ancestral halls. This shows that Huizhou architecture combines some religious customs and embodies the concept of the unity of humanity and nature. Fujian-style architecture is mainly popular in Taiwan, and over time, it began to combine some Western architectural styles.

Sichuan-style architecture is mainly popular in regions such as Sichuan and Yunnan. Sichuan is the birthplace of Taoism; therefore, Sichuan-style architecture pays more attention to introversion and does not favor bright colors. The main color used is gray, reflecting the simple Taoist style. Jin-style architecture is mainly popular in Shanxi, Gansu, and Ningxia. Owing to the complex terrain of these places, including mountains, highlands, and plains, various Jin-style buildings have been created, such as Jinzhong and Jinxi kilns and Beiheyuan.

Su-style architecture is mainly popular in Jiangsu and Zhejiang and includes rich layers, elegant colors, and quiet scenery. It is typical southern architecture and includes exquisite brick carvings. An inscription is generally hung outside the door or in the lobby courtyard, which demonstrates the integration of cultures. As one of China's six major architecture factions, Wanpai architecture is a very influential style in China. Among them, Huipai is considered a part of Wanpai. Wanpai architecture is characterized by blue tiles and white walls, and its style is similar to that of Huizhou.

Residential houses in southern China are generally square, which has a great relationship with lighting in terms of geographical environment, and the distance between modern buildings is scientifically calculated based on the angle of sunlight. Siheyuan can be regarded as a typical building in Beijing. The design is solemn and grand, with extremely profound cultural connotations. Fengshui is emphasized in architectural design, which is closely related to early ancient history and culture. Courtyards are surrounded on four sides, which promotes relationships among neighbors. The mutual communication has strong humanistic characteristics.

In summary, most of the architectural styles in China have regional differences, which are closely related to factors such as geographical environment, climate, temperature, and humidity, and lead to the formation of architectural styles in different regions. In addition to typical residential buildings, various forms of historical and cultural relics also exist. Furthermore, the details in some architectural designs adopt patterns typically found in nature, reflecting the integration of architecture and nature. The continuous development of architecture has gradually affected the style of classical architecture. In particular, some rare architectural styles have been gradually disappearing; therefore, combined with the actual situation, we should comprehensively consider the protection of precious ancient architectural culture.

9.2.5 Traffic Culture

Transportation symbolizes human civilization and progress. In ancient times, humans only knew how to use brute force to perform a series of production activities. With the continuous progress of civilization, primitive humans gradually began to invent, create some simple transportation facilities, and realize the improvement of social production by producing simple tools. In the early days of China, the cold climate and dry weather in the north resulted in some vehicles being built as early inventions and creations, including horse-drawn carriages, sleds, and stair-climbing vehicles, to facilitate the transportation of goods, such as grain. In the south, where the climate is warm and most cities are coastal, early inventions were often adapted to local conditions, with local wood used to build ships and develop water transportation.

With the development of transportation facilities in various places, people gradually began to build roads that were convenient for vehicles and horses. Roadside pavilions were placed to provide supplies to passersby. With gradual development, pedestrians who stopped at these pavilions often wrote poems, which they exchanged with others, thereby promoting the progress of ancient Chinese civilization. People also began to build arched bridges across rivers. The invention of the bridge promoted communication between people on both sides of the river bank. However, construction not only facilitated the development of people on both sides of the strait but also led those located in a mountainous area to view themselves as superior. In ancient China, humans invented the sedan chair approximately 4000 years ago. This type of transportation is mainly used to carry people or livestock and can be used to show the dignity of the person sitting in the sedan chair. These chairs are also often used by people who are inconvenient and during marriage.

The abovementioned transportation methods are mainly typical in ancient China. With the continuous advancement of science and technology, automobiles, trains, subways, airplanes, and ships have gradually become the mainstream transportation methods, and the old-fashioned underdeveloped transportation facilities have gradually retreated. However, the foundation of China's ancient transportation methods has allowed for the creation of today's convenient transportation, promoted the progress of maritime civilization, further responded to the call to development of the Maritime Silk Road, and facilitated cultural and other exchanges and progress nationwide.

9.2.6 Religious Culture

China is a multiethnic and multireligious country, with religions such as Buddhism, Taoism, and Islam. Buddhism was introduced to China from India in the first century A.D. After a long period of propagation, a Sinicized Buddhism was formed for the Chinese nation. Differences in time periods, routes, regions, ethnic cultures, and social and historical backgrounds have resulted in three major systems of Buddhism in China: Chinese Buddhism, Tibetan Buddhism, and Theravada Buddhism in Yunnan. Taoism is a religion native to China, with a history spanning 1700 years. Islam was introduced to China in the seventh century AD, and the Tang, Song, and Yuan dynasties were important periods for the spread of Islam in China. After long-term development, Islam has not only taken root among ethnic minorities such as the Hui and Uyghurs in China but also a Sinicized Islam has formed. Presently, Islam in China is mostly practiced among the Hui, Uyghur, Kazak, Tajik, and other ethnic minorities.

9.3 Differences in the Spatial Distribution of Cultural Industries and the Causes Thereof

9.3.1 Evaluation of Cultural Industries Development

9.3.1.1 Data Source

China's cultural industries has continued to develop. Overall, the tertiary industry represented by the cultural industries accounts for an increasing proportion of the national economy and has contributed greatly to China's economic development. However, the geographical environment and political and historical factors have resulted in the development of the cultural industry being unbalanced across regions.

Table 9.2 Missing data and handling of cultural industry in 2017

Variables	Description of missing data
Number of educational institutions in the cultural sector	Guizhou, Tibet, and Qinghai have missing data, and the median is used to replace missing values
Number of employees in educational institutions in the cultural sector	Guizhou, Tibet, and Qinghai have missing data, and the median is used to replace missing values
Number of graduates from cultural and educational institutions	Jilin, Guizhou, Tibet, and Gansu have missing data, and the median is used to replace missing values
Total assets of educational institutions in the cultural sector	Guizhou, Tibet, and Gansu have missing data, and the median is used to replace missing values
Area of scenic spots (square kilometers)	Shanghai data are missing, and the median is used to replace missing values
Number of tourists (10,000 person-times)	Shanghai has missing data, and the median is used to replace missing values
Total capital income of scenic spots (10,000 yuan)	Inner Mongolia and Shanghai have missing data, and the median is used to replace missing values

The statistical yearbooks of 31 provinces, autonomous regions, and municipalities outside Taiwan; the 2017 China Cultural Industry Statistical Yearbook; and the Chinese Culture Database were used in this study. During the data collection process, some data were found to be missing; therefore, we used median values to replace the missing values (Table 9.2).

9.3.1.2 Basic Theory

(i) Diamond Model Theory

The diamond model, proposed by Porter (1997), is mainly used to measure competitive advantages among countries. Porter argues that a country's competitiveness is mainly affected by four factors: production factors, demand conditions, related and supporting industries, and strategic factors. By adding government and opportunity as two auxiliary factors, a diamond model with six vertices is formed (Fig. 9.3).

(ii) The basic principle of the entropy method

The entropy value method is an objective assignment method, which uses the amount of information provided by the entropy value of each index to determine the weight of the index, and an index with a relatively large degree of change has a greater weight. Using the entropy method, the weight of each index is calculated by using the value coefficient, of the index information. The higher the value coefficient, the greater the importance to the evaluation.

The entropy value method is commonly used to judge the degree of dispersion of an indicator. Entropy is a measure of uncertainty in information theory. The greater the uncertainty, the greater the entropy, and vice versa. The less the information in

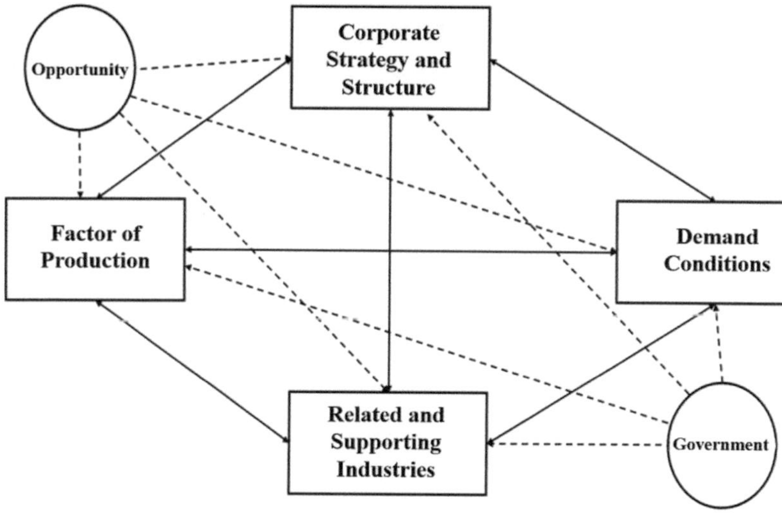

Fig. 9.3 Porter's diamond theory model

the data, the greater the entropy. Entropy values can be used to judge the randomness of events and the degree of dispersion of a certain indicator. The greater the degree of dispersion of the indicators, the greater the impact on the comprehensive evaluation.

Samples and indicators n in the cultural industry development model $m$can form an initial matrix: $X = (x_{ij})m \times n$. For a certain indicator x_j, if the degree of dispersion of its data is large, the effect of the indicator will be greater.

The steps of the entropy value method are as follows.

a. Select 31 provinces and 41 indicators in China, among which x is the value of the first i th ($i = 1,2, \ldots, n$) province and the j th ($j = 1,2, \ldots, m$) indicator.
b. Normalization of indicators. Evaluation indicators are typically divided into positive, negative, and moderate indicators. To reflect the advantages and disadvantages of data indicators, positive processing must be performed on each indicator.
c. Standardization of indicators. Owing to the different dimensions of data between indicators, to avoid calculation errors caused by different dimensions, the data should be standardized. The specific processing method is as follows:

Positive indicators:

$$x'_{ij} = \frac{x_{ij} - \min\{x_{ij}, \ldots, x_{nj}\}}{\max\{x_{1j}, \ldots, x_{nj}\} - \min\{x_{ij}, \ldots, x_{nj}\}} \tag{9.1}$$

Negative indicators:

$$x'_{ij} = \frac{\max\{x_{ij}, \ldots, x_{nj}\} - x_{ij}}{\max\{x_{1j}, \ldots, x_{nj}\} - \min\{x_{ij}, \ldots, x_{nj}\}} \tag{9.2}$$

Then, it x'_{ij} is the value of the i th province j, which is still recorded as x_{ij}.

d. Use the entropy value method to determine the weight of each index.

Calculate the e_j entropy value of the index item:

$$e_j = -k \sum_{i=1}^{n} p_{ij} \ln(p_{ij}) \tag{9.3}$$

Among them, $k = 1/\ln(n)$, p_{ij} calculate the information entropy redundancy for the proportion of the i country under the j indicator.

Calculate the weight of each indicator:

$$w_j = \frac{d_j}{\sum_{j=1}^{m} d_j} \tag{9.4}$$

Calculate the composite score for each region:

$$s_i = \sum_{j=1}^{m} w_j \cdot p_{ij} \tag{9.5}$$

(iii) Construction of the indicator system

Based on the diamond model theory proposed by Porter (1997), and data sources such as the Statistical Yearbook of Chinese Culture and Statistical Bulletin of Culture and Tourism Development of the Ministry of Culture and Tourism, this section follows scientific principles. The final index evaluation system includes five first-level indicators, 12 s-level indicators, and 41 third-level indicators. The specific indicators are shown in Table 9.3 (Ze, 2013).

Using R language software and following the steps of the entropy value method, the weight and entropy values of the third-level indicators are determined, as shown in Table 9.4.

The weights of the second- and first-level indicators can be obtained based on the weights of the third-level indicators, as shown in Table 9.5.

Following the steps of the entropy value method, the data of each province and city are multiplied by the entropy value of the corresponding index; then, the multiplied data are weighted and summed according to the weight; and finally, the cultural industry index system of each province is calculated to obtain the first-level index (the results are rounded to two decimal places). The ranking of the first-level indicators is shown in Table 9.6.

Table 9.3 Index system for cultural industry evaluation

First-level indicators	Second-level indicators	Third-level indicators
Production factors	Human resources	Number of employees in public libraries
		Number of employees in mass cultural institutions
		Number of cultural market operators
		Number of cultural industry employees
	Infrastructure	Number of cultural market operators (units)
		Number of cultural property institutions
		Number of museum institutions
		Building area of public buildings used by public libraries (10,000 square meters)
		Construction area of public libraries per 10,000 people (square meters)
		Building area of mass cultural facilities per 10,000 people (square meters)
	Cultural resources	Total collection of public libraries (10,000 volumes)
		Cultural relics collection (piece/set)
		Area of scenic spots (square kilometers)
		Number of museum collections (pieces/sets)
Related products	Art industry	Number of performing arts organizations
		Number of employees in performing arts groups
		Number of performances by art performance groups (10,000 performances)
	Tourism	Number of tourists (10,000 person-times)
		Total capital income of scenic spots (10,000 yuan)
	Radio and television industry	Comprehensive population coverage of radio programs (%)
		Comprehensive population coverage of TV programs (%)
		Number of cable broadcasting and TV subscribers (10,000 households)
		Annual production time of radio programs (hours)
		Annual production time of TV programs (hours)

(continued)

Table 9.3 (continued)

First-level indicators	Second-level indicators	Third-level indicators
		Radio and television practitioners (persons)
	Culture and education	Number of educational institutions in the cultural sector
		Number of employees in educational institutions in the cultural sector
		Number of graduates from cultural and educational institutions
	Advertising	Advertising business unit
		Number of advertising practitioners
Consumer demand	Cultural industry needs	Total circulation of public libraries (10,000 person-times)
		Number of borrowed books and periodicals (10,000 volumes)
		Number of cultural activities organized (times)
		Number of exhibitions held (pieces)
Government actions	Government input	Public expenditure on culture, sports, and media (100 million yuan)
		Financial appropriation income of art performing groups (thousand yuan)
Industrial strength	Industry income	Total income from radio and television (10,000 yuan)
		Advertising revenue (10,000 yuan)
		Operating profit of cultural market operation (thousand yuan)
	Total assets	Total assets of performing arts groups (1000 yuan)
		Total assets of educational institutions in the cultural sector

9.3.2 Development Level and Pattern of Cultural Industries

Based on the comprehensive evaluation of the cultural industries of 31 provinces, autonomous regions, and municipalities other than Hong Kong, Macao, and Taiwan in Sect. 9.3.1, Table 9.6 shows that the cultural industries of Zhejiang, Guangdong, and Jiangsu have the highest comprehensive scores of 0.20587, 0.16048, and 0.15174, respectively. The comprehensive score of Qinghai's cultural industry is the lowest among the 31 provinces, autonomous regions, and municipalities, at 0.00434. Provinces, autonomous regions, and municipalities directly under the central government have the second-lowest scores, with a score of 0.00891. The scores indicate

Table 9.4 Weights and entropy values of the third-level indicators of cultural industries in 31 provinces and cities

Third-level indicator	Entropy value	Weight
Number of employees in public libraries (persons)	0.958532	0.010070
Number of employees in mass cultural institutions	0.965408	0.008400
Number of employees in cultural market operations (persons)	0.912542	0.021238
Number of cultural industry employees	0.927334	0.017646
Number of cultural market operators (units)	0.927006	0.017725
Number of cultural property institutions	0.926130	0.017938
Number of museum institutions	0.933710	0.016097
Area of public housing buildings used by public libraries (10,000 square meters)	0.941494	0.014207
Construction area of public libraries per 10,000 people (square meters)	0.983309	0.004053
Building area of mass cultural facilities per 10,000 people (square meters)	0.960250	0.009653
Total collection of public libraries (10,000 volumes)	0.924921	0.018232
Cultural relics collection (piece/set)	0.880666	0.028978
Area of scenic spots (square kilometers)	0.836092	0.039802
Number of museum collections (pieces/sets)	0.883167	0.028371
Number of performing arts organizations (number)	0.863001	0.033268
Number of employees in performing arts groups (person)	0.896549	0.025121
Number of performances by art performance groups (10,000 performances)	0.767887	0.056365
Number of tourists (10,000 person-times)	0.866753	0.032357
Total capital income of scenic spots (10,000 yuan)	0.765446	0.056958
Comprehensive population coverage of radio programs (%)	0.999974	0.000006
Comprehensive population coverage of TV programs (%)	0.999991	0.000002
Number of cable broadcasting and TV subscribers (10,000 households)	0.927570	0.017589
Annual production time of radio programs (hours)	0.944237	0.013541
Annual production time of TV programs (hours)	0.946202	0.013064
Radio and television practitioners (persons)	0.938586	0.014913
Number of educational institutions in the cultural sector	0.902607	0.023650
Number of employees in educational institutions in the cultural sector	0.918364	0.019824

(continued)

Table 9.4 (continued)

Third-level indicator	Entropy value	Weight
Number of graduates from cultural and educational institutions	0.920921	0.019203
Advertising business unit	0.831255	0.040977
Number of advertising practitioners	0.877905	0.029649
Total circulation of public libraries (10,000 person-times)	0.880890	0.028924
Number of borrowed books and periodicals (10,000 volumes)	0.884805	0.027973
Number of cultural activities organized (times)	0.940946	0.014340
Number of exhibitions held (pieces)	0.936878	0.015328
Public expenditure on culture, sports, and media (100 million yuan)	0.955368	0.010838
Per capita cultural expenses (yuan)	0.950682	0.011976
Total income of radio and television (10,000 yuan)	0.798799	0.048858
Advertising revenue (10,000 yuan)	0.738943	0.063393
Operating profit of cultural market operation (1000 yuan)	0.708620	0.070757
Total assets of performing arts groups (1000 yuan)	0.894508	0.025617
Total assets of educational institutions in the cultural sector	0.863712	0.033095

Table 9.5 Weights of first- and second-level indicators of the cultural industry in 31 provinces and cities

First-level indicator	First-level index weight	Second-level indicator	Second-level index weight
Production factors	0.252411	Human resources	0.057353
		Infrastructure	0.079674
		Cultural resources	0.115384
Related products	0.396488	Art industry	0.114754
		Tourism	0.089315
		Radio and television industry	0.059100
		Culture and education	0.062678
		Advertising	0.070626
Consumer demand	0.086566	Cultural industry demand	0.086566
Government actions	0.022814	Government input	0.022814
Industrial strength	0.241721	Industry income	0.183009
		Total assets	0.058712

Table 9.6 Final scores of the first-level indicators of the cultural industry in 31 provinces and cities

Province	Production factors 0.252411	Related products 0.396488	Consumer demand 0.086566	Government actions 0.022814	Industrial strength 0.241721	Score
Beijing	0.11771	0.08151	0.02234	0.02281	0.15217	0.10127
Tianjin	0.02803	0.05193	0.01082	0.00670	0.05334	0.04165
Hebei	0.04427	0.12127	0.03285	0.00270	0.00878	0.06428
Shanxi	0.06357	0.13938	0.01688	0.00255	0.00888	0.07498
Inner Mongolia	0.04184	0.06150	0.01395	0.01045	0.00845	0.03843
Liaoning	0.05143	0.07086	0.02418	0.00269	0.02123	0.04836
Jilin	0.02976	0.04758	0.00969	0.00293	0.00092	0.02750
Heilongjiang	0.05107	0.04932	0.01778	0.00000	0.00046	0.03410
Shanghai	0.11739	0.10344	0.04133	0.02271	0.01554	0.07849
Jiangsu	0.17121	0.16351	0.06136	0.01333	0.15752	0.15174
Zhejiang	0.14398	0.24767	0.08657	0.01346	0.26282	0.20587
Anhui	0.09560	0.15809	0.03370	0.00056	0.17241	0.13141
Fujian	0.04852	0.10075	0.01944	0.00395	0.02652	0.06038
Jiangxi	0.06009	0.14533	0.01713	0.00039	0.04692	0.08562
Shandong	0.16764	0.16261	0.07381	0.00624	0.03341	0.12140
Henan	0.12225	0.17735	0.04566	0.00162	0.02814	0.11197
Hubei	0.10501	0.11116	0.03094	0.00424	0.06374	0.08876
Hunan	0.09474	0.16717	0.02579	0.00694	0.06728	0.10885
Guangdong	0.16546	0.14999	0.05874	0.02069	0.22212	0.16048
Guangxi	0.03394	0.04902	0.02466	0.00017	0.06524	0.04591
Hainan	0.00531	0.02239	0.00000	0.00253	0.01265	0.01334
Chongqing	0.05077	0.09928	0.02322	0.00136	0.01837	0.05866
Sichuan	0.20111	0.15474	0.05700	0.00725	0.02379	0.12297
Guizhou	0.03561	0.03709	0.01512	0.00144	0.00967	0.02737
Yunnan	0.09048	0.06674	0.02424	0.00163	0.01328	0.05465
Tibet	0.01079	0.01689	0.00260	0.01699	0.00269	0.01068
Shaanxi	0.14392	0.07025	0.01990	0.00602	0.00127	0.06635
Gansu	0.03188	0.04094	0.01326	0.00226	0.00834	0.02749
Qinghai	0.00048	0.00963	0.00237	0.00752	0.00010	0.00434
Ningxia	0.01357	0.01244	0.00593	0.00158	0.00001	0.00891
Xinjiang	0.01578	0.04124	0.04114	0.00402	0.00021	0.02404

that the development level of the cultural industry in the lower reaches of the Yangtze River in eastern China is higher than that of other regions in terms of geographical environment, whereas the development level of cultural industries in the northwestern region of China is relatively weak. Therefore, it can be roughly concluded that China's cultural industry presents an overall pattern in which the development level of the cultural industry gradually increases from west to east and decreases from south to north. The development structure of China's cultural industry is intricate.

Jiangsu, Guangdong, and Zhejiang are important places for economic development in China. Jiangsu and Zhejiang are often collectively referred to as the Jiangsu-Zhejiang area, which is known as the land of fish and rice. From the perspective of the geographical environment, these three places are all coastal areas, with developed maritime transportation, suitable for exchanges and development with other countries, reflecting China's current "One Belt, One Road" development strategy. The local economic development is relatively better than that of other cities, and the geographical environment is rich in language, food, clothing, and other cultural elements. Developed transportation has further promoted the development of the local cultural industry. The combination of these complex factors has led to the culture of this eastern coastal city, where the cultural industry is booming.

Qinghai, Tibet, and Ningxia are among the regions located in northwestern China, in the border areas of China. They did not take the lead in China's early development, and have steep local terrains and high altitudes. The early transportation methods and geographical environment were relatively closed. They are isolated from each other; therefore, the economy, culture, and cultural industry are relatively less developed. In recent years, China has gradually implemented the western development strategy, thus providing financial support to the western region and strengthening the infrastructure of the cultural industry, which has improved the cultural industry and promoted economic development in these regions.

As an emerging industry in the twenty-first century, the cultural industry has continued to grow and develop in China with the "Belt and Road" strategy and supply-side reform policy. As a representative of national soft power and soft science, the development of the cultural industry is an important development strategy to promote regional transformation, accelerate industrial upgrading, and realize economic development in poor areas. The above analysis helps reveal the basic status of cultural industry ratings in provinces, autonomous regions, and municipalities directly under the central government, which shows differences in the cultural industry development in various regions. From another perspective, the cultural industry competitiveness ranking of each province is relatively close to the GDP ranking. By understanding the relative development status of the cultural industry in provinces, autonomous regions, and municipalities directly under the central government, we can grasp the overall status of cultural industry development in China. In summary, for future development of the cultural industry in China, the cultural industry development in all provinces and cities of China must be considered comprehensively to finally realize that the cultural industry drives local economic development, thereby improving China's comprehensive strength.

9.3.3 Influencing Factors of Cultural Industry Development

Through the analysis described in the previous two sections, we can build a comprehensive evaluation system of cultural industry indicators based on the diamond model and entropy method, and calculate the comprehensive cultural industry scores in 31 provinces, autonomous regions, and municipalities nationwide. The comprehensive score is high, whereas the comprehensive scores of the cultural industry in regions such as Qinghai and Tibet are low. China's cultural industry level generally presents a pattern of gradual increases from north to south and from west to east. Many factors could explain the differences in cultural industry scores across various regions in China, such as geographical environmental, political, and historical factors.

Differences in the geographical environment of each region result in differences in regional cultural industry scores (Cabra & Guerrero, 2022). Jiangsu, Zhejiang, and Guangzhou are located in the coastal areas with developed maritime traffic, which is convenient for export trade with other countries. Compared with Qinghai, Tibet, and other regions, they have obvious advantages in geographical environment. For Tibet, owing to multiple factors such as terrain and landforms, the climate is relatively cold and dry. Qinghai is located on a plateau and has a plateau continental climate with large temperature differences. In comparison, Jiangsu and Zhejiang do not have as good of an environment, and they are close to the inland, without developed water transportation. Therefore, the unique natural and geographical factors affect the development level of the cultural industry across regions, and the level is relatively high in areas with a good geographical environment. The development of characteristic cultural industries in areas with a relatively fragile geographical environment promotes the development of regional cultural industries.

Regarding political factors, China currently has several relevant cultural industry policies, such as the promulgation of many relevant cultural industry laws, strengthening the rational regulation of cultural industries, deepening national system reforms, and continuously improving the cultural industry structure. Although the various policies issued by the state are aimed at the whole country, the sequence of development may show some differences in each region. For example, in China's early development, the development of the east was relatively ahead, while the development of the west was relatively behind. Moreover, provincial and municipal governments will still be affected by the local environment and other factors when they implement national policies at the local level, which will create a certain lag in local cultural industry development. These factors will affect the provinces and autonomous regions to some extent. The cultural industry development levels in municipalities directly under the central government show some differences. Therefore, from a political perspective, China should proceed based on the overall situation and make reasonable arrangements for different regions to achieve coordinated development nationwide.

From the perspective of historical factors, with continuous development and progress over time, China's excellent culture has gradually been constructed. Due to the accumulation of historical humanities, a strong humanistic style has been formed. For example, the northwest region is located on the Loess Plateau, and the local area has formed the unique northwestern characteristics of Northern Shaanxi culture, cave architecture, and northwestern clothing, whereas the water towns of China are well-known Huizhou architecture and Jiangsu and Zhejiang cuisines, and in southern China, early local earth buildings were built according to local conditions. The rich resources of aquatic products resulted in a local diet mostly comprised of fish and shrimp. Compared with other regions, the maritime transportation culture is more developed, thereby forming unique local coastal customs. In northern China, the climate is relatively cold and dry. Vegetables are not common in the winter, and are therefore often pickled, and the diet is mostly salty. Therefore, we should make good use of the historical civilizations in different regions. By so doing, we can create a cultural industry with certain local characteristics, help regions transform, and achieve sustainable regional economic development.

From a regional economy perspective, variations in the GDP and economic levels of each region lead to certain differences in cultural industry development and cultural infrastructure construction. Economically developed areas generally have higher levels of cultural industry development and relatively complete infrastructure, whereas urban and rural areas with a relatively less developed economy have lower cultural industry development levels and relatively less developed infrastructure, which leads to unbalanced cultural industry development across various regions. Furthermore, the development of the cultural industry has a greater impact. To balance or improve economic development in poverty-stricken areas, we should increase financial support from the perspective of the state, invest in high-tech support for areas with weak cultural industry development, strengthen regional cultural infrastructure, promote the development of less developed areas, and then help regional economies progress.

In summary, cultural industry development is not only affected by the economic, political, and geographical environment and historical civilization but by other factors as well. The influence of multiple factors results in differences in cultural industry development levels across regions. To better promote the development of China's cultural industry, in the context of supply-side structural reforms, we should combine economic, political, geographical environment, historical, and cultural factors to develop the cultural industry in various regions, especially for resource-exhausted cities with resource-development industries. Promoting cultural industry development can help regional economies transform and develop, thereby narrowing the gaps between provinces, autonomous regions, and municipalities directly under the central government, and realizing the balanced development of tertiary industries such as the national cultural industry, to improve the overall strength of China's economic development.

9.4 Strategic Spatial Layout of China's Culture and Cultural Industries

9.4.1 Strengthening the Regional Layout of Cultural Industries

As an emerging industry in China, the cultural industry plays a substantial role in promoting China's economic development and realizing the harmonious and healthy development of humanity and nature. The cultural industry is a special product formed by combining the spirit of history and natural factors. The development level of a country's cultural industry represents the strength of China's cultural soft power, and can also reflect China's comprehensive strength. Cultural industry development in cities has clear regional characteristics. Regional evolution has resulted in unique local customs and cultures. Combining these spiritual connotations with specific objects creates cultural products and spiritual inheritance with local characteristics. The development of a cultural product can often affect the development of the cultural industry within other industries, and its geographical location analysis shows an agglomeration state. As it affects the development of other industries, a complex cultural industry chain and cultural industry network are formed.

Currently, some problems persist in China's cultural industry, such as the lack of a sound development system in the early stages, the unbalanced development of various sectors of the cultural industry, and the people's early concept of cultural industry development. The problems should be comprehensively analyzed to establish an accurate concept of development and transformation, eliminate the simple reproduction mode of the early stages that relies on geological resources, and recognize and understand the connotation of cultural industry development. Considering the unbalanced development of various industries, we should analyze the development of relatively backward aspects of the cultural industry in combination with geographical factors and help support their development. Based on the overall development status of China's cultural industry, the overall industrial structure should be further optimized to accelerate the industrial upgrading and transformation of less developed sectors. Regarding the imbalance in cultural industry development across various regions, we should consider the overall construction of a comprehensive cultural industry layout and strengthen the infrastructure for the cultural industry in various regions.

Therefore, a reasonable layout for the development of the cultural industry should be considered from multiple perspectives. Regional cultural industry can only be improved by providing technical support to regions with a relatively less developed industry. The current situation in China indicates the need to train talent in the technology field and improve cultural undertakings based on the power of science and technology. From the geographical environment perspective, in economically developed regions such as Beijing and Shanghai, the development of transportation, financial industry, information technology, and other industries is relatively advanced compared with that in the northwestern and northeastern regions, providing them with certain advantages in cultural industry development. Some relatively less developed cities only relied on industry in the early development process, and the industrial structure was relatively simple, which eventually led to a lower level of cultural industry development than that in areas such as Beijing, Tianjin, and Hebei. Considering this complicated situation, China should consider the development of various regions. According to each specific situation, we should seek new development methods and promote the harmonious development of regional cultural industry, thereby driving regional economic development. From a cultural perspective, the continuous evolution of history has resulted in various regions developing different cultural characteristics and landscapes. We should make good use of local customs in the development process, and combine these cultural customs with the cultural industry to construct a unique cultural environment. A characteristic cultural industry will promote the transformation of regional economic development.

9.4.2 Coordinating the Development of Urban and Rural Cultural Industries

The coordinated development of urban and rural cultural industries is currently a prominent problem facing China's cultural industry in terms of the economic gap between urban and rural areas (Su & Heshmati, 2014). The early separate development of cities and the countryside has resulted in different development models. Cities continue to use resources to develop and build industrial bases, whereas the countryside uses land resources for simple agricultural production. With the urban development model, the development process requires many skilled workers, resulting in a significant decrease in the rural population and a sharp increase in the urban population. Development has gradually become unbalanced between urban and rural areas, resulting in the rural economy failing to keep up with the economy of the northern cities. A clear dual development pattern exists between urban and rural areas, and the gap is widening.

The main gaps between rural areas and cities in cultural industry development include unbalanced cultural industry investment between rural and urban areas and cities, substantial differences in the construction of cultural infrastructure between rural and urban areas, and rural areas being relatively behind in the construction of spiritual civilization. Cities have invested heavily in the cultural industry, and most have built libraries, movie theaters, and cultural industry parks, of which both the variety and quantity are gradually increasing. The number of available library books increased by approximately 4.3% between 2015 and 2016 and approximately 2.3% between 2016 and 2017. The above data indicate that, although the number of books, libraries, and performing arts groups in China's overall cultural industry are increasing annually, comparing these numbers between cities and rural areas shows a large gap in construction, and these increases are mainly attributable to the contributions of cities. Therefore, these factors can reflect the large regional differences between urban and rural areas in the cultural industry overall.

In terms of the infrastructure construction of rural and urban cultural industries, in the urban environment, cultural centers, cinemas, concert halls, and cultural performance halls have increased in number with industry development, and the gradual development of science and technology has promoted the cultural industry. It also gradually drives the construction of these infrastructures. With improved infrastructure construction, urban spiritual civilization for the cultural industry is also gradually improving, while urban areas have mainly developed agricultural life and maintained traditional production methods. The development of cultural and other industries is relatively scattered. Therefore, cultural knowledge urgently needs to be replenished to improve the status quo of cultural industry development. An obvious duality exists in the cultural industry between the city and the countryside. Compared with rural areas, in urban areas, residents' spiritual literacy is higher, the construction of cultural infrastructure is more complete, and the proportion of investment in cultural industry construction is greater.

Therefore, to address these differences, we should comprehensively consider and coordinate the cultural industry development between urban and rural areas. The development gap between urban and rural areas is caused by the combination of time and other factors. As a result of accumulation, this problem would be difficult to solve in a short period. Therefore, it is necessary to consider a long-term perspective, improve the atmosphere of the rural cultural industry, and strengthen the construction of rural cultural industry infrastructure. Furthermore, we should conduct cultural publicity activities, organize cultural performances, and help improve the cultural level in rural areas. Simultaneously, urban and rural cultural industry development should be coordinated, and the state should fully coordinate support by providing funding and implementing policies according to the specific differences between urban and rural areas. During the development process, we should make overall plans, narrow the gap between urban and rural areas as much as possible, develop a reasonable layout for urban and rural cultural industry development, and focus on the future. In terms of the differences between rural and urban areas, it is necessary to apply scientific and rational development strategies and utilize resources in rural areas, such that narrowing the gap does not bring adverse consequences. In addition, the urban and rural market environments must be handled correctly, and the cultural industry should be developed in ways that are conducive to the development of urban and rural areas considering the actual situation and overall market environment, to realize the overall planning of urban and rural cultural industries.

9.4.3 Featured Industries Contribute to Poverty Alleviation

The report of the Nineteenth National Congress of the CPC noted that China's cultural industry has achieved relatively fruitful results. For the future development of the cultural industry, we must establish cultural confidence, take the unique traditional culture of the Chinese nation as the basis, bear in mind China's unique cultural genes, and recognize China's current existence. The principal contradiction of the Chinese nation gives full play to the deep connotation of socialism with Chinese characteristics and helps realize the great rejuvenation of the Chinese nation. The current problems in the development of China's cultural industry are mainly the need to improve China's cultural industry system and policies, rationally arrange the layout of China's cultural industry, narrow the gap between cultural industries in various regions, and balance various sectors within the cultural industry.

For various provinces, cities, and urban and rural areas, the large gap in culture and cultural industries remains a relatively serious problem that must be addressed. Under China's "One Belt, One Road" strategy and supply-side reform, the cultural industry and cultural undertakings of the Internet have risen rapidly, accounting for an increasing market share compared with the traditional cultural industry development model. As the world is grown, the development of the Internet cultural industry has become the mainstream. Presently, China is at the end of its "13th Five-Year Plan." Promoting resource dependence among, industrial transformation in, and the upgrading of poverty-stricken areas and targeted poverty alleviation are China's outstanding tasks. Therefore, the promotion of regional cultural industry is the key issue in alleviating poverty.

Poverty-stricken areas generally do not have financial support, high-tech talent, or uniform industrial development organizations. Therefore, cultural industry activities are scattered and relatively difficult to develop, and cultural industry development in less developed cities and counties requires national cooperation. The market and local areas are generally coordinated and unified, such as the state providing support through financial assistance and tax relief; however, the state's investment is focused on precision and appropriate investment, to support the cultural industry to the greatest possible extent. In addition, the state should encourage China's cultural industry experts to help poor and less developed areas develop scientifically. Sending high-quality talent to less developed areas is greatly beneficial for regional cultural industry development. Thus, some less developed areas could enrich their own cultural industries and promote their economic development.

In China's border areas of Hubei, Hunan, Chongqing, and Guizhou, the geographical environment is rich in resources and products, geographical indications, regional cultural industries, and ethnic cultural tourism characteristics. Accordingly, the geographical advantages of local resources should be transformed into cultural industry advantages. For specific areas, such as developing the tea cultural industry in Zhangjiajie, Yongshun County, and Guzhang County, and the traditional Chinese medicine industry in regions such as Cili County, Xianfeng County, and Longshan County, we recommend taking advantage of the geographical environment. Longshan County, Yongding District, Fenghuang County, and Xiangtan County, a region with rich fruit products, should develop fruit products with local characteristics. In addition, the local handicraft culture is strong; therefore, key handicraft markets should be developed in specific cities and counties. In recent years, the state has carried out large-scale development in the western region, with the help of the geographical advantages of natural grasslands, to create characteristic cultural advantages relevant to the local nomadic people. The combination of local herder culture and other cultural factors has had an impact on local art shows, dance performances, and exhibitions. Therefore, we should focus on the regional and cultural characteristics of the grassland area of Inner Mongolia, increase investment in infrastructure construction, give full play to the locally formed grassland culture, rationally develop and utilize grassland resources, improve the competitiveness of the regional culture, and realize the grassland culture..

Regional culture is the basis for cultural industry development and growth. The current problems in the regional cultural industry are mainly the result of imperfect policies, relatively loose industrial organization, and no specific cultural brands being build. Based on the above analysis, we should become better at mining local characteristic culture, focus on cultivating one side's characteristic culture, play a role in market operation for characteristic cultures that can realize industrialization, protect local characteristic cultures as needed, and properly develop them during protection. For example, establishing local museums and cultural centers has turned the characteristic cultural industry into a pillar of the local key cultural industry. For the future development of the characteristic cultural industry, regions should establish a good concept of civilization construction, innovate the local cultural industry, and create local characteristics with obvious differences. Local governments should actively respond to cultural industry development plans promulgated by the state, and under the condition of supply-side reform, carry out in-depth development of regional cultural characteristics, help poor and less developed areas improve technological and economic development, and realize the development of a characteristic cultural industry to drive regional economic development goals.

9.5 Summary

This chapter primarily explores the differences in the geographical distribution of culture, specifically from perspectives of the role of culture and cultural industry in geographical development, differences in the spatial distribution of culture and cultural industry and their causes, and strategies for developing China's culture and cultural industry. The spatial layout is analyzed from these perspectives.

Regarding the role of culture and cultural industry in geographical development, this chapter mainly introduces the development history of human civilization, analyzes the impact of human activities on the geographical environment over time, and explores the importance of culture and the cultural industry to the geographical environment. We should regulate our own behavior, protect our living environment, strengthen the construction of ecological civilization, and promote China's economic development with the goal of developing culture and the cultural industry.

Regarding the differences in the spatial distribution of culture, this chapter presents an analysis of the differences in the distribution of culture in different industries in

China from five perspectives: language, clothing, food, architecture, and transportation. From the perspective of the spatial distribution of language, using descriptive analysis, we conclude that Mandarin is the most widely distributed and common language among many dialects. From the perspective of the spatial distribution of clothing, we conclude that Han clothing accounts for the largest proportion of national clothing, which is mainly related to geography. In terms of the environment, the Han ethnicity comprising the largest proportion of the population has much to do with it. From the perspective of the spatial distribution of food and buildings, we conclude that the food and buildings in various places have much to do with the geographical environment. Local materials are usually combined with the geographical environment to form unique food and architectural cultures. From the perspective of the spatial distribution of traffic, we conclude that traffic culture between the northern and southern regions was quite different in the early days, owing to the different environments. The southern region is good at shipbuilding, whereas the northern region is good at using vehicles and other tools to carry out activities on land.

With regard to the differences in the spatial distribution of the cultural industry, we collected cultural industry data from 31 provinces, autonomous regions, and municipalities, in addition to Hong Kong, Macao, and Taiwan, based on the diamond model and basic theory of the entropy value method, to construct a cultural industry evaluation index system. The comprehensive cultural industry scores in Jiangsu, Guangdong, and Zhejiang are relatively high, whereas those in Qinghai and Tibet are relatively low. China's cultural industry presents an overall pattern in which the development level gradually increases from west to east, but gradually decreases from south to north.

With regard to the strategic spatial layout of China's culture and cultural industry, this chapter examines the current development of the cultural industry and proposes a scientific and rational layout of China's regional culture and cultural industry and financial support for regions where the development is relatively weak. With the help of the cultural industry, we should continue to improve the development policy of the industry. Considering the gap in cultural industry development between urban and rural areas, we should develop overall plans, increase the construction of rural infrastructure, focus on discovering special products in areas where the cultural industry is not well developed, and transform special cultural products into regional pillar industries to build a cultural brand with regional characteristics, eliminate cultural industry development gaps, and realize economic progress.

Exercise

1. Briefly describe the role of culture and cultural industries in geographical development?
2. What are the differences in the spatial distribution of culture?
3. What are the reasons for the differences in the spatial distribution of culture?
4. What are the differences in the spatial distribution of culture and cultural industry and the reasons thereof?

5. Briefly describe the strategic spatial distribution of culture and cultural industries in China?
6. Briefly describe the main performance of the difference in geographical distribution of culture between north and south of China, and analyze its causes?
7. Analyze the development trend of culture and the cultural industry in the context of globalization and suggest measures to promote the development of the cultural industry.
8. Briefly describe the role of cultural industries in protecting and inheriting traditional culture, and make suggestions in that regard.
9. Summarize the important position of culture in the development of contemporary society and how cultural industries can promote economic development?
10. Analyze the present situation, problems and countermeasures and suggestions for the development of China's cultural industry from the perspective of cultural industry policy.

Extended Reading

The development of Chinese culture dates back thousands of years and has gone through many stages of change and development. Chinese culture originated in the Yellow River Basin in Neolithic Age, about 5000 years ago. Through the archaeology of cultural remains such as pottery, bronze, iron, jade, charred textile fragments, and rice siliceous bodies unearthed, it is evident that thousands of years ago, ancient people's smelting and casting technology, agriculture, pottery, and textile production had already begun to take shape, as well as etiquette and the concept of the family, also gradually took shape. During the Spring and Autumn Period and the Warring States Period, Confucianism, Taoism, Mohism, Legalism, and many other doctrines emerged successively, resulting in the flourishing of "a hundred schools of thought," forming the important foundation of Chinese philosophical and cultural thought, and also the most brilliant and starry era of thought and culture. In this era, moral concepts such as benevolence, righteousness, propriety, wisdom, and faith were widely spread, which profoundly influenced later generations.

After the Spring and Autumn Period and the Warring States Period, the Qin Dynasty unified the world and ended the long period of war between the vassals, and at the same time implemented a series of powerful measures to consolidate its rule, including writing the same text, setting the same track for all vehicles, and standardizing weights and measures. These measures not only promoted economic prosperity and development but also achieved cultural unification, facilitated exchanges and integration among multiple ethnic groups, and laid the foundation for the political, economic, and cultural development of later generations. Subsequently, the Han Dynasty inherited the development foundation of the Qin Dynasty and further developed art forms such as stone carving and calligraphy, adding to the ancient culture. During the Ming and Qing dynasties, the literati and writers set off a wave of reflection on and inheritance of traditional culture, with rapid development in the fields of literature, history, and science and the birth of many unique cultural and artistic forms, such as essays, novels, and operas. Literary masterpieces such as The

Compendium of Materia Medica, Dream of the Red Chamber, and Journey to the West were also created during this period.

The development of modern Chinese culture was a process of continuous change, innovation, and integration. Chinese culture has undergone profound changes and development in modern times, as it began to connect with the international community, introduced and studied foreign cultures, then rationally scrutinized and fused Chinese and Western cultures, and finally, critically inherited traditional cultures and absorbed and fused advanced cultures. Meanwhile, with the advancement of globalization and the progress of science and technology, Chinese culture is gradually moving toward the world stage, showing the deep heritage and unique charm of oriental culture, and in the collision with the world culture, constantly absorbing the cultural essence from all over the world, taking the essence and abandoning the dregs, which not only enriches the connotation of Chinese culture but also infuses it with new vitality.

References

Bian, M. (2023). Research on the measurement and coupling coordination of the development level of material civilization and spiritual civilization under the background of Chinese path to modernization. *Academic Journal of Management and Social Sciences, 5*(2) (Article 2). https://doi.org/10.54097/ajmss.v5i2.07

Cabra, J. F., & Guerrero, C. D. (2022). Regional creativity: Cultural and socio-economic differences. *Journal of Creativity, 32*(2), 100022. https://doi.org/10.1016/j.yjoc.2022.100022

Porter, M. E. (1997). *The competitive advantage of nations: Creating and sustaining superior performance*. Free Press.

Su, B., & Heshmati, A. (2014). Analysis of the determinants of income and income gap between urban and rural China. *China Economic Policy Review*. https://doi.org/10.1142/S1793969013500027

Yao, L., Maneejuk, P., Yamaka, W., & Liu, J. (2023). Quantifying the competitiveness of cultural industry and its impacts on Chinese economic growth. *Sustainability, 15*(1), 79. https://doi.org/10.3390/su15010079.

Ze, Y. (2013). Design of evaluation system of city cultural industry competitiveness based on diamond model. *Science and Technology Management Research*. https://www.semanticscholar.org/paper/Design-of-Evaluation-System-of-City-Cultural-Based-Ze/ec65afc35746bdb217397cf1f39fec5cfa560754

Chapter 10
Resource and Environment Carrying Capacity and Its Regional Differences

Abstract Measuring resource and environmental carrying capacity and eco-efficiency is conducive to a full understanding of the coordination between social development and resource and environmental protection. This chapter analyzes the role of environmental protection and ecological civilization construction in geographic development, the measurement and characterization of environmental capacity and resource and environmental carrying capacity, the measurement and characterization of eco-efficiency, the spatial layout of the ecological civilization construction and sustainable development strategy, and the path of transformation and policy choices for underdeveloped regions. It also proposes the development of the economy, the protection of the ecological environment, and adherence to green, low-carbon recycling and sustainable development. The objectives of this chapter are as follows: 1. To understand the role of environmental protection and ecological civilization construction in geographical development. 2. To understand the measurement and representation of environmental capacity, resource and environmental carrying capacity, and ecological development efficiency. 3. To become familiar with the spatial layout of China's ecological civilization construction and sustainable development strategy.

10.1 Environmental Protection and Ecological Civilization

Currently, China's resources are excessively consumed, environmental pollution is prominent, and ecosystems are seriously degraded. Thus, we must respect the laws of nature, love the environment, rationally develop and utilize natural resources, and follow the path of sustainable development. The construction of ecological civilization must adhere to green, low-carbon, circular, and sustainable development and protect the ecological environment while developing the economy. This will not only provide high-quality ecological products for the present but also leave more ecological assets for future generations.

X. Deng and M. Song, *Development Geography*,
https://doi.org/10.1007/978-981-96-6917-2_10

10.1.1 Improving People's Well-Being

Over the past 40 years of reform and opening up, China's economic development has made remarkable achievements. The living standards of the Chinese people have greatly improved, and they have begun to pay more attention to their and their families' health, living standards, and home environment.

Chinese people require a good ecological environment to improve their lives. The quality of the ecological environment directly determines people's quality of life and happiness index; therefore, the ecological environment must be improved, as damage to it will have a detrimental effect on people.

10.1.2 Laying a Solid Foundation for Survival

The quality of the ecological environment is related to the existence and development of human civilization. The ecological environment is a public product that no one can leave and that can never be replaced. A good ecological environment is the natural basis and prerequisite for people to live happily and a necessity that money cannot replace. In real life, problems such as air, river, and soil pollution are not new, and the losses caused by environmental pollution are substantial. Faced with increasingly severe environmental challenges, people have begun to worry about their safety in various respects, such as drinking water, food, and air safety.

10.1.3 Improving the Image of the Region

All regions can coordinate development between ecological protection and the improvement of people's livelihoods, strive to find a transformation and upgrading path that changes ecological benefits into economic benefits and ecological advantages into development advantages, and continuously move toward a high degree of unity between beautiful regional scenery and people's prosperity and happiness. Accordingly, efforts must be made to increase local residents' income, further improve people's living conditions, and improve people's quality of life. Most people should consume high-quality ecological products, such as those produced from the resources of green water and mountains, to ensure that ecological achievements can benefit poor people and development achievements can be shared by everyone.

10.2 Measurement of Environmental Capacity and Resource and Environmental Carrying Capacity

10.2.1 Environmental Capacity and Resource and Environmental Carrying Capacity

10.2.1.1 Environmental Capacity

Environmental capacity refers to the maximum number of organisms that an ecosystem can sustain within the limits of maintaining the adaptation, regeneration, and renewal capabilities of living organisms in a balanced state under the condition of human survival and undamaged natural ecosystems, or the maximum amount of pollution that the environmental system can accommodate in a given area.

10.2.1.2 Environmental Carrying Capacity

Environmental carrying capacity refers to the population size and economic scale that can be supported by available environmental resources during a certain period while remaining in a relatively stable state. In the context of the survival and development of human society, the environmental system can provide support for various activities.

10.2.1.3 Resource and Environmental Carrying Capacity

Resource and environmental carrying capacity is the collective name of "resource carrying capacity" and "environmental carrying capacity" and refers to a comprehensive carrying capacity concept that includes both resource and environmental elements. It refers to the ability of regional resources and the environmental system to remain stable under the premise that the resource structure meets the needs of regional sustainable development, and environmental functions can continue to maintain balance within a certain period, region, and level of human economic and social activities.

10.2.2 Resource and Environmental Carrying Capacity Evaluation

With China's rapid economic development, the problems of resource shortages and environmental pollution have become increasingly prominent. Simultaneously, with improved living standards, people have increasingly higher requirements for an esthetically pleasing environment. Saving resources and protecting the environment has become an important topic of sustainable development in China, and research on resource and environmental carrying capacity has attracted extensive scholarly attention. The rise of research theories and methods has promoted the development of research on resource and environmental carrying capacity models.

10.2.2.1 The Evaluation Index System

The construction of the evaluation index system of regional resources and environmental carrying capacity is based on the aspects of economy, society, environment, and resources. According to China's resources and environmental conditions and drawing on the evaluative concepts and frameworks of existing studies, we select 20 indicators from those aspects to construct an evaluation index system of resource and environmental carrying capacity (Table 10.1).

10.2.2.2 Evaluation Model Based on Entropy Weight TOPSIS (Technique for Order Preference by Similarity to Ideal Solution)

(i) Standardized evaluation matrix construction

Based on the determined indicators, an initial evaluation matrix of resource and environmental carrying capacity was established. Among them, m is the number of evaluation targets, $i = 1, 2 \ldots m$; and n is the number of evaluation indicators, $j = 1, 2 \ldots n$. The matrix involves m evaluation objectives and n evaluation indicators. The matrix form of the initial evaluation index is expressed in Eq. 10.1:

$$X = \begin{bmatrix} x_{11} & x_{12} & \cdots & x_{1n} \\ x_{21} & x_{22} & \cdots & x_{2n} \\ \vdots & \vdots & \vdots & \vdots \\ x_{m1} & x_{m2} & \cdots & x_{mn} \end{bmatrix} \tag{10.1}$$

where X is the initial evaluation matrix and x_{mn} is the initial value of the jth index of the ith evaluation target. Standardize the collected data and, as much as possible, eliminate the deviation caused by the attributes of indicators such as benefit (positive), cost (reverse), dimension, and order of magnitude. According to the different properties of the selected indicators, the following formulas are used to standardize the forward and reverse indicators, respectively:

Positive indicators:

$$r_{ij} = \frac{x_{ij} - \min(x_{ij})}{\max(x_{ij}) - \min(x_{ij})} \tag{10.2}$$

Table 10.1 Resource and environmental carrying capacity evaluation index system

Standard layer	Index layer	Index unit	Indicator type	Code	Indicator description
Economic and social index	GDP per capita	Yuan/ person	Positive index	S1	GDP/resident population
	Output value of the tertiary industry as a percentage of GDP	%	Positive index	S2	Output value of the tertiary industry/GDP
	Urbanization level	%	Positive index	S3	Urban resident population/total resident population
	Per capita disposable income of residents	Yuan/ person	Positive index	S4	–
	Per capita consumption expenditure of residents	Yuan/ person	Positive index	S5	–
	Technology market turnover as a percentage of GDP	%	Positive index	S6	–
	Urban registered unemployment rate	%	Inverse index	S7	–
Resource index	Cultivated land area per 10,000 people	Hectare/ 10,000 people	Positive index	R1	Cultivated land area/ resident population
	Construction land area per 10,000 people in the city	Hectare/ 10,000 people	Positive index	R2	
	urban population density	People/ square kilometer	Inverse index	R3	–
	Per capita water resources	Cubic meter/ person	Positive index	R4	Total water resources/ resident population
	Water consumption per 10,000 yuan of GDP	Cubic meter/ 10,000 yuan	Inverse index	R5	Comprehensive water consumption/GDP
	Energy consumption per 10,000 yuan of GDP	kg of standard coal/ 10,000 yuan	Inverse index	R6	Energy consumption/ GDP
Environmental index	Forest cover rate	%	Positive index	E1	–

(continued)

Table 10.1 (continued)

Standard layer	Index layer	Index unit	Indicator type	Code	Indicator description
	Green coverage rate of urban built-up areas	%	Positive index	E2	–
	Investment in environmental protection as a percentage of GDP	%	Positive index	E3	Total investment in environmental pollution control/GDP
	Comprehensive utilization rate of industrial solid waste	%	Positive index	E4	Comprehensive utilization of industrial solid waste/production of industrial solid waste
	SO_2 emissions per 10,000 yuan of GDP	kg/10,000 yuan	Inverse index	E5	–
	NO_x emissions per 10,000 yuan of GDP	kg/10,000 yuan	Inverse index	E6	–
	Wastewater discharge per 10,000 yuan of GDP	kg/10,000 yuan	Inverse index	E7	–

Contrarian indicator:

$$r_{ij} = \frac{\max(x_{ij}) - x_{ij}}{\max(x_{ij}) - \min(x_{ij})} \tag{10.3}$$

where x_{ij} is the initial value of the jth index of the ith evaluation target; $min(x_{ij})$, $max(x_{ij})$ are the minimum and maximum values of the jth index in each evaluation target; and r_{ij} is the standardized value of the jth index of the ith evaluation target. After processing, the following standardized evaluation matrix is obtained:

$$R = \begin{bmatrix} r_{11} & r_{12} & \cdots & r_{1n} \\ r_{21} & r_{22} & \cdots & r_{2n} \\ \vdots & \vdots & \vdots & \vdots \\ r_{m1} & r_{m2} & \cdots & r_{mn} \end{bmatrix} \tag{10.4}$$

(ii) Entropy weight method to determine index weight

The determination of the weight is relatively subjective, and the information of multi-index variables often overlaps. Therefore, the entropy weight method can use the original data information to analyze the correlation degree of the indicators and, according to the amount of information carried by each indicator, to make the determination of the weight more objective. Thus, we use the entropy weight method to determine the weight of each indicator.

First, calculate the proportion of the ith evaluation target index value under the jth index to the following index:

$$p_{\mathrm{ij}} = \frac{r_{\mathrm{ij}}}{\sum_{i=1}^{m} r_{\mathrm{ij}}} \tag{10.5}$$

Second, calculate the entropy value of the jth index as follows:

$$e_j = -\mathrm{k}\sum_{i=1}^{m} p_{\mathrm{ij}}\ln(p_{\mathrm{ij}}) \tag{10.6}$$

Among them, $\mathrm{k} = 1/\ln(\mathrm{m})$, $\mathrm{e_j} \geq 0$ and $\ln 0 = 0$. Then, calculate the information entropy redundancy as follows:

$$d_j = 1 - \mathrm{e}_j \quad (\mathrm{j} = 1, 2, \ldots \mathrm{n}) \tag{10.7}$$

Finally, calculate the weight of each indicator as follows:

$$w_j = \frac{d_j}{\sum_{i=1}^{n} d_j} \quad (\mathrm{j} = 1, 2, \ldots, \mathrm{n}) \tag{10.8}$$

(iii) Evaluation matrix construction based on entropy weight

Based on the standardized evaluation matrix of China's resource and environmental carrying capacity, combined with the entropy weight, a standardized evaluation matrix Y based on entropy weight is constructed. The calculation process is shown in Eq. 10.9:

$$Y = \begin{bmatrix} y_{11} & y_{12} & \cdots & y_{1\mathrm{n}} \\ y_{21} & y_{22} & \cdots & y_{2\mathrm{n}} \\ \vdots & \vdots & \vdots & \vdots \\ y_{\mathrm{m}1} & y_{\mathrm{m}2} & \cdots & y_{\mathrm{mn}} \end{bmatrix} = \begin{bmatrix} r_{11}\cdot w_1 & r_{12}\cdot w_2 & \cdots & r_{1\mathrm{n}}\cdot w_n \\ r_{21}\cdot w_1 & r_{22}\cdot w_2 & \cdots & r_{2\mathrm{n}}\cdot w_n \\ \vdots & \vdots & \vdots & \vdots \\ r_{\mathrm{m}1}\cdot w_1 & r_{\mathrm{m}2}\cdot w_2 & \cdots & r_{\mathrm{mn}}\cdot w_n \end{bmatrix} \tag{10.9}$$

(iv) Positive and negative ideal solution determination

Set y_j^+ as the maximum value of the jth index in the evaluation data among all evaluation objectives to obtain the positive ideal solution (i.e., the most preferrable solution to choose). y_j^- is the minimum value of the jth index in the evaluation data among all evaluation objectives, which is the negative ideal solution (i.e., the least preferrable solution). See the Eqs. 10.10 and 10.11 for the calculation method as follows:

$$y_j^+|j = 1, 2, \ldots, n = \left\{\max(y_{ij})|i = 1, 2, \ldots, m\right\} \quad (10.10)$$

$$y_j^-|j = 1, 2, \ldots, n = \left\{\min(y_{ij})|i = 1, 2, \ldots, m\right\} \quad (10.11)$$

(v) Distance calculation

To calculate distance, we adopt the Euclidean metric formula. Let D_i^+ be the distance between the *i*th evaluation target and y_j^+ and D_i^- be the distance between the *i*th evaluation target. The calculation methods are shown in Eqs. 10.12 and 10.13.

$$D_i^+ = \sqrt{\sum_{j=1}^{n}(y_{ij} - y_j^+)^2} \quad (10.12)$$

$$D_i^- = \sqrt{\sum_{j=1}^{n}(y_{ij} - y_j^-)^2} \quad (10.13)$$

(vi) Calculate the closeness between the evaluation target value and the ideal solution

Assuming T_i that the carrying capacity of the *i*th evaluation target resource environment is close to the optimal carrying capacity, it is called closeness, and its value range is [0,1]. The larger T_i is, the closer it is to the optimal level of resource and environmental carrying capacity. When $T_i = 0$, the carrying capacity of resources and the environment is the smallest. When $T_i = 1$, the carrying capacity of resources and the environment is the largest. The size of China's resource and environmental carrying capacity can be measured by the degree of closeness to the ideal solution. If the degree of closeness is greater, the carrying capacity is greater. In contrast, if the degree of closeness is smaller, the carrying capacity is smaller. Equation 10.14 shows the calculation method.

$$T_i = \frac{D_i^-}{D_i^+ + D_i^-} \quad (10.14)$$

10.2.2.3 Empirical Research on Resource and Environmental Carrying Capacity in China

(i) Evaluation objectives and data sources

Owing to the availability of index data, 30 provinces, autonomous regions, and municipalities directly under the central government in mainland China (excluding Tibet, Hong Kong, Macao, and Taiwan) were selected as the research objects, and 2009–2017 was selected as the research period. Owing to the lack of individual index data

Table 10.2 Resource and environmental carrying capacity evaluation index weight

Index	S1	S2	S3	S4	S5	S6	S7	R1	R2	R3
Weight	0.0549	0.0398	0.0288	0.0551	0.0519	0.2562	0.0425	0.0775	0.0583	0.0225
Index	R4	R5	R6	E1	E2	E3	E4	E5	E6	E7
Weight	0.1275	0.0040	0.0169	0.0558	0.0164	0.0440	0.0327	0.0062	0.0048	0.0042

for Shanghai between 2009 and 2011, the Shanghai sample from 2009 to 2011 was removed from the evaluation targets, leaving 276 evaluation targets.

(ii) Calculation process

Equations 10.2 and 10.3 are used to standardize the original evaluation index matrix composed of original data, and then Eqs. 10.5–10.8 are used to calculate the weight of the evaluation index of China's resource and environmental carrying capacity (Table 10.2).

Determine the positive and negative ideal solutions according to Eqs. 10.9–10.11, and then use Eqs. 10.12 and 10.13 to obtain the distance between the resource and environmental carrying capacity of China and 30 provinces (autonomous regions and municipalities) between 2009 and 2017 and the positive and negative ideal solutions. The results are shown in Tables 10.3 and 10.4.

Finally, according to the distance between the resource and environmental carrying capacity and the positive and negative ideal solutions in Tables 10.3 and 10.4, Eq. 10.14 can be used to obtain the closeness of the resource and environmental carrying capacity of each evaluation target. The results are shown in Table 10.5.

(iii) Results analysis

The overall level of China's resource and environmental carrying capacity is increasing. During the study period, 2011 was the only year in which China's resource and environmental carrying capacity declined; all other years increased compared with the previous year. This is mainly because in 2011, poor performance was reported for the per capita arable land area, per capita water resources, proportion of environmental protection investment in GDP, and comprehensive utilization rate of industrial solid waste.

From the perspective of the overall levels in the eastern, central, and western regions, the eastern region has the highest level of resource and environmental carrying capacity, followed by the western region, and the lowest in the central region. The gap between the resource and environmental carrying capacity levels in the central, western, and eastern regions is widening, whereas that between the central and western regions has not changed much.

Table 10.3 Resource and environmental carrying capacity and positive ideal solution distance in China (2009–2017)

Regions	2009	2010	2011	2012	2013	2014	2015	2016	2017
Nationwide	0.30	0.30	0.30	0.29	0.29	0.29	0.29	0.28	0.28
Beijing	0.19	0.18	0.18	0.17	0.16	0.16	0.16	0.16	0.16
Tianjin	0.30	0.30	0.29	0.29	0.29	0.28	0.27	0.27	0.27
Hebei	0.32	0.32	0.32	0.31	0.31	0.31	0.31	0.31	0.31
Shanxi	0.32	0.31	0.31	0.31	0.31	0.31	0.31	0.31	0.30
Inner Mongolia	0.30	0.30	0.30	0.29	0.29	0.29	0.29	0.30	0.30
Liaoning	0.30	0.30	0.30	0.29	0.30	0.30	0.29	0.28	0.28
Jilin	0.30	0.30	0.30	0.30	0.30	0.30	0.30	0.29	0.28
Heilongjiang	0.29	0.29	0.30	0.29	0.28	0.29	0.29	0.29	0.29
Shanghai	–	–	–	0.27	0.27	0.27	0.27	0.27	0.27
Jiangsu	0.31	0.30	0.30	0.30	0.30	0.30	0.29	0.29	0.29
Zhejiang	0.30	0.30	0.30	0.30	0.30	0.30	0.29	0.29	0.29
Anhui	0.31	0.31	0.31	0.30	0.30	0.30	0.30	0.29	0.29
Fujian	0.31	0.30	0.30	0.30	0.30	0.30	0.30	0.29	0.30
Jiangxi	0.31	0.30	0.31	0.30	0.30	0.30	0.30	0.29	0.30
Shandong	0.31	0.31	0.31	0.31	0.31	0.30	0.30	0.30	0.30
Henan	0.32	0.32	0.32	0.32	0.32	0.31	0.31	0.31	0.31
Hubei	0.30	0.30	0.30	0.30	0.29	0.28	0.27	0.26	0.26
Hunan	0.31	0.31	0.31	0.31	0.31	0.30	0.30	0.30	0.30
Guangdong	0.30	0.30	0.30	0.30	0.29	0.30	0.29	0.29	0.29
Guangxi	0.31	0.30	0.31	0.30	0.30	0.30	0.30	0.30	0.29
Hainan	0.30	0.30	0.30	0.30	0.29	0.30	0.30	0.29	0.30
Chongqing	0.30	0.30	0.30	0.30	0.30	0.29	0.30	0.29	0.30
Sichuan	0.31	0.31	0.31	0.30	0.30	0.30	0.29	0.29	0.29
Guizhou	0.31	0.31	0.31	0.31	0.31	0.30	0.30	0.30	0.29
Yunnan	0.31	0.30	0.31	0.30	0.30	0.30	0.30	0.30	0.29
Shaanxi	0.30	0.30	0.29	0.28	0.27	0.26	0.26	0.26	0.25
Gansu	0.30	0.30	0.30	0.29	0.29	0.29	0.29	0.28	0.28
Qinghai	0.29	0.29	0.28	0.28	0.28	0.27	0.27	0.26	0.25
Ningxia	0.31	0.31	0.31	0.31	0.31	0.31	0.31	0.31	0.31
Xinjiang	0.31	0.30	0.30	0.30	0.30	0.30	0.30	0.29	0.29

Table 10.4 Resource and environmental carrying capacity and negative ideal solution distance in China (2009–2017)

Regions	2009	2010	2011	2012	2013	2014	2015	2016	2017
Nationwide	0.05	0.05	0.05	0.05	0.06	0.06	0.06	0.06	0.06
Beijing	0.19	0.20	0.21	0.24	0.25	0.26	0.27	0.27	0.28
Tianjin	0.07	0.07	0.07	0.08	0.08	0.09	0.10	0.10	0.10
Hebei	0.04	0.04	0.05	0.04	0.04	0.04	0.05	0.05	0.05
Shanxi	0.04	0.04	0.04	0.05	0.05	0.05	0.05	0.06	0.05
Inner Mongolia	0.08	0.09	0.09	0.09	0.10	0.10	0.10	0.10	0.10
Liaoning	0.06	0.06	0.06	0.07	0.07	0.07	0.07	0.08	0.08
Jilin	0.07	0.07	0.07	0.07	0.08	0.08	0.08	0.08	0.08
Heilongjiang	0.10	0.10	0.09	0.10	0.10	0.10	0.10	0.10	0.10
Shanghai	–	–	–	0.11	0.11	0.12	0.12	0.12	0.12
Jiangsu	0.06	0.06	0.06	0.07	0.07	0.08	0.08	0.08	0.08
Zhejiang	0.08	0.08	0.08	0.08	0.09	0.09	0.09	0.10	0.10
Anhui	0.05	0.05	0.05	0.05	0.06	0.06	0.06	0.06	0.07
Fujian	0.07	0.08	0.07	0.08	0.08	0.08	0.08	0.09	0.09
Jiangxi	0.06	0.07	0.07	0.08	0.07	0.07	0.07	0.08	0.07
Shandong	0.06	0.06	0.06	0.06	0.06	0.07	0.07	0.07	0.07
Henan	0.04	0.04	0.04	0.04	0.04	0.05	0.05	0.05	0.05
Hubei	0.05	0.05	0.05	0.06	0.06	0.07	0.08	0.08	0.08
Hunan	0.06	0.06	0.05	0.06	0.06	0.06	0.06	0.07	0.07
Guangdong	0.07	0.08	0.07	0.08	0.08	0.08	0.08	0.09	0.09
Guangxi	0.06	0.07	0.06	0.07	0.07	0.07	0.08	0.08	0.08
Hainan	0.08	0.08	0.08	0.08	0.08	0.08	0.08	0.08	0.08
Chongqing	0.05	0.06	0.06	0.06	0.06	0.07	0.07	0.07	0.07
Sichuan	0.05	0.05	0.05	0.05	0.05	0.05	0.06	0.06	0.06
Guizhou	0.05	0.05	0.05	0.05	0.05	0.06	0.06	0.06	0.06
Yunnan	0.06	0.06	0.06	0.06	0.07	0.06	0.07	0.07	0.07
Shaanxi	0.05	0.05	0.06	0.06	0.08	0.08	0.09	0.09	0.09
Gansu	0.05	0.05	0.05	0.06	0.07	0.07	0.07	0.07	0.07
Qinghai	0.13	0.11	0.11	0.13	0.10	0.12	0.10	0.10	0.12
Ningxia	0.07	0.06	0.07	0.07	0.07	0.07	0.07	0.07	0.07
Xinjiang	0.07	0.07	0.07	0.08	0.08	0.08	0.08	0.08	0.08

Table 10.5 Proximity to carrying capacity of resources and environment in China (2009–2017)

Regions	2009	2010	2011	2012	2013	2014	2015	2016	2017
Nationwide	0.14	0.15	0.14	0.16	0.16	0.17	0.17	0.18	0.18
Eastern mean	0.21	0.22	0.22	0.24	0.24	0.25	0.25	0.26	0.27
Central mean	0.16	0.17	0.16	0.17	0.18	0.18	0.19	0.20	0.20
Western mean	0.17	0.18	0.18	0.19	0.20	0.21	0.20	0.21	0.22
Beijing	0.50	0.52	0.54	0.59	0.61	0.62	0.62	0.63	0.64
Tianjin	0.19	0.19	0.20	0.21	0.22	0.24	0.26	0.27	0.28
Hebei	0.12	0.12	0.13	0.12	0.12	0.12	0.13	0.13	0.14
Shanxi	0.12	0.12	0.12	0.14	0.14	0.14	0.14	0.17	0.14
Inner Mongolia	0.22	0.23	0.23	0.24	0.25	0.25	0.25	0.24	0.25
Liaoning	0.16	0.17	0.17	0.19	0.19	0.19	0.19	0.21	0.22
Jilin	0.19	0.20	0.19	0.20	0.21	0.20	0.20	0.22	0.23
Heilongjiang	0.25	0.25	0.24	0.25	0.26	0.25	0.26	0.26	0.26
Shanghai	–	–	–	0.29	0.29	0.30	0.31	0.30	0.31
Jiangsu	0.16	0.17	0.17	0.18	0.20	0.20	0.21	0.22	0.22
Zhejiang	0.20	0.21	0.21	0.22	0.23	0.23	0.24	0.25	0.25
Anhui	0.14	0.14	0.14	0.15	0.17	0.17	0.17	0.18	0.18
Fujian	0.19	0.21	0.19	0.22	0.22	0.22	0.22	0.24	0.22
Jiangxi	0.17	0.20	0.18	0.20	0.19	0.19	0.20	0.21	0.20
Shandong	0.15	0.16	0.16	0.17	0.17	0.18	0.18	0.18	0.19
Henan	0.11	0.11	0.11	0.12	0.12	0.13	0.13	0.14	0.15
Hubei	0.14	0.15	0.15	0.16	0.18	0.21	0.22	0.23	0.24
Hunan	0.15	0.16	0.14	0.15	0.16	0.16	0.18	0.18	0.19
Guangdong	0.20	0.22	0.20	0.21	0.21	0.21	0.22	0.23	0.24
Guangxi	0.17	0.18	0.17	0.19	0.20	0.19	0.21	0.20	0.22
Hainan	0.20	0.21	0.21	0.20	0.22	0.20	0.20	0.22	0.21
Chongqing	0.15	0.17	0.17	0.17	0.17	0.19	0.18	0.19	0.19
Sichuan	0.13	0.14	0.13	0.15	0.15	0.15	0.16	0.16	0.17
Guizhou	0.13	0.13	0.13	0.14	0.15	0.16	0.16	0.16	0.17
Yunnan	0.16	0.17	0.16	0.17	0.18	0.18	0.18	0.19	0.20
Shaanxi	0.14	0.14	0.16	0.18	0.22	0.23	0.25	0.26	0.26
Gansu	0.15	0.15	0.15	0.17	0.19	0.19	0.19	0.20	0.20
Qinghai	0.14	0.15	0.14	0.16	0.16	0.17	0.17	0.18	0.18
Ningxia	0.21	0.22	0.22	0.24	0.24	0.25	0.25	0.26	0.27
Xinjiang	0.16	0.17	0.16	0.17	0.18	0.18	0.19	0.20	0.22

10.2.3 Dynamic Model of Resource and Environmental Carrying Capacity System

10.2.3.1 Principles

System dynamics was founded by Professor J. W. Forrester of the Massachusetts Institute of Technology in the 1950s. Its early research objects were mainly centered on the management of industrial enterprises, studying issues such as changes in output and employee conditions, enterprise development, and stock market instability. Its basic principle is that, by observing the actual system, collecting information about the state of the object system, and then using the relevant information to make decisions, The result of the decision is to take action, which then impacts the actual system. This change, in turn, provides new information to the observer, thereby forming a feedback loop in the system.

10.2.3.2 System Dynamics Modeling

Most extant research evaluates the carrying capacity of China's historical period by establishing an index system, which has important reference value for improving China's resource and environmental carrying capacity. However, few studies have examined the dynamic changes in resource and environmental carrying capacity at the national level. Accordingly, we introduce the system dynamic model of "economic and social development-resource and environmental carrying capacity" to simulate and predict the future trends in China's resource and environmental carrying capacity under different development scenarios in the process of economic development and propose the principles of China's economic and social development and resource and environment protection and policy suggestions.

10.2.3.3 System Analysis of Resource and Environmental Carrying Capacity

(i) Modeling purpose

We hope that by establishing a system dynamics model, we can fully and clearly reflect the function system of resource and environmental carrying capacity (Wang et al. 2023), determine the system elements and their interrelationships, and understand the impact of economic and social development on China's resource and environmental carrying capacity channels. In addition, we hope to be able to predict the development trends in elements of the economic and social system and understand their impact on China's resource and environmental carrying capacity; simulate different economic and social development situations and analyze the effects of economic and social development on China's resource and environmental carrying capacity and the actual scenario model that should be adopted in the study. Then, we aim to perform a policy simulation for optimizing economic and social development and propose policy measures to improve China's resource and environmental carrying capacity based on the simulation results.

(ii) System boundary determination

To establish a simulation model based on system dynamics theory, it is necessary to first determine the system boundary (i.e., the research scope of the model). For the research purposes of the model, the influencing factors closely related to the system behavior are covered.

Internally, the complex feedback relationship in the system dynamics model forms a closed loop. We aim to simulate the relationship and development trends among the economy, population, atmospheric environment, and resources under the constraints of resources and the environment in China, to provide a theoretical reference for policy formulation.

The geographical boundary of the model is China's land territory. The time boundary is from 2009 to 2030, and the simulation step is one year. Data from 2009 to 2017 are used for model testing. The simulation base year is 2009, and the period from 2018 to 2030 is used to predict the resource and environmental carrying capacity of China. The research data are mainly collected from the China Statistical Yearbook, China Environmental Statistical Yearbook, China Environmental Yearbook, and China Urban and Rural Construction Statistical Yearbook.

(iii) Subsystem division

The structure of China's resource and environmental carrying capacity system is complex. The system is limited by factors such as economic development, population growth, atmospheric environment quality, energy consumption, and pollutant emission reduction measures. Complex constraints exist among various system elements. According to the interrelationships of the internal structure of the system and the integrity and hierarchy of the system, China's resource and environmental carrying capacity system can be further divided into economic and social, resource, and environmental subsystems, which together promote the coordinated environmental and economic development.

(iv) General structural framework

The key step in establishing a system dynamics model is to determine the causal feedback relationship of the system. The system behavior of each subsystem is closely related to the internal feedback of the system and also restricted by other subsystems. The causal feedback loop of China's economic, social, resource, and environmental systems is mainly based on the concepts described below.

First, GDP growth, especially industrial development, will increase air pollutant emissions, leading to increasingly serious problems with environmental pollution. Therefore, it is necessary to improve the economical and intensive use of energy and reduce pollutant emissions.

Second, an increased population requires more resources, and China's demand for construction land area and water consumption increases accordingly. GDP growth has promoted the migration of the rural population to cities and towns, and the ongoing growth of urban construction land has continuously improved China's urbanization level. Simultaneously, as the demand for land for other purposes increases, the cultivated land area in China continues to decrease.

Third, environmental pollution is forcing adjustments to China's energy and economic structures. The energy consumption structure is gradually optimized, energy use efficiency is continuously improved, and industrial structure upgrades reduce pollutant production. Improved emission reduction technologies and increased investment in environmental protection continuously increase pollutant treatment capacity. Combined with the actual situation in China, a causal feedback relationship diagram of the economy, society, resources, and the environment is established.

10.2.3.4 System Dynamics Model Flowchart Establishment

Based on the analysis of China's resource and environmental carrying capacity system and the dynamic feedback mechanism of each subsystem, a system flowchart of China's resource and environment carrying capacity is established. The main state variables include GDP, total population, urbanization rate of the registered population, urbanization rate of the permanent population, national construction land area, cultivated land area, and comprehensive water consumption. The main rate variables include GDP growth, total population growth, urbanization rate growth of the registered population, urbanization rate growth of the resident population, growth of national construction land area, reduction of cultivated land area, and increase of comprehensive water use. The main auxiliary variables include land development intensity, SO_2 emissions, NO_x emissions, actual urbanization rate, industrial output value, and industrial energy consumption. Constants include urban resident population coefficient, land area, citizenization index of the agricultural transfer population, and greening factor of the urban built-up area. The system operation behavior is regulated by changing the parameters. Vensim software is used to simulate China's resource and environmental carrying capacity system.

10.2.3.5 Explanation of the Main Variables of the Model

In establishing the system dynamics model, the initial and constant values of the state variables need to be determined, and the reasonable assignment of these variables can improve the model simulation accuracy. These variables can be determined through statistical data and simulation experiments: within the historical fluctuation range of variables, select appropriate parameters to debug the model. In addition, some necessary auxiliary variables are included in the modeling process to ensure the completeness of the model structure.

(i) The initial value of the state variable is determined

Based on relevant statistics from the China Statistical Yearbook, China Environmental Statistical Yearbook, China Environmental Yearbook, and China Urban and Rural Construction Statistical Yearbook, historical data for the nine years between 2009 and 2017 are used to establish the logic among the variables in the system and quantitative relationships (Table 10.6).

(ii) Model parameter value determination

According to the structural characteristics of the resource and environmental carrying capacity system, and considering the interaction and change characteristics among variables, the model constant values are determined in three ways.

First, we check the information. Some variable values can be obtained by consulting relevant statistical yearbooks. For example, land area can be obtained by searching the China Statistical Yearbook, and the data are reliable. In addition, some variable values can refer to other research reports or literature to obtain corresponding empirical values.

Second, we conduct regression analyses. A clear quantitative relationship exists between some variables; therefore, regression analysis can be used for calculations based on historical data. The parameters calculated by the regression analysis method mainly include the greening factor of urban built-up areas, the growth coefficient of the national construction land area, and the urban resident population coefficient.

Third, we produce table functions. Parameters with variable variation rules or unclear quantitative relationships among variables can be provided in the form of table functions, mainly including GDP growth, industrial and agricultural water use, birth rate, death rate, proportion of total industrial output value in GDP, and urban area.

Table 10.6 Initial values of the state variables of the dynamic model of China's environmental carrying capacity system

Variable	Value	Unit	Variable	Value	Unit
Year	2009	Year	GDP	34,908.1	10^9 yuan
Urban built-up area	53,665	km^2	Total population	133,450	10^4 Per
Green area of urban built-up area	18,598.4	km^2	Urbanization rate of registered population	33.74	%
Cultivated area	135,385	10 km^2	Urbanization rate of permanent residents	48.34	%
Comprehensive water consumption	5965.2	10^8 m^3	National construction land area	355,777	km^2

Note The national construction land area in 2009 is based on estimates from other years

10.2.3.6 Validity Test of Resource and Environmental Carrying Capacity Model

(i) Model structure check

By studying China's historical statistical data, we can check whether the relationship between variables in the model and the system flow chart is consistent with the actual situation, and analyze whether the variables are reasonably set to ensure the model's reliability and accuracy. After a comprehensive analysis of China's economic and social development and the status of resources and the environment, the system structure analysis function of the Vensim software is used to test the relationship between the variables in the system model and the system flow chart. The parameter values of the model are obtained mainly through the analysis of China's historical statistical data and consulting relevant statistical yearbooks and research reports. The structural test function of the model provided by the Vensim software is used to test the model, and after repeated debugging, the dynamic model of China's resource and environmental carrying capacity system is continuously improved and brought close to the actual situation.

(ii) Model history check

The system behavior can be difficult to accurately describe if the simulation results of the model cannot match the historical records of the actual system. The period used in this study for system historical inspection is from 2009 to 2017, and the simulated values of the variables are obtained using Vensim software simulation.

The variables selected for the historical test are the main variables that can reflect the functions and elements of each subsystem: SO_2 emissions, NO_x emissions, arable land area, urban per capita construction land area, land development intensity, comprehensive water consumption, GDP, and permanent population urbanization rate. The validity of the model is evaluated by calculating the error rate between the key variable simulation results and the real historical data. Table 10.7 shows the calculation results. Clearly, the error rate of the main variables is lower, and the validity of the model is better. This strongly demonstrates that the established dynamic model of China's resource and environmental carrying capacity system is correct, can better show the real situation of China's resource and environmental carrying capacity system, and has strong validity and reliability.

Table 10.7 Fitting error rate of main variables of the model (error rate unit: %)

Index	2009	2010	2011	2012	2013	2014	2015	2016	2017
GDP (billion yuan)	34,908	38,621	42,304	45,627	49,167	52,755	56,396	60,184	64,311.23
	34,908	38,621	42,304	45,627	49,167	52,755	56,396	60,185	64,311.2
	0	0	0	0	0	0	0	0	0
Urbanization rate of permanent residents (%)	48.34	49.95	51.27	52.57	53.73	54.77	56.1	57.35	58.52
	48.34	49.95	51.07	52.17	53.17	54.23	55.31	56.4	57.54
	0	0	– 0.39	– 0.76	– 1.04	– 0.99	– 1.41	– 1.66	– 1.67
Urban per capita construction land area (square meters/person)	106.55	105.29	105.02	110.57	111.02	114.12	114.1	112.43	114.29
	103.35	104.54	106.48	108.27	109.73	111.21	112.6	113.95	115.19
	– 3	– 0.71	1.39	– 2.08	– 1.16	– 2.55	– 1.31	1.35	0.79
Cultivated land area (ten square kilometers)	135,384.6	135,268.3	135,238.6	135,158.4	135,163.4	135,057.3	134,998.7	134,920.9	134,881.2
	135,385	135,268	135,206	135,146	135,087	135,033	134,976	134,918	134,859
	0	0	– 0.02	– 0.01	– 0.06	– 0.02	– 0.02	0	– 0.02
Land development intensity (%)	–	–	–	–	3.9	3.97	4.02	4.07	4.12
	3.71	3.76	3.81	3.87	3.91	3.96	4.02	4.07	4.12
	–	–	–	–	0.26	– 0.25	0	0	0
Comprehensive water consumption (billion cubic meters)	5965.2	6022	6107.2	6131.2	6183.4	6094.9	6103.2	6040.2	6043.4
	5965.2	5965.25	6025.26	6110.39	6133.02	6191.99	6106.2	6107.28	6035.44
	0	– 0.94	– 1.34	– 0.34	– 0.81	1.59	0.05	1.11	– 0.13

(continued)

Table 10.7 (continued)

Index	2009	2010	2011	2012	2013	2014	2015	2016	2017
SO_2 emissions (10,000 tons)	2214.4	2185.1	2217.9	2117.6	2043.9	1974.4	1859.1	1102.86	875.4
	2344.49	2252.14	2093.7	2084.7	1966.42	1923.23	1885.55	1102.86	875.4
	5.87	3.07	− 5.6	− 1.55	− 3.79	− 2.59	1.42	0	0
NO_x emissions (10,000 tons)	1692.7	1852.4	2404.3	2337.8	2227.4	2078	1851	1394.31	1258.83
	1832.26	1900.31	2263.42	2285.54	2148.99	2044.1	1889.75	1394.31	1258.83
	8.24	2.59	− 5.86	− 2.24	− 3.52	− 1.63	2.09	0	0

10.2.4 Scenario Analysis of China's Development Under the Constraints of Resource and Environmental Carrying Capacity

10.2.4.1 Scenario Setting for Different Development Models in the Future

(i) Scenario description

China's economic growth has long been overly dependent on resource consumption, which has resulted in serious environmental pollution. Although China's economic development has shifted from a stage of high-speed growth to one of high-quality development, and resource utilization efficiency has continuously improved while environmental protection has steadily increased, China's resource and environmental problems remain severe.

To solve the current resource and environmental problems, coordinate economic and social development with resources and the environment, and take the path of sustainable development, a feasible economic and social development plan must be designated. Therefore, the scenario analysis method can be used to analyze and assist in decision-making, using four development scenarios, namely of status quo continuation, performance-driven, environmental protection, and coordinated development, to systematically simulate China's future resource and environmental carrying capacity.

(ii) Status quo continuation model

Historical development trends are used as the standard to maintain the current economic and social development model, continue to operate according to the previous development trend, and continue the existing resource utilization, environmental protection, production, and lifestyle.

(iii) Performance-driven model

With rapid GDP growth as the goal, this model considers the scale and speed of economic development; however, it does not consider resource consumption, environmental pollution, or people's living conditions.

(iv) Eco-friendly model

With the goal of protecting the ecological environment, this model focuses on the economical and intensive use of resources and the reduction of pollutant emissions, without emphasizing the scale and speed of economic development.

(v) Coordinated development model

To achieve sustainable development, we need not only healthy and rapid economic development but also economical resource use and ecological environment protection.

10.2.4.2 Scenario Parameter Setting

(i) Status quo continuation model: In the dynamic model of China's resource and environmental carrying capacity system, the values of each parameter are kept constant, and the simulation is conducted according to the current development trend.
(ii) Performance-driven model: The Outline of the Thirteenth Five-Year Plan for National Economic and Social Development (2016–2020) predicted that the GDP in 2020 would double that of 2010, with an average annual increase of more than 6.5%. During the period, China's economy maintained rapid growth. We set the GDP growth rate from 2018 to 7.5%, increase the national construction land area growth coefficient to 1.5, and the urban built-up area growth coefficient to 1, and keep other parameters of the model unchanged. Based on this, the future development situation is simulated.
(iii) Eco-environmental protection model: Economic development will inevitably consume large amounts of resources while simultaneously discharging pollutants into the natural environment. To protect both resources and the environment, we should greatly reduce the speed of economic development, save and optimally utilize resources and energy, reduce the use of fossil energy such as coal, and develop new energy. The GDP growth rate is set to 3% from 2018, industrial energy saving coefficient is increased to 0.3, population growth rate is 5.32‰ before 2020 and then adjusted to decrease by 0.532‰ per year based on 5.32‰, industrial SO_2 removal rate is increased to 95%, industrial NO_x removal rate increased to 80%, national construction land area growth coefficient decreased to 1, urban built-up area growth coefficient decreased to 0.5, and other parameters of the model remained unchanged. Based on this, the future development situation was simulated.
(iv) Coordinated development model: Maintain a moderate economic development speed, save and intensively utilize resources and energy simultaneously, moderately reduce the use of fossil energy such as coal, develop new energy sources, and enable China's economy to achieve healthy development. Set the 2018–2020 GDP growth rate to 6.3% and the 2021–2030 GDP growth rate to 5%. The industrial energy saving coefficient increases to 0.2, the industrial SO_2 removal rate is 92%, the industrial NO_x removal rate is 78%, and the parameters for other models remain unchanged. Based on this, the future development situation is simulated. Table 10.8 shows the values of the control variables of the four development models.

10.2.4.3 Scenario Simulation and Analysis of Resource and Environmental Carrying Capacity

We adjust the models in turns according to the requirements of each model to obtain the scale of China's economic and social development and resource and environment quality (2018–2030) under different models (Table 10.9). The numbers 1–4 after the

Table 10.8 Values of simulation variables for each development model system

Control variable	Status quo continuation	Performance-driven	Eco-friendly	Coordinated development
GDP growth rate	6.86%	7.5%	3%	6.3%, 5%
Industrial energy saving factor	0	0	0.3	0.2
Industrial SO_2 removal rate	90.825%	90.825%	95%	92%
Industrial NO_x removal rate	68.816%	68.816%	85%	78%
National construction land area growth coefficient	1.38	1.5	1	1.09
Urban built-up area growth coefficient	0.77469	1	0.5	0.7446
Population growth rate	5.32‰ (decrease by 0.48363‰ every year after 2020	5.32‰ (decrease by 0.48363‰ every year after 2020	5.32‰ (decrease by 0.532‰ every year after 2020	5.32‰ (decrease by 0.48363‰ every year after 2020

indicators in Table 10.9 represent the four modes of continuation of the status quo, performance-driven, eco-environmental protection, and coordinated development, respectively. For example, the urbanization rate of permanent residents1 represents the urbanization rate of permanent residents under the status quo continuation model.

(i) Status quo continuation model

During the simulation period, China's economic development maintained rapid growth, and the total economic volume increased significantly. The GDP in 2020 will be 78.4751 trillion yuan, which is at least double that of 2010. The urbanization rate of the permanent population in 2020 will be 61.51%. The urbanization rate is 82.26%; the urbanization rate of the registered population will be 46.07% in 2020 and 67.23% in 2030; and the actual urbanization rate will be 53.33% in 2020 and 74.29% 2030. According to the goals of the Thirteenth Five-Year Plan (2016–2020) for National Economic and Social Development and National Population Development Plan (2016–2030), in 2030, the GDP will reach 152.364 trillion yuan, and the rapid economic development can better meet people's material needs.

(ii) Performance-driven model

During the simulation period, China's economy developed rapidly, and the total economic volume increased rapidly. The GDP in 2020 will be 79.8936 trillion yuan, at least double that of 2010. The urbanization rate of the permanent population will be 61.78% in 2020 and 85.43% in 2030. The urbanization rate of the registered population will be 46.34% in 2020 and 70.47% in 2030. The actual urbanization rate will

Table 10.9 Key variable prediction results

Index/year	2018	2019	2020	2021	2022	2023	2024	2025	2026	2027	2028	2029	2030
Urbanization rate 1	58.77	60.10	61.51	63.02	64.64	66.36	68.21	70.18	72.28	74.53	76.94	79.51	82.26
Urbanization rate 2	58.77	60.22	61.78	63.45	65.25	67.18	69.26	71.49	73.89	76.47	79.24	82.23	85.43
Urbanization rate 3	58.77	59.35	59.95	60.56	61.19	61.85	62.52	63.21	63.92	64.65	65.41	66.19	66.99
Urbanization rate 4	58.77	59.99	61.28	62.65	63.81	65.03	66.31	67.65	69.06	70.54	72.09	73.72	75.43
GDP1	68,722.9	73,437.3	78,475.1	83,858.5	89,611.2	95,758.5	102,328	109,347	116,848	124,854	133,430	142,583	152,364
GDP2	69,134.5	74,319.6	79,893.6	85,885.6	92,327	99,251.6	106,695	114,698	123,300	132,547	142,488	153,175	164,663
GDP3	66,240.5	68,227.7	70,274.6	72,382.8	74,554.3	76,790.9	79,094.6	81,467.5	83,911.5	86,423.9	89,021.7	91,692.4	94,443.1
GDP4	68,362.8	72,669.7	77,247.8	81,110.2	85,165.8	89,424	93,895.2	98,590	103,520	108,695	114,130	119,837	125,829
$SO_2$1	956.68	983.28	1011.70	1042.08	1074.54	1109.22	1146.29	1185.90	1228.22	1273.45	1321.78	1373.43	1428.62
$SO_2$2	959.00	988.26	1019.71	1053.52	1089.86	1128.93	1170.93	1216.09	1264.62	1316.80	1372.89	1433.19	1498.01
$SO_2$3	556.67	560.94	565.35	569.89	574.56	579.37	584.33	589.44	594.70	600.12	605.70	611.45	617.37
$SO_2$4	770.03	786.98	805.00	820.20	836.16	852.92	870.52	888.99	908.40	928.77	950.16	972.62	996.20
NOx1	1347.01	1375.88	1406.72	1439.67	1474.89	1512.53	1552.74	1595.72	1641.64	1690.71	1743.15	1799.19	1859.07
NOx2	1349.53	1381.28	1415.40	1452.08	1491.52	1533.91	1579.48	1628.47	1681.13	1737.75	1798.61	1864.03	1934.36
NOx3	900.14	904.24	908.46	912.80	917.28	921.89	926.64	931.53	936.57	941.76	947.10	952.61	958.28
NOx4	1057.97	1072.85	1088.67	1102.02	1116.03	1130.74	1146.19	1162.41	1179.45	1197.33	1216.11	1235.83	1256.53
Land intensity 1	4.182	4.245	4.313	4.386	4.463	4.546	4.634	4.729	4.829	4.937	5.053	5.176	5.307
Land intensity 2	4.262	4.338	4.419	4.506	4.600	4.700	4.808	4.925	5.050	5.184	5.329	5.484	5.651
Land intensity 3	3.927	3.948	3.968	3.990	4.012	4.034	4.057	4.081	4.106	4.132	4.158	4.185	4.213
Land intensity 4	3.988	4.034	4.083	4.135	4.179	4.225	4.273	4.324	4.377	4.433	4.492	4.553	4.618

be 53.59% in 2020 and 77.50% in 2030. In 2030, the GDP will reach 164.663 trillion yuan, and the rapid economic development will meet residents' material needs. However, over-exploitation of land resources, rapid urbanization of the resident population, a large gap between supporting facilities and corresponding services in the urbanization process, and the increasingly serious environmental pollution caused by this are unrelated to people's concern for the ecological environment.

(iii) Eco-friendly model

During the simulation period, China's economic development maintained low-speed growth, and the total economic volume increased slowly. The GDP in 2020 was 70.2746 trillion yuan, which failed to double that of 2010. The urbanization rate of the permanent population in 2020 was 59.95%, and the permanent population in the urbanization rate was 66.99%. The urbanization rate of the registered population was 44.48% in 2020 and 51.66% in 2030. The actual urbanization rate was 51.75% in 2020 and 58.86% in 2030. A gap remains between the goals of the Thirteenth Five-Year Plan for National Economic and Social Development (2016–2020) and National Population Development Plan (2016–2030). In 2030, the GDP will reach 94.443 trillion yuan, and the slow economic growth cannot keep up the people's pursuit of better lives.

(iv) Coordinated development model

During the simulation period, China's economic development maintained moderate growth, and the total economic volume continued to increase. The GDP in 2020 will be 77.2478 trillion yuan, which is at least double that of 2010. The urbanization rate of the permanent population in 2020 will be 61.28%. The urbanization rate of the registered population is 75.43% and will be 45.84% in 2020 and 60.27% in 2030. The actual urbanization rate will be 53.09% in 2020 and 67.39% in 2030. According to the goals of the Thirteenth Five-Year Plan (2016–2020) for Economic and Social Development and National Population Development Plan (2016–2030), in 2030, the GDP will reach 125.829 trillion yuan, and people's growing need for better lives will be continuously met through economic development.

The first three models all have some drawbacks. In the status quo continuation model, resource shortages and environmental problems brought about by economic development continue to worsen, failing to meet the national planning requirements. In the performance-driven model, the extensive economic development mode will lead to excessive resource consumption and more severe environmental pollution, which will break the dynamic balance between the economy, resources, and the environment and cause resource shortages and environmental pollution to restrict the healthy development of China's economy and society. In the eco-friendly model, natural resources and the environment are fully protected; however, economic development is slow, the urbanization process is delayed, and people's living standards cannot be effectively improved. In view of the problems existing in the status quo continuation, performance-driven, and eco-friendly models, the coordinated development model focuses on the dual needs of China's economic and social development and resource and environmental protection and considers the protection of resources and the environment along with economic and social development. Currently, this is a feasible development model, which is in line with the requirements for China's sustainable development.

10.3 Characteristics of Ecological Development Efficiency

10.3.1 Ecological Development Efficiency Measurement

10.3.1.1 Measurement Methods

Ecological efficiency is the corresponding relationship between input and output when the resource and environment conditions on which humans rely meet their needs and create social wealth. Regarding the measurement of ecological efficiency, existing studies have adopted methods such as the economic and environmental impact ratio method, fuzzy comprehensive evaluation method, material flow analysis (MFA), ecological footprint method, and data envelopment analysis (DEA). Of these, the DEA method is the most widely used. The DEA method is a systematic analysis method for effectively evaluating comparable decision-making units of the same type based on multiple input and output indicators and using linear programming methods.

According to the existing literature, several DEA methods are used to measure ecological efficiency, such as the CCR (Charnes, Cooper, and Rhodes) model, BBC (Banker, Charnes, and Cooper) model, directional distance function method, DEA model based on relaxation measure, DEA window analysis, super-efficiency DEA method, and three-stage DEA method. Wang et al. (2023) used the super SBM (Slack-Based Measure) model, the group-frontier Malmquist-Luenberger (GML) productivity index, and meta-frontier Malmquist-Luenberger (MML) productivity index, to evaluate the total factor ecological productivity and its driving force. In this study, the super-efficiency SBM is used to measure the efficiency of ecological development in China from 2009 to 2017, and the Malmquist index model is used for dynamic analysis.

10.3.1.2 Measurement Index and Data Sources

(i) Evaluation index system construction

In the development process, a country or region must invest in both natural and social resources to obtain economic development. However, this will also cause some environmental problems. Therefore, input indicators are divided into natural resource and social resource input indicators, and output indicators are divided into undesired and expected output indicators. We select four natural resource input indicators, two social resource input indicators, five undesired output indicators, and one target of expected output to build an evaluation index system for ecological development efficiency (Table 10.10).

(ii) Description of research samples and data sources

Owing to the availability of index data, 30 provinces, autonomous regions, and municipalities directly under the central government in mainland China (excluding Tibet, Hong Kong, Macao, and Taiwan) were selected as the research objects, and the static and dynamic methods were used to measure the level of interprovincial ecological efficiency in China. The research period spans nine years, from 2009 to 2017. The data on the urban construction land area in Shanghai from 2009 to 2011 are missing, and the trend extrapolation method is used to obtain substitute values. The unutilized amount of industrial solid waste (including general industrial solid waste and hazardous waste) is obtained by subtracting the amount of industrial solid waste utilized from the amount of industrial solid waste produced to better reflect the true level of the index data. The investment amount of fixed assets and the regional GDP index are processed with 2009 as the base period, and the influence of price factors is removed.

Table 10.10 Ecological development efficiency measurement index system

Indicator type	Indicator category		Indicator name	Index unit
Input indicators	Resource class	Natural resource input	Urban construction land area	Square kilometer
			Cultivated area	Square kilometer
			Energy consumption	10,000 tons of standard coal
			Water consumption	100 million cubic meters
		Social resource investment	Number of employed persons	Million people
			Investment in fixed assets	100 million yuan
Output indicators	Environment	Undesired output	Wastewater discharge	Tons
			SO_2 emissions	Tons
			NO_x emissions	Tons
			smoke and dust emissions	Tons
			Unutilized amount of industrial solid waste	Tons
	Economics	Expected output	GDP	100 million yuan

10.3.1.3 Measuring Results of Ecological Development Efficiency Level

Taking 30 provinces, autonomous regions, and municipalities directly under the central government in mainland China (excluding Tibet, Hong Kong, Macao, and Taiwan) as the research object, according to the input–output index system constructed above, super-efficiency from 2009 to 2017 was calculated using MaxDEA 8 software. Simultaneously, to better analyze the change trend in China's ecological development efficiency, according to the input-oriented Malmquist index model, DEAP 2.1 software was used to obtain the ecological efficiency value (ML index) and its decomposition value for each region: technological progress and technical efficiency variety.

(i) Super-efficiency performance of regional ecological development efficiency

Timing characteristics

China's overall ecological development efficiency level generally rises first and then declines. From the perspective of the overall levels of China's eastern, central, and western regions, the production efficiency values generally increased first and then declined. Regarding ecological efficiency values, that of the eastern region reached a maximum of 2.99 in 2011, that of the central region reached a maximum of 0.43 in 2011, and that of the western region reached a maximum of 0.63 in 2010.

Spatial features

Table 10.11 shows that five regions, namely Hainan, Tianjin, Beijing, Shanghai, and Qinghai, had ecological efficiency values greater than or equal to 1 from 2009 to 2017. The number of regions accounts for one-sixth of the total number of regions. From 2009 to 2017, Hainan and Tianjin had the highest ecological efficiency values. Specifically, Hainan had the highest ecological efficiency values in 2009, 2010, 2013, 2015, and 2016, and Tianjin had the highest ecological efficiency values in the other years. The lowest ecological efficiency value from 2009 to 2017 was in Xinjiang. From 2009 to 2017, the differences between the highest value and the lowest value of regional ecological efficiency were 7.25, 10.26, 16.6, 11.03, 8.03, 7.81, 5.02, 4.34, and 4.64, respectively.

(ii) Regional eco-efficiency index (ML Index) and its decomposition

The comprehensive growth index of the ecological development efficiency of the eastern, central, and western regions during the study period was greater than 1, which shows that the overall ecological development efficiency is increasing. The ecological development efficiency growth rate is the fastest in the central region, followed by the eastern region, and the slowest in the western region (Table 10.12).

Table 10.13 shows that, since 2010, the comprehensive growth index of China's ecological development efficiency has been greater than 1 and has risen in a fluctuating pattern, which shows that China's ecological development efficiency is generally on the rise, with an increasing growth rate. It also shows that, under the severe

Table 10.11 Super-efficiency values of regional ecological development efficiency (2009–2017)

Year	2009	2010	2011	2012	2013	2014	2015	2016	2017
Beijing	1.48	1.5	1.56	1.55	1.56	1.55	1.56	1.5	1.72
Tianjin	2.06	2.19	16.91	11.33	6.74	8.08	4.22	3.63	4.9
Hebei	0.56	0.53	0.52	0.5	0.49	0.48	0.47	0.46	0.46
Liaoning	0.44	0.44	0.45	0.45	0.45	0.45	0.47	0.59	1
Shanghai	1.18	1.22	1.2	1.22	1.19	1.17	1.16	1.16	1.16
Jiangsu	0.76	0.7	0.67	0.65	0.64	0.64	0.62	0.62	0.62
Zhejiang	0.91	1.01	0.86	0.78	0.67	0.67	0.67	0.67	0.66
Fujian	1.04	1.02	1.02	1.01	0.66	0.65	0.64	1	1
Shandong	0.62	0.61	0.59	0.57	0.6	0.58	0.55	0.55	0.56
Guangdong	1.02	1.02	1.04	1.05	1.06	1.03	0.67	0.65	0.64
Hainan	7.6	10.6	8.06	8.7	8.31	6.47	5.28	4.61	3.64
Eastern mean	1.61	1.89	2.99	2.53	2.03	1.98	1.48	1.40	1.49
Shanxi	0.45	0.45	0.45	0.43	0.42	0.4	0.38	0.37	0.46
Jilin	0.4	0.41	0.42	0.43	0.44	0.44	0.42	0.43	0.43
Heilongjiang	0.36	0.35	0.36	0.35	0.35	0.37	0.36	0.37	0.36
Anhui	0.36	0.36	0.37	0.37	0.36	0.36	0.35	0.35	0.35
Jiangxi	0.42	0.42	0.42	0.42	0.41	0.42	0.39	0.39	0.38
Henan	0.53	0.51	0.51	0.5	0.49	0.49	0.47	0.47	0.47
Hubei	0.42	0.4	0.4	0.39	0.41	0.4	0.41	0.41	0.4
Hunan	0.44	0.47	0.48	0.49	0.5	0.51	0.51	0.53	0.51
Central mean	0.42	0.42	0.43	0.42	0.42	0.42	0.41	0.42	0.42
Inner Mongolia	0.53	0.48	0.49	0.49	0.49	0.47	0.53	0.54	0.53
Guangxi	0.42	0.41	0.42	0.4	0.4	0.39	0.37	0.36	0.35
Chongqing	0.44	0.44	0.45	0.49	0.5	0.49	0.48	0.48	0.48
Sichuan	0.43	0.43	0.44	0.43	0.43	0.43	0.4	0.39	0.38
Guizhou	0.39	0.4	0.38	0.36	0.36	0.35	0.34	0.33	0.3
Yunnan	0.38	0.37	0.36	0.37	0.39	0.37	0.35	0.35	0.34
Shaanxi	0.67	1	1.01	0.54	0.51	0.51	0.49	0.48	0.45
Gansu	0.37	0.36	0.34	0.34	0.34	0.32	0.32	0.33	0.36
Qinghai	2.21	2.25	2.26	2.32	2.22	2.28	2.36	2.15	2.05
Ningxia	0.47	0.48	0.49	0.5	0.49	0.51	0.56	0.59	0.62
Xinjiang	0.35	0.34	0.31	0.3	0.28	0.27	0.26	0.27	0.26
Western mean	0.61	0.63	0.63	0.59	0.58	0.58	0.59	0.57	0.56
National average	0.92	1.04	1.44	1.26	1.07	1.05	0.87	0.83	0.86

Table 10.12 Annual average Malmquist index and its decomposition value by region (2009–2017)

Regional index	Technical efficiency	Skill improvement	Pure technical efficiency	Scale efficiency	Total factor productivity
Beijing	1.000	1.096	1.000	1.000	1.096
Tianjin	1.000	0.957	1.000	1.000	0.957
Hebei	0.998	1.044	0.996	1.002	1.042
Liaoning	1.051	1.053	1.038	1.012	1.106
Shanghai	1.000	1.064	1.000	1.000	1.064
Jiangsu	0.997	1.060	1.000	0.997	1.056
Zhejiang	0.991	1.057	1.000	0.991	1.047
Fujian	1.000	1.038	1.000	1.000	1.038
Shandong	0.985	1.043	1.000	0.985	1.027
Guangdong	0.986	1.039	1.000	0.986	1.025
Hainan	1.000	0.889	1.000	1.000	0.889
Eastern mean	1.001	1.029	1.003	0.998	1.030
Shanxi	1.003	1.034	1.012	0.992	1.038
Jilin	1.010	1.063	1.009	1.000	1.073
Heilongjiang	0.998	1.048	0.995	1.003	1.046
Anhui	1.001	1.058	1.000	1.000	1.059
Jiangxi	0.987	1.056	0.986	1.001	1.042
Henan	1.004	1.039	1.003	1.001	1.043
Hubei	0.994	1.054	0.995	0.999	1.048
Hunan	1.026	1.023	1.027	0.999	1.049
Central mean	1.003	1.047	1.003	0.999	1.050
Inner Mongolia	1.002	1.040	1.001	1.002	1.042
Guangxi	0.988	1.048	0.987	1.001	1.036
Chongqing	1.013	1.053	1.010	1.003	1.067
Sichuan	0.987	1.039	0.988	0.999	1.025
Guizhou	0.967	1.015	0.966	1.001	0.981
Yunnan	1.007	1.018	1.007	1.001	1.025
Shaanxi	0.979	1.047	0.979	1.000	1.025
Gansu	0.987	1.027	1.014	0.973	1.014
Qinghai	1.000	0.947	1.000	1.000	0.947
Ningxia	1.041	0.975	1.004	1.037	1.015
Xinjiang	0.964	1.031	0.961	1.004	0.995
Western	0.994	1.021	0.992	1.002	1.015
Average	0.999	1.031	0.999	1.000	1.030

Table 10.13 Annual Malmquist index and its decomposition of ecological development efficiency in China (2009–2017)

Area index	Technical efficiency	Technical progress	Pure technical efficiency	Scale efficiency	Total factor productivity
2009–2010	1.000	1.011	1.000	1.000	1.011
2010–2011	0.987	1.051	0.995	0.991	1.037
2011–2012	0.984	1.045	0.988	0.996	1.029
2012–2013	0.977	1.045	0.986	0.991	1.021
2013–2014	1.000	1.003	0.998	1.002	1.003
2014–2015	1.018	1.003	1.010	1.008	1.021
2015–2016	1.020	1.054	1.007	1.012	1.075
2016–2017	1.005	1.037	1.008	0.996	1.042
Average	0.999	1.031	0.999	1.000	1.030

environmental pollution situation, China's environmental protection and governance have achieved results, providing a breakthrough for sustainable development.

Since 2009, China's ecological development efficiency has shown a sinusoidal growth trend. This is similar to the results obtained by Wang and Kong (2021) in their study on the change trend in the environmental efficiency of construction land in various regions of China. Judging from the factors affecting the growth of China's ecological development efficiency, technological progress is the main reason for its growth, whereas technical efficiency, pure technical efficiency, and scale efficiency are the main factors restricting its growth in some years.

10.3.2 Convergence Analysis of Ecological Development Efficiency

The analysis of the ecological development efficiency in each region in the previous sections indicates that large ecological development efficiency gaps exist between the eastern, central, and western regions overall and between provinces. To further analyze the differences in the efficiency of ecological development in China's eastern, central, and western regions, as well as within China, a convergent analysis of the efficiency of ecological development is conducted.

According to the convergence hypothesis of neoclassical growth theory, regional ecological efficiency convergence means that areas with lower ecological efficiency have higher growth rates than areas with higher ecological efficiency. The empirical research methods mainly include σ convergence and β convergence. σ convergence reflects the change trend in regional ecological efficiency differences, and the standard deviation and coefficient of variation are often used to test the σ convergence hypothesis. Let e_i be the ecological efficiency value of the ith province and city, σ be the standard deviation of regional ecological efficiency, C_v represent the coefficient

Table 10.14 Eco-efficiency standard deviation from 2009 to 2017

Year	National	Eastern	Central	Western
2009	1.3494	2.0401	0.0547	0.5397
2010	1.8754	2.93	0.0541	0.5664
2011	3.2469	5.1052	0.0518	0.5723
2012	2.4431	3.7626	0.0531	0.5773
2013	1.8125	2.7579	0.0539	0.5481
2014	1.7544	2.6636	0.0537	0.5691
2015	1.1444	1.6649	0.0549	0.5955
2016	0.984	1.3954	0.0602	0.5333
2017	1.025	1.4487	0.0571	0.5065
Average	1.7372	2.6409	0.0548	0.5565

of variation of ecological efficiency, and n be the number of regions. The formulas for calculating the standard deviation and coefficient of variation are as follows:

$$\sigma = \sqrt{\frac{1}{n}\sum_{i=1}^{n}(e_i - \overline{e}_i)^2} \tag{10.15}$$

$$C_v = \frac{\sigma}{\overline{e_i}} \tag{10.16}$$

If the standard deviation and coefficient of variation of the regional ecological efficiency gradually decrease, then, σ convergence is determined to be present.

10.3.2.1 Standard Deviation of Ecological Development Efficiency

The change trends in the standard deviation are shown in Table 10.14. The change trends in the standard deviation of the ecological development efficiency nationwide and the eastern region are similar—relatively large and generally increasing first and then declining—and the difference in the standard deviation between the two has narrowed. The standard deviation of the efficiency of ecological development in western China increased first and then declined. We observe σ convergence in eco-efficiency nationwide and in the eastern region, but not in the central and western regions. The coefficient of variation is then used for further analysis.

10.3.2.2 Coefficient of Variation of Eco-efficiency

The change trend of the variation coefficient in Table 10.15 indicates that the variation coefficient of national ecological development efficiency is larger than that of the

Table 10.15 Coefficient of variation of ecological efficiency from 2009 to 2017

Year	National	Eastern	Central	Western
2009	1.4609	1.27	0.1295	0.8915
2010	1.805	1.5466	0.1284	0.8952
2011	2.2527	1.708	0.1215	0.9058
2012	1.9426	1.4883	0.1257	0.971
2013	1.6907	1.3562	0.1276	0.9406
2014	1.6682	1.3458	0.1267	0.9798
2015	1.3174	1.1229	0.1335	1.014
2016	1.1794	0.9941	0.1452	0.9356
2017	1.19	0.974	0.1359	0.9104
Average	1.6119	1.3118	0.1304	0.9382

eastern, western, and central regions, and the variation coefficient of the central region is relatively small and stable, generally increasing first and then declining. The coefficient of variation of ecological development efficiency in the central and western regions increased slightly in fluctuations. We observe σ convergence in eco-efficiency nationwide and in the eastern region, but not in the central and western regions.

10.4 Convergence of Regional Economic Development

The analysis in the previous section shows σ convergence in eco-efficiency at the national level and in the eastern region, but not in the central and western regions. The status of regional resource and environmental carrying capacity and ecological efficiency will affect regional economic development levels. This section further analyzes the convergence of economic development in 30 provinces, autonomous regions, and municipalities directly under the central government (excluding Tibet, Hong Kong, Macao, and Taiwan) in mainland China. The economic development level is characterized by per capita GDP. The historical data from 2009 to 2017 are selected, and the influence of price factors is eliminated with 2009 as the base period (Table 10.16).

Table 10.16 Economic development level from 2009 to 2017 (Note: GDP per capita: yuan/person; Xizang, Hong Kong, Macao and Taiwan are not included; The price deflator is based on 2009)

Year	2009	2010	2011	2012	2013	2014	2015	2016	2017
Beijing	65,339	68,322	71,771	75,429	79,471	83,806	88,804	94,756	101,198
Tianjin	61,253	67,980	75,859	82,784	89,399	95,422	102,273	110,509	114,855
Hebei	24,503	26,881	29,724	32,368	34,807	36,814	39,100	41,507	43,952
Shanxi	21,472	23,450	26,359	28,876	31,282	32,652	33,518	34,855	37,128
Inner Mongolia	39,627	45,313	51,584	57,331	62,291	66,962	71,945	76,850	79,640
Liaoning	35,044	39,709	44,472	48,630	52,849	55,902	57,697	56,306	58,792
Jilin	26,565	30,154	34,290	38,391	41,562	44,248	47,018	50,630	53,627
Heilongjiang	22,444	25,248	28,346	31,181	33,666	35,570	37,805	40,248	42,937
Shanghai	68,083	72,064	76,511	81,109	86,088	91,697	98,470	105,047	112,388
Jiangsu	44,119	49,350	54,570	59,923	65,518	71,030	76,913	82,674	88,295
Zhejiang	43,575	47,230	51,330	55,295	59,600	64,013	68,748	73,298	78,079
Anhui	16,413	19,359	21,932	24,503	26,863	29,079	31,295	33,732	36,254
Fujian	33,378	37,740	42,074	46,521	51,282	55,885	60,392	64,873	69,464
Jiangxi	17,273	19,558	21,876	24,196	26,533	28,979	31,450	34,086	36,845
Shandong	35,794	39,702	43,805	47,860	52,196	56,412	60,566	64,514	68,880
Henan	20,534	23,302	26,122	28,705	31,266	33,965	36,614	39,363	42,314
Hubei	22,659	25,977	29,407	32,611	35,781	39,137	42,358	45,533	48,943
Hunan	20,387	22,780	25,594	28,302	30,918	33,625	36,235	38,910	41,790
Guangdong	38,976	42,504	46,470	49,858	53,841	57,608	61,500	65,211	69,035
Guangxi	15,979	19,221	21,423	23,655	25,864	27,855	29,848	31,749	33,676
Hainan	19,146	22,082	24,506	26,434	28,792	30,962	33,084	35,333	37,439
Chongqing	22,840	26,505	30,492	34,333	38,232	42,101	46,330	50,766	55,000
Sichuan	17,289	20,246	23,269	26,116	28,618	30,925	33,107	35,439	38,125
Guizhou	11,062	12,686	14,631	16,549	18,522	20,488	22,538	24,730	27,062
Yunnan	13,498	15,056	17,011	19,107	21,291	22,884	24,728	26,716	29,071
Shaanxi	21,921	25,067	28,490	32,080	35,505	38,835	41,704	44,639	47,933
Gansu	13,259	14,794	16,617	18,610	20,587	22,342	24,068	25,798	26,564
Qinghai	19,412	22,144	24,912	27,732	30,461	32,979	35,379	37,888	40,314
Ningxia	21,653	24,266	26,946	29,674	32,233	34,391	36,808	39,377	42,013
Xinjiang	19,810	21,650	23,984	26,573	29,093	31,528	33,402	35,371	37,327

Table 10.17 Standard deviation of economic development level from 2009 to 2017

Year	National	Eastern	Central	Western
2009	15,284	16,115	3197	7681
2010	16,131	16,630	3511	8797
2011	17,186	17,499	4059	10,048
2012	18,200	18,475	4583	11,142
2013	19,299	19,457	4878	12,031
2014	20,476	20,627	5096	12,928
2015	21,899	22,208	5424	13,909
2016	23,396	24,252	5918	14,871
2017	24,611	25,637	6163	15,349
Average	19,609	20,100	4759	11,862

10.4.1 σ Convergence Analysis

10.4.1.1 Standard Deviation of Economic Development

The change trend in standard deviation in Table 10.17 shows that the standard deviation of the economic development level of the eastern region from 2009 to 2017 was larger than that of China overall, the western region, and the central region, whereas the standard deviation of the central region was much smaller than that at the national level and in the eastern and western regions. The national, eastern, central, and western economic development levels have approximately the same standard deviation change trend, which is expanding annually, and the national and eastern changes are closer. From the perspective of standard deviation, σ convergence cannot be proven at the national level or in the eastern, central, and western regions. The coefficient of variation is then used for further analysis.

10.4.1.2 Coefficient of Variation

The change trend in the variation coefficient in Table 10.18 shows that the variation coefficient of national economic development is larger than that of the western, eastern, and central regions, whereas the variation coefficient of the central region is relatively small and stable. Similarly, the change trend of the coefficient of variation for economic development level at the national level and in the eastern, central, and western regions is approximately the same and generally declines in fluctuations. We can observe σ convergence in the economic development levels at the national level and in the eastern, central, and western regions. Regional economic development is restricted by the resource and the environmental carrying capacity. Although the resource and environmental carrying capacity and efficiency of ecological development differ across regions, the growth potential is relatively large in less

Table 10.18 Coefficient of variation of economic development level from 2009 to 2017

Year	National	Eastern	Central	Western
2009	0.5373	0.3778	0.1525	0.3905
2010	0.5092	0.3562	0.1480	0.3919
2011	0.4890	0.3431	0.1518	0.3956
2012	0.4728	0.3352	0.1549	0.3931
2013	0.4616	0.3273	0.1513	0.3862
2014	0.4557	0.3243	0.1471	0.3830
2015	0.4551	0.3268	0.1464	0.3826
2016	0.4556	0.3360	0.1492	0.3810
2017	0.4505	0.3348	0.1451	0.3697
Average	0.4763	0.3402	0.1496	0.3860

economically developed areas. Therefore, in the long run, inter-regional economies the level of development is convergent.

10.4.2 Absolute *β* Convergence Analysis

Absolute β-convergence of economic development means that the growth rate of economic development in areas with low economic development levels is faster than that in areas with higher economic development levels, and finally reaches a certain stable state, which requires different areas to have the same "basic characteristics" (i.e., have the same external factors). Absolute β convergence can be expressed using Eq. 10.17:

$$\ln(I_{i,t+T}/I_{i,t})/\mathrm{T} = \mathrm{a} + \mathrm{b}\ln(I_{i,t}) + \mu_{it} \tag{10.17}$$

where $I_{i,t}$ and $I_{i,t+T}$ represent the value of the economic development level of the ith region in period t and period $t+T$, respectively; $\ln(I_{i,t+T}/I_{i,t})/\mathrm{T}$ represents the average annual growth rate of the economic development level in period t and period $t+T$; a is a constant term; b is the coefficient of the economic development level in the base period; and μ_{it} is a random error term. If b is significantly negative, then, the growth rates of the economic development levels in each region have an inverse relationship with the initial values, and the regions with lower economic development levels have a tendency to "catch up" with the regions with higher economic development levels; thus, there is absolute β convergence. Conversely, if b is significantly positive, then, there is no absolute β convergence. To effectively use the sample data and make the time series of econometric regression continuous, set $\mathrm{T} = 1$. Therefore, the absolute β convergence of China's economic development level growth can be expressed as Eq. 10.18:

$$\ln(I_{i,t+T}/I_{i,t}) = a + b\ln(I_{i,t}) + \mu_{it} \quad (10.18)$$

According to Eq. 10.17, the test results of absolute β convergence regression of economic development are obtained, as shown in Table 10.19. When conducting the analysis, the Hausman test is first applied to determine whether to choose a fixed- or random-effects model. Table 10.19 shows that the national overall and the eastern, central, and western regions have passed the Hausman test at the 1% significance level; thus, a fixed-effect model is selected for all regions and the national level.

From a national perspective, the estimated value of the coefficient of the economic development level at the beginning of the period is – 0.0914, which is significant at the 1% significance level, indicating that the economic development level at the beginning of the period is inversely proportional to its growth rate. There is an absolute β convergence in economic development at the national level. The economic development levels of the 30 provinces, autonomous regions, and municipalities directly under the central government (excluding Tibet, Hong Kong, Macao, and Taiwan) tend to remain in a steady state. The estimated coefficients of the initial economic development levels of the eastern, central, and western regions are – 0.0761, – 0.0944, indicating that absolute β convergence exists among all 11 eastern provinces and the 8 central provinces, as well as among all 11 provinces in the west.

Table 10.19 Absolute β convergence regression results of economic development at the national level and in the eastern, central, and western regions

Regions	Model	a	b	Hausman test	P-value	Conclusion
National level	FE	1.0439*** (45.31)	– 0.0914*** (– 41.61)	167.90	0.0000	Fixed effect
	RE	0.7463*** (23.62)	– 0.0631*** (– 20.96)			
Eastern	FE	0.9031*** (35.71)	– 0.0761*** (– 32.76)	28.93	0.0000	Fixed effect
	RE	0.6103*** (8.73)	– 0.0492*** (– 7.67)			
Central	FE	1.0603*** (23.76)	– 0.0944*** (– 21.78)	10.13	0.0015	Fixed effect
	RE	1.0917*** (16.56)	– 0.0974*** (– 15.24)			
Western	FE	1.1499*** (35.99)	– 0.1034*** (– 33.05)	75.62	0.0000	Fixed effect
	RE	0.9320*** (19.27)	– 0.0821*** (– 17.36)			

Data source The data in the table are calculated and sorted using EViews software
***Indicates significance at the 1% level, and t-statistics are in parentheses

10.5 Ecological Civilization Construction and Sustainable Development

Land space is a precious resource and the home we rely on for survival and development. To strengthen environmental protection and pollution control, improve the level of ecological civilization construction, and promote sustainable development, China has planned the development of land space and divided the main functional areas.

10.5.1 Land Space Strategic Pattern in China

"National Main Functional Area Planning" helped form the main target of the main functional area in 2020. The planning task is for long-term development, and it will be implemented in due course according to the actual situation revised. It determines the main functional zone strategy of China's land space and sets the direction and tone for the scientific development of land space.

10.5.1.1 Constructing an Urbanization Strategic Pattern with "Two Horizontal and Three Vertical" as the Main Body

The construction takes the land bridge passage and the passage along the Yangtze River as the two horizontal axes and the coastal, Jingha Jingguang, and Baokun passages as the three vertical axes, with the urbanized areas optimized and developed by China as the main support, and other cities on the axis. The strategic pattern of urbanization in which urbanization areas are an important component.

10.5.1.2 Constructing an Agricultural Strategic Pattern with "Seven Districts and Twenty-Three Belts" as the Main Body

An agricultural strategic pattern is constructed with the Northeast Plain, North China Plain, Yangtze River Basin, Fenwei Plain, Hetao irrigation area, South China, and Gansu Xinjiang as the main body, basic farmland as the basis, and other agricultural areas as important components.

10.5.1.3 Constructing an Ecological Security Strategic Pattern with "Two Screens and Three Belts" as the Main Body

The Qinghai-Tibet Plateau ecological barrier, Loess Plateau-Sichuan-Yunnan ecological barrier, northeast forest belt, northern sand control belt, southern hilly and mountainous areas, and important systems of major rivers are used as the framework, and other key national ecological function areas are used as important supports. The distribution of national prohibited development areas is an important component of the strategic pattern of ecological security.

Facing profound changes in the global situation and China's national conditions, the State Council issued the "National Land Planning Outline (2016–2030)" (hereinafter referred to as the "Outline"). The base period of the "Outline" is 2015, the medium-term target year is 2020, and the long-term target year is 2030. Solid promotion is necessary for the overall strategy of regional development, strategy of main functional areas, and three major strategies (i.e., "Belt and Road" construction, coordinated development of Beijing-Tianjin-Hebei, and development strategy of the Yangtze River Economic Belt), to further improve the pattern of land and space development.

The "Outline" not only plans the main goals for 2020 but also determines the main goals for 2030 accordingly, while simultaneously clarifying the attributes of each indicator (anticipatory or binding). As shown in Table 10.20, land development intensity, arable land retention, total water consumption, proportion of excellent water quality in the seven key river basins in China, and water quality compliance rate of important rivers and lakes water function areas are the binding indicators, and other indicators are expected indicators.

Table 10.20 Index value of the main goal

Serial number	Indicator name	2020	2030	Attributes
1	Urban space (10,000 km^2)	10.21	11.67	Predictable
2	Land development intensity (%)	4.24	4.62	Restrictive
3	Amount of cultivated land (100 million mu)	18.65	18.25	Restrictive
4	Forest cover rate (%)	> 23	> 24	Predictable
5	Total water consumption (billion cubic meters)	6700	7000	Restrictive
6	Grassland comprehensive vegetation coverage (%)	56	60	Predictable
7	Wetland area (100 million mu)	8	8.3	Predictable
8	Road and railway network density (km/km^2)	≥ 0.5	≥ 0.6	Predictable
9	Proportion of excellent water quality in the seven key river basins in China (%)	> 70	> 75	Restrictive
10	Water quality compliance rate of important rivers and lakes water function zones (%)	> 80	> 95	Restrictive
11	Newly controlled soil erosion area (10,000 kn^2)	32	94	Predictable

10.5.2 Types of National Main Functional Areas

The "National Main Functional Zone Planning" defines "development" and proposes six development concepts: development according to the suitability of natural conditions, distinction of main functions, development according to the resource and environmental carrying capacity, control of development intensity, adjustment of spatial structure, and provision of ecological products. Table 10.21 shows the classification criteria and specific categories of the main functional areas.

Restricted development areas can be divided into two categories. The first is the main production area of agricultural products, which is characterized by more cultivated land and better conditions for agricultural development. For national sustainable development, the primary task is to enhance the comprehensive agricultural production capacity. The second is restricting the development of key ecological function areas, which are characterized by important ecological functions or fragile ecosystems and poor resource and environmental carrying capacity and do not meet the needs of the general public. To meet the requirements of large-scale and high-intensity industrialization and urbanization development, the primary task is to enhance the production capacity of ecological products.

10.5.3 National Ecological Function Zoning in China

10.5.3.1 National Ecological Function Zoning Plan

(i) Partition method

According to the natural attributes of the ecosystem and types of dominant service functions, ecosystem service functions can be divided into three categories: ecological regulation, product provision, and human settlement security.

Based on these three major categories, ecological functions are divided into nine types according to the importance of ecosystem service functions. Ecological adjustment functions include five types: water conservation, biodiversity protection, soil conservation, windbreak and sand fixation, and flood regulation and storage. Product supply functions include two types: agricultural products and forest products. Human settlement security functions include population and economically dense metropolises, of which there are two types of clusters and key town clusters.

Ecological function areas are divided according to the types of ecological functions and their spatial distribution characteristics, as well as the spatial differentiation characteristics of ecosystem types, terrain differences, and combinations of land use.

(ii) Zoning plan

Compared with the 2008 edition of *National Ecological Function Zoning*, the division standards of ecological function categories and types remain unchanged; however,

Table 10.21 Classification of main functional areas

Serial number	Classification criteria	Main functional area category	Meaning	Classification basis
1	Development content	Urbanization	Areas whose main function is to supply industrial and service products, but also supply agricultural and ecological products	Based on the type of main product provided
		Area	The main function is to supply agricultural products, but also supply ecological, service, and industrial products	
		Agricultural products	The main function is to supply ecological products, but also supply certain agricultural, service, and industrial products	
2	Development method	Main producing area	Urbanized areas with a relatively developed economy, relatively dense population, high development intensity, and more prominent resource and environmental problems should be optimized for industrialization and urbanization	Based on the resource and environmental carrying capacity, current development intensity, and future development potential of different regions, it is divided based on whether it is suitable for large-scale and high-intensity industrialization and urbanization
		Key ecological functional areas	Urbanized areas that have a certain economic foundation, strong resource and environmental carrying capacity, high development potential, and good conditions for agglomerating population and economy, and focus on industrialization and urbanization	
		Optimizing the development area	The main production areas of agricultural products and the key ecological function areas that also limit the development of large-scale and high-intensity industrialization and urbanization	

(continued)

Table 10.21 (continued)

Serial number	Classification criteria	Main functional area category	Meaning	Classification basis
		Key development areas	Protected areas of natural and cultural resources at all levels and types established in accordance with the law, as well as other key ecological function areas that prohibit industrialization and urbanization and require special protection	
3	Level	Restricted development area	Taking the development-prohibited areas as an example, it includes national natural reserves, world cultural and natural heritage sites, national scenic spots, national forest parks, and national geological parks	It is divided based on whether the functional area is at the national or provincial level
		Prohibited development area	Taking the prohibited development areas as an example, it includes various natural and cultural resource protection areas at the provincial level and below, important water sources, and other provincial governments that need to determine prohibited development areas	

the number of ecological function areas has increased by 26 . Table 10.22 presents the National ecological function zoning system.

10.5.3.2 Types and Overview of Ecological Functional Areas

National Ecological Function Zoning (Revised Edition) (No. 61, 2015) classifies the national ecological function areas according to the dominant ecosystem service functions and analyzes the spatial distribution characteristics and problems and protection directions of various ecological function areas to form national terrestrial ecological function zones (Table 10.23).

Table 10.22 National ecological function zoning system

Ecological Function category (3 categories)	Ecological Function type (9 categories)	Examples of ecological functional areas
Ecological regulation	Water conservation	Micang Mountain - Daba Mountain Water Conservation Functional Area
	Biodiversity Conservation	Xiaoxing'anling Biodiversity Conservation Functional Zone
	Soil conservation	Soil Conservation Functional Zone of Loess Hills and Gullies in Northern Shaanxi
	Windbreak and sand fixation	Horqin Sandy Land Windbreak and Sand Fixation Functional Area
Product offer	Flood storage	Wanjiang Wetland Flood Regulation and Storage Functional Area
	Agricultural products provided	Sanjiang Plain Agricultural Products Supply Functional Area
Habitat security	Forest product offering	Small Khingan Mountains Mountain Forest Product Supply Functional Area
	Metropolitan agglomeration	Yangtze River Delta Metropolitan Agglomeration Functional Area

10.6 Transformation and Policy in Underdeveloped Areas

The "Two Mountains Theory" proposed by General Secretary Xi Jinping is the guiding ideology for the economic and social development of various regions in China in the new era. Underdeveloped regions, especially resource-based cities, should thoroughly implement this guiding ideology, promote the transformation and upgrading of industrial structure, strengthen cities through the comprehensive implementation of ecological industries, eliminate the development path in the past, and make regional development glow with strong vitality through green transformation.

10.6.1 Implementing Development Concepts

The five development concepts of innovation, coordination, green, openness, and sharing are the basic development concepts for realizing the two centenary goals and the great rejuvenation of the Chinese nation, as well as the basic policy for regional

Table 10.23 Types of national terrestrial ecological functional zones

Dominant ecosystem service function		Ecological function area (a)	Area (10,000 square kilometers)	Area ratio (%)
Ecological adjustment	Water conservation	47	256.85	26.86
	Biodiversity Conservation	43	220.84	23.09
	Soil conservation	20	61.40	6.42
	Windbreak and sand fixation	30	198.95	20.80
	Flood storage	8	4.89	0.51
Product	Agricultural products provided	58	180.57	18.88
	Forest product offering	5	10.90	1.14
Supply	Metropolitan agglomeration	3	10.84	1.13
	Key urban agglomerations	28	11.04	1.15
Total		242	956.29	100.00

Note This division does not include Hong Kong Special Administrative Region, Macao Special Administrative Region, or Taiwan Province, whose combined total area is 37,100 square kilometers

transformation and development. In transforming and developing underdeveloped areas, the five development concepts should be conscientiously implemented.

First, adhere to innovative development by adhering to concept innovation, technological innovation, and idea innovation to find innovation points, explore new directions, and implement the innovation-driven development strategy. Second, insist on coordinated development by coordinating the development of urban and rural areas, giving full play to the advantages of traditional pillar industries, and vigorously developing other industries to make emerging industries bigger and stronger. Third, adhere to green development by effectively combining regional transformation and development with resource conservation and environmental protection and providing more and higher-quality ecological products while developing the economy. Fourth, adhere to open development by improving the level of China's interprovincial ecological efficiency and opening up to the outside world and developing a high-level export-oriented economy. Fifth, adhere to shared development to share benefits with everyone across the nation from the achievements of economic, political, cultural, social, and ecological civilization.

10.6.2 Promoting the Adjustment of Industrial Structure

The adjustment of industrial structure in underdeveloped areas should promote industrial diversification horizontally while also promoting the extension of industrial chains vertically. In the development process, we should choose emerging industries suitable for our own development according to our own development reality (Qian & Zhou, 2023; Hou, 2023), which is key to the green transformation of economic development.

10.6.3 Promoting Intensive Use of Resources, Regional Energy Conservation, and Emission Reduction

Green consumption is a way to approach consumption needs while meeting the development requirements of the new era. It considers the health of consumers, saves resources, and protects the environment. It advocates the consumption of green products that are healthy, environmentally friendly, and low in resource consumption. It strives to avoid wasting resources in the consumption process, and its utilization is more efficient. Residents are important consumers of products and services, and the greening of their consumption preferences will promote further changes in the direction of enterprise production and operation, thereby promoting the transformation and upgrading of industrial structures. In the green transformation of underdeveloped areas, in addition to promoting green production methods, green lifestyles should also be vigorously promoted to encourage all residents to save and intensively use natural resources, thereby reducing pollution.

10.6.4 Strengthening Environmental Protection Policies

To further promote the construction of ecological civilization and the smooth transformation of green development in various regions nationwide, China has introduced several policies and measures to protect the environment and control pollution. In the development process, all regions should conscientiously apply the concept of green development, implement various control policies, and protect the ecological environment. The effectiveness of environmental protection policies largely depends to on whether regional environmental protection policies are implemented correctly and remain in place. However, in the regional development process, some environmental protection policies will inevitably not be implemented.

10.7 Summary

This chapter mainly explores the geographical distribution differences in the ecological environment, specifically from the perspectives of the role of environmental protection and ecological civilization construction in geographical development, measurement and representation of environmental capacity and resource and environmental carrying capacity, measurement and representation of ecological development efficiency, and construction of ecological civilization and its feasibility. The spatial layout of China's sustainable development strategy and the transformation path and policy choices for underdeveloped areas are analyzed in five aspects.

This chapter mainly discusses the role of environmental protection and ecological civilization construction in geographical development from the perspectives of improving people's well-being, consolidating the foundation of survival, enhancing the image of the region, and creating economic benefits. Saving resources and protecting the environment are China's basic national policies. We should implement the five development concepts; adhere to the path of green, low-carbon, circular, and sustainable development; and protect the ecological environment while developing the economy. This will not only provide high-quality ecological products for the present but also leave a legacy of more ecological assets for future generations.

The measurement and characterization of environmental capacity and resource and environmental carrying capacity starts with the concepts of environmental capacity, environmental carrying capacity, and resource and environmental carrying capacity; then, the evaluation index system of China's resource and environmental carrying capacity is constructed from the aspects of economy, society, resources, and environment. The entropy weight TOPSIS method is used to evaluate the resource and environmental carrying capacity level of 30 provinces, autonomous regions, and municipalities directly under the central government (excluding Tibet, Hong Kong, Macao, and Taiwan).

The measurement and characterization of the efficiency of ecological development is mainly conducted from two aspects: measurement of the efficiency of ecological development and analysis of the convergence of the efficiency of ecological development. First, the super-efficiency SBM is used to measure the efficiency of ecological development in China from 2009 to 2017, and the Malmquist index model is used for dynamic analysis. The results show that, except for a few areas in China where the ecological efficiency values increased or remained unchanged between 2009 and 2017, the ecological efficiency values in most areas have declined to varying degrees and are quite different among different regions. Based on the factors affecting the growth of China's ecological development efficiency, technological progress is the main reason for its growth. Second, the standard deviation and coefficient of variation of China's ecological development efficiency are calculated, and the results show σ convergence in ecological efficiency at the national level and in the eastern region, but not in the central and western regions.

The σ convergence of economic development was analyzed by calculating the standard deviation and variation coefficient of China's economic development level, and the absolute β convergence of economic development was examined. The results show the existence of both σ convergence and absolute β convergence.

Regarding the strategic spatial layout of ecological civilization construction and sustainable development, we analyze China's development strategy for promoting environmental protection and ecological civilization construction based on three aspects: the strategic pattern of China's land space, types of national main functional areas, and national ecological function zoning.

Regarding the transformation path and policy choices of underdeveloped regions, the following four aspects are analyzed: implementing the five major development concepts to remove obstacles to regional development, promoting industrial restructuring to facilitate green economic transformation; promoting economical and intensive use of resources to promote regional energy conservation and emission reduction, strengthening the implementation of environmental protection policies, and vigorously implementing pollution control.

Exercise

1. Briefly discuss the meaning of environmental capacity and environmental carrying capacity of resources.
2. Briefly describe the selection of indicators and measurement methods for eco-efficiency assessment.
3. Please select a case study area and carry out an empirical analysis of the convergence of economic development.

Extended Reading

Abernethy, V. D. (2001). Carrying capacity: The tradition and policy implications of limits. *Ethics in Science and Environmental Politics, 2001*, 9–18.
Daily, G. C., & Ehrlich, P. R. (1992). Population, sustainability, and earth's carrying capacity. *BioScience, 42*(10), 761–771. https://doi.org/10.2307/1311995
Siegel, F. R. (2021). *The earth's human carrying capacity: Limitations assessed.* Springer International Publishing. https://doi.org/10.1007/978-3-030-73476-3_1

References

Hou, J. (2023). Green-production transitions and hazardous industrial discharge: A regional case study from China. *Science Progress, 106*(1), 00368504231152747. https://doi.org/10.1177/00368504231152747

Qian, J., & Zhou, Y. (2023). Can an emission trading policy promote green transformation of regional economies? Evidence from China. *Journal of Water and Climate Change, 15*(1), 171–191. https://doi.org/10.2166/wcc.2023.540

Wang, D., & Kong, X. (2021). Evaluation of carrying capacity on resource and environment based on improved TOPSIS method. *IOP Conference Series: Earth and Environmental Science, 811*(1), 012004. https://doi.org/10.1088/1755-1315/811/1/012004

Wang, Y., Wang, J., Duan, X., & Wang, L. (2023). Assessment and simulation of water environment carrying capacity in a river basin using system dynamics model. *Polish Journal of Environmental Studies, 32*(3), 2893–2907. https://doi.org/10.15244/pjoes/161326

Chapter 11
Impacts of and Adaptation to Climate Change

Abstract The impacts of global climate change vary based on geographic location, resource and energy endowment, and socio-economic development levels. Outcomes differ significantly across regions and latitudes. Development Geography emphasizes the regional development impacts and adaptation strategies related to climate change, examining how climate change affects economic systems, factor inputs, and social welfare. Internationally, climate change research focuses on impacts and adaptation, both crucial and interdependent. Adaptation measures are essential for mitigating adverse effects on ecosystems and societies and promoting sustainable regional development. The impacts of climate change are particularly severe in major countries such as the United States, China, Russia, and the United Kingdom. These impacts can influence global economic and social development through interconnected mechanisms such as trade. Less developed regions, with weaker economic resilience, are more vulnerable to climate change, struggling to cope with such crises. This chapter analyzes climate change impacts from a Development Geography perspective across six dimensions: agriculture, water resources, health, energy, disasters, poverty, and employment. The objectives are to understand the regional impacts of climate change, the importance of adaptation measures, and the need for sustainable regional development to ensure healthy progress. The objectives of this chapter are as follows: 1. To assess the direct and indirect impacts of climate change on various sectors of regional economy, including agriculture, water resources, human health, disaster risk, poverty, and employment. 2. To understand the regional disparities in the extent and severity of these impacts, recognizing that different regions may be more or less vulnerable to climate change due to their unique geographical, economic, and social characteristics. 3. To highlight the urgency of addressing climate change and its impacts on regional development, emphasizing the need for adaptive strategies and policies to mitigate potential harm and foster resilience. 4. To provide a framework for analyzing and evaluating the effectiveness of adaptation measures, aiming to inform policymakers and stakeholders about the best practices for addressing climate change impacts in diverse regions.

X. Deng and M. Song, *Development Geography*,
https://doi.org/10.1007/978-981-96-6917-2_11

11.1 Background of Impacts of and Adaptation to Climate Change

Since the Industrial Revolution, the rapid development of regions has resulted in a large amount of greenhouse gas emissions and led to global climate change, which in turn has caused extreme weather events such as storms, floods, droughts, and high temperatures to intensify and increase, and regional development and climate change have become prominent environmental issues currently affecting the survival and development of humankind. As global warming continues to intensify, the sea level is rising, extreme weather is occurring frequently, the warmth and rainfall of the balmy monsoon has turned into extremely destructive tropical storms, the North and South Poles glaciers are melting, and the rivers that have regular dry periods are suddenly flooded and inundated and the ensuing diseases and pests are aggravated, water resources are diminishing, agricultural yields are decreasing, and hundreds of thousands of people are being displaced. Over the past 60 years or so, climate change, characterized mainly by warming around the globe, is continuing to undergo significant changes that are already having a dramatic impact on global nature and human society and regional development.

The United Nations Intergovernmental Panel on Climate Change (IPCC) has issued five assessment reports on global climate change, in particular, noting that the important role of global human activities in altering greenhouse gases and in controlling the process of global climate change formation and development has become prominent. Numerous studies domestically and internationally have confirmed that climate change has had a profound and direct impact on the survival and development of global human civilization, and that it is one of the fundamental factors and forces that ultimately lead to the survival and rise and fall of human civilization in all regions around the world. There is a close and complex correlation between various social events in human society and history, such as demographic fluctuations and human migrations, economic fluctuations, social changes, dynastic changes and even the rise and fall of civilizations, and global climate change. Due to the differences in social patterns, regional environments, and socio-economic conditions in different regions, the responses to climate change are also very different. Therefore, we should respond to climate change in different regions from the perspective of regional development and analyze specific problems to ensure the sustainable development of the regions.

A system of interdependence between human activities and climate change has been formed, and we must develop adaptive strategies from the perspective of climate adaptation and regional diversity. This should be done by reducing environmental vulnerability, increasing resilience, strengthening adaptability, making more progress in adaptation measures in space and time, building consensus in the international community, applying rational approaches to make action on adaptation to climate change more feasible, and establishing regional and global climate governance mechanisms centered on the building of a climate regime. Six elements of human survival and development, namely agriculture, water resources, health, energy, disasters, and poverty and employment, are closely related to the impact of climate change. The most direct response to climate change is the anomaly of temperature and precipitation, which directly affects the total amount of water resources and agricultural production. Second, disasters caused by climate change will exacerbate poverty and people's health problems, which, through a series of transmission mechanisms, will affect aspects such as employment and energy utilization. Therefore, the six areas of agriculture, water resources, health, energy, disasters, and poverty and employment were selected for this study.

11.2 Impacts of Climate Change and Regional Development

Presently, due to the complex and volatile international and domestic market environments and the combined effects of multiple factors, regional development is facing serious challenges and dilemmas, such as overcapacity and slow upgrading in traditional industries, high thresholds and long lead times in new industries, and the fact that traditional industries, which are predominantly secondary in nature, bring about a large amount of greenhouse gas emissions, which, in turn, lead to climate change and indirectly affect the sustainable development of the region. In view of this, this section discusses the impacts and challenges of adapting to climate change in four categories of regions with different levels of economic and social development, namely developed countries, developing countries, least developed countries (LDCs), and island states. The impacts of climate change are mainly in six aspects, namely agriculture, water resources, health, energy, disasters, and poverty and employment. Agricultural production is related to the food security problem of human beings. Water resources directly affect the survival of human beings, according to the observation data from the main hydrological stations of the main stream of the Yellow River in the past 50 years. The quantitative analysis concludes that the climate change is the main driving force for decreasing the actual amount of incoming water in the main stream of the upper reaches of the Yellow River, which is about 75%. The disaster, poverty, and health problems affect the quality of life of human beings, and the projections are that by 2050, increasing temperatures will result in 1.6 to 2.2 million new cases of kidney stones per year in the United States, with up to a 30% increase in certain regions. Energy and employment are related to human production, among other things. The relationships among climate change and agriculture, water resources, health, energy, disasters, and poverty and employment are shown in Fig. 11.1. The figure shows that these six issues involve all aspects of human life, and solving them is an important scientific way to realize the harmonious coexistence of human-earth relations and a necessary path for the sustainable development of human beings. The issue of climate change has attracted global attention; for example, the Intergovernmental Panel on Climate Change (IPCC, 2014) warns that if the current rate of greenhouse gas emissions continues, the temperature of the earth's surface is likely to rise by 0.2 °C every 10 years, which will have a huge impact on agricultural production,

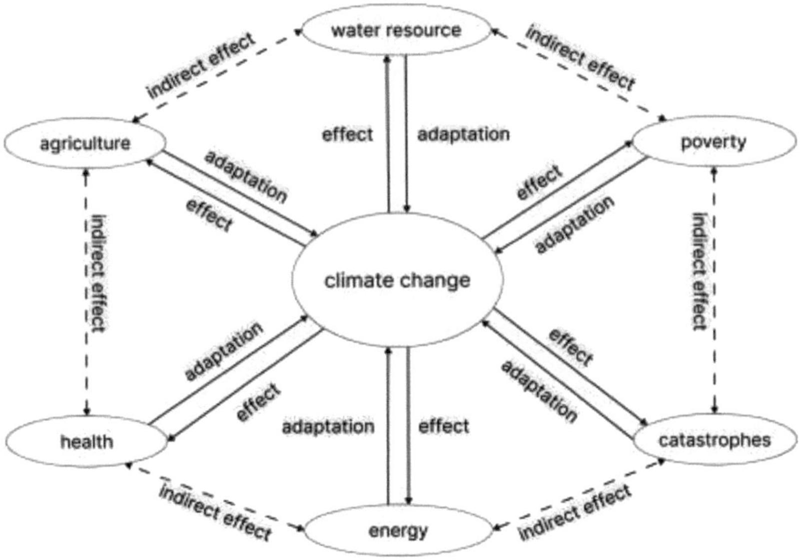

Fig. 11.1 Framework map of regional development and climate change impacts

and other areas (Houghton, 1995). For example, climate change affects the growth and development of crops, an increase in average temperature shortens the number of days of rice fertility by an average of 7.6 days, and an increase in temperature leads to a shortening of the fertility period of first-season and early rice. Meanwhile organizations such as the Human Dimensions of Global Change Program and the Future Earth Research Framework have studied these six aspects to varying degrees.

11.2.1 Impacts of Climate Change on Agriculture

Regional development mainly relies on and is reflected in industrial development, which is the dominant factor in activating the regional economy and the main pillar of regional economic development, with economic growth injecting vitality into regional development. As one of the three major industries, agriculture is greatly significant for the national economy and people's livelihood. In the context of climate change, how agriculture will be affected is one of the core issues of development geography. As the population continues to grow, the demand for food has increased. However, due to the increasing scarcity and depletion of natural capital (including soil, water, and biodiversity) required for food cultivation and production, yields are difficult to maintain, and climate change has brought about several negative impacts on food production (Chen et al., 2015). Higher temperatures, wetter climates, and increased atmospheric carbon dioxide levels will increase the incidence of a wide range of weeds, pests, and epidemics, and extreme temperatures combined with reduced rainfall will prevent crops from growing (Iglesias et al., 2012). Meeting food demand and adapting agriculture to climate change in this situation will be a great challenge.

Global climate change is expected to have a severe impact on food security in different parts of the world, which, in turn, will hinder regional development. The projected distribution of marine species diversity and reduction of marine biodiversity in sensitive areas of the globe in the mid-twenty-first century and beyond will pose significant challenges to the productivity of marine fisheries and other marine ecosystem services in all regions. Simulations of global catch potential have shown that climate change is likely to result in an average increase of 30–70% in catch potential at high latitudes and a decline of about 40% in the tropics over the next 40 years (Cheung et al., 2010). For wheat, rice, and maize in tropical and temperate high-altitude regions of the globe, global climate change is projected to have a huge impact on food production if adaptation measures are not taken, although marine fisheries in very few regions may benefit (IPCC, 2014). The rapidly increasing demand for food production will pose significant risks to China's and the global food security system (IPCC, 2014), as the level of high global average temperature increase is at least 4 °C above late 20th-century levels.

Global climate change is already having a more prominent impact on global agricultural production, which, in turn, is one of the major constraints on regional development, especially for less developed regions. On the issue of addressing climate change, different regions and countries have attached great importance to the mitigation of agricultural greenhouse gas emissions, while different countries have long neglected agricultural adaptation to climate change. Currently, developed countries are more mature in terms of the theoretical basis and technical methods of climate change adaptation, whereas less developed countries are less adapted to climate change. Table 11.1 presents main impacts or threats of climate change on agricultural production in different countries or regions. This chapter provides an overview of the progress of climate change adaptation in countries around the world, discusses the urgency of climate change adaptation in agriculture, summarizes the current major adaptation technologies, distills the current status and gaps in agricultural adaptation technologies, and focuses on the challenges facing climate change adaptation in agriculture in various countries.

In terms of regional development and agricultural adaptation to climate change in developed countries, the Nordic region is typically representative of regional development and adaptation to climate change. Northern Europe is the region that is most affected by climate change, with the southern regions of Northern Europe expected to experience the greatest changes in the spring, whereas Norway, southern Finland, and east-central Sweden are expected to experience the greatest changes in the fall. Average temperatures are expected to increase by 1–3 °C in spring, summer, and fall, varying by region with the seasons, but overall, the largest increase in temperature is in the northern regions. Winter temperatures are expected to increase by 2–8 °C, with the largest increases (6–8 °C) in Norway, northern Sweden, and the northern half of Finland. Average precipitation is expected to remain stable or increase across the region, with a 30% increase in spring in northern Sweden. Climate change will increase the potential for food production in the Nordic countries, and improved

Table 11.1 Impacts or threats of climate change on agricultural production in different countries or regions

Country or regional development level	Main impacts or threats to agricultural production
Developed country	Snowmelt ice water and cooler soil temperatures will limit the opportunities presented by an extended growing season, and delayed spring tillage will create an undisturbed, favorable environment for pests and weeds
Developing country	Significant increase in the frequency and intensity of extreme precipitation events (droughts and floods), impacts of increased droughts on the northward shift of the distribution area of crops in certain regions, changes in rainfall, acceleration of the water cycle, intensification of the conflict between supply and demand, and a sharp increase in the cost of agricultural production and irrigation utilization in certain regions
Less developed country	Parts of the region (e.g., West and Central Africa) are highly vulnerable to climate change, which contributes to the increasing frequency of extreme events such as devastating droughts and floods. The lack of rainfall and high summer temperatures strain water resources for agriculture in southern Africa, which adversely affects the growth and development of local crops

Note Adopted from the Third National Assessment Report on Climate Change (2015)

conditions in northern Europe are likely to support a shift from spring sowing to winter cereals, which will have higher yields (Trnka et al., 2011). Warmer and wetter conditions, more frequent extreme weather events, and longer growing seasons may allow current production crops to expand northward and provide opportunities to breed new crop varieties. However, the greatest negative impact on yields will occur if crops are subjected to drought during the growing season and extreme precipitation during the harvest period. Excessive snow and ice melt and cooler soil temperatures will limit the opportunities presented by an extended growing season, and delayed spring tillage will create undisturbed, favorable environments for pests and weeds. Without effective adaptation actions, extreme events could lead to yield losses in Northern Europe, presenting a range of challenges for Nordic agriculture.

As a largely agricultural country, China's agricultural production occupies an important position in the world, and global climate change can have a direct or indirect impact on China's agricultural production, making research on adaptation responses to China's agricultural development in the context of global climate change imperative. Overall, China's basic research on agricultural adaptation to climate change is still relatively weak. Extreme events such as sea level rise and hurricane heat waves have had a great impact on Chinese agriculture. Most of the impacts of climate change on crop yields in developing countries have been negative, and the severity of the negative impacts and the loss-causing ratio is likely to increase further in the future. Climate change contributes to changes in crop growth and development, cropping areas and systems, yields, production costs, and management practices. Climate change causes an increase in heat, which expands crop cultivation to areas at higher latitudes and altitudes. In the past 50 years, except for certain localized areas in South China and Southwest China, the vast majority of China's high-altitude areas have already shown a trend of rapid temperature increase. Precipitation in sensitive areas has decreased, and there is a trend of uneven distribution of marine species in sensitive areas, Climate change has caused changes in the planting areas of relevant crops in China, which should be adjusted to adapt to regional development and climate change. Climate warming, the release and decomposition of nitrogen-containing microorganisms from soil organic matter, will accelerate the water cycle, increase the loss of organic matter and nitrogen-containing microorganisms from the soil, shorten the release and utilization cycle of fertilizers, accelerate soil degradation and organic matter erosion, and greatly weaken the defensive capacity of agroecosystems to withstand natural disasters. In arid areas, soil erosion is severe, increasing irrigation cost and investment. For example, increased precipitation variability in northeastern China has resulted in a significant increase in the frequency and intensity of extreme precipitation events (droughts and floods), and droughts have led to soil salinization and desertification in some areas, which have reduced the quality of the environment for agricultural production, lengthened the crop-growing seasons, and exacerbated pests, diseases, and weeds, making them more difficult to control and more costly to invest in. Meanwhile, the northward shift in crop growth and distribution in the region, the impact of the intensification of drought on the northward shift in the distribution of crops in certain regions, changes in rainfall, acceleration of the water cycle, the intensification of the contradiction between supply and demand, and the sharp increase in the cost of agricultural production and irrigation utilization in certain regions also have had an impact.

In terms of adaptation to climate change in agriculture in less developed countries, the West and Central Africa region has abundant water resources; however, they are unevenly distributed throughout the region, with less than three percent of cultivated land having sufficient water for irrigation. The region is highly vulnerable to climate change and farmers across the region are suffering from climate variability and weather risks. Climate change is contributing to the increasing frequency of extreme events such as devastating droughts and floods hitting smallholder farmers. Poor rainfall and high summer temperatures have strained water resources for agriculture in Southern Africa, adversely affecting the growth and development of local crops. In 2016, the number of people affected by food insecurity in the Southern African region increased dramatically because of declining crop yields caused by El Niño weather. Cereal yields bounced back significantly to record levels in 2017, resulting in food insecurity. The number of people who are food insecure has fallen by nearly 90%. In the case of Mali, agriculture is the backbone of China's economy, accounting for about 42% of China's GDP and about 72% of the Union's member states' labor force engaged in agricultural production. Agriculture therefore has great potential to reduce poverty in China. However, climate change, through changes in temperature and precipitation, could seriously affect agricultural production, leading to reduced or even no food production. At the same time, Mali's agricultural production is under-utilized in terms of resources: high cost of irrigation infrastructure, limited technical capacity, and lack of financial resources. However, some of these constraints can be gradually overcome with the small-scale irrigation promotion program developed by the Malian government. Comoros is one of the many island countries located mainly in Africa, west of Madagascar and east of Mozambique in the Red Sea and the Indian Ocean, and is known as the "Moon Country" of Africa. The average production of spices such as vanilla and cloves has always been among the highest in the world. Comoros is the smallest and poorest country in Africa, where the modernization of agricultural technology is quite low. Climate change has made agriculture in China extremely challenging. Climate change has led to prolonged droughts in the island, with concentrated rains from January to March, increasing the risk of pests and other crop diseases, and strong winds, floods, hurricanes, and sea level rise threatening infrastructure, significantly impacting China's agricultural economy.

11.2.2 Impacts of Climate Change on Water Resources

Resource endowment is a key constraint for regional transformational development and socio-economic growth, which plays a role in constraining sustainable regional development. As an irreplaceable and strategic resource for human social development, water resources are vulnerable to human activities and climate change (Iglesias et al., 2018), which can alter the spatial and temporal distribution of water resources by directly affecting precipitation. In the context of the unusual complexity of global climate change, the contradiction between the supply and demand of water resources is becoming increasingly prominent. As the global climate changes in an exceptionally complex way, the changes in temperature and precipitation will be greatly affected, and the spatial and temporal changes in the distribution of water resources and uneven distribution will be further exacerbated. Water scarcity affects more than 40% of the global population, and water-related disasters account for 70% of all deaths from natural disasters. Prolonged drought and water scarcity, hydrological uncertainty, and extreme weather events (floods and droughts) are recognized as the greatest threats to global prosperity and social stability. It is therefore important to develop climate change-adapted water resource management strategies promptly to adapt to the needs of regional development.

The impacts of climate change on water resources pose different forms of challenges in different regions of the world. Developed countries have made detailed deployments in dealing with the impacts and challenges of water resource adaptation to climate change in terms of water resource strategies, policies, and planning (Tayfur et al., 2016), and the Netherlands is one of the more representative countries in this regard. With approximately 30% of China below sea level and 66% of China in flood-prone zones, Dutch socio-economic systems are highly vulnerable to global climate change. According to the prediction of relevant scholars, the sea level height in the Netherlands may rise again by 35–85 cm by the end of this century. The highest part of the sea level in the Netherlands may also rise again by 1.3 m in the next 150 years, and coping with the inundation crisis has always been a challenge for the Netherlands. According to the KNMI (Koninklijk Nederlands Meteorologisch

Instituut) Climate Change Scenarios 2006 for the Netherlands, it is predicted that by 2100, global warming and subsidence will result in a sea level rise of 2.5–5 m in the Netherlands, and that the threat of flooding in coastal cities will increase to more than 10 times the current level. The inundation potential of the Dutch delta cities has increased sixfold since the twentieth century, and according to 2010 statistics, about 23% of the dykes in the Netherlands are insufficient to cope with the water levels during flood surges.

China has experienced regional development and climate change impacts on water resources and made some achievements. With its vast land area, uneven distribution of water resources and significant spatial differences in water disasters, frequent floods in parts of the southeast, and droughts in the inland northwest that have already affected people's productivity and life, China's regional water resources will be further affected by climate change. Water resources in the northern part of China account for only 16%, whereas those in the southern account for 84%. In the context of global warming, the runoff of China's major rivers is on a decreasing trend. The per capita runoff and water resources of the major rivers in the north (especially Hai River, Yellow River, and Liao River in the northeast) have been significantly reduced. Per capita runoff and water resources in northern China are small, and the inter-annual temperature variations between years are large, unevenly distributed, and not fully matched with the distribution of regional productivity. The region prone to severe droughts and floods, and climate change is making it more difficult to conserve, develop, and utilize water resources. The amount of available water resources is limited. Regional water shortages are also occurring in the southern region, and future climate change will further increase the frequency of floods and droughts in China, making it difficult to effectively alleviate water shortages in some parts of China.

Scarcity of freshwater resources and pollution of the water environment are a major threat to the regional sustainable development of underdeveloped countries and small island developing states (SIDS) in terms of population growth, public health, and ecological environment. Therefore, it is an important research topic from the perspective of development geography. In terms of water resource response to climate change, the socio-economic development of in less developed regions has long been plagued by water scarcity, and both ecological and economic development have been constrained. Water resource adaptation to climate change policies, strategies, and actions are weak. In the Mediterranean region, water resources are a constraint to development. Exacerbated water scarcity will be the most devastating impact of climate change due to over-drying and over-heating of the land surface and high levels of heat gain and distribution. Environmental and urban pressures in the Mediterranean region challenge the use of water for food production. Agricultural management should reconsider ways to provide water for crop production in hot and arid climates while contributing to the sustainability and quality of freshwater supplies. Water is also one of several key issues facing Africa today and in the future. Agriculture in Africa is particularly dependent on precipitation; however, frequent natural disasters (Quandt et al., 2017), such as droughts and floods, have largely limited agricultural development on the continent.

Small island states share many similarities, such as vulnerability to natural disasters and climate extremes, extreme openness of their economies, and low adaptive and risk-sharing capacity, which will increase their vulnerability and reduce their resilience to climate variability and change. The main issues facing island states are broadly similar, such as freshwater quality and quantity, integrated watershed planning and management, and water supply reliability. Under most climate change scenarios, the water resources of small island states are likely to be severely jeopardized. Most SIDS have limited water supply capacity, and their water resources are particularly vulnerable to changes in and distribution of future rainfall.

11.2.3 Impacts of Climate Change on Human Health

Health is the basis for human survival and regional development. From the perspective of development geography, sustainable regional development depends on high-quality elements of labor capacity, and the sustainable development of human health depends on the harmony of human-earth and inter-regional relations, which is an important factor in ensuring stable regional economic growth, and is also a major area of research in development geography.

As living standards continue to improve, health issues are gradually receiving attention, and climate change may have a significant impact on human health. The impacts of weather-related disasters on people's health go beyond mortality and include injuries, mental health problems, spread of disease, food and water insecurity, and limited access to health care and other basic services (Table 11.2). Half of the world's population does not have access to basic health services, and 800 million people spend 10% or more of their household expenditure on health care. Therefore, it is particularly critical to develop regional development strategies for health adaptation to climate change. By the middle of this century, global climate change is expected to directly affect human health primarily by exacerbating existing health and environmental problems around the globe, and the World Health Organization has predicted that "the incidence of non-melanoma skin cancers could increase by 3.5–6.0% after 2050." In contrast to the current lack of international baseline data predicting climate change, throughout the twenty-first century, climate change is expected to potentially lead to further deterioration and exacerbation of environmental and health conditions in many countries and regions, and the World Health Organization has predicted that deaths from hyperthermia will triple globally by 2020, particularly in low-income developing countries.

In terms of human health adaptation to climate change in developed countries, the fundamental cause of dengue transmission is temperature. As temperatures continue to rise, the incubation period of the virus in mosquitoes continues to shorten, the frequency of mosquito bites increases rapidly, and the range that mosquitoes can reach is likely to expand. With climate change, dengue fever is likely to spread to higher latitudes or higher altitudes (Campbell-Lendrum et al., 2015; Ebi et al., 2016). A total of 6942 cases of dengue fever were reported on the French island of

Table 11.2 Several climate change-related impacts and threats to human health events

Impunity	Events
High temperature and heat wave	Climate change-related health effects of heat waves have resulted in deaths from cardiovascular and other respiratory diseases
	During a heat wave in the European summer of 2003, an annual increase of 70,000 deaths due to cardiovascular and respiratory diseases was recorded (Robine et al., 2008)
El Niño	In Venezuela and Colombia in South America, malaria outbreaks have been associated with the El Niño phenomenon, which, in turn, has affected human health
	The occurrence of dengue fever in the Asia–Pacific region is related to the El Niño phenomenon (Tong et al., 2004)
	Malaria epidemics in East Asia and South America, cholera outbreaks in coastal Bangladesh, and viral pneumonia among women in California, USA, have also been shown to be associated with El Niño (Ebi et al., 2001)
Disease infection	Waterborne diseases and diseases transmitted by insects, snails, or other cold-blooded animals have a significant impact (Randolph, 2013)
	The effects of climate change may lengthen the seasons of inter-infection and transmission of infectious diseases transmitted by the important vector Schistosoma haematobium and further altering their geographic and transmission ranges
	Global climate change is expected to further expand and contribute to the transmission of infectious schistosomiasis in China and other countries and regions through the snail schistosome
Other threats or impacts	Sea level rise: Destruction of homes, medical facilities, and other essential services
	Impact on freshwater availability: The lack of safe natural freshwater resources can have a serious impact on children's hygiene and seriously increase the risk of diarrhoeal epidemics; in extreme cases, water shortages can lead directly to drought and severe famine
	Impact of floods: Summer floods severely contaminate freshwater supplies, increasing the risk of human contamination of freshwater and infection by diseases affecting the communication system, and creating a good breeding ground for aquatic insects and birds, such as flies and mosquitoes, that may carry a wide range of diseases;
	Decline in the production quality of food varieties: Increased prevalence of malnutrition and undernutrition (Wheeler et al., 2013), leading to 3.1 million deaths per year

Réunion in 2018. The southern part of the island was the most affected area, with cases distributed in various small settlements. Two vectors, Aedes aegypti and Aedes albopictus, are present on Reunion Island, but the Aedes albopictus populations are relatively high and Aedes aegypti larvae are only found in the western coastal areas of the island.

Health service accessibility and health outcomes in China vary by region. While life expectancy and other health indicators in Beijing and Shanghai have surpassed those of high-income countries such as South Korea and the United States, health indicators in places such as Guizhou, Tibet, and Xinjiang are comparable to those of low-income countries. Women and children in remote rural areas especially lack quality health care, and the incidence of malnutrition, infectious diseases, and anemia is much higher in rural areas compared to urban areas. Rising temperatures, changes in precipitation, and increased levels of CO_2 in the atmosphere can pose significant threats to human health. For example: high-temperature heat waves can easily cause cardiovascular and cerebrovascular diseases, respiratory diseases, etc. The changes in humidity resulting from climate change contribute to the spread of schistosomiasis. The distribution range and density of snails, which can be altered by humidity, rainfall, and other factors, also play a role in facilitating malaria. North China Plain and southern and northeastern China are particularly sensitive to climate change and require enhanced monitoring and implementation of preventive and control measures. Climate change also affects the distribution and reproduction rates of insect vectors, potentially contributing to the spread of dengue fever in Hainan, China. Furthermore, physiological and psychological impacts stemming from air, soil, and water pollution, as well as water scarcity, adversely affect people's short- and long-term health status, including increasing risks of cancers, respiratory illnesses, cardiovascular diseases, and diarrheal diseases.

Low- and middle-income countries are the most severely affected by the health impacts of climate change. Weak health systems and poor access to basic public services have a weaker capacity to adapt to climate change. Improving health services for women and children in low-income households is a critical challenge for Nigeria. One-third of Africa's 250 million children under the age of five are stunted, a situation that seriously threatens the region's social development prospects. The challenge is to establish or strengthen mechanisms to detect, verify and respond rapidly and effectively to unexpected outbreaks and epidemics. Small island developing states (SIDS) are mostly producers of nutrient-rich pulses, seeds, fruits, vegetables, and nuts. The impacts of climate change, scarcity, and degradation of land and marine resources have led to local food scarcity, dietary changes, and a high dependence on imported foods (often ultra-processed products containing high levels of salt, sugar, and fat), resulting in "extremely high levels of overweight and obesity." For example, adult obesity rates in SIDS are nearly 60% higher than the global average (20.9% and 13.2%, respectively), with almost one in two adults obese in many small island states in Polynesia and Micronesia, and the prevalence of hunger and malnutrition in SIDS is higher than globally (17.5% and 10.7%, respectively). In the Pacific, 75% of adult deaths are caused by non-communicable diseases (World Health Organization, 2017), and five SIDS have obesity rates of more than 50% among adult women. The challenges of food security and nutrition caused by climate change in SIDS should be addressed.

11.2.4 Impacts of Climate Change on Energy Supply and Demand

The intensity of energy consumption is a characterizing indicator of the size of the regional economy and a reflection of the degree of regional development. Energy is at the heart of development, and an adequate supply of energy enables business activities, innovation, and new industries and can further drive economic growth. However, 30% of the world's energy is consumed by the food system. An estimated one-third of produced food is lost or wasted, and an estimated 38% of the energy consumed by the food system is wasted. More than one-tenth of the world's population (840 million people) does not have access to modern electricity services. About 3 billion people rely on traditional biomass for cooking and heating, with negative impacts on health, the environment, and economic development. CO_2 emissions from fossil fuel combustion and industrial processes accounted for roughly 78% of the increase in total GHG emissions during the period 1970–2010 (IPCC, 2014). Globally, economic development and population growth remain the two most important drivers of increased CO_2 emissions from fossil fuel combustion. Increased coal use has reversed the long-term trend of decarbonization of the world's energy supply. With global warming, changes in regional weather patterns and extreme weather events projected to affect energy demand, production, and transmission, the energy sector faces significant challenges from climate change.

In terms of developed countries, Japan has a scarcity of fossil energy and relies heavily on overseas imports. Climate change will affect changes in global ocean currents, causing great damage to the marine transportation industry. The occurrence of the Fukushima nuclear accident has further aggravated the energy security situation, with the energy self-sufficiency rate dropping from 20% in 2010 to about 8% in 2016, whereas the cost of electricity consumption has risen, putting pressure on economic and social development and national life. In addition, Japan is still under great pressure to reduce emissions. According to the Paris Agreement, Japan's greenhouse gas emissions are to be cut by 26% in 2030 and 80% in 2050 compared to 2013.

As the climate continues to change, the economy develops, and the population increases, China, as the second-largest emitter of greenhouse gases, will take on more and more emission reduction tasks. Climate change will continue to influence the changes in our energy industry structure. China's energy structure is characterized by a coal-based energy production and consumption structure, which is different from the world's energy composition dominated by oil and natural gas, which makes China's CO_2 intensity of energy consumption more than 30% higher than the world's average. Climate change has prompted China to adjust its energy policy and industrial structure to reduce energy demand. The Chinese government has taken a series of actions to adjust the industrial structure, optimize the energy structure, conserve energy, improve the efficiency of energy use, control greenhouse gas (GHG) emissions, and increase carbon sinks, and has achieved positive results. China's carbon intensity level decreased by about 46% in 2017 compared to 2005, exceeding the 2020 target of a 40–45% reduction.

There are about 621 million people without electricity in less developed Africa. Particularly in countries such as the Democratic Republic of Congo, Libya, Malawi, and Sierra Leone, up to 90% of Africa's population is without electricity. Energy access in the African region was only 42.8% in 2016, lagging far behind the rest of the world, and insufficient energy supply has been the biggest obstacle to Africa's infrastructural development (Nepal et al., 2019). Island states have the unique problem of geographic remoteness and isolation, and their small size and limited financial, institutional, and natural resources contribute to their underdevelopment. Their domestic markets are extremely limited in size and their exports are constrained by their distance from the mainland and lack of resources. Island states are heavily dependent on very expensive imported energy, as long-distance transportation costs add to the cost of products, including oil. The need to import fuel more frequently due to the low oil reserve capacity of island countries further increases the purchase cost of gasoline and diesel. The volatility of fuel prices more seriously affects the energy security of island states, making them more vulnerable. Table 11.3 introduces the significant challenges that the energy sector faces from climate change from three perspectives: partial impact, emission reduction measures, and policy framework.

11.2.5 Impacts of Climate Change on Disaster Risks

Disaster risk due to climate change is one of the key concerns of development geography. Natural disasters pose a great threat to regional development, and for a region, the impact of a large natural disaster may not be recovered for a long time, seriously hindering the sustainable development of the region. Scientific assessment of the risk of natural disasters in a region and the adoption of preventive measures are of far-reaching significance to regional development. The risk posed by climate-related impacts arises from the interaction between climate-related hazards (including hazard events and trends) and the vulnerability and exposure of human and natural systems, including the resilience of these systems. Natural hazards hurt the poor and vulnerable the most, and their impacts continue to grow with climate change. The rate and magnitude of warming and other changes in the climate system, accompanied by ocean acidification, increase the weight of losses and even the risk of irreversible harmful effects in some cases. Some risks are region-specific, such as floods in South Asia, severe droughts in Africa, and devastating hurricanes in the Caribbean, leading to large numbers of deaths, injuries, and property and economic losses, while others are global in nature. Climate change will also lead to the outbreak and spread of infectious diseases, plagues, etc., seriously affecting human health and survival, among other issues; therefore, it is essential to develop strategies for regional development impacts and challenges of natural disaster risk adaptation to climate change.

In the context of global warming, coastal flooding due to sea level rise and extreme water levels caused by high tide levels and storm surges have a huge impact on coastal socio-economics and the natural environment. Eighty percent of global deaths due to flooding from 1975 to 2016 resided within 100 km of the coast (Hu et al., 2018).

Table 11.3 Energy sector faces significant challenges from climate change

Partial impact	Electric power plant	Plumbing	Cords	Renewable energy	Nuclear energy
	Thermal power plants are affected by increased ambient temperatures, resulting in less efficient heat conversion. Less cooling water and higher water temperatures may lead to a drop in power operations or temporary power outages	Oil and gas pipelines are subject to sea level rise in coastal areas and thawing permafrost in cold climates. Design for risk, infrastructure construction standards, and structural upgrades should be considered	Extreme weather events, especially strong winds, can damage power lines. Standards can be revised to change power line routes to evacuate them from high-risk areas	Changes in regional weather patterns may affect the water cycle of the SPH (Smoothed Particle Hydrodynamics). The increase in cloudy weather in some regions has had an impact on solar technology, and the increase in the number and intensity of storms could damage equipment	Water scarcity and extreme weather events can disrupt the operation of critical equipment and processes, which can threaten nuclear power plants
Emission reduction measure	Carbon Capture and Storage	Increase efficiency	Fuel switching	Alternative energy	Demand reduction
	Carbon Capture and Storage (CCS) technology for fossil fuel plants reduces emissions. Carbon dioxide storage capacity is large and all parts of this technology have been demonstrated. Bioenergy combined with CCS can extract carbon dioxide from the air	Improvements in energy efficiency can be achieved through the refurbishment of existing plants, the adoption of efficient new processes to improve transmission and distribution, and the upgrading of fossil fuel extraction and conversion technologies	Switching to low-carbon fuels (e.g., coal to gas) can reduce emissions. Moving from a coal plant with the world's average efficiency to a state-of-the-art gas plant can halve emissions if methane emissions are controlled	Increased use of renewable energy sources, such as solar, wind, and biofuels. Increased use of nuclear energy. Hydropower is currently the largest single renewable resource, but solar, wind, and biofuels will see growth	Reducing consumer demand is a key mitigation strategy. The extent of demand reduction determines the scale of the mitigation challenge facing the energy sector. Potential constraints from the "rebound effect" need to be considered

(continued)

Table 11.3 (continued)

Policy framework	**Largest greenhouse gas emitting industries**	**Regulatory framework**	**Technology investment**	**Carbon pricing**
	The energy sector is the largest emitter of greenhouse gases. Reaching the 2°C target implies that the overall energy system will need to rapidly curb the growth of emissions and reduce warming to 0 °C by the end of the century, and may also require "negative emission" technologies, such as carbon sequestration	Governments can promote the use of emission reduction measures by creating an attractive fiscal and regulatory framework	New technologies can be used to improve efficiency, power generation, mining, storage, transportation, and distribution	A major challenge for governments and regulators is to ensure that carbon price incentivizes additional investment in low-carbon technology

Flood risk is a constant challenge for the Netherlands. Located in the Rhine delta, the Netherlands is a typical example of a low-lying country, with 25% of China's land located below sea level, more than half of the land requiring artificial dykes to ensure safety, about 10 million people living within flood protection zones, and a large number of industrial enterprises located in the zone. The high standard of dykes continues to be put to the test, as in 1995 when the Rhine River suffered its second-largest flood in recorded history. Climate change is extremely hazardous, with a 46% increase in the frequency of weather-related disasters between 2007 and 2016 (compared to the 1990s), which cause more than 60,000 deaths per year, mainly in developing countries. China's major climate disaster risks include droughts (mainly in the northern agricultural and pastoral zones of Ningnan, Longdong, and northern Shaanxi, northern Jinbei and southern Inner Mongolia, and the Shandong Peninsula), floods (in the Yangtze River Delta, Pearl River Delta, and lower Liaohe River areas, among others), frosts (which have a significant impact on agricultural production), low summer temperatures (affecting summer agriculture in northeast China), typhoons (in coastal areas), snowstorms (a major disaster in China's pastoral areas), storms (a major disaster in China's pastoral areas), and cold disaster (mainly affecting South China). Of the various types of natural disasters that occur in China, those related to extreme weather and climate events account for more than 70%.

Disasters have largely hurt the ability of poor and vulnerable communities to develop. From 1995 to 2014, 89% of storm-related fatalities occurred in low-income countries, even though only 26% of storms in these countries. Climate change is taking a huge toll on Africa. Rainfall fluctuations have increased in the Sahel and other dry lands, and water resources continue to run out. Natural disasters such as droughts in East Africa, floods in South Africa, and hurricanes are increasing in frequency and intensity. On July 25, 2019, there was a locust outbreak in the deserts of Yemen and the Horn of Africa. Increased summer breeding of desert locusts due to heavy rains could pose a serious threat to agricultural production areas in Yemen, Sudan, Eritrea, parts of Ethiopia, and the northern regions of Somalia over the next three months. This will have a potentially adverse impact on seasonal agricultural yields and local economies, affecting food security and sustainable livelihoods in the countries concerned. Small island states are particularly vulnerable to climate change and are at risk from sea-level rise. Small island states will be disproportionately affected by sea-level rise. Low-lying small island states face the prospect of total submersion as a result of sea-level rise. Countries such as Maldives, and Pacific Island countries such as Kiribati, the Marshall Islands, and Tuvalu are vulnerable to these impacts. Sea level rise is projected to worsen the closer you get to the equator, and countries such as the Maldives could face extinction. Sea level rise in the Maldives is projected to be 70–125 cm, depending on the degree of global warming. Sea level rise causes more severe storm surges, inundation, salinization of freshwater sources and soils by salt water intrusion, subsidence, land loss, and erosion. This hampers economic development. The infrastructure of many SIDS is concentrated in coastal areas. Consequently, climate change is already affecting strategic sectors of their economies, thus threatening ecosystems. Ocean acidification and changes in water temperature are also expected to severely affect marine biodiversity and ecosystems, threatening fish and coral community life.

11.2.6 Impacts of Climate Change on Poverty and Employment

Poverty and employment are important indicators of the sustainable development of a country or region, and they are also a practical manifestation of the impact of regional development and climate change from the perspective of development geography. Globally, the key to realizing stable poverty eradication in less developed countries or regions lies in stabilizing employment and strengthening industries. This process is directly or indirectly affected and constrained by climate change, therefore, studies should be conducted on the impact of regional development and climate change on poverty and employment from the perspective of development geography.

As of 2015, half of the 736 million people living in extreme poverty globally these are in five countries, namely Nigeria, India, Ethiopia, Democratic Republic of Congo, and Bangladesh. These are also the most populous countries in South Asia and sub-Saharan Africa, two regions that will be disproportionately affected by the effects of climate change and together account for 85% (629 million) of the world's poor. The risk of catastrophe posed by violent conflict and climate poses a challenge to poverty eradication (Bank, 2016). In the Arab region, extreme poverty was previously less than 3%. However, the conflicts in Syria and Yemen have led to an increase in poverty rates in the region, exposing more people to hunger and homelessness. Poverty eradication must be tailored to each country with appropriate strategies based on the latest data and analysis and the needs of the people. To sustainably reduce poverty, countries across the globe need to grow in an inclusive, labor-intensive way. Good, inclusive work is the surest route out of poverty. A productive private sector that operates on a level playing field and provides employment opportunities for all sectors of society is key to ending extreme poverty and promoting shared prosperity.

Climate change and poverty are major challenges facing human society in the twenty-first century. Climate poverty refers to the phenomenon of poverty triggered or exacerbated by climatic or climate change factors. Addressing climate poverty and boosting employment is of strategic significance for countries and regions around the world in addressing climate change. The impact of climate change on poverty has a double effect: first, it exacerbates the poverty level of the existing poor population, and second, it increases the number of new poor people. Worldwide, countries at different stages of development use different measures to adapt to climate change. The UK is representative of the developed world. one in five Britons were in poverty during 2016–2017. Of those in poverty, more than four million were in persistent poverty. Comparing the 2011–2012 and 2016–2017 periods, child poverty is on the rise and rising faster than the child population. Worker poverty is more severe than at any time in the past 20 years. The job crisis is a major challenge for developing countries, hindering the eradication of extreme poverty and the promotion of shared prosperity. As a typical case of developing countries, more than 700 million of India's 1.353 billion people are rural, but the labor force is only 400 million. The Indian government has not been able to solve the problems of poverty and unemployment well, and its challenges include the following:

(i) The population growth rate is too fast. India's excessive population growth rate means the rapid growth of the labor force; however, the employment opportunities provided by economic development are far from meeting the demand, leading to serious employment problems, and poverty will continue to emerge.

(ii) Social demand and talent training model are contradictory. In India's institutions of higher learning, the setting of specialized courses is not compatible with the needs of the labor market.

(iii) Unbalanced industrial structure. 90% of India's population lives in rural areas, indicating that the main labor force lives in rural areas, and more than 50% of the population is engaged in agricultural labor; however, the slow development of agriculture has failed to effectively absorb labor to create wealth. Underdeveloped regions are also significantly affected by climate change. Droughts and floods tend to be extreme and devastating, with reduced or even no crop harvests, which can have a dramatic impact on farmers' incomes and assets, resulting in the inability of farmers to purchase seeds, fertilizers, and the means of production needed to resume production the following year, seriously affecting people's productivity and even their survival. the maize harvest in Malawi in 2005 was at its lowest level since records began. Persistent drought and flooding reduced maize production by 29%, and more than 5 million people faced food shortages. As rural incomes plummeted, poverty grew because farmers lacked the funds to plant properly during the 2006 season. Many small island developing states (SIDS) are among the most vulnerable and sensitive to climate change in the world and endure the greatest impact of hazards. Increasing flooding, rising ocean and land temperatures, ocean acidification, bank erosion, and damage to infrastructure from extreme weather events are among the direct damages caused by sea level rise. In addition to these direct threats and damages, the impacts of climate change on SIDS have resulted in significant losses to their socio-economic sectors, such as fisheries and tourism, in addition to a significant increase in the economic costs of responding to climate change.

11.3 Adaptation to Climate Change and Regional Development

Development geography focuses on the regional impacts of climate change and, in recent years, on regional adaptation to climate change. Specifically, it includes agriculture, water resources, health, energy, disasters, and poverty and employment. Mitigating and adapting to climate change at the regional level in the above aspects is significant for regional development. Formulating scientific and reasonable adaptive strategies for regional development to adapt to climate change is the first task in coping with climate change and an important prerequisite for realizing sustainable regional development. Therefore, in response to the impacts and challenges encountered in the six aspects of agriculture, water resources, health, energy, disasters, and poverty and employment in the previous section, in developed, developing, underdeveloped, and island countries, this section proposes corresponding adaptive strategies as well as national and regional response.

Regions and countries worldwide are pursuing regional development and adaptation to climate change. The adapting process is influenced by many factors that determine the choices faced by individuals, businesses, and governments. The nature of the risks associated with climate change varies from place to place and from time to time, as does adaptive capacity. The state of human development, technological and scientific capacity, and financial resources all play a role in defining adaptive capacity. In some respects, the incremental risks posed by climate change are a matter of degree. Today's social and economic policies for adapting countries and populations, investments in infrastructural defenses against floods and cyclones, and systems for regulating watershed management are no different from those that will be needed to deal with future threats. However, the magnitude of these threats poses both quantitative and qualitative challenges. Some countries and populations are better prepared to deal with them than others.

11.3.1 Strategies for Adaptation of Agriculture to Climate Change

Industrial development, as the main pillar of regional development and economic growth, is directly and inextricably linked to regional development. Agriculture, industry, and services are all areas of industrial research covered in the introduction to development geography, which has become more focused on agriculture in the context of global climate change. Agriculture is closely related to human survival and food security. Agricultural development is greatly affected by climate change, which is reflected in temperature and precipitation changes (Challinor et al., 2014); therefore, due to the different geographic locations of different countries, their climate types are also quite different. At the same time, there are certain differences in the level of development of different countries, and their agricultural production technology is also quite different. Therefore, it is necessary to consider the actual situation of specific countries, in accordance with the principle of adapting to local conditions and time, and adopt corresponding applicable strategies to cope with climate change.

In terms of developed countries' strategies for regional development and agricultural adaptation to climate change, Europe emphasizes improving water use efficiency as a means of coping with climate change. Europe emphasizes the importance of improving water use efficiency as a key measure to cope with climate risks, as well as the importance of extension services. Short-term adaptation measures include advancing planting dates and sowing new varieties of existing crops, which can be adapted relatively quickly. Long-term adaptation measures are further categorized into technical measures (e.g., introduction of new varieties), management measures (e.g., change in cropping patterns, increase in soil organic matter content, landscaping, water), and infrastructural modifications (e.g., improvement of buildings, access roads, drainage, and irrigation systems). At the technical level, more heat-tolerant varieties or new varieties better suited to specific climates should be introduced. At the management level, cropping patterns and in the day of sowing of crops should be changed. The organic matter content of the soil should be increased. The northern regions have acid-free drained soils with high organic matter content that makes blueberry cultivation possible. At the infrastructural level, the drainage and irrigation systems should be improved, and the European Water Framework should enforce full cost recovery for water services. Buildings and roads should be reinforced; in the Arctic region, thawing permafrost affects the stability of buildings and roads. The population should be informed about soil needs, texture, pH, fertilizer needs, storage knowledge, etc.

China's modernization requires the supporting role of a good agricultural production environment. The formation of a sound system to help the agricultural industry to adapt to climate change in terms of policies, concepts, methods, and technologies; therefore, it is necessary to establish scientific, technological, and production systems to cope with climate change in the field of agriculture. China's main response measures are threefold:

(i) Strengthen the construction of agricultural infrastructure and promoting the development of water-saving agriculture. Climate change has led to seasonal droughts and shortages of irrigation water; therefore, the development of water-saving agriculture is urgent. In northern and western China, water shortage is currently the most important factor limiting agricultural development. Therefore, it is important to develop and establish a long-term mechanism for efficient water-saving agricultural inputs based on open-source technology, water-saving irrigation technology, agronomic drought-resistant technology, and engineered water conservation technology according to local conditions.
(ii) Enhance the comprehensive management level of agricultural adaptation to climate change. Through scientific and reasonable agricultural management techniques, maximize yield potential, reduce excessive use of nutrients, and reduce the impact of agriculture on the environment. Accelerate the promotion of good seeds, implement water and fertilizer management systems, and strengthen the construction of integrated technology systems such as pest control.
(iii) Appropriately adjust the layout of agricultural industries. By adjusting the layout of the agricultural industry, we can improve the output and quality of the main grain-producing areas, improve the current planning of agricultural land, and enhance the resilience and recoverability of the agricultural production system to adapt to changes in the climate.

In the case of less developed countries, the development of irrigated agriculture is one of the main courses of action to improve food and income security. To guide the development of irrigation in China, Mali introduced the Small-Scale Irrigation Promotion Program 2012–2021 (Neighborhood Irrigation Program), thus financing projects such as irrigated vegetable gardens and inland valley floors. Small-scale irrigation in Mali takes different forms: (i) Village Irrigation Schemes (VISs): VISs are established along rivers to facilitate the cultivation of crops such as rice. (ii) Controlled Flood Recession (CFR): rice is cultivated such that it is dependent on the level of water reached. These systems have low investment costs; however, their feasibility is questioned because they depend on rainfall and floods. (iii) Inland valley bottoms: There are four different types of irrigation systems for inland valley bottoms. The main ones include river diversions, micro-dams, and well-equipped inland valley bottoms (these systems are equipped with earth or stone embankments that keep a layer of water longer and prevent erosion). (iv) Water retention works in wadis and oases: The main objective is to recharge the water table or to feed the animals. (v) Irrigated vegetable gardens: These comprise small plots situated around a borehole or a wellhead. Water extraction can be done manually or using a pump. These systems are also used for off-season crops. Agricultural development in Comoros must adopt relevant adaptation strategies to climate change, thus addressing the hunger of the local population.

The main responses in Comoros can be categorized into three points: (i) Farmers are the main players in local agriculture, and it is important to improve their relevant knowledge. Ensure that farmers have the necessary capacities, tools, and technologies to meet the challenges of climate change and increase the resilience of the agricultural system to climate change and various types of shocks, which to date has 200 vulnerable farming communities throughout Comoros. Train farmers in composting, soil restoration, commercial agriculture, banana and potato cultivation, as well as cattle and goat breeding. (ii) Science and technology are the first productive forces, and the improvement of relevant technologies will ensure food production. Increase farm yields in Comoros by providing climate-resistant seeds, introducing good livestock breeds, improving agricultural and breeding techniques, and improving water reserves, use, and management; (iii) Establish an agrometeorological observation system. Strengthen the service capacity of the Comorian National Civil Aviation and Meteorological Service (CNACM) and set up an agrometeorological observation system to provide farmers with accurate real-time meteorological information so that they can take timely measures to respond to climate change, thus ensuring sustainable and adaptive development of agriculture.

Adaptation strategies to climate change in agriculture in developed countries (Europe)

Plant breeding is the most common and fundamental action taken to adapt to the challenges of climate change. If longer growing seasons and unique light conditions come with climate change, future new crop varieties can be better utilized (Uleberg et al., 2014) and accelerated maturation and yield reductions can be minimized. Therefore, breeding research programs and directed breeding to support farm-autonomous adaptation (Himanen et al., 2013) are recommended adaptation strategies. For different adaptation purposes, corresponding adaptation strategies are given (Wiréhn, 2018).

- Planting new crops or crop rotations were implemented to take full advantage of the extended growing season and change climatic conditions. Arable crops and varieties were expanded northward: legumes, more productive forages, vegetables, cereals, peas, broad beans, rapeseed, soybeans, sunflowers, and maize.
- Increased fertilizer application and higher pesticide use in response to increased CO_2 levels in the atmosphere. To cope with accelerated grain phenology and shorter utilization of the nutrient period, (i) use long-season varieties; (ii) use spring-sown crops (spring sowing has less impact than winter sowing in Denmark); (iii) select new varieties.

- Prevent stress in times of drought by (i) improving soil properties through measures such as better drainage; (ii) focusing on crop rotation; (iii) developing irrigation systems and reservoirs (especially for potatoes, sugar beet, vegetables, and fruits); (iv) determining current and future abstraction needs and available water resources; and (v) using winter-sown crops.
- To cope with increased rainfall, (i) improve soil properties; (ii) focus on crop rotation; (iii) adopt intercropping; and (iv) increase and improve drainage systems.
- Seeding and harvesting problems due to excess moisture—use perennial crops.
- Increase soil buoyancy to prevent yield loss after heavy rains and waterlogged farmland. (i) Use cropping systems that enhance soil structure and filtration; (ii) improve the size and management of major drainage systems; (iii) ensure cooperation between administrative units and utilities; and (iv) breed flood resistant varieties.
- Prevention of nitrogen leaching with (i) reduced intensive tillage in the fall and (ii) lime application.
- Pest and disease risks, with increased demand for crop protection and pest control products.

11.3.2 Strategies for Adaptation of Water Resources to Climate Change

From the perspective of development geography, the shortage of resources often leads regional development into a short-board effect, in which the resources that are the short-board become the resistance that restricts regional development. Currently, the shortage of water resources is a key constraint to economic and social development. Climate change will have a direct impact on water resources, and the development of applicable regional countermeasures for water resource adaptation to climate change is crucial for the sustainable development of the region (Garrote et al., 2015). The following is an analysis of water resource adaptation strategies to climate change for a typical developed country (the Netherlands), developing countries (China and the Mediterranean region), less developed regions (Africa), and island countries.

In terms of adaptation strategies for water resources in developed countries, the Netherlands has developed new measures and is now a pioneer in the fight against climate change. The Netherlands no longer sees people and water as opposites. Water is no longer a threat but an opportunity, and it is pursuing a strategy of "human-water symbiosis": on the one hand, it is responding to the rise in sea level by constructing over-water dwellings with polystyrene coated on the outside of the foundations to ensure that the dwellings will float on the surface of the water and will not sink. On the other hand, the Netherlands is building a water channel for the IJssel River. In addition, Nijmegen, a border city with Germany in the southeast of the Netherlands, has also begun constructing a large-scale project to take advantage of the Waal River running through the city to dig an artificial stream for the river. The project plans to build an artificial island at the confluence of the old and new tributaries of the Waal River, with office buildings, parks, homes, and stores, similar to Manhattan in New York. Therefore, although in some countries, due to sea level rise, coastal areas are facing the danger of being submerged; however, the idea of "people and water coexistence" has greatly influenced the mainstream concept of the Netherlands throughout China, as people's lives are beginning to be closely related to water.

In terms of water management and allocation, public utilities are responsible for water management in the Netherlands. Due to the evolution of the Dutch management system, water management has become localized and centralized. As a country with a decentralized territory and a complex water system, the Dutch water management institutions can be divided into three levels, national, provincial, and water board and municipal. Each level has its own legislature and decision-making executive and performs the corresponding duties, and different departments are responsible for implementing different policies and tasks in accordance with the legislative authorization.

In terms of China's water resources adaptation strategy to climate change, China is experiencing an increasingly serious conflict between water supply and demand, and combining conventional water resources allocation methods with emergency management measures to improve China's water resource management and regulation capacity to proactively respond to climate change. Based on multiple considerations, including the implementation of the strictest water resource management, efficient utilization, rational allocation, optimal dispatching, strengthening protection, extensive open source, timely water transfer, and scientific management, we propose strategies and measures to improve the capacity of water resource management and regulation level. This is mainly reflected in the following five aspects.

(i) Strictly manage water resources and rationally control water demand growth. Control water use efficiency, decomposition of the total water resource control implementation procedures, implementation of the most stringent water resource management system, and establishment of the control of pollution "three red lines" assessment index system. The establishment of a scientific and technological water management system requires the improvement of laws and regulations related to water resource management to cope with climate change and the establishment of sound national policies and related laws and regulations on water resource management.

(ii) Widely implement water-saving technologies and improve the efficiency of water use in river basins. China's agriculture accounts for about 70% of water consumption, and agricultural water conservation is the key to water conservation methods. Construct a water-saving irrigation system in the basin, adjust the agricultural planting structure, and research and develop new water-saving technologies, including dry farming, non-sufficient irrigation, and biological water-saving technologies. Form an integrated system of water conservation in the basin, including comprehensive water conservation techniques for irrigation districts (adapted to arid and semi-arid areas), technically and economically appropriate industrial water conservation and emission reduction techniques, and urban water recycling techniques.

(iii) Improve the implementation of emergency management measures and enhance the ability to implement measures for emergency dispatching. Global climate change has increased the probability of extreme events such as droughts and heavy rains associated with temperature and rainfall, and it is necessary to improve the emergency response capacity of water conservancy for flood prevention and drought mitigation. Flood prevention and mitigation measures are facing a shift from the previous response direction of controlling floods as the main objective to scientific and modernized management of floods, improving flood early warning and forecasting capabilities, identifying the occurrence and development patterns of extreme events in a timely manner, and establishing a scientifically reasonable disaster evaluation index system. It is necessary to formulate a set of reasonable emergency response scheduling plans for extreme events to provide a guide to action in response to the occurrence of extreme events. The establishment of a standardized system is indispensable to ensure the effective implementation of emergency dispatching.

(iv) Enhance the capacity of sewage treatment and seawater and rainwater utilization. Compared with the severe drought in the northwest, the south, especially the coastal area, has abundant rainfall. However, the sewage treatment rate in the eastern coastal areas is not high, and the seawater utilization capacity should be improved. In the Yellow River Basin, the problem of over-exploitation of water resources is prominent, with over-exploitation of surface and underground water sources. In the past 20 years, the runoff water consumption rate of the Yellow River Basin has reached 70%, and the amount of groundwater resources extracted in the basin has reached 14 billion m^3, and even the amount of groundwater resources extracted in some basins has seriously exceeded the standard, resulting in the large-scale formation of groundwater landing funnels. Unconventional water resources, such as brackish water, rainwater, and recycled water, which are more widely distributed, have failed to improve utilization efficiency. There is an urgent need to strengthen the capacity of non-conventional water resource utilization and adjust the pattern of water resource utilization to achieve the goals of multi-source complementarity and abundance-demerit transfer. Improve the scheduling rules for water resources in reservoirs to rationalize them scientifically, and raise the level of utilization of water resources such as rainwater and floodwater.

(v) Strengthen the construction of water conservancy facilities. The establishment of water conservancy engineering measures is an effective means of coping with flood control and drought relief, and can significantly alleviate the outstanding contradiction between the supply and demand of water resources. Increase the infrastructure construction of regional water conservancy projects such as reservoirs, river embankments, and flood storage areas. Accelerate the construction progress of projects such as the South-to-North Water Diversion Project and consider the construction of necessary water storage facilities in areas with greater resource potential to improve the ability of water resources to cope with climate change.

In terms of the adaptation strategies of water resources in less developed countries, to solve the problem of insufficient water supply caused by climate change and achieve the goal of sustainable development in the context of climate change, there is an urgent need to strengthen the national capacity for water resource management and to develop a modern scientific management model for water resources in African river basins. The Niger River Basin in West and Central Africa covers nine countries (Nigeria, Burkina Faso, Guinea, Chad, Niger, Cameroon, Mali, Côte d'Ivoire, and Benin). The impacts of climate change on the water cycle in the basin are highly controversial, with some predictions of more arid impacts on the Niger River Basin, and other predicting it to be wetter than before. The uncertainty of such impacts makes investment in increasing the IWRM (Integrated Water Resources Management) capacity of countries in the basin even more important, which will give the basin the desired development in the current climate change environment. There are six main adaptation strategies for the countries covered by the Niger River Basin.

(i) Improve reference information and data related to decision-making to support national water resource strategy and decision-making. Scientific decisions cannot be made without accurate data, and the first thing to do when formulating national climate change adaptation strategies is to obtain reliable reference information and data.
(ii) Ensure the implementation of decision-making. Identify the needs of relevant organizations for information sharing, enhance the sustainability of water storage infrastructure, mitigate the impacts of climate change, and accelerate the development of carbon energy to ensure the effectiveness and impact at the grassroots level.
(iii) Strengthen Infrastructure. Develop multi-purpose infrastructure targeting power generation, irrigation, transportation, and minimum flows to reduce infrastructure construction, operation, and maintenance costs and improve infrastructure operational efficiency.
(iv) Establish run-of-river hydropower plants that provide low-carbon energy. The natural geographic advantage of being adjacent to the Niger River is that it makes the surrounding area rich in hydroelectric energy resources. By establishing run-of-river hydroelectric power stations, this resource advantage is fully utilized, and the consumption of fossil energy is reduced, which is a low-carbon way of utilizing energy.
(v) Optimize the function of reservoirs. Reservoirs have important roles such as flood control and irrigation, which are highly important for the sustainable development of the basin. Optimizing water storage in reservoirs to improve water redistribution and to maintain the amount of dry water is an indispensable part of climate change adaptation.

(vi) Improve the risk resilience of poor communities. Equip the rural poor with small-scale water storage facilities to reduce the impacts of natural disasters such as drought, and improve their risk resilience and adaptive capacity.

Water resource adaptation to climate change in small island states is generally a sustainable development strategy. More recent studies have identified key adaptation areas including water resource basin management; island and reef protection; agriculture and forest management; biodiversity conservation; security of energy storage and extraction; further development of renewable energy sources and optimization of energy consumption; construction of seawalls, drainage systems, water tanks, and logging bans, and relocation and reinforcement of existing infrastructure. Adaptation strategies include:

(i) Some small island states (e.g., Seychelles, Tuvalu, Barbados, and Maldives) have already begun investing in climate change adaptation strategies, including desalination of seawater to compensate for current and projected water shortages.
(ii) Establishing the response of terrestrial upland and inland ecosystems to changes in mean temperature and rainfall, as well as to temperature and rainfall extremes.

Adaptation Strategies for Water Resources to Climate Change in Developed Countries (Netherlands)

Water resource management in the Netherlands includes three features: natural river basin management, emphasis on the role of law in guaranteeing the operation of the river basin management system, and emphasis on public participation.

(i) Natural river basin management

In Dutch water management, the focus is on the integrated management of water resources and the use of coordinated approaches. Coordination takes the form of consultations, administrative decisions, and court judgments. Conflicts between regions, sectors, institutions, and individuals in the development, utilization, management, and protection of water resources are first resolved by consultation between the parties, and if an agreement is reached, the agreement is signed and complied with. Inter-provincial conflicts over important rivers are to be determined by the Ministry of Water Resources or its subsidiary bodies, and if the parties concerned do not agree, they may request a higher administrative review by the authority that issued the administrative decision. In the Netherlands, administrative adjudication is an important tool for dealing with and harmonizing conflicts in water resource management. Integrated water resource management requires the unified management of surface water and groundwater and of water quantity and quality to ensure the sustainable use of water resources.

(ii) Emphasis on legal safeguards for the operation of river basin management systems

The Dutch government has stipulated that the development of natural resources and the improvement of shipping conditions, both of which are to be developed in harmony, must be consistent with the water quality improvement measures needed to meet the harmonious development of recreation, nature, and landscape. The government, provinces, and territories must ensure that the project meets water-related needs. Water-related national development plans and policy documents should include specific steps on how to manage water resources. Water resource management legislation affects the entire field of water resource management and can be seen in existing relevant legislation. It complements existing laws and regulations, coordinates the overall management of water resources, and is a specialized regulation for quantitative surface water management.

(iii) Emphasis on public participation

Public participation is fully emphasized in water resource management, that is, all sectors related to water are involved in the management and take coordinated action. In the "Water Parliament" management system, water users, together with local governments and experts, become members of a basin committee to consult and make decisions on the development, utilization, and protection of water resources in the basin.

11.3.3 Strategies for Adaptation of Human Health to Climate Change

The health status of the labor force has long been one of the key concerns of development geography, and as the level of economic and social development continues to rise, health issues are gradually becoming a global concern. Climate change can lead to the emergence of widespread health problems (Watts et al., 2015). Therefore, the development of regional human health development strategies adapted to climate change is practically related to everyone and is an important prerequisite for the realization of comprehensive human development.

In terms of health adaptation strategies to climate change in developed countries, climate change causes variable rainfall, stagnant water, etc., leading to mosquito breeding and global warming, which provides conditions for the creation and spread of dengue fever. The public health response to the high incidence of the dengue transmission phenomenon includes:

(i) Stronger vector control, focusing on areas where dengue cases have been reported.
(ii) Enhanced case surveillance.
(iii) Enhanced human source material security measures for cell, tissue, and organ donations, including screening for viral genomic testing for each donation.
(iv) Social mobilization, including mobilization of local representatives in cities and urban areas, organizing awareness-raising campaigns in schools and door-to-door informing the population on how to avoid contact with and remove mosquito breeding sites. Production of specific guidelines on dengue control and distribution to local and regional elected officials and communities.
(v) Risk-specific communication aimed at raising awareness of dengue among the public and health workers, including posters at China's points of entry, verbal communication on flights to and from Reunion Island, weekly press releases, and local television and radio campaigns.

China's strategy for addressing the impacts of climate change on health includes five areas, namely establishing mechanisms, upgrading capacity, strengthening research, increasing publicity and strengthening international cooperation. These are explained below:

(i) Establishing and improving mechanisms for actively responding and adapting to climate change. Formulate important national strategies and policies to cope with the current climate change environment. The health sector plans and summarizes the adaptation work to cope with the impacts of climate change on an annual basis and further strengthen the leadership and organizational mechanisms and to enhance the formulation and implementation of work plans and policies in the health field.
(ii) Improve adaptive capacity. Improve the national infectious disease monitoring and early warning information reporting system and build an integrated monitoring network covering urban and rural areas. Use various models to build early warning systems for different diseases so that healthcare, disease control and decision-making command organizations can respond quickly to sudden-onset diseases in a relatively short time.
(iii) Increase research related to the impact of climate change on the health aspects of living things. Prioritize the impact of climate change on human health and increase the scope and intensity of pilot studies. Conduct pilot studies on monitoring and control, and design and field validate climate change adaptation technologies.
(iv) Increase publicity and raise public awareness. Actively carry out publicity campaigns on how to deal with climate change, especially extreme weather, and raise the importance of national and people's awareness about self-protection. Enhance public awareness of self-protection and health education and promote communication between policy makers and scientists.
(v) Strengthen international cooperation and communication. Establish communication and coordination between China and the international community in all relevant fields, and strengthen research and technical cooperation.

Climate change threatens the food safety and hygiene, China's high population density, higher urban population density, and increasing international exchange of goods, services, and people will increase the difficulty of controlling major infectious diseases that threaten public health and its transmission risk. Since 2013–2018, the number of people infected with H7N9 avian influenza in China has continued to increase, and it has been particularly severe in the mid- and low-latitude, coastal areas (Wang et al., 2017). Therefore, it is necessary for national and local governments to take appropriate preventive measures. Depending on the outbreak monitoring work, they should guide provinces to strengthen monitoring to ensure timely and effective detection and response to the outbreak, emphasize prevention and control work, and guide provinces to take advantage of the current low incidence period to review the prevention and control efforts made to facilitate the implementation of long-term measures and reduce the risk of infection in poultry by starting at the source. The agriculture sector should implement mass vaccination of poultry using the bivalent influenza A H5 and H7 vaccines, increasing vigilance and strengthening local government control. The government has urged provinces to emphasize infection prevention and control and remain vigilant to ensure timely and effective case detection and management, pay attention to the impacts of climate change on the health of the population, and assess and develop emergency plans for people's health affected by climate change. Improve public healthcare facilities in areas vulnerable to climate change, improve epidemiological characterization, laws, adaptation strategies, and technological research on climate change-related diseases, particularly infectious diseases and outbreaks. Establish a mechanism for psychological interventions after extreme weather and climate disasters. Conduct regular risk assessments to prioritize seasonal and regional control and strengthen surveillance and prevention of vector-borne diseases under climate change.

In less developed countries, to improve healthcare, Nigerian states have used financing to strengthen its healthcare system (Eichler et al., 2013). Today there are more women receiving antenatal care, more skilled and competent birth attendants, and more vaccinated children, which has greatly improved local health. Specific measures also include the following four areas.

(i) Adjustment of human resources for health. in the face of the inequitable distribution of health professionals to urban centers and the shortage of some categories of health human resources, the Nigerian government has helped to focus on human rights in service delivery through the development of the Human Rights Plan (HRP).

(ii) Provision of essential medicines for medical treatment, the National Drug Policy has been approved and a strategic plan has been developed to properly integrate traditional medicine into the Nigerian health system and to promote the African Regional Strategy for the promotion of the role of traditional medicine in the health system.

(iii) Building an information-based health system and developing a decision support system. Nigeria provides timely and quality information for decision-making through the National Health Information System (NHIS) to develop functional databases of basic indicators that consider the health MDGs.

(iv) Basic national health research will provide information and inputs for decision-making in the selection of cost-effective and efficient health interventions. Emphasis will be placed on promoting research in areas such as reducing disease risk factors and improving health systems and health promotion.

Health Adaptation Strategies for Climate Change in Developing Countries (Philippines)

Chronic low routine immunization coverage and hesitant fear of vaccination are the root causes of disease outbreaks. Therefore, to adapt to climate change, strategies in the health sector are:

(i) Strengthen the surveillance of acute rashes and fevers to ensure rapid detection and confirmation of measles cases and prompt public health action. Ensure that the various protocols for situation reporting, sample collection, case confirmation and rapid laboratory testing for geno-typing, and epidemiologic linkage of confirmed cases are updated and strengthened by all concerned.

(ii) Develop a risk communication plan.

(iii) Strengthen outbreak preparedness; disseminate health information on avoiding measles for the benefit of the public.

(iv) Strengthen the entire immunization system by increasing the coverage of routine vaccinations; increase the coverage mass vaccinations, thereby reducing the number of people contracting measles.

(v) Plan and implement preventive supplementary immunization campaigns to improve the immunity of the population.

(vi) Ensure the development of a national action plan for the elimination of measles and its implementation efforts. Develop rational and scientific regional guidelines to prevent and respond to measles and rubella outbreaks.

11.3.4 Strategies for Adaptation of Energy Supply and Demand to Climate Change

Energy is a necessity for economic and social development and an important strategic reserve resource, and its importance to regional development has made the issue of energy a focus of attention in development geography. Energy issues mainly include energy access and security, and the impact of climate change on energy security is particularly important. The pollution caused by energy utilization also requires attention. Solving the energy problem is a major issue for regional development, and the applicable strategies for climate change adaptation in different countries and regions are analyzed below.

In terms of energy adaptation strategies in developed countries. The Japanese government has made significant efforts to address climate change at the policy level, and as part of the climate change adaptation policy system, the Energy Basic Plan has been released to the fifth issue (July 2018), which outlines Japan's development goals for 2030 and formulates the basic policy. The so-called development goal is to reduce energy import dependence and achieve an energy self-sufficiency rate of 24% by the target year. The basic policy is "3E + S," i.e., prioritizing the guarantee of a stable supply of energy (Energy security), and improving economic efficiency, (Economic efficiency), cost reduction, and environmental suitability (Environmental protection), taking energy safety as a prerequisite. The Energy Basic Plan also proposes building an energy-saving society, accelerating the development of renewable energy, restarting nuclear power generation under the premise of safety, efficient and clean utilization of fossil fuels, building a hydrogen society, and popularizing emerging energy technologies to promote Japan's energy transition.

Some important types of energy (e.g., hydrogen energy) have also received sufficient attention, with the government's energy ministry releasing the Hydrogen Energy Basic Strategy in December 2017, which plans to take the lead internationally in realizing a hydrogen energy society. The Hydrogen Basic Strategy proposes to build a stable, low-cost commercial hydrogen supply chain, develop key technologies for internationalization of the hydrogen supply chain, accelerate the promotion of hydrogen-powered electric vehicles, plan for hydrogen and improve the hydrogen supply chain, and overcome key scientific issues in order to realize the industrial use of hydrogen. In addition to specific policies such as research and development of key technologies for the utilization of hydrogen energy, we have set development targets for 2030, including the establishment of a commercialized hydrogen supply chain and the achievement of 800,000 hydrogen fuel cell vehicles.

The nuclear accident in Fukushima in 2011 dealt a huge blow to Japan's nuclear energy industry, leading to a nationwide shutdown of all nuclear power plants until late 2016 when some of them were restarted with government support. In July 2017, the Japanese government adopted the Basic Principles for the Utilization of Nuclear Energy, which propose the rational and appropriate use of nuclear energy to clarify the system of responsibility and implement risk management and control; thus contributing to the environment, isotope technology, and strengthening basic nuclear energy research and development.

Taking photovoltaic power generation as an example, Japan has set up R&D programs such as "high-performance/high-reliability photovoltaic power generation cost reduction technology," "photovoltaic power generation system efficiency improvement and maintenance management technology development," and "photovoltaic power generation system recycling technology development." The R&D projects cover the entire life cycle of photovoltaic technology. Universities, enterprises, and research institutes are organized to carry out joint R&D, and enterprises are actively promoted to apply, demonstrate, and promote the technology. Below are the characteristics of Japan's energy policy:

(i) Diversification and clean and efficient utilization of energy sources. Japan has taken the lead in proposing a hydrogen energy development strategy at the national level, and has incorporated "clean and efficient" into the basic guidelines of energy policy.
(ii) Science and technology innovation is important in achieving energy policy goals, and collaborative innovation is an important organizational form of science and technology innovation. The Japanese government attaches great importance to science and technology, and the formulation of energy policy also reflects this. The government-led establishment of the new energy industry technology comprehensive development organization strongly links the government, scientific research institutions, and enterprises, and becomes a bridge between the three, which is conducive to the implementation of policies and transformation of scientific and technological achievements.

In China, the general direction of energy adaptation to climate change is to emphasize the strengthening of China's cooperation in all aspects of energy, build and improve the legal security system of energy system adaptation to climate change, strengthen the scientific and technological support of energy system adaptation to climate change, develop an integrated energy system, and develop corresponding adaptation strategies from these aspects.

(i) Strengthen China's cooperation in all aspects of energy. From the perspective of the long-term development of China's energy strategy, strengthen cooperation with other countries around the world. International cooperation among countries must pay attention to the urgent problems of developing countries, which are more sensitive to the adverse effects of climate change in the context of global climate change. In terms of clean energy, low-carbon technologies are adopted, energy efficiency and energy conservation measures are improved, smart grids are developed, and renewable energy is promoted (Yang, 2016). A new international cooperation system for clean energy is gradually being established, and a carbon emission market is being set up around the world to promote the development of clean energy technologies and realize the transformation of global energy use. In addition, it is important to strengthen inter-regional cooperation in the field of new energy, strengthen the dialogue between governments under the synergy of regional organizations, increase trust among countries, and expand the field of cooperation. China's various departments of coal, chemical industry, and electric power break the boundaries of the industry, and cooperate with each other. China should take advantage of the promotion of the "One Belt, One Road" strategy, and it is expected that the goal of intra-continental, inter-continental, and global networking can be achieved gradually. The actual energy development strategy of each country will be formulated, the cooperation between the government and the enterprises will be strengthened, and the importance of low carbon energy will be raised.
(ii) Establish a sound legal protection system. At the legal level, the absolute and relative caps on the use of renewable energy should be regulated, and national

laws and regulations should be based on the needs of the national development strategy, with clear aggregate targets and the establishment sound policies. Environmental and social policies related to the development of renewable energy should be regulated. In addition, the rights of the public to make decisions and the administrative behavior toward renewable energy development should be regulated, including citizens' participation. Furthermore, the citizens' right to participate in policymaking and administrative behavior, including both citizen participation and supervision should be regulated. The legal obligations of electric power companies, including both the construction of the electric power system and the service of electric power supply should be stipulated, emphasizing both the legal and social responsibility, and the local renewable energy legislation should be strengthened.

(iii) Strengthen the support of science and technology system. Including energy internet development (smart energy system, integrated energy system), new energy system development/integrated energy technology, smart grid technology, integrated data mining, and cloud computing technology.

From 2014 to 2018, the energy sector in Egypt underwent a remarkable transformation. By implementing strong social protection mechanisms and increasing funding for social programs, the Egyptian government was able to drastically reduce energy subsidies, with subsidy spending now down to 3.3% of GDP. The government has enacted several laws and regulations to promote private sector participation in the renewable energy, electricity, and gas sectors, and has built 1500 megawatts of new solar power capacity. Energy Africa has adopted strategies to adapt to climate change.

(i) Grid extension and transmission network expansion. Expanding renewable energy capacity through grid extension and transmission network expansion, adopting innovative off-grid generation technology solutions (Eberhard et al., 2014), and developing regional power pools to improve the efficiency of energy services.

(ii) Increase the development and deployment of a range of new technologies, solar storage solutions (Timilsina et al., 2012), smart meters, mobile payments for utilities, satellite mapping and imaging, high-voltage DC transmission, and solar home systems and mini-grids.

(iii) Promotion of renewable energy projects. Grid-connected solar projects with private investment were operational within two years, with competitive tariffs. In Zambia, electricity prices reached the lowest level in Africa ($0.06 per kWh) through an auction. The program has now been expanded to countries such as Ethiopia.

(iv) Improve climate resilience and promote sustainable transportation. To reduce greenhouse gas emissions and facilitate access to work, Senegal built an 18.3 km fully segregated bus rapid transit system in Dakar, which has the potential to be replicated and scaled up across the region.

National Energy Adaptation Strategies to Climate Change in Different Regions

(i) Dominica in the Caribbean—Hydroelectric Power Generation

Dominica in the Caribbean relies on hydroelectricity as an important part of its energy supply. Research and development on energy efficiency and programs appropriate for small island states, such as solar and wind power, can contribute to adaptation and mitigation strategies and the realization of a sustainable development vision.

(ii) Seychelles—Consideration of Climate Change in the Development of National Policies

Due to the special characteristics of island countries, the government of Seychelles must consider the impacts of climate change in its infrastructural development process, and the government's energy committee considers the potentially catastrophic consequences of global climate change when preparing relevant projects and national energy policies that come from international organizations and the government and proposes efficiency improvement measures for renewable energy production and consumption to control greenhouse gas emissions.

(iii) New Zealand—Wind power, hydroelectric power generation

Climate-induced reductions in water supply have led to reductions in water supplied to hydroelectric turbine units and cooling water for thermal power plants. In New Zealand, increased westerly wind speeds promote wind power generation and allow precipitation to overflow into the main hydroelectric catchment of the South Island and increase winter rains in the Waikato catchment. Warmer climates increase snowmelt, the ratio of rainfall to snowfall, and river flows in winter and early spring, which contribute to hydropower generation in times of peak energy demand for heating.

(iv) Strengthening energy cooperation

SIDS should strengthen international and interregional cooperation in the field of energy and develop partnerships that emphasize national energy security from a sustainable development perspective to expand access to energy services. Greater access to energy services means greater risk tolerance and a more stable domestic energy supply.

11.3.5 Strategies for Adaptation of Natural Disasters to Climate Change

Natural disasters pose a great threat to regional development, and development geography focuses on the impact of natural disasters in several dimensions. The occurrence of natural disasters not only affects developed countries or regions but also has a stronger impact on developing countries or regions. Disasters caused by climate change are often large-scale, and global warming can cause glaciers to melt and cause sea level rise, which is often devastating for some island countries. Therefore, exploring regional development applicability strategies for disaster adaptation to climate change from the following aspects can help enhance the ability to adapt to climate change.

In terms of the adaptive strategies of developed countries to cope with climate change. By the end of the twentieth century, the Netherlands had built a high-standard flood defense engineering system, consisting of an early warning system, a safety inspection system, and an evaluation system together with a strict management system, which has strengthened the Netherlands' flood control capability. Some scholars have proposed resilience strategies for flood management in the Netherlands based on spatial planning, one of which is the zonal flood retention strategy for the purpose of flood retention and peak shaving, and the other is the green river strategy that utilizes the natural reserves or agricultural areas along the river. The zonal flood retention strategy is to simulate the development process of various floods exceeding the protection standard and estimate the economic losses caused by different levels of floods through model simulation methods, then, apply the zoning thinking to further divide all the perimeter zones along the coast and classify a part of them as flood retention zones, and determine the inundation sequences, depths, and frequencies of a series of different zones to achieve a very low flood frequency in the other zones (<1/10000) and a higher standard of safety than before. The goal of the zoning flood control strategy is to achieve flood control and peak shaving in the event of a flood disaster; therefore, the overall planning of the entire area should follow certain principles: the planning of land use should prioritize the flood control function as the first element to be considered to meet the criterion of safety first and eliminate non-natural disaster factors that may cause serious consequences of flooding in the area. The idea of zoning should be considered, according to the frequency of regional flooding, divided into small blocks with different frequency levels to form a large enclosure. In this way, even if a flood occurs, it will not affect the entire enclosure, and the flooding will stop at one or several small blocks, thus greatly reducing the disaster losses. Based on the principle of loss minimization, higher frequency blocks can be set up in areas with less economic damage, such as agricultural areas or natural ecological areas, whereas high frequency should be avoided as much as possible in blocks in populated and economically developed areas. The green river strategy is to create a floodway next to the original river channel, which is usually in a spacious, narrow area near the river. As the primary tool for controlled flood diversion, the diversion channel can increase flood flows and reduce flood levels in the main channel. Green rivers are called green because they are selected according to the principle of greenness. Floodways are usually located in nature reserves or agricultural areas along the river, where inundation losses are low and recovery is high, and due to the green vegetation cover, the river channel remains green and waterless all year round, except during flooding periods.

Green rivers include three types, namely natural development, biodiversity, and multifunctional types when implemented. (i) The natural development type is simpler in terms of the scale of the project, usually only needing to construct dykes and their supporting facilities for flood relief, and additional auxiliary settings hardly need to be constructed. The labor input is also low, and there is no need to employ specialized managers for active short-cycle maintenance work; therefore, the capital investment cost is low. (ii) In the biodiversity type, more labor input is required, and the production and development of plants and animals need to be considered in the construction and later maintenance of the project. A rich variety of plants and animals will help reduce the vulnerability of the river and improve the self-recovery ability of the river ecosystem. (iii) Multi-functional type is a combination of disaster prevention, recreation, and ecological functions, not only for flood prevention but also for public use for physical and mental relaxation and recreational sports.

In terms of China's disaster risk adaptation strategy to climate change, China faces the challenge of serious disaster risks, such as increased aridification in Northwest China, increased extreme droughts in North China, and more frequent droughts and floods in the Yangtze River Basin, all of which are closely related to climate change and human activities, and the frequency and intensity of these climatic hazards are in turn linked to climate change at larger scales. The main adaptation strategies are as follows:

(i) Establish an early warning system. Disaster prevention and mitigation departments release disaster monitoring and forecasting information in a timely manner and establish relevant disaster and secondary and derivative disaster monitoring and forecasting and early warning linkage mechanisms with other relevant government departments such as transportation, water conservancy, and fire protection. Establish a disaster coordination system to realize rapid disaster intelligence transmission and emergency response through the network.

(ii) Establish social resilience. Socially marginalized groups are particularly vulnerable to natural disasters through a combination of factors, namely geography, financial, socio-economic, cultural, and gender status, and in terms of access to services, decision-making and justice. Inclusive community resilience programs strengthen the resilience of these vulnerable groups by promoting a community-led perspective on risk.

(iii) Improve drainage systems. Improving drainage systems can play a role in flood mitigation or avoidance. In areas prone to flooding, special attention should be paid to the maintenance and improvement of the drainage system, especially in areas where the drainage system is old and cannot cope with today's drainage, and where major works are needed to improve the drainage system.

(iv) Establish disaster relief stockpiles. The establishment of flood and cyclone shelters can reduce the impact of disasters on people, plan and construct specialized warehouses for the storage of disaster relief supplies, and further strengthen the construction of material reserves, basically covering China's disaster-prone areas, and improve the agreement reserve system of disaster relief supplies.
(v) Carry out disaster emergency assessment technology. Actively carry out the construction of technical support capacity for disaster emergency assessment. Relying on the construction of a four-tier disaster reporting and management system, the research and development of disaster process simulation models, environmental disaster reduction satellite resources and other satellite remote sensing data, and the use of unmanned aerial vehicles and disaster site collection, will comprehensively enhance the capacity of emergency response assessment for major disasters, thereby minimizing the impact of disasters on human beings.
(vi) Expand the resilience of cities. Rapidly urbanizing areas require substantial and well-planned infrastructure investments to meet growing resource demands, promote economic growth, and foster social development. We should strengthen urban adaptive planning, increase investment in urban adaptive construction, and better cope with climate hazards through improved urban disaster adaptation patterns and sustained functional operation.

In terms of disaster risk response strategies for climate change adaptation in less developed countries, some countries in Africa are faced with desert locust epidemics and need to take urgent action to prevent the spread of the epidemic and save crops. The main measures are as follows:

(i) Adopt desert locust emergency action. Emergency desert locust control action in Yemen is required to protect crops, mitigate the risk of infestation in Yemen, and prevent locust swarms from invading neighboring countries. In Yemen, desert locust survey and control operations are hampered by insecurity in locust-breeding areas and insufficient capacity in terms of equipment and operating funds. Conduct a much-needed locust response campaign and implement a series of urgent measures to support monitoring and control operations in breeding areas.
(ii) Initiate the process of purchasing additional equipment. The process of purchasing additional equipment needs to be initiated to increase the capacity of the Ministry of Agriculture to control locusts in outbreak areas in a timely manner. The main beneficiaries of the operation will be farmers, livestock breeders, and nomads residing in desert locust-infested areas, whose livelihoods will be unaffected against the background of the expected damage to agricultural production and pasture caused by the intensification of locust infestation. Crops will be saved, thus preventing crop losses. Crop losses could lead to increased hunger and poverty in rural areas plagued by protracted conflict.
(iii) Conduct regular ground survey monitoring. Given the severity of the current locust epidemic, it is important for all countries to monitor field conditions through regular ground surveys and to take the necessary control measures

when major outbreaks are detected. The spread of pest and disease epidemics can be better prevented through regular monitoring, and losses can be kept within tolerable limits.

(iv) Build an information service system to provide forecasts. Operate the Desert Locust Information Service System (DLIS) to receive and analyze data from countries affected by locust epidemics to assess the current locust outbreak, provide forecasts, and issue alerts and warnings as necessary to inform the international community of the latest situation and threat of the outbreak.

Small island states are increasingly adapting to climate change and disaster risk reduction through infrastructure protection, coastal polders, coral protection to help reduce the impact of storms and the introduction of salt-tolerant plants and fish species to help combat invasions caused by storms and sea-level rise. Kiribati, for example, is scaling up climate change adaptation measures across China, focusing on the most disaster-prone sectors. These include improved water supply, coastal management protection, strengthened laws to reduce coastal erosion, and population settlement planning to reduce individual risk. The main countermeasures are twofold. The first is to improve immigration policies. International migration is particularly important for low-lying small island states whose entire territories face the prospect of being inundated by rising sea levels. Migration from low-lying islands in the Pacific to other Pacific countries at higher elevations, such as Fiji and Vanuatu, is expected to increase. The Government of Kiribati in particular has stepped up its efforts to develop a combination of strategies, including the introduction of an advance voluntary migration pathway, especially for labor migrants, and the purchase of 20 km^2 of land in Fiji to prepare for potential population movements. Since historical relations differ from country to country, there is a need for well-developed migration policies. Second, early warning mechanisms need to be improved. The risks posed by such disasters could be reduced by improving early warning mechanisms and establishing early warning systems. Collective action among participating countries, regions, and international organizations, as well as wide dissemination of information, communication and response, and building the capacity of Governments and local communities to prevent and mitigate disasters, are required to ensure effective cooperation in risk knowledge and disaster monitoring and warning services. The establishment of the last line of defense for early warning systems at the community level is critical to the success of the overall early warning system.

Disaster Risk Adaptation Strategies for Climate Change in Developed Countries (UK)

The United Kingdom (UK) Climate Impacts Program (UKCIP) has conducted detailed regional and industry studies to explore the challenges associated with adaptation. A flood risk management strategy is being developed based on risk assessments of sea level rise and increased rainfall. Projected changes in climate type, storm type, and rainfall type are expected to result in increased flood risk. In contrast to the Netherlands, the UK flood protection system is designed to cope with the largest floods expected to occur every 100 to 200 years. With rising sea levels and predictably more frequent storms and rain, the flood protection strategy is currently being revised. Estimates from the insurance industry suggest that the number of households at risk of flooding will rise from 2 million in 2004 to 3.5 million in the distant future if the flood protection infrastructure is not reinforced. Roughly half of China's flood infrastructure is in good condition. The Environment Agency, a government agency, has proposed at least \$8 billion to toughen the Thames to strengthen flood defense structures in the near term The annual cost of flood management and coastal erosion is currently around \$1.2 billion per year, and the massive floods of 2007 have prompted renewed calls for increased spending.

11.3.6 Strategies for Adaptation to Climate Change in Terms of Anti-poverty and Employment

Poverty and employment are often widespread social problems and are important concerns in development geography, and climate change tends to exacerbate the problem of poverty and lead to increased unemployment, which has a negative impact on regional development. In this context, developed, developing, and underdeveloped countries, as well as island nations, have proposed adaptive coping strategies for regional development in response to climate change, as described below.

In terms of developed countries (UK), poverty and employment adaptation strategies to climate change, the main response measures (IPCC, 2014; UNFCCC, 2014) are as follows:

(i) Reorganize the social security system. The social security system should be more oriented to the more vulnerable households, such as providing more support to low-income and elderly households, for example, by increasing the minimum wage. In terms of housing, the state should provide more low-cost housing and guaranteed commercial housing for low-income families and strictly prohibit the rapid increase of housing prices to improve the living conditions in poor urban areas.

(ii) Develop a clean growth strategy. The British federal government, through the development of the growth strategy, has further proposed the development of a clean growth strategy to reduce carbon dioxide emissions at the strategic level, with a two-pronged approach from both the supply side and the consumption side, and a total of nearly 50 measures have been proposed to realize the emissions reduction target. In addition, it will increase the number of social jobs and further reduce the number of people living in poverty, and ultimately achieve the goal of adapting to climate change. The UK has developed a clean growth strategy to increase productivity, create quality jobs, solve the employment problems of some poor people, and protect the natural environment. It is predicted that by 2030, the growth rate of the low-carbon economy will reach 11% and the number of jobs will reach 430,000 jobs.

In terms of poverty and employment adaptation strategies to climate change in India, an effective way and important means to lift poor areas out of poverty is to increase the employment rate. The Indian government has implemented several policies and measures to address the employment issue, and the employment rate has increased, thus improving the poverty situation of part of the population. The following seven points are the specific implementation strategies.

(i) Combine market guidance and legal protection. This two-pronged approach is conducive to increasing employment opportunities for farmers. In market guidance, the Indian government on the one hand increases the demand for relevant jobs; on the other hand, through the training of farmers to improve their employment competitiveness and labor productivity, the policy is biased toward labor-intensive agriculture. At the legal level, the state provides employment security in rural areas, unemployment benefits and other aspects of the provisions of the specific content of the formation of the "National Rural People's Employment Security Act." The bill also specifies the government's responsibilities, such as providing sufficient employment opportunities for families performing manual labor. The bill was a milestone in eliminating absolute poverty in rural areas and among farmers.

(ii) Implement family planning policies to control population growth. To control population growth, the Indian government has implemented a nationwide family planning policy starting from the fourth five-year plan of China, with a view to slowing down the rate of population growth and reducing the pressure of competition for jobs.

(iii) Reform the education system. Reform the existing higher education system to align with national and social development needs, establishing relevant programs that match professional fields with employment opportunities and are market-oriented. This will help improve the employment rate of graduates and significantly reduce the employment challenges faced by college graduates. India's education department has introduced vocational education into the university curriculum and organized the establishment of a special management committee, which has set up a relatively complete vocational education system involving the compilation and review of teaching materials, curriculum teaching, teacher training, and supervision and management. India has also implemented the Rural Service Center Scheme and the Rural Engineering Scheme, providing self-employment assistance to university graduates and encouraging them to go to rural areas for employment and entrepreneurship.

(iv) Guarantee employment rights of women in India. As women in India are often discriminated against by job providers when seeking employment due to historical and religious reasons, there is a need for specific initiatives to safeguard women's right to employment. To this end, the government has launched the Jawahar program, which requires job providers to allocate a specific percentage of their jobs to women, thus protecting women's legal rights and improving their social and economic status to a certain extent.

(v) Rationally distribute rural land. Rationally distributing land, which is the basis of agricultural production, can be an institutional guarantee for successfully overcoming the problem of poverty. The Indian government has realized a green revolution in land by abolishing the outdated agricultural and land systems and implementing a nationwide "landholding ceiling" system. This has increased land resources and the viability of agricultural operations and enabled China to become self-sufficient in food grains. Currently, the lessors of land are small landholders. Therefore, to combat poverty, particularly rural poverty, it is necessary to carry out a reasonable reform of the land system.

(vi) Increase the severity of punishment for corruption. Corruption is the biggest obstacle to the implementation of poverty eradication policies, and the halfway implementation of many poverty policies has been marred by corruption. Developing countries are more vulnerable to corruption due to their low level of economic development. In other words, in developing countries, to fight corruption is, to a certain extent, to fight poverty. Through the governance of corruption, the policy of poverty governance can be better implemented.

(vii) Rationally distribute wealth. A reasonable distribution of national wealth should be based on national realities and within the limits of China's productivity level. The implementation of universal free education, free medical care, and other policies that do not match the level of development of China is beyond the scope of public affordability, will only make the poor poorer. India's comprehensive free health care system has overloaded both the financial system and the health care system, seriously affecting people's lives.

In terms of poverty and employment adaptation strategies to climate change in less developed countries, rational and effective public intervention policies can break the vicious cycle of "climate disaster-employment-poverty." The Government of Malawi has proposed a measure that has been effective in realizing government interventions: the transfer of productive inputs back to farmers' accounts through transfer payments. Farmers purchased about 300,000 tons of fertilizer and 10,000 tons of maize seed, as well as large quantities of fertilizer, at lower prices, saving them about two thirds of what they would have spent at normal prices. The Government established a national agency while encouraging and organizing private enterprises as suppliers of subsidized materials for efficient distribution. It is estimated that, excluding the effects of rainfall variability, maize yields throughout Malawi were 600,000–700,000 tons higher than normal in 2007 as a result of the interventions, and that incomes increased more significantly than costs. Higher crop yields have reduced the need for food imports into Malawi, alleviating economic pressures on China and creating more jobs (Chirwa et al., 2013). Malawi needs to do more to increase government accountability (Brown, 2011), and Malawi's experience highlights the important role of public intervention policies to increase employment opportunities and reduce vulnerability to climate risks by creating an enabling environment for poverty reduction (Harrigan, 2008), in addition to increasing investment in the provision of basic services to the poor.

Island Countries (Philippines) Poverty and Employment Adaptation Strategies to Climate Change

To adapt to the employment pressure and poverty that climate change brings to China, the Philippines has adopted the following strategies:

(i) Macroeconomic stability and smooth economic growth based on free enterprise, emphasizing the importance of economic growth in poverty alleviation. Only with economic growth will there be more job opportunities, the state be able to invest more in eradicating poverty and employment challenges, and the society be able to develop.
(ii) Modernization of agriculture and fisheries to ensure that demand is focused on food security and growth of the agricultural sector to address regional poverty and unemployment. The modernization of agriculture and fisheries should rely on scientific and technological progress and institutional innovation to transform agricultural and fishery production methods to form a modern agricultural and fishery industrial system that is green and safe, core competitive, prosperous, of high-quality, and efficient.
(iii) Integrated human development and protection of vulnerable groups to meet the need for more social protection and provide accurate and efficient basic services for the poor. To alleviate poverty, we need to better understand the needs of the poor, tailor our approach to the local context and the time, and use a more precise approach.
(iv) Good and efficient governance, which emphasizes the fundamental role of government in poverty alleviation efforts, whether national or local. Regardless of location, the most fundamental force in poverty eradication is the government. All poverty governance must be government-led, and good and efficient governance is fundamental to alleviating poverty.
(v) Developing a green economy to avoid the effects of climate change.

11.4 Summary

This chapter examines the intricate relationship between regional development and climate change, exploring the various ways in which a warming planet is altering the social, economic, and environmental landscapes of different regions. The chapter begins by examining the impacts of climate change on agriculture, highlighting how shifts in temperature and precipitation patterns can affect crop yields, livestock health, and food security. It then focuses on water resources to discuss how changes in precipitation and glacial melt can lead to both water scarcity and flooding, with severe consequences for regional economies and populations.

The chapter also considers the health implications of climate change, examining how extreme weather events, air pollution, and the spread of vector-borne diseases pose significant challenges to public health systems. Furthermore, it explores the link between climate change and disaster risk, emphasizing the need for better preparedness and response mechanisms to mitigate the damage caused by natural disasters.

The economic consequences of climate change are also discussed, with a focus on its impacts on poverty and employment. The chapter argues that climate change exacerbates existing inequalities and can trap vulnerable communities in a vicious cycle of poverty and environmental degradation. The chapter concludes by outlining the importance of adaptation strategies and policies in addressing the challenges posed by climate change, emphasizing the need for a multi-stakeholder approach that engages governments, businesses, and communities in building resilience.

Exercise

1. What are the specific impacts of climate change on regional development? Please provide examples.
2. In the face of the challenges posed by climate change, how should different regions develop adaptation strategies based on their own characteristics?
3. What are the impacts of climate change on regional agriculture, industry, tourism, and other industries? How can these impacts be addressed?
4. How can inter-regional cooperation in addressing climate change be realized to achieve co-development?
5. In the context of climate change, how can we balance the relationship between regional development and ecological and environmental protection?

Extended Reading

Huitema, D., Adger, W. N., Berkhout, F., Massey, E., Mazmanian, D., Munaretto, S., Plummer, R., & Termeer, C. C. J. A. M. (2016). The governance of adaptation: choices, reasons, and effects. Introduction to the Special Feature. *Ecology and Society*, *21*(3).

This article discusses the governance of adaptation to climate change, distinguishing between the different types of adaptation. It also distinguishes between mitigation, which focuses on reducing interference with the carbon cycle as the primary cause of human-induced climate change, and adaptation, which focuses on adjusting to the consequences of climate change impacts on its natural and social systems.

Chang, J.-J., Wei, Y.-M., Yuan, X.-C., Liao, H., & Yu, B.-Y. (2020). The Nonlinear Impacts of Global Warming on Regional Economic Production: An Empirical Analysis from China. *Weather, Climate, and Society*, *12*(4), 759–769.

This article estimates the impacts of global warming on economic production, at both the global and regional levels, and explores regional development under climate change.

References

Bank, T. W. (2016). *Poverty in a rising Africa.* World Bank Publications.

Brown, D. (2011). Making the linkages between climate change adaptation and spatial planning in Malawi. *Environmental Science and Policy, 84*(8), 940–949.

Campbell-Lendrum, D., et al. (2015). Climate change and vector-borne diseases: What are the implications for public health research and policy? *Philosophical Transactions of the Royal Society B-Biological Sciences, 370*, 1665.

Challinor, A. J., et al. (2014). A meta-analysis of crop yield under climate change and adaptation. *Nature Climate Change, 4*, 287–291.

Chen, Y., et al. (2015). Progress and prospects of climate change impacts on hydrology in the arid region of Northwest China. *Environmental Research, 139*, 11–19.

Cheung, W. W. L., et al. (2010). Large-scale redistribution of maximum fisheries catch potential in the global ocean under climate change. *Global Change Biology, 16*(1), 24–35.

Chirwa, E., & Dorward, A. (2013). *Agricultural input subsidies: The recent Malawi experience.* Oup Catalogue.

Climate Change, I. P. C. C. (2014). *Impact, adaptation, and vulnerability.* Cambridge University Press.

Eberhard, A., Town, C., & Kolker, J. (2014). South Africa's renewable energy IPP procurement program: Success factors and lessons. Public-Private Infrastructure Advisory Facility (Vol. 2, pp. 1–56).

Ebi, K. L., & Nealon, J. (2016). Dengue in a changing climate. *Environmental Research, 151*(11), 115–123.

Ebi, K. L., et al. (2001). Association of normal weather periods and El Ni?o Events with hospitalization for viral pneumonia in females: California, 1983–1998. *American Journal of Public Health, 91*(8), 1200–1208.

Eichler, R., et al. (2013). Performance-based incentives to improve health status of mothers and newborns: What does the evidence show? *Journal of Health Population and Nutrition, 31*, 36–47.

Garrote, L., et al. (2015). Quantitative assessment of climate change vulnerability of irrigation demands in Mediterranean Europe. *Water Resources Management, 29*(2), 325–338.

Harrigan, J. (2008). Food insecurity, poverty and the Malawian starter pack: Fresh start or false start? *Food Policy, 33*(3), 237–249.

Himanen, S. J., et al. (2013). CropResponses to climate and socioeconomic change in Northern Regions. *Regional Environmental Change, 13*(1), 17–32.

Houghton, R. A. (1995). Changes in the storage of Terrestrial carbon since 1850. In R. Lal, J. M. Kimble, E. Levine & V. A. Stewart (Eds.), *Soils and global change* (Vol. 1995, pp. 45–65). Lewis Publisher CRC.

Hu, P., et al. (2018). Flood-induced mortality across the globe: Spatiotemporal pattern and influencing factors. *Science of the Total Environment, 643*, 171–182.

Iglesias, A., et al. (2012). From climate change impacts to the development of adaptation strategies: Challenges for agriculture in Europe. *Climatic Change, 112*(1), 143–168.

Iglesias, A., Santillan, D., & Garrote, L. (2018). On the barriers to adaption to less water under climate change: Policy choices in Mediterranean Countries. *Water Resources Management, 32*(15), 4819–4832.

IPCC. (2014). In *Climate change 2014: Impact, adaptation, and vulnerability.* Cambridge University Press.

Nepal, R., & Paija, N. (2019). Energy security, electricity, population and economic growth: The case of a developing South Asian resource-rich economy. *Energy Policy, 132*(5), 771–781.
Quandt, A., Neufeldt, H., & McCabe, J. (2017). The role of agroforestry in building livelihood resilience to floods and drought in semiarid Kenya. *Ecology and Society, 22*(3).
Randolph, S. (2013). Is expert opinion enough? A critical assessment of the evidence for potential impacts of climate change on tick-borne diseases. *Animal Health Research Reviews, 14*(2), 133–137.
Robine, J. M. et al. (2008). Death toll exceeded 70,000 in Europe during the summer of 2003. *Comptes Rendus Biologies, 331*(2), 0–178.
Tayfur, G., Onoz, B., Cancelliere, A., et al. (2016). Editorial: Water resources management in a changing world: Challenges and opportunities. *Water Resources Management, 30*(15), 5553–5557.
Timilsina, G. R., Kurdgelashvili, L., & Narbel, P. A. (2012). Solar energy: Markets, economics and policies. *Renewable and Sustainable Energy Reviews, 16*(1), 449–465.
Tong, S., Hu, W., & Mcmichael, A. J. (2004). Climate variability and ross river virus transmission in Townsville Region, Australia, 1985–1996. *Tropical Medicine and International Health, 9*(2), 298–304.
Trnka, M., et al. (2011). Expected changes in Agroclimatic conditions in central Europe. *Climatic Change, 108*, 261–289.
Uleberg, E., Hanssen-Bauer, I., Oort, B., et al. (2014). Impact of climate change on agriculture in Northern Norway and potential strategies for adaptation. *Climatic Change, 122*, 27–39.
UNFCCC. (2014). Warsaw international mechanism for loss and damage associated with climatic change impacts (Decision 2/CP19).
Wang, X., et al. (2017). Epidemiology of Avian influenza a H7N9 virus in human beings across five epidemics in Mainland China, 2013–17: An epidemiological study of laboratory-confirmed case series. *The Lancet Infectious Diseases, 17*(8), 822–832.
Watts, N., et al. (2015). Health and climate change: Policy response to protect public health. *The Lancet, 386*(10006), 1861–1914.
Wen, L., & Yan, F. (2018). Regional differences and influencing factors in the CO_2 emissions of China's power industry based on the panel data models considering power-consuming efficiency factor. *Environment, Development and Sustainability, 20*(5), 1987–2007.
Wheeler, T., & von Braun, J. (2013). Climate change impacts on global food security. *Science, 341*(6145), 508–513.
Wiréhn, L. (2018). Nordic agriculture under climate change: A systematic review of challenges, opportunities and adaptation strategies for crop production. *Land Use Policy, 77*, 63–74.
World Health Organization. (2017). Tracking universal health coverage: 2017 global monitoring report.
Yang, J. (2016). Environmental and climate change co-benefits analysis of wind power generation in China. *Energy Procedia, 88*, 76–81.

Chapter 12
Global Convergence of Regional Socio-economic Development

Abstract As globalization deepens, development disparities between and within developed and less developed countries and regions are exhibiting new characteristics. Development Geography has become essential for understanding these changes and the underlying mechanisms. Currently, regional development gaps are showing diverse features, and scholars are increasingly focusing on the convergence of regional socio-economic development. Long-term imbalances and steady-state stages coexist between economies and regions. Therefore, it is crucial to explore the dynamic evolution of regional development differences, convergence trends, and influencing factors to understand regional development disparities or convergence. This chapter reviews domestic and international literature on regional development convergence and integrates knowledge from economics, management, sociology, and geography. This chapter examines the convergence of socio-economic development regionally and globally in the context of globalization. The objectives of this chapter are as follows: 1. **To analyze the key factors** influencing the level and pace of regional economic development, with a particular focus on human capital investment, industrial development, and globalization. 2. **To examine convergence models** for regional economic development and explore the extent to which less developed regions can catch up with more developed ones over time. 3. **To conduct empirical studies** and analyze the relative and absolute differences in per capita income and GNI across different regions and countries, providing insights into the patterns and trends of regional economic development. 4. **To identify policies and strategies** that can facilitate the convergence process, enabling less developed regions to achieve sustainable and inclusive growth.

X. Deng and M. Song, *Development Geography*,
https://doi.org/10.1007/978-981-96-6917-2_12

12.1 Convergence of Regional Social Development

12.1.1 Regional Social Development Convergence Profiles

Since the twenty-first century, along with the wave of globalization and industrial transformation and upgrading, developed and underdeveloped countries have presented different development models and mechanisms, and the problems of unbalanced regional development and localized poverty have become more and more prominent. The fundamental goal of social development is to raise the level of well-being of all human beings, reduce poverty, and promote sustainable development in all regions (Gohou and Soumaré, 2012; Howell, 2005). Resident income is a visual reflection of the living standard and poverty level of the population, and the Gini coefficient, calculated based on the distribution of resident income, serves as a global and universal indicator of income disparity and the balance of social development (Mussini and Mauro, 2013; Portnov & Felsenstein, 2010). In the context of economic globalization and industrial transformation and upgrading, several questions remain, such as whether the living standards and social development levels of residents in developed and less developed countries are rising in tandem, whether globalization contributes to narrowing the imbalance in regional development and reducing poverty, and whether industrial upgrading may exacerbate the income gap, thereby triggering a higher level of inequity in social development (Monfort et al., 2018; González & Figueiredo, 2019). Neoclassical growth theory provides a theoretical perspective, which suggests that over time, the economic growth rate of low-income regions is expected to outpace that of high-income regions, thus causing the regional per capita incomes to converge. However, globally, whether the level of social development in developed and less developed countries is converging or diverging remains a controversial topic that has attracted the attention of many economists and sociologists.

The research on regional social development convergence mainly focuses on the state of convergence and its influencing factors. Scholars have studied α convergence, β-conditional convergence, β absolute convergence, and club convergence on income disparity at the city, regional, national, and global levels. The results show that in the United States, there has been a trend of convergence in residents' income between 1988 and 1999 (Lynch, 2003). However, in China, the characteristics of income gap convergence vary across different economic zones; for example, the Southwest Comprehensive Economic Zone has stable convergence, whereas the Northwest, Central-South, and Central-North Economic Zones show a divergent trend. There is significant conditional and absolute β convergence in the urban–rural income gap in prefecture-level cities; however, the speed of convergence varies according to the size of the city, with an overall downward trend. The ChangZhuTan urban agglomeration also exhibits absolute β convergence in a specific period. In addition, income gap convergence may show nonlinear changes; for example, China's interprovincial income gap converges under a widening trend in low-income areas, whereas high-income areas converge under a narrowing trend. In addition to income indicators, the Gini coefficient is a key indicator of the degree of income inequality. Generally, a Gini coefficient below 0.2 indicates a highly even distribution of income, 0.2–0.29 indicates a relatively even distribution, 0.3–0.39 indicates a relatively reasonable distribution, 0.4–0.59 indicates a large disparity, and 0.6 or higher indicates a huge disparity. 0.4 is regarded as the "warning line" for the disparity of income distribution. From 2003 to 2012, China's Gini coefficient exceeded 0.4, but declined year by year, reflecting the gradual improvement of social well-being.

Regional social development is influenced by various factors, with globalization, the level of economic development, and industrial structure considered important drivers. Globalization, especially economic globalization, plays a role in global social development in terms of trade, the establishment of multinational corporations, international finance, and foreign exchange investment, and the role of trade and capital flows brought about by economic globalization on social development has been demonstrated in 47 countries around the world (Ezcurra and Rodríguez-Pose, 2013). According to Kuznets' inverted U-shaped theory, economic growth affects social development to different degrees depending on the stage of economic development. Rapid economic growth in a region helps drive a significant increase in the income level of residents, and the impact of economic growth on income disparity has been confirmed in China. There are also significant differences in the extent to which industrial restructuring, especially industry and services, has contributed to social development in different countries, with developed countries placing greater emphasis on services and less developed countries paying more attention to the development of industry. Differences in the level of productivity development are an important cause of income disparity. Globalization, the level of economic development, and industrial structure play a role in widening or narrowing the income gap from different perspectives, which in turn directly affects the enhancement of the well-being of human society. In addition, the mobility of human resources and capital brought about by globalization directly contributes to the strengthening of social ties among countries, forming social development communities such as the European Union, Association of Southeast Asian Nations (ASEAN), and the African Union. The formation of such communities intensifies interactions. The spatial interaction of social development has been proven effective in narrowing the gap between countries in 41 countries along the "Belt and Road" route. Rey and Montouri (1999) have also pointed out that spatial interaction must be taken into account in the study of regional income convergence in the United States. The spatial interaction role and social development community, among other factors, play an important role in the realization of social development and further affect its convergence mechanism.

Studies have generally focused on the impact of a single industry, urban and rural income, and regional poverty, and related systematic studies are relatively few. The literature is also dominated by a single country or region, and there are fewer comparative studies on income disparity between developed and less developed countries globally. The research methodology used in this study uses more traditional convergence analysis and cross-sectional data regression, and spatial interaction and its spillover effects are considered. In this section, we focus on global social development as the main object of study, focusing on exploring its convergence and driving mechanisms and considering spatial interactions and their spillover effects. Differences in the convergence mechanisms and influencing factors of social development between developed and less developed countries are also discussed in this section. The study takes the well-being of human society as represented by the Gini coefficient as the characterizing variable of the level of social development, the level of globalization in different dimensions as the main explanatory variable, the level of economic development and the level of industrialization as the main control variables, and the multivariate variables together explore the trend of change in social development globally.

12.1.2 Regional Social Development

In the wave of globalization in the twenty-first century, regional development links have become stronger and interactions among countries more frequent. Globalization has had a profound impact on economies, particularly in terms of capital and labor mobility, cultural exchanges, trade, technology sharing, and institution-building. Among these, trade globalization and financial globalization have had a particularly significant impact on regional social development.

According to Keynesian multiplier theory, export trade is not only an important engine of economic growth but also contributes to the growth of residents' incomes in accordance with the foreign trade multiplier in the sustained growth of the economy. When a country's social productivity exceeds its domestic demand, demand from other countries becomes a key driver of its own economic growth. The expansion of export trade not only expands the domestic production market but also contributes to the growth of the population's income in accordance with the multiplier effect. In addition, consumer preferences for different products affect market size and trade prices, which in turn have an impact on a country's income distribution. For example, trade globalization may help narrow the gap between the rich and the poor if high-income earners prefer imported products and low-income earners prefer local products. Conversely, it may exacerbate the polarization between the rich and the poor.

However, the international movement of capital is aimed at cost reduction, the pursuit of high profits, new technologies, and overseas markets. According to the theory of international capital flows, capital-exporting countries develop overseas markets and increase exports and profits through capital flows; however, they may also face risks such as reduced fiscal revenues, declining employment opportunities, and social instability. As for capital inflow countries, the influx of international capital can raise the level of investment in their countries, introduce advanced technology, increase residents' employment and income, and solve the problems of insufficient capital and lagging economic development. However, if foreign capital over-controls China's economic lifeblood or leads to excessive international debt, it may lead to the predatory exploitation of resources and environmental problems and thus hinder economic development and aggravate social welfare inequality.

This section analyzes the means and standard deviations of economic, trade, and financial globalization for 46 countries worldwide based on globalization data published by the KOF Swiss Institute for Economic Research (e.g., Schedules 12.1).

This study provides an in-depth analysis of the changes in the mean and standard deviation of the indicators of economic, trade, and financial globalization for 46 countries worldwide through a comparative analysis of developed and developing countries (see Schedule 12.1 for the list of countries). Overall, the trade globalization index of these 46 countries significantly outperforms the financial globalization index, and financial globalization is more volatile and shows a stronger trend of change (Fig. 12.1). From a time-series perspective, the global economic crisis had a significant impact on the financial globalization index of each country between 2007 and 2013, leading to a steep decline in the index. With the economic recovery in 2014, economic globalization, trade globalization, and financial globalization all reached their highest values during the study period. It is noteworthy that the standard deviation of economic globalization for 46 countries worldwide showed a downward trend during the period from 2006 to 2013, which implies that the differences among countries in terms of economic globalization are gradually decreasing (Fig. 12.2). In addition, it is also found that the standard deviation of trade globalization is significantly larger than that of financial globalization, suggesting that countries are more differentiated in terms of trade globalization than financial globalization. Overall, before 2013, the economic globalization index showed a downward trend to a lesser extent due to the economic crisis. However, in 2014 and 2015, the degree of economic globalization remained high. Although the trade globalization index is generally higher than the financial globalization in 46 countries worldwide, the variability among countries remains high.

12.1.2.1 Analysis of Economic Globalization in Developed Countries

First, we conducted an in-depth analysis of the changes in the means and standard deviations of economic, trade, and financial globalization in 24 developed countries worldwide between 2006 and 2015. During this period, the mean value of the economic globalization index of these developed countries varied within the range of 80.72–85.22, showing that economic globalization fluctuated at a relatively high level and peaked in 2014. Looking at the detailed indicators, the trade globalization index is generally higher than the financial globalization index for these countries (Fig. 12.3). The standard deviation of the economic globalization index for these 24 developed countries fluctuates between 4.18 and 5.10, which implies that the levels of economic globalization in these countries vary within a relatively stable and low range. However, an in-depth analysis of the refined indicators reveals that the standard deviation of the financial globalization index of these countries is significantly

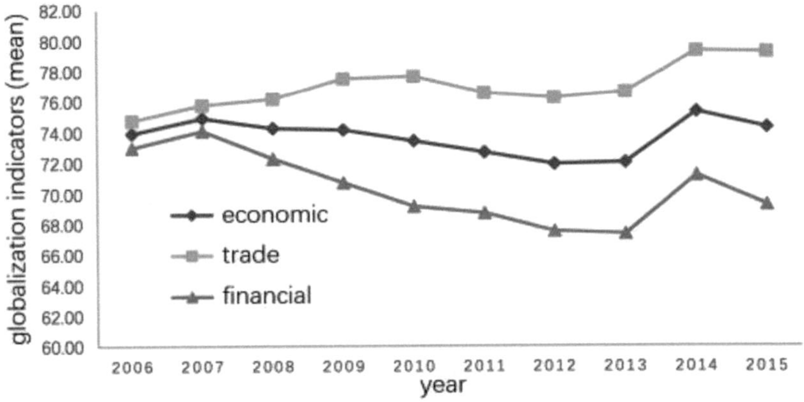

Fig. 12.1 Changes in the mean values of economic globalization, trade globalization, and financial globalization for 46 countries worldwide

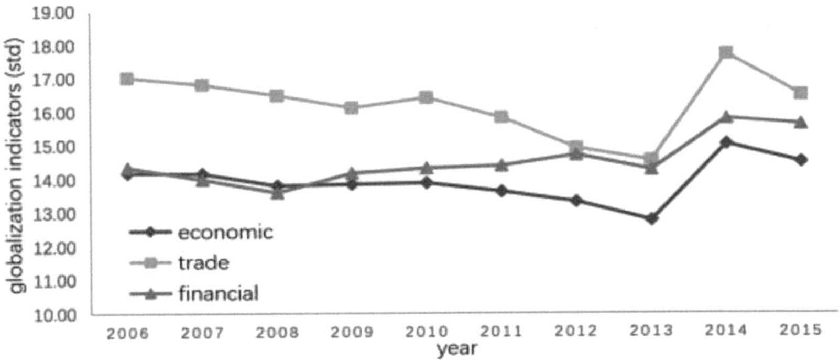

Fig. 12.2 Changes in the standard deviation of economic globalization, trade globalization, and financial globalization for 46 countries worldwide

higher than that of trade globalization, which suggests that there is more variability among these developed countries in terms of financial globalization than in terms of trade globalization (Fig. 12.4).

12.1.2.2 Analysis of Economic Globalization in Less Developed Countries

The 22 less developed countries also adopt the changes in the mean and standard deviation of economic, trade, and financial globalization. The data show that the mean value of the economic globalization index of the 22 less developed countries around the world fluctuated within the range of 61.70 to 64.53 from 2006 to 2015, a level significantly lower than that of developed countries, and the index moved within

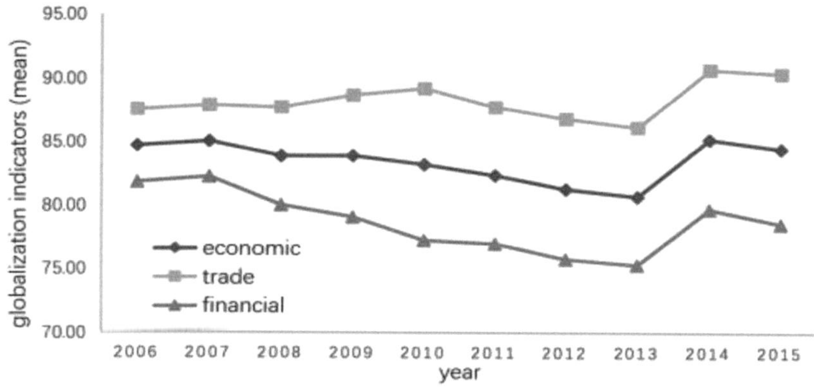

Fig. 12.3 Changes in the mean values of economic, trade, and financial globalization in 24 developed countries

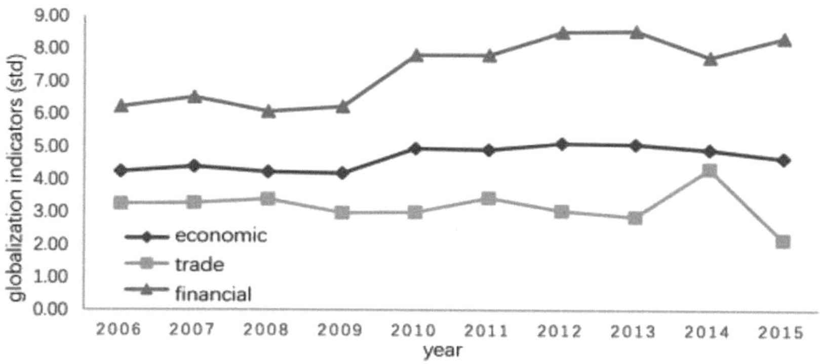

Fig. 12.4 Changes in the standard deviation of economic, trade, and financial globalization in 24 developed countries

a relatively low range, with the highest value reached in 2014. In terms of specific indicators, the trade globalization index of less developed countries is generally higher than the financial globalization index (Fig. 12.5). Further analysis of standard deviation changes reveals that the standard deviation of the economic globalization index for these 22 less developed countries varies within a range of 11.51 to 14.93, a range of fluctuation that is significantly higher than that of developed countries. At the level of refined indicators, the standard deviation of the trade globalization index of less developed countries is higher than that of financial globalization in most cases, which suggests that less developed countries exhibit greater variability in trade globalization compared to financial globalization (Fig. 12.6).

In summary, not only are the overall indices of economic, trade, and financial globalization in developed countries higher than those in less developed countries; however, there is also relatively less variability across countries. This difference is

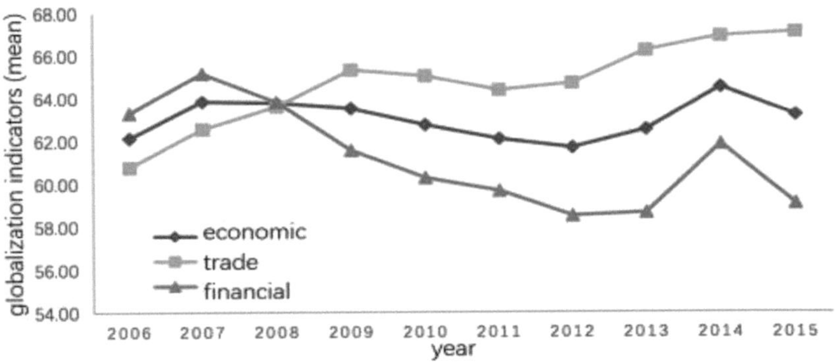

Fig. 12.5 Changes in the means of economic globalization, trade globalization, and financial globalization in 22 less developed countries

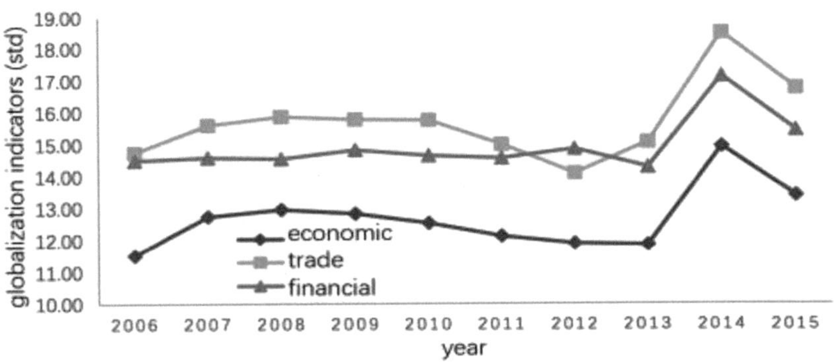

Fig. 12.6 Changes in standard deviation of economic globalization, trade globalization, and financial globalization in 22 less developed countries

closely related to the integrity of the industries of developed countries and the high level of economic development.

12.1.3 Convergence Analysis of Regional Social Welfare

Based on Nobel Prize-winning economist Amartya Sen's welfare index and the American Social Health Association (ASHA)'s social welfare evaluation index, we constructed a comprehensive social welfare index and conducted an empirical study of regional social development convergence. Amartya Sen proposed a welfare index formula for measuring social development.

$$S = RY \times (1 - G) \tag{12.1}$$

where S is the social welfare index, RY is the growth rate of real income per capita and G is the Gini coefficient. This index pays attention to the growth of economic aggregate and the fairness of income distribution when evaluating social welfare; however, this index ignores the consideration of basic human needs, and the empirical results are not satisfactory. The ASHA has proposed the use of the ASHA index to measure social welfare, which is calculated by dividing the product of the employment rate, the average life expectancy, the literacy rate, and the growth rate of GNP per capita by the product of the birth rate and the infant and child mortality rate. This index measures social welfare from social, medical, cultural, and economic perspectives and is considered to better reflect the level of social welfare and quality of life of a country's population. Based on the Amartya Sen welfare index and the ASHA index, considering regional poverty, this subsection proposes a social welfare index (SWI) based on seven dimensions: growth rate of per capita gross domestic product (RG), Gini coefficient (G), proportion of the poor population (RP), rate of employment (RE), rate of growth of life expectancy (RL), rate of births (RB), and infant and child mortality rate (RD). The literacy rate is excluded in the calculation because of missing data. The formula is as follows.

$$SWI = \frac{RG \times (1 - G) \times (1 - RP) \times RE \times (1 + RL) \times (1 - RD)}{1 + RB \times (1 - RD)} \tag{12.2}$$

The Gini coefficient, a breakdown of the social welfare index, is a measure of income disparity that is commonly used internationally, and its value reflects the variability of social well-being in individual countries.

12.1.3.1 α Convergence Analysis of Regional Social Welfare

We used key data such as GDP per capita growth rate, Gini coefficient, poverty headcount ratio (based on the national poverty line), employment rate, life expectancy, birth rate, and infant and child mortality rate for 46 countries between 2006 and 2015 obtained from the World Bank statistics (https://data.worldbank.org/), and using a scientific method of calculation, we have derived a social welfare index. We further conducted a statistical analysis of α convergence and β convergence to reveal the trend of social welfare levels in these countries. It is worth noting that the absolute difference index may have some errors in measuring the differences in social development. Therefore, to more accurately assess whether the dispersion of the social welfare indices of the 46 countries has gradually declined over time, we adopted a relative difference indicator, the coefficient of variation. This indicator can more accurately reflect the relative changes in social welfare levels in different countries, thus providing a more accurate basis for judgment.

By analyzing the coefficient of variation, we can observe the trend of changes in the degree of dispersion of the social welfare index in 46 countries. If the coefficient of variation shows a gradual downward trend over time, it indicates that the differences in social welfare among these countries are gradually narrowing and social

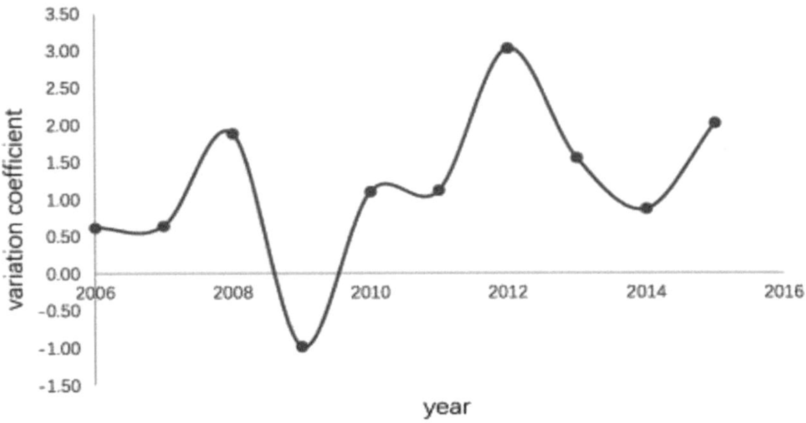

Fig. 12.7 Change in the coefficient of variation of the social welfare index for 46 countries worldwide

development is more balanced. By contrast, if the coefficient of variation remains stable or increases, it may mean that the differences in social welfare levels have not significantly improved and may even have increased.

The coefficients of variation of the social welfare index for all 46 countries show significant fluctuation characteristics and do not exhibit a clear convergence trend (Fig. 12.7). To gain a more detailed insight into the differences in this trend among countries at different levels of development, we analyzed the changes in the coefficient of variation of the social welfare index in 24 developed countries and 22 less developed countries separately. Among developed countries, changes in the relative variance of the social welfare index are characterized by large fluctuations, especially in 2008 and 2013, when the absolute value of the coefficient of variation was relatively high. This indicates that the differences in social development in developed countries show a fluctuating trend (Fig. 12.8). In contrast, the change in the relative variance of the social welfare index in less developed countries shows an upward and downward trend within a small range. It is noteworthy that the coefficient of variation of social welfare in less developed countries is significantly smaller than that of developed countries, which indicates that the degree of variation in social welfare is smaller in less developed countries than in developed countries (Fig. 12.9).

12.1.3.2 β Convergence Analysis of Regional Social Welfare

The study found that the fixed effects model is more applicable in the β-absolute convergence analysis (Prob > chi2 = 0.0000, rejecting the original hypothesis) by conducting a Hausman test on the panel data from 2006 to 2015. Therefore, we chose the fixed-effects model for the β-absolute convergence analysis of the growth rate of the social welfare index in 46 countries worldwide.

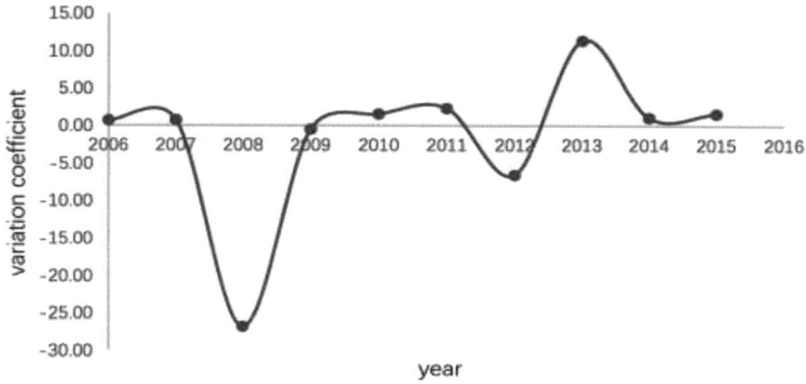

Fig. 12.8 Changes in the coefficient of variation of the social welfare index in 24 developed countries

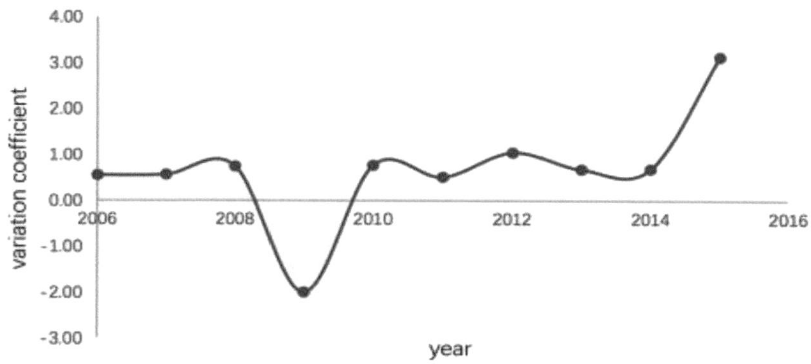

Fig. 12.9 Changes in the coefficient of variation of the social welfare index in 22 less developed countries

In the analysis of the initial value of global social development measured using the social welfare index, we find that the convergence coefficient was negative, with an absolute value of 0.8131. However, the within-group regression R2 of the model was only 0.3943 and the model did not pass the F-test, suggesting that the model was not significant. Therefore, we cannot conclude that social development in the 46 countries shows a trend of convergence based on this model alone (Table 12.1). The data for the 24 developed countries also reveal a negative coefficient of convergence for the initial value of social development, with an absolute value of 0.7220. However, the within-group regression R2 of the model is only 0.3484, also failing the F-test indicating that the model is not significant. Therefore, a tendency of convergence in the social development of the developed countries is not proven (Table 12.1). Similarly, for the 22 less developed countries worldwide, the coefficient of convergence of the initial value of social development is also negative, with an absolute value of 0.9542.

Table 12.1 β-absolute convergence analysis results (*Note* * $p < 0.10$, ** $p < 0.05$, *** $p < 0.001$)

Explanatory variables		Model (growth rate of GINI coefficient)			
46 countries worldwide		24 developed countries worldwide		22 less developed countries worldwide	
Constant	0.0047***	Constant	0.0016	Constant	0.0096***
Initial value	– 0.8131***	Initial value	– 0.7220***	Initial value	– 0.9542***
R^2 within group	0.3943	R^2 within group	0.3484	R^2 within group	0.4654
F-value	1.0700	F-value	0.6100	F-value	1.1300

Although the within-group regression R2 of the model reaches 0.4654, the model still fails the F-test, and thus also fails to demonstrate a convergence trend in social development in less developed countries (Table 12.1).

In summary, global social development, characterized by the social welfare index, did not show a clear convergence trend between 2006 and 2015. This result may be influenced by various factors, including but not limited to economic policies, social systems, cultural differences, and the degree of globalization in each country.

12.1.3.3 Study on the Driving Mechanisms of Regional Social Welfare

The spatial interaction of social development in regions across the world and the Cobb–Douglas production function together constitute the basic theoretical framework for analyzing the driving mechanism of global social development. In this study, the social welfare index of 46 countries worldwide from 2006 to 2015 is taken as the main object of investigation, and the level of economic globalization is selected as the key explanatory variable. To explore the impact of economic globalization more comprehensively, we further refine it into two dimensions: trade globalization and financial globalization. Meanwhile, the level of industrialization (measured by the share of value added of the service sector in GDP), the level of consumption (total consumption as a percentage of GDP), and the level of investment (fixed asset investment as a percentage of GDP) are introduced as control variables to comprehensively analyze the driving mechanism of social welfare.

In this study, we conducted Hausman tests on panel data for 46 countries, 24 developed countries, and 22 less developed countries worldwide. The results show that the fixed effects model is more suitable for analyzing the driving mechanism (Prob > chi2 = 0.0000, rejecting the original hypothesis). We constructed four-panel regression models: Model I incorporates the level of economic globalization, the level of industrialization of services, the level of investment, and the level of consumption as control variables; Model II focuses on the impact of the level of trade globalization; Model III focuses on the role of the level of financial globalization; and Model IV integrates the impact of the level of trade and financial globalization.

Table 12.2 Results of panel regression analysis of social welfare development in 46 countries worldwide (*Note* * $p < 0.10$, ** $p < 0.05$, *** $p < 0.001$)

Explanatory variables	Coefficient			
	Model I	Model II	Model III	Model IV
Constant	0.1588***	0.1666***	0.1520***	0.1563***
Economic globalization	0.0001			
Trade globalization		– 0.0001		– 0.0001
Financial globalization			0.0001	0.0001
Industrialization level (service sector)	– 0.1359***	– 0.1328***	– 0.1313***	– 0.1278***
Investment level	0.0923***	0.0934***	0.0890***	0.0890***
Consumption level	– 0.1261***	– 0.1273***	– 0.1270***	– 0.1281***
R^2	0.3055	0.3057	0.3069	0.3074
F-value	3.49***	3.50***	3.40***	3.30***

Through the simulation analysis of the four models, we conclude that for all countries around the world, the driving effect of economic globalization, trade globalization, and financial globalization on social welfare development is not significant. On the contrary, the level of industrialization of services and the level of consumption have a significant negative driving effect on social welfare development, whereas the level of investment has a significant positive driving effect (Table 12.2). For developed countries, the level of economic globalization, financial globalization, and industrialization of services are also not significant drivers of social welfare development. However, the level of trade globalization and the level of investment are significant positive drivers of social welfare development in developed countries, whereas the level of consumption is a significant negative driver (Table 12.3). For the less developed countries, economic globalization, trade globalization, financial globalization, and the level of consumption are not significant drivers of their social welfare development. However, the level of service industrialization has a significant negative driving effect on social welfare development in less developed countries, while the level of investment has a significant positive driving effect (Table 12.4).

12.1.4 Convergence Analysis of Regional Income Disparities

12.1.4.1 Descriptive Statistical Analysis of Regional Income Disparities

With regard to the Gini coefficients (https://data.worldbank.org/) of 46 countries from 2006 to 2015, which were compiled by the World Bank, the variability of social well-being in the 46 countries was statistically analyzed based on six dimensions, namely, the very small value, the very large value, the mean value, the standard deviation, the variance and the coefficient of variation (Table 12.5). The results of

Table 12.3 Results of panel regression analysis of social welfare development in 24 developed countries (*Note* * $p < 0.10$, ** $p < 0.05$, *** $p < 0.001$)

Explanatory variables	Coefficient			
	Model I	Model II	Model III	Model IV
Constant	0.2009***	0.1580***	0.2411***	0.1740***
Economic globalization	0.0004			
Trade globalization		0.0008**		0.0010**
Financial globalization			– 0.0001	– 0.0003
Industrialization level (service sector)	– 0.0372	-0.0444	– 0.0424	– 0.0530
Investment level	0.0759***	0.0896***	0.0826***	0.1032***
Consumption level	– 0.2973***	– 0.2929***	– 0.3024***	– 0.2954***
R^2	0.4400	0.4492	0.4379	0.4520
F-value	5.60***	5.73***	5.64***	5.77***

Table 12.4 Results of panel regression analysis of social welfare development in 22 less developed countries (*Note* * $p < 0.10$, ** $p < 0.05$, *** $p < 0.001$)

Explanatory variables	Coefficient			
	Model I	Model II	Model III	Model IV
Constant	0.1144***	0.1162***	0.1036***	0.1109***
Economic globalization	– 0.0001			
Trade globalization		– 0.0002		– 0.0002
Financial globalization			0.0001	0.0001
Industrialization level (service sector)	– 0.1929***	– 0.1825***	– 0.1928***	– 0.1796***
Investment level	0.1009***	0.1008***	0.0970***	0.0987***
Consumption level	– 0.0227	– 0.0256	– 0.0239	– 0.0270
R^2	0.2694	0.2740	0.2695	0.2754
F-value	3.17***	3.40***	2.90***	2.94***

the analysis show that the average value of the Gini coefficient for the 46 countries was below 0.4 during the decade, which indicates that the distribution of income is relatively balanced in most of the countries, with small income gaps and good overall social well-being. However, some countries' Gini coefficients exceeded 0.5 during this period, indicating large disparities in income distribution. Observed from a time-series perspective, the average value of the Gini coefficient of 46 countries worldwide declined by 0.0143 between 2006 and 2015, with a rate of change of 3.98%. Meanwhile, the absolute difference (i.e., standard deviation) of the Gini coefficient also decreased by 0.0173, with a rate of change of 19.64%. This trend shows that both the mean and standard deviation of the global Gini coefficient have shown a downward trend, and the absolute difference has changed more significantly than the mean. This further confirms that global well-being in social development has been steadily increasing from year to year, while differences among countries have been narrowing from year to year.

The changes in the mean value of the Gini coefficient of 46 countries, 24 developed countries, and 22 less developed countries worldwide differed significantly. The year 2013 was the abrupt change point of the change in the mean value of the Gini coefficient, and the overall Gini coefficient of the 46 countries showed a downward trend, with different trends in the changes in the developed and less developed countries. Specifically, except for 2013, the less developed countries showed a decreasing trend overall, and the mean value of the Gini coefficient in 2006 was 0.4167. The change in 2015 was 0.3803, a decrease of 9.58%. The developed countries' Gini coefficient in 2006–2015 had a change interval of [0.3068, 0.3155], with a change of 0.87%, and the developed countries' Gini coefficient is obviously smaller than that of less developed countries. However, it shows a slow growth trend, and the income gap tends to widen (Fig. 12.10).

Between 2006 and 2015, the standard deviation of the Gini coefficient for 46 countries, 24 developed countries, and 22 less developed countries worldwide showed significant differences in changes. 2011 became a turning point in the changes in the standard deviation of the Gini coefficient due to the combined effects of the shrinking of the global stock market, the debt crisis in the United States, and the war in the Middle East and North Africa. Overall, the absolute differences in the Gini coefficients of 46 countries worldwide show a downward trend; however, the changes differ between developed and less developed countries (Fig. 12.11). Specifically, the standard deviation of the Gini coefficient in less developed countries decreased from 0.0918 in 2006 to 0.0824 in 2015, with an average annual rate of change of 0.00094, which suggests that the differences in the Gini coefficient among less developed countries are gradually narrowing. In contrast, the standard deviation of

Table 12.5 Differences in the gini coefficient of income distribution among 46 countries worldwide (2006–2015)

Year	Number of countries observed	Minimum value	Maximum value	Mean value	Standard deviation	Variance	Coefficient of variation
2006	46	0.2440	0.5750	0.3594	0.0881	0.0080	0.2450
2007	46	0.2440	0.5580	0.3572	0.0850	0.0070	0.2381
2008	46	0.2370	0.5550	0.3536	0.0820	0.0070	0.2391
2009	46	0.2480	0.5180	0.3504	0.0787	0.0060	0.2247
2010	46	0.2480	0.5310	0.3493	0.0772	0.0060	0.2210
2011	46	0.2460	0.5620	0.3486	0.0792	0.0060	0.2271
2012	46	0.2470	0.5610	0.3484	0.0764	0.0060	0.2194
2013	46	0.2460	0.5260	0.3510	0.0744	0.0060	0.2120
2014	46	0.2460	0.5070	0.3460	0.0728	0.0050	0.2105
2015	46	0.2540	0.5080	0.3451	0.0708	0.0050	0.2052

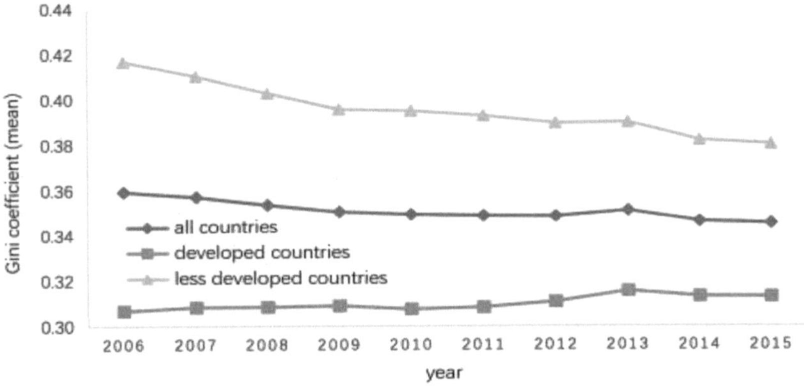

Fig. 12.10 Changes in the global mean Gini coefficient for 46 countries, 24 developed countries, and 22 less developed countries

the Gini coefficient of developed countries maintains fluctuations within a small range, with a change interval of [0.0340, 0.0383], showing that the differences in the Gini coefficient among developed countries are relatively small.

Overall, it seems that the global Gini coefficient has shown a decreasing trend during the period from 2006 to 2015, the income gap between countries has been narrowing year by year, and global social well-being has shown an increasing trend. It is worth noting that 2013 was a turning point for the mean value of the Gini coefficient, whereas 2011 was a turning point for the standard deviation of the Gini coefficient. The Gini coefficients of developed countries are significantly lower than those of less developed countries, and although there is a slow upward trend within a small range, the differences among countries are volatile. This suggests that developed countries

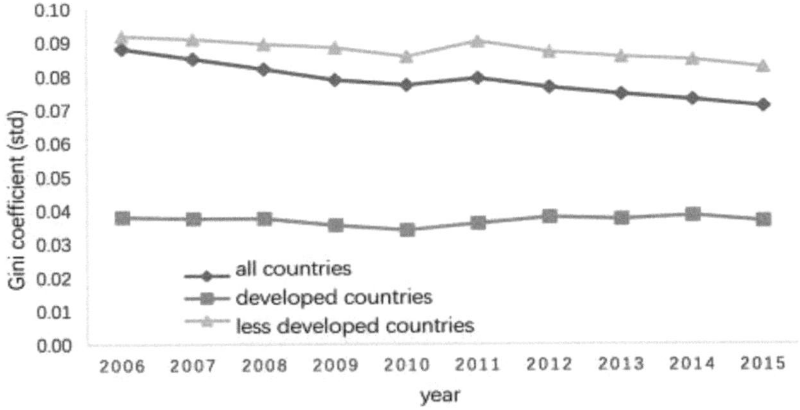

Fig. 12.11 Change in the standard deviation of the Gini coefficient for 46 countries, 24 developed countries, and 22 less developed countries, globally

have a more reasonable distribution of income, keep income disparities at a low level, have a more balanced social development, and have better social well-being. The Gini coefficients of the less developed countries, although larger, show a clear downward trend, and the differences among countries are narrowing year by year, reflecting the positive progress made by the less developed countries in narrowing income gaps, enhancing social well-being, and reducing regional imbalances in social development.

12.1.4.2 α Convergence Analysis of Regional Income Disparities

In view of the possible errors of absolute difference indicators in measuring social development differences, this study turns to the relative difference indicator, the coefficient of variation, to verify more accurately whether the dispersion of the Gini coefficient in 46 countries worldwide has gradually declined over time (Fig. 12.12). The results of the analysis show that the coefficient of variation of the Gini coefficient of the 46 countries worldwide does show a decreasing trend overall from the perspective of relative differences. Between 2006 and 2010, the coefficient of variation decreased significantly from 0.245 to 0.221. However, there was a sudden change in 2011; however, between 2012 and 2015, the relative variance remained on a downward trend. To gain a deeper understanding and comparison of social development differences among countries at different levels of development, we further analyzed the trends in the coefficients of variation of the Gini coefficients of 24 developed countries and 22 less developed countries. The fluctuation of the relative coefficient of variation of the Gini coefficient in developed countries is within the range of [0.1106, 0.1237]. Specifically, the overall relative variances of the Gini coefficients of developed countries showed a declining trend during the period from 2006 to 2010 and reached their lowest point in 2010. However, between 2010 and 2012, this difference again showed an upward trend, followed by fluctuations, suggesting that despite the relatively small differences in social development in developed countries, there have been significant fluctuations in recent years (Fig. 12.13). However, the change in the relative difference in the Gini coefficient in the less developed countries also shows more pronounced fluctuations. From 0.2204 in 2006 to 0.2167 in 2015, despite the existence of fluctuations, the overall trend suggests that the differences in social development in less developed countries are gradually decreasing (Fig. 12.14). This finding is important for understanding the dynamics of social development and its differences in countries at different levels of development.

12.1.4.3 β Convergence Analysis of Regional Income Disparities

(i) **β-absolute convergence analysis**

In this study, Hausman tests were conducted for 46 countries, 24 developed countries, and 22 less developed countries worldwide using panel data from 2006 to 2015.

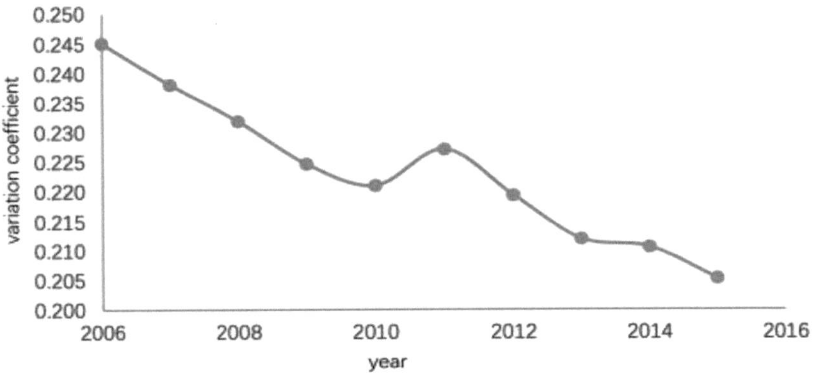

Fig. 12.12 Change in the coefficient of variation of the Gini coefficient for 46 countries worldwide

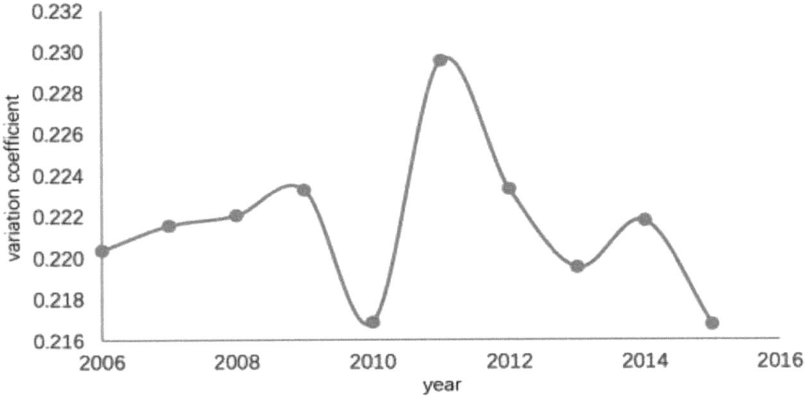

Fig. 12.13 Evolution of the coefficient of variation of the Gini coefficient for 24 developed countries

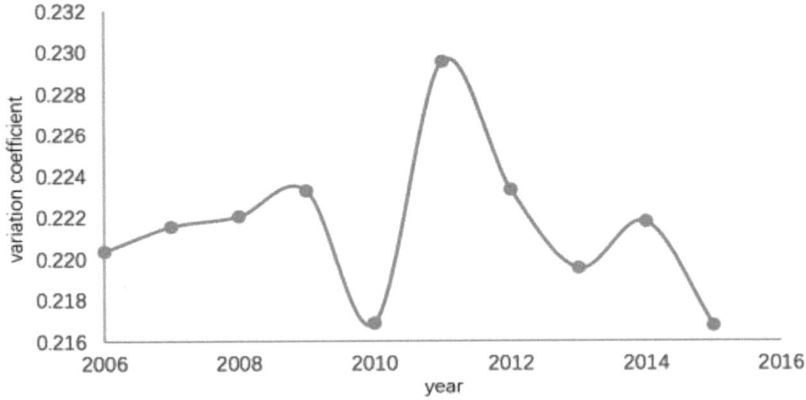

Fig. 12.14 Changes in the coefficient of variation of the Gini coefficient for 22 less developed countries

The test results (Table 12.6) show that the fixed effects model is more suitable for β-absolute convergence analysis (Prob > chi2 = 0.0000, original hypothesis rejected). The results of the analysis indicate that global social development shows a clear convergence trend during this period. Specifically, the convergence coefficients of the initial values of global social development are all negative, which implies that countries with higher initial Gini coefficients have relatively lower growth rates of social development, thus achieving convergence. The absolute value of the convergence coefficient is 0.3524, indicating moderate strength of convergence. The within-group regression R2 of the model is 0.2198, indicating that the model has some goodness of fit in explaining the phenomenon of global social development convergence. Further, a comparative analysis of the convergence dynamics between developed and less developed countries was conducted. The social development of developed countries also shows a convergence pattern, and their initial value of the convergence coefficient is also negative. The absolute value of the convergence coefficient is 0.4926, which is higher than the overall global level, showing a faster convergence of the growth rate of social development in developed countries. The within-group regression R2 of the model is 0.2658, indicating that the model is more capable of explaining the phenomenon of convergence in developed countries. Social development in less developed countries also shows convergence, with a negative convergence coefficient for the initial value. The absolute value of the convergence coefficient is 0.312, which is slightly lower than the global level, implying that the rate of convergence of the growth rate of social development in less developed countries is slightly slower. The within-group regression R2 of the model is 0.2095, showing the fit of the model in explaining the phenomenon of convergence in less developed countries.

In summary, regional social development globally has shown a certain convergence trend between 2006 and 2015. In this trend, the growth rate of social development in developed countries converged faster than that in less developed countries, showing differences in the convergence of social development in countries at different levels of development.

(ii) **β-conditional convergence analysis**

In this study, the Hausman test was first conducted on the panel data of 46 countries, 24 developed countries, and 22 less developed countries worldwide from 2006 to 2015. The test results show that the fixed effects model is more suitable for β-conditional convergence analysis (Prob > chi2 = 0.0000, rejecting the original hypothesis).

Table 12.6 β-absolute convergence analysis results (*Note* * $p < 0.10$, ** $p < 0.05$, *** $p < 0.001$)

Explanatory variables		Model (growth rate of GINI coefficient)			
46 countries worldwide		24 developed countries worldwide		22 less developed countries worldwide	
Constant	0.1223***	Constant	0.1532***	Constant	0.1199***
Initial value	– 0.3524***	Initial value	– 0.4926***	Initial value	– 0.3120***
R^2 within group	0.2198	R^2 within group	0.2658	R^2 within group	0.2095
F-value	2.9800***	F-value	3.3000***	F-value	2.9900***

Subsequently, based on the β-absolute convergence regression analysis model, we selected the Gini coefficient as a key indicator of regional social development and introduced the economic globalization indicator representing the level of globalization, as well as the level of economic development (GDP per capita) as a control variable for the conditional convergence analysis, and carried out the β-conditional convergence regression analysis. In the β-conditional convergence analysis of the Gini coefficient growth rate, we constructed four models. Model I is the original β-absolute convergence model, Model II adds the level of economic globalization as a control variable based on Model I, Model III adds the level of economic development as a control variable based on Model I, and Model IV combines the control variables of Models II and III, taking into account both the level of economic globalization and the level of economic development. From the data in Table 12.7, we observe that global social development shows convergence under all four models. This convergence is reflected in the fact that the convergence coefficients of the initial values are all negative, indicating that countries with higher initial Gini coefficients will have relatively lower growth rates, thus achieving convergence. Further analysis reveals that in Model II, the effect of the level of economic globalization on the convergence of global social development is negative, implying that as the level of economic globalization increases, the rate of convergence of global social development may accelerate. Similarly, in Model III, the level of economic development has a negative effect on global social development convergence. In Model IV, which combines the level of economic globalization and the level of economic development, we find that both factors have a negative impact on global social development convergence.

To deeply analyze the specific impact of trade globalization and financial globalization on the convergence of regional social development, we conducted a β-conditional convergence regression analysis by selecting 46 countries worldwide as our research subjects, using the trade globalization index and the financial globalization index as the main explanatory variables, the level of economic development as a control variable, and the Gini coefficient as an explanatory variable to measure the level of social development. In the β-conditional convergence analysis of the Gini

Table 12.7 Results of β-conditional convergence analysis (economic globalization) (*Note* * $p < 0.10$, ** $p < 0.05$, *** $p < 0.001$)

Explanatory variables	Coefficient			
	Model I	Model II	Model III	Model IV
Constant	0.1223***	0.1680***	0.2617***	0.3146***
Initial value	– 0.3524***	– 0.3562***	– 0.4074***	– 0.4134***
Economic globalization		– 0.0006***		– 0.0006***
Level of economic development			– 0.0125***	– 0.0130***
R^2	0.2198	0.2399	0.2432	0.2650
F-value	2.98***	3.16***	3.14***	3.46***

coefficient growth rate, we constructed five models. Model I pays special attention to the effect of the level of trade globalization on the convergence of social development, while Model II focuses on the role of the level of financial globalization. Models III and IV combine the level of trade globalization, the level of financial globalization, and the level of economic development for comprehensive analysis, respectively. Model V comprehensively incorporates the three control variables of trade globalization level, financial globalization level, and economic development level. After the estimation of the five models, we find that global social development shows a convergence trend, which is reflected in the fact that the convergence coefficients of the initial values are all negative. Specifically, in Models I, III, and V, the effect of the level of trade globalization on the convergence of global social development is shown to be negative, implying that the increase in the level of trade globalization helps to promote the convergence of social development. Similarly, in Model IV, the level of financial globalization has a negative impact on global social development convergence, showing the positive role of financial globalization in driving social development convergence.

Furthermore, the results of Model V, which combines all the control variables, show that the level of trade globalization, financial globalization, and economic development all have a negative impact on the convergence of global social development. This suggests that the globalization of trade and finance, as well as the increase in the level of economic development, are important drivers of social development convergence (Table 12.8).

In conducting the β-conditional convergence analysis of the growth rate of the Gini coefficient in 24 developed countries worldwide, four different models were developed to explore the impact of different factors on the convergence of social development. Model I is the baseline model, i.e., the original β-absolute convergence model. Model II adds the level of economic globalization as a control variable to Model I. Model III introduces the level of economic development as a control variable to Model I. and Model IV combines the control variables of Models II and III, taking

Table 12.8 Results of β-conditional convergence analysis (trade globalization and financial globalization) (*Note* * $p < 0.10$, ** $p < 0.05$, *** $p < 0.001$)

Explanatory variables	Coefficient				
	Model I	Model II	Model III	Model IV	Model V
Constant	0.1658***	0.1359***	0.2970***	0.2872***	0.3158***
Initial value	– 0.3760***	– 0.3427***	– 0.4273***	– 0.3995***	– 0.4186***
Trade globalization	– 0.0005***		– 0.0004**		– 0.0003**
Financial globalization		– 0.0002*		– 0.0002**	– 0.00030**
Level of economic development			– 0.0120***	– 0.0134***	– 0.0128***
R^2	0.2355	0.2273	0.2570	0.2536	0.2647
F-value	2.98***	2.99***	3.33***	3.29***	3.44***

into account both the level of economic globalization and the level of economic development. The data in Table 12.9 indicate that the social development of developed countries shows convergence under all four models. This convergence is reflected in the fact that the convergence coefficients of the initial values are all negative, indicating that the developed countries with higher initial Gini coefficients will have relatively lower growth rates, thus gradually achieving convergence. Further analysis reveals that the effect of the level of globalization on the convergence of social development in developed countries is negative. This implies that as the level of globalization increases, the rate of convergence of social development in developed countries is likely to accelerate. However, contrary to our expectations, the effect of the level of economic development on the convergence of social development in developed countries is not significant. This may be because developed countries have already reached a certain height in terms of the level of economic development; therefore, its effect on social development convergence is no longer as significant as it is in less developed countries.

To sum up, the level of globalization is one of the important factors affecting the convergence of social development in developed countries, whereas the effect of the level of economic development is relatively weak (Table 12.9).

In the β-conditional convergence analysis of the growth rate of the Gini coefficient for the 24 developed countries, five different models were constructed to explore the impact of different factors on the convergence of social development. Model I incorporates the level of trade globalization as a control variable, Model II takes into account the level of financial globalization, Model III combines the level of trade globalization and the level of economic development, Model IV integrates the level of financial globalization and the level of economic development, and Model V covers all three control variables, namely the level of trade globalization, the level of financial globalization, and the level of economic development. The results of the analysis show that all five models show a convergence of social development in developed countries, i.e., the growth rate of the Gini coefficient gradually converges over time. This convergence is reflected in the fact that the convergence

Table 12.9 Results of β-conditional convergence analysis for developed countries (economic globalization) (*Note* * $p < 0.10$, ** $p < 0.05$, *** $p < 0.001$)

Explanatory variables	Coefficient			
	Model I	Model II	Model III	Model IV
Constant	0.1532***	0.2140***	0.2360***	0.2988***
Initial value	– 0.4926***	– 0.48879***	– 0.5081***	– 0.5047***
Economic globalization		– 0.0007***		– 0.0008***
Level of economic development			– 0.0074	– 0.0076
R^2	0.2658	0.2924	0.2707	0.2975
F-value	3.30***	3.53***	3.36***	3.58***

coefficients for the initial values are all negative, indicating that developed countries with initially higher Gini coefficients will have relatively lower growth rates, thus achieving convergence. In examining the effect of the level of globalization on convergence, we find some interesting phenomena. In Models I, III, and V, the effect of the level of trade globalization on the convergence of social development in developed countries is not statistically significant. However, in Models II, IV, and V, the level of financial globalization has a significant negative impact on the convergence of social development in developed countries, which implies that an increase in the level of financial globalization may accelerate the process of convergence of social development in developed countries.

In Model V, which combines all control variables, we further observe the effects of the level of trade globalization and the level of financial globalization on the convergence of social development. The results show that the effect of the level of financial globalization on the convergence of social development in developed countries remains negative, whereas the effect of the level of trade globalization on the convergence of social development is not significant at the 5% level. This suggests that financial globalization may play a more important role in driving social development convergence in developed countries (Table 12.10).

Finally, we conducted a β-conditional convergence analysis to analyze the convergence of the growth rate of the Gini coefficient in 22 less developed countries in depth. Specifically, we constructed four models: Model I serves as the benchmark and represents the original β-absolute convergence model. Model II adds the level of economic globalization as a control variable based on Model I. Model III introduces the level of economic development as a control variable based on Model I. Model IV combines the control variables of Models II and III, taking into account the level of economic globalization and the level of economic development. The results of the analysis show that all four models indicate a convergence of social development in less developed countries, i.e., the growth rate of the Gini coefficient converges over time. This convergence is reflected in the fact that the convergence coefficients

Table 12.10 Results of β-conditional convergence analysis for developed countries (trade globalization and financial globalization) (*Note* * $p < 0.10$, ** $p < 0.05$, *** $p < 0.001$)

Explanatory variables	Coefficient				
	Model	Model II	Model III	Model IV	Model V
Constant	0.1799***	0.1962***	0.2933***	0.2519***	0.2783***
Initial value	– 0.4854***	– 0.4990***	– 0.5033***	– 0.5095***	– 0.5070***
Trade globalization	– 0.0003		-0.0004		– 0.0002
Financial globalization		– 0.0005***		– 0.0005***	– 0.0005**
Level of economic development			– 0.0095	– 0.0051	– 0.0063
R^2	0.2708	0.2952	0.2786	0.2976	0.2992
F-value	3.31***	4.04***	3.42***	3.54***	3.54***

for the initial values are all negative, implying that countries with higher initial Gini coefficients will have relatively lower growth rates, thus achieving convergence. In Model II, we find that the level of economic globalization has a negative effect on the convergence of social development in less developed countries at the 10% significance level. This suggests that as the level of economic globalization increases, the rate of social development convergence in less developed countries may accelerate. The results of Model III show that the level of economic development has the same negative effect on social development convergence in less developed countries. This implies that an increase in the level of economic development helps drive social development convergence in less developed countries.

In Model IV, which combines the level of economic globalization and the level of economic development, we find that both factors negatively affect social development convergence in less developed countries. This suggests that, in the context of globalization, an increase in the level of economic development is one of the key factors contributing to the convergence of social development in less developed countries (Table 12.11).

To explore the convergence of the growth rate of the Gini coefficient in 22 less developed countries in depth, we conducted a β-conditional convergence analysis. Five different models were constructed, each incorporating different control variables to observe their effect on convergence. Model I considers the level of trade globalization as a control variable. Model II incorporates the level of financial globalization. Model III combines the level of trade globalization and the level of economic development. Model IV includes the level of financial globalization and the level of economic development. Model V combines the level of trade globalization, the level of financial globalization, and the level of economic development. From the data in Table 12.12, all five models show that the social development of less developed countries shows a convergence pattern, that is, countries with a higher initial value will have a relatively slower rate of social development, thus achieving convergence. It is particularly noteworthy that the convergence coefficients are all negative, further validating the existence of convergence. However, in examining the effect of the level

Table 12.11 Results of β-conditional convergence analysis for less developed countries (economic globalization) (*Note* * $p < 0.10$, ** $p < 0.05$, *** $p < 0.001$)

Explanatory variables	Coefficient			
	Model I	Model II	Model III	Model IV
Constant	0.1199***	0.1155***	0.2561***	0.2996***
Initial value	– 0.3120***	– 0.3170***	– 0.3768***	– 0.3850***
Economic globalization		– 0.0005*		– 0.0006**
Level of economic development			– 0.0128***	– 0.0134***
R^2	0.2095	0.2254	0.2389	0.2573
F-value	2.99***	3.07***	3.17***	3.38***

Table 12.12 Results of β-conditional convergence analysis for less developed countries (trade globalization and financial globalization) (*Note* * $p < 0.10$, ** $p < 0.05$, *** $p < 0.001$)

Explanatory variables	Coefficient				
	Model I	Model II	Model III	Model IV	Model V
Constant	0.1598***	0.1263***	0.2822***	0.2777***	0.2989***
Initial value	– 0.3430***	– 0.3030***	– 0.3997***	– 0.3688***	– 0.3908***
Trade globalization	– 0.0004*		-0.0004		– 0.0004
Financial globalization		– 0.0002		– 0.0002	– 0.0002
Level of economic development			– 0.0120**	– 0.0140***	– 0.0131***
R^2	0.2251	0.2131	0.2506	0.2469	0.2566
F-value	3.13***	2.87***	3.29***	3.23***	3.35***

of globalization on convergence, we find some interesting results. In Models I, III, and V, the effect of the level of trade globalization on the convergence of social development in less developed countries is not significant at the 5% significance level. Similarly, in Models II, IV, and V, the effect of the level of financial globalization on the convergence of social development in less developed countries is not significant. In Model V, we combined three control variables, namely the level of trade globalization, the level of financial globalization, and the level of economic development. The results show that the effect of the level of economic development on the convergence of social development in less developed countries remains negative, which means that the increase in the level of economic development helps promote convergence. However, the effects of the level of trade globalization and the level of financial globalization on social development convergence in less developed countries remain non-significant at the 5% level.

In summary, although the level of globalization has an important impact on socio-economic development at the global level, its impact is not significant in the process of social development convergence in less developed countries. In contrast, the level of economic development remains the key factor influencing the convergence of social development in less developed countries.

12.1.4.4 Analysis of the Driving Mechanisms of Regional Social Development

This study constructs an analytical framework for the driving mechanisms of global social development based on the spatial interaction of social development in regions worldwide and the Cobb–Douglas production function. In this framework, the Gini coefficient is chosen as an explanatory variable to measure the degree of inequality in the level of social development. The level of globalization is used as an explanatory variable, which is further subdivided into trade globalization and financial globalization to fully reflect the multidimensional impact of globalization on social development. Considering the close relationship between land and capital and the structure of production, the level of industrialization of industry and that of services are selected as control variables to reveal the potential impact of industrial structure on social development. Meanwhile, given that the level of economic development is the cornerstone of social development, GDP per capita is included in the model to control the differences in the level of economic development of different countries.

The study uses averaged cross-sectional data from 46 countries, 24 developed countries, and 22 less developed countries worldwide from 2006 to 2015. To explore the spatial autocorrelation phenomenon in the models in depth, Moran's I test was performed on the error term of each model and Lagrangian function tests for spatial lag and spatial error were conducted.

By constructing four OLS regression models, this study finds that economic globalization has no significant impact on global social development, whereas trade globalization has a significant positive effect on social development, and financial globalization shows a negative effect. In the model combining the trade globalization, financial globalization, industrial industrialization level, service industrialization level, and economic development levels, trade globalization and economic development level have significant positive effects on global social development, whereas financial globalization, industrial industrialization level, and service industrialization level may exacerbate income disparity and negatively affect social development. In terms of the degree of influence, financial globalization and trade globalization have comparable effects on social development.

In addition, the error terms of Models I, II, and IV exhibit a certain degree of spatial autocorrelation and pass the Lagrangian function test with spatial lag. Given that Model IV contains all the variables of Model II, spatial econometric regression analyses of Models I and IV for the 46 countries were finally conducted to reveal more accurately the spatial impact mechanisms of globalization on global social development (Table 12.13).

In the two-effects fixed model, since the spatial coefficient is not significant in both models of economic globalization and trade globalization, we focus on the analytical results of the individual fixed-effects model and the time fixed-effects model. In these two models (Table 12.14), we find that both economic globalization and trade globalization have significant negative effects on the Gini coefficient. The negative correlation between the Gini coefficient and the level of global social development implies that economic globalization and trade globalization have a significant driving effect on global social development over a long period. Meanwhile, our study also shows that the level of economic development also has a significant negative effect on the Gini coefficient in both the individual fixed effects model and the time fixed effects model. This suggests that an increase in the level of economic development contributes to the reduction of income disparity, which in turn enhances global social well-being. The individual fixed-effects model shows that industrialization level of industry and that of services have a significant positive effect on the level of social development; however, the time fixed-effects model shows a significant negative

Table 12.13 Results of OLS regression analysis of social development in 46 countries (*Note* * $p < 0.10$, ** $p < 0.05$, *** $p < 0.001$)

Explanatory variables	Coefficient			
	Model I	Model II	Model III	Model IV
Constant	0.0558	0.1303	0.0973	0.2474
Economic globalization	– 0.0006			
Trade globalization		– 0.0021***		– 0.0031***
Financial globalization			0.0018**	0.0031***
Industrialization level (industry)	0.8938***	0.7282***	0.897***	0.5984**
Industrialization level (services)	1.18***	1.12***	1.10***	0.9587***
Level of economic development	– 0.0619***	– 0.0494***	– 0.0794***	– 0.0629***
R^2	0.4659	0.5509	0.5096	0.6781
Adj R^2	0.4137	0.5071	0.4617	0.6378
Moran's I (Z-value)	6.75***	4.46***	– 1.24	7.48***
LM_Lag	15.53***	15.95***	0.6270	15.58***
LM_error	0.0050	0.5010	2.71	0.7080

effect. This may be because the individual fixed-effects model focuses more on the impact of individual changes on social development, taking into account the differences of different individuals, whereas the time fixed-effects model reflects more on the long-term impact of time changes on global social development. It is worth noting that the spatial regression coefficients in the individual fixed-effects models of economic globalization and trade globalization are all positive at the 1% significance level. This suggests that there is a significant positive correlation effect between countries, that is, the level of social development in one country is influenced to some extent by the level of social development in neighboring countries. Geographical proximity has a significant positive effect on the level of social development of countries, and neighboring geographical locations are conducive to the flow of capital, the transmission of information, and the exchange of technology. The process of globalization has, to a certain extent, strengthened the impact of this geographical proximity.

Therefore, when exploring the impact of globalization on the level of global social development, it is necessary to consider spatial factors. The use of spatial panel regression models enables a more rational analysis of the relationship between globalization and the level of global social development.

The study conducted Hausman tests on the panel data of 24 developed countries from 2006 to 2015 and found that random effect model is more suitable for driving mechanism analysis (Prob > chi2 = 0.1245, original hypothesis accepted). The results of the analysis of 24 developed countries worldwide show that economic globalization, trade globalization, and financial globalization do not have a significant effect on the level of social development in developed countries. The level

Table 12.14 Spatial panel regression results for social development in 46 countries (*Note* * $p<0.10$, ** $p<0.05$, *** $p<0.001$)

Explanatory variables	Economic globalization trade globalization					
	Individual fixed effects	Time fixed effects	Double effects fixed	Individual fixed effects	Time fixed effects	Double effects fixed
Economic globalization	– 0.0007***	– 0.0005**	– 0.0007***			
Trade globalization				– 0.0009***	– 0.0015***	– 0.0008***
Financial globalization				0.0001	0.0013***	0.0001
Industrialization level (industry)	– 0.2176***	0.2478***	– 0.2095***	-0.2118***	0.2484***	– 0.1996***
Industrialization level (services)	– 0.1388***	0.4734***	– 0.0907*	– 0.1019**	0.5177***	– 0.0820
Level of economic development	– 0.0309***	–0.0244***	– 0.0372***	– 0.0275***	–0.0328***	– 0.0369***
Spatial regression coefficient	0.1255***	0.6692***	0.0537	0.1151***	0.5119***	0.0514
R^2 within group	0.2711	0.0230	0.2622	0.2974	0.1458	0.2880
R^2 between groups	0.2281	0.6913	0.2299	0.3051	0.7358	0.2811
Overall R^2	0.2290	0.6311	0.2305	0.3048	0.6916	0.2812
Log-likelihood	1355	779.01	1367	1362	806.29	1372

of economic development has a significant negative effect on the Gini coefficient, that is, the level of economic development has a significant positive effect on the level of social development in developed countries, and a high level of economic development contributes to the reduction of the income gap and the enhancement of social well-being. The industrialization level of industry does not have a significant effect on the level of social development in developed countries, whereas the industrialization level of services has a positive effect on the widening of the income gap (Table 12.15).

We conducted the Hausman test on the panel data of 22 less developed countries from 2006 to 2015 and found that the fixed effects model is more suitable for driving mechanism analysis (Prob > chi2 = 0.0111, rejecting the original hypothesis). The results of the analysis of 22 less developed countries show that economic globalization and trade globalization have a significant negative impact on the Gini coefficient of less developed countries, that is, they have a significant positive impact on the level of social development, and financial globalization has a non-significant impact. The level of economic development, the level of industrialization, and the industrialization level of services have a significant negative effect on the Gini coefficient,

Table 12.15 Panel regression results for social development in 24 developed countries (*Note* * $p < 0.10$, ** $p < 0.05$, *** $p < 0.001$)

Explanatory variables	Coefficient			
	Model I	Model II	Model III	Model IV
Constant	0.2870***	0.2843***	0.2726***	0.2896***
Economic globalization	– 0.0003**			
Trade globalization		– 0.0002***		– 0.0002***
Financial globalization			-0.0002	-0.0001
Industrialization level (industry)	0.0701***	0.0627***	0.0738***	0.0684***
Industrialization level (services)	0.2338***	0.2368***	0.2348***	0.2335***
Level of economic development	– 0.0114***	– 0.0115***	– 0.0113***	– 0.0114***
R^2	0.4174	0.4052	0.4166	0.4170

that is, the level of economic development, the industrialization level of industry and that of services have a significant positive effect on the level of social development of less developed countries, and an increase in the level of economic development, the industrialization level of industry, and the industrialization level of services can help reduce the income disparity and enhance social well-being (Table 12.16).

Table 12.16 Panel regression results for social development in 22 less developed countries (*Note* * $p < 0.10$, ** $p < 0.05$, *** $p < 0.001$)

Explanatory variables	Coefficient			
	Model I	Model II	Model III	Model IV
Constant	0.9705***	0.9251***	0.9378***	0.9086***
Economic globalization	– 0.0007**			
Trade globalization		– 0.0009***		– 0.0009***
Financial globalization			0.0001	0.0001
Industrialization level (industry)	– 0.02358***	– 0.2193***	– 0.2629***	– 0.2228***
Industrialization level (services)	– 0.2452***	– 0.2106***	– 0.2709***	– 0.2100***
Level of economic development	-– 0.0386***	– 0.0344***	– 0.0380***	– 0.0335***
R^2	0.3622	0.3898	0.3486	0.3915
F-value	201.67***	187.30***	157.48***	116.35***

12.1.5 Conclusions

Based on the theoretical framework of Development Geography, this section provides an in-depth discussion of the convergence trend of social development at the global scale and its external driving mechanisms. Through a comparative analysis of the differences in social development patterns between developed and less developed countries over the past decade, the study finds that global social development has shown a significant convergence trend. The development of globalization, industrial progress, and the growth of national economies are the main drivers of this trend. However, global social development presents unprecedented opportunities as well as many challenges. To gain a deeper understanding of this convergence trend, the study employs income disparity as a key indicator to comprehensively analyze the convergence status and driving mechanism of regional social development. In terms of the convergence situation, we explored α convergence, β absolute convergence, and conditional convergence, and conducted a club convergence analysis for developed and less developed countries. In the study of driving mechanisms, we use economic globalization, trade globalization, and financial globalization as the core explanatory variables, combined with the industrialization level of industry and services and economic development status as the control variables, and analyze the impacts of these factors on global social development and the spatial interactions between them.

Between 2006 and 2015, the level of global social development steadily increased, showing a clear convergence trend. Developed countries converged relatively quickly, with their social development levels fluctuating in a higher range, smaller differences between countries, and higher social well-being. However, the less developed countries, although starting from a lower base, have shown a significant upward trend in their level of social development, with inter-country differences narrowing from year to year, again showing a trend of convergence. This demonstrates the commitment of less developed countries to narrowing income gaps, enhancing social well-being, and reducing regional inequalities in social development during this decade. Overall, global social well-being has shown a trend of convergence based on steady improvement, and controlling income disparities and promoting welfare equalization are key paths to achieving harmonious social development. Countries should seize the opportunity to strengthen international cooperation and exchanges to further enhance social well-being on the existing basis and realize the sustainable development of national societies. However, global social development still faces many challenges, with more than 20% of countries having a Gini coefficient that has long exceeded the warning line and Panama even consistently exceeding 0.5, with a wide gap in income distribution. Thus, the harmonious development of the global society still requires more effort.

Globalization, economic development, and the level of industrialization are important forces driving global social development, in which there are significant differences in the driving mechanisms between developed and less developed countries. From a global perspective, economic globalization and its subsectors, such as trade globalization and financial globalization, not only affect the level of global social development but also significantly contribute to its convergence. In particular, economic globalization, trade globalization, and financial globalization have a dampening effect on global social development convergence, with financial globalization having a particularly strong impact on developed countries. As far as the level of social development itself is concerned, economic globalization and trade globalization have had a significant positive contribution to less developed countries' however, this effect has not been significant in developed countries. The profound impact of economic globalization on social development is manifested through capital and human mobility and information connectivity. Developed countries generally have a high level of globalization; however, financial globalization varies more among countries than trade globalization and becomes an important factor affecting the convergence of social development. On the contrary, the index of trade globalization in less developed countries is significantly higher than that of financial globalization, indicating that trade globalization has played a key role in raising the level of social development in these countries. In the wave of globalization in the twenty-first century, less developed countries should grasp the opportunities of trade globalization, optimize the structure of import and export commodities and markets, promote the development of strategic industries, special industries, and service trade with a competitive advantage, and seek to diversify their export markets. This will not only strengthen China's long-term comprehensive competitiveness and promote the stability and long-term development of trade but also further enhance the level of social development. However, we cannot ignore the challenges posed by globalization. Economic globalization may exacerbate inequalities in regional social development; therefore, its rational use is the key to achieving sustainable global social development. The level of economic development is the cornerstone of global social development, which not only has a significant positive contribution to the level of social development but also inhibits the convergence of social development to a certain extent. Regions with higher levels of economic development usually have higher resident incomes, more reasonable income distribution mechanisms, smaller income gaps, and better social welfare, thus presenting higher levels of social development. At the national level, raising the level of economic development is an important way to raise the level of social development. However, from the global perspective, rapid economic growth may also lead to a further widening of the differences in the level of social development among countries, which is a challenge that needs to be seriously addressed. In addition, there are significant positive spatial interactions between regional social development, and the interaction between geographically neighboring countries is particularly significant. Geographical proximity facilitates the flow of capital, the transfer of information, the sharing of knowledge, and the exchange of technology, and globalization further reinforces this interactive effect. Geographical proximity has a significant positive impact on the level of social development of a region, and the level of social development of one country affects the level of social development of neighboring countries to a certain extent. Geographical proximity facilitates the flow of financial resources, the transfer of information, the sharing of knowledge, and the exchange of technology (Rosenthal & Strange, 2004), and globalization further reinforces this effect to some extent. With the increasingly close global ties, countries should make full use of their geographical advantages to strengthen exchanges and cooperation with neighboring countries to jointly enhance global social welfare and promote sustainable social development. By strengthening international cooperation, we can better cope with the challenges posed by globalization and jointly promote the harmony and progress of the global community.

12.2 Convergence of Regional Economic Development

12.2.1 Regional Economic Development Convergence Profiles

Against the background of globalization and industrial transformation, global economic development in the twenty-first century has entered a new stage, with interregional linkages becoming increasingly prominent, the industrial division of labor and layout becoming increasingly refined, and the pattern of economic development among different economies experiencing a huge transformation. The growth mechanisms and interaction mechanisms of developed and less developed countries in the global economic network are distinctive (Blecker et al., 2017; Royuela & García, 2015). The level of economic development not only determines the status and role of a country or region in the global economic and trade system but also is an important reflection of national strength and international competitiveness, which is mainly measured by core indicators such as the gross national product and national income. After going through the stages of industrial revolution, industrial structure upgrading, and global integration, the trend of convergence or divergence of the level of economic development of countries has become the focus of research. While the global economy continues to grow, how do its internal differences change? Does it tend to converge or diverge? These are urgent questions for economists and management scientists (Borozan, 2017). Based on neoclassical economic growth theory and endogenous economic growth theory, research on the convergence and divergence of economic growth has increasingly received attention, and the research methodology has gradually shifted from the traditional regression of cross-sectional data to a more rigorous analysis of time-series panel data.

Convergence, driving mechanisms, and spatial interactions of economic development are the core issues of global and regional economic development research. In the study of convergence, scholars have explored the convergence trend of economic development at the national, regional, and global levels, and the results show that different countries and regions show heterogeneity in time and space. For example, during 1989–1999, the economic development of 145 European regions showed obvious convergence (Dall'Erba, 2008), whereas during 1995–2000, 256 European regions had club convergence, and this convergence was heterogeneous in terms of different steady states and speeds of convergence. When spatial factors are considered, regional economic growth in Anhui Province, China, exhibits absolute β-convergence during the period 1996–2012. By contrast, the 15 Asian economies did not show a trend of convergence in their economic development over the past 50 years but have shown absolute convergence in the last 20 years. China's regional development has gone through multiple stages of balanced, unbalanced, coordinated, integrated, and synergistic development, and economic growth has been characterized by region-wide convergence. The inter-regional economic development gap is shrinking, and the overall economic development of the central and western regions shows a converging trend, accompanied by the phenomenon of random convergence of local means. However, the economic development gap in certain central regions is still widening. Overall, the economic growth rate of less developed regions is gradually increasing, the economic growth of developed countries is relatively stable but with differences in the steady state, and the internal structure is being optimized and transformed. In terms of the driving mechanism of economic development, industrial development and globalization are the most important influencing factors. Globalization was reflected not only in trade, transnational corporations, international finance, and foreign exchange investment but also in social globalization, such as international tourism, migration, culture, information, talent, transportation, and interpersonal network relations and political globalization, such as the integration of institutional systems. The role of factor mobility brought about by globalization on economic development has been confirmed in the BRIC[1] countries (Coulibaly et al., 2017). Policy effects, such as the "harmonious and balanced expansion" policy proposed by the EU to promote economic development, also have an impact on the β convergence of economic growth (Yin et al., 2003). The adjustment of industrial structure, especially the development of manufacturing and service industries, has had different degrees of impetus to the economic development of countries, which has also received attention from a wide range of scholars (Herman, 2016; Lauridsen, 2017). In terms of the spatial interaction of economic development, globalization has promoted the global mobility of factors of production such as labor and capital, thus strengthening the spatial influence among economies. For example, the economic growth of Europe's 256 regions not only relies on endogenous factor accumulation and mobility but is also influenced by the growth paths of neighboring regions (Fischer et al., 2006). The regional economic development of Japan's 47 first-level administrative districts has also been shown to be the result of a combination of economic convergence and economic agglomeration. In China, dynamic externalities formed by spatial interactions are an important factor in regional economic development and convergence, both at the prefecture city and city-ring levels.

The convergence and driving mechanisms of regional economic development have received extensive attention at both the national and global levels. However,

[1] BRICS initially referred to a group of rapidly developing nations expected to emerge as leading economic powers of the twenty-first century if their growth trends continued. The name, derived from the initials of its founding members—Brazil, Russia, India, China, and South Africa—has since become the title of a formal intergovernmental organization aimed at promoting deeper economic and geopolitical cooperation among its members.

most previous studies have focused on the impact of factors such as industry, trade, finance, and investment on regional economies, and relatively few have systematically explored the impact of globalization. Most empirical studies have focused on a single economy, whereas comparative studies of developed and less developed countries at the global level are scarce. In addition, traditional research methods tend to use convergence and econometric analyses, with less consideration of the role of spatial interactions and their spillover effects in economic development. Therefore, this section takes the global level of economic development as the core research object and analyzes its regional convergence, driving mechanisms, and spatial interaction effects in depth. In particular, we conduct a systematic analysis of developed and less developed countries, using the level of globalization as the main explanatory variable and industrialization development as a control variable to shed light on the changing trends of global economic development in a comprehensive manner.

12.2.2 *Impacts of Climate Change on Agriculture*

12.2.2.1 Descriptive Analysis of Differences in GDP Per Capita

Figure 12.15 shows the global distribution of GDP per capita based on the World Bank data (https://data.worldbank.org/) for 120 countries worldwide (e.g., Exhibit 12–1) from 2004 to 2016. Table 12.17 shows the minimum, maximum, mean, standard deviation, and variance of GDP per capita in 120 countries.

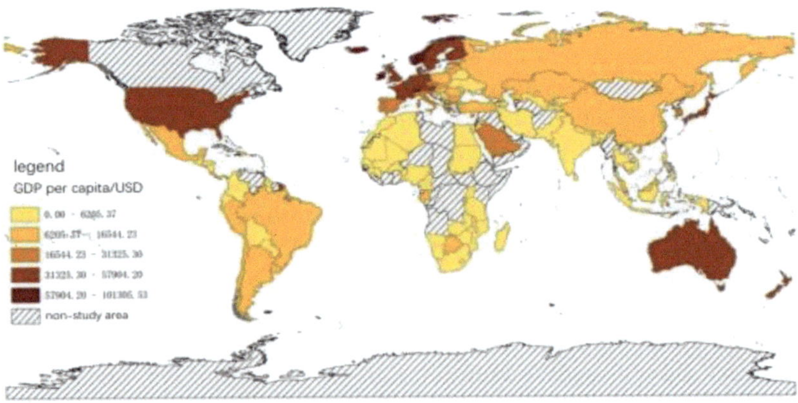

Fig. 12.15 Global distribution of GDP per capita

Table 12.17 Global GDP per capita differences for 120 countries, 2004–2016

Year	Number of countries observed	Minimum value	Maximum value	Mean value	Standard deviation	Variance
2004	120	128	75,716	10,651	15,497	1.45
2005	120	152	80,290	11,511	16,508	1.43
2006	120	167	89,740	12,429	17,587	1.41
2007	120	172	106,019	14,314	20,101	1.40
2008	120	198	114,294	15,604	21,391	1.37
2009	120	212	103,199	13,947	18,966	1.36
2010	120	234	104,965	14,717	19,614	1.33
2011	120	250	115,762	16,369	21,860	1.34
2012	120	252	106,750	16,015	21,100	1.32
2013	120	257	113,625	16,499	21,726	1.32
2014	120	275	118,824	16,672	21,905	1.31
2015	120	306	100,428	14,926	19,383	1.30
2016	120	282	101,306	15,012	19,522	1.30

Changes in the absolute differences (standard deviation) between the GDP per capita of the 120 countries show some volatility and are characterized by clear phases (Fig. 12.16). The mean value of the GDP per capita of the 120 countries, shows a general trend of increasing and then decreasing. Since the standard deviation is affected by changes in the mean during the calculation process, the increase in absolute differences can be partially attributed to the increase in GDP per capita. Therefore, to remove this partial effect on the standard deviation, the change in GDP per capita and its relationship with the absolute difference (standard deviation) in GDP per capita is further examined and verified.

The mean value of GDP per capita for the 120 countries was $10,651 in 2004, which increased to $15,012 in 2016, an overall increase of $4,361, with a growth rate of 40.94%.The standard deviation of the absolute difference in GDP per capita for the 120 countries was $15,497 in 2004, which increased to $195,222 in 2016, an overall increase of $4,025 and a growth rate of 25.97%, which is a substantial change.

The change of the absolute difference (standard deviation) between the GDP per capita of the120 countries shows a stage characteristic. From 2004 to 2008, the standard deviation increased year by year, from $15,497 to $21,391, with an average growth rate of 38.03%. After 2008, following the global economic crisis, the absolute difference decreased rapidly, especially in 2009, from $21,391 in 2008 to the lowest point of $18,966. After 2010, the global economy gradually recovered, and the absolute difference in GDP per capita began to rise, reaching $21,860 in 2011, and then experienced a slow decline from 2011 to 2012. Then, from 2013 to 2014, the standard deviation of GDP per capita showed a slow rise, reaching $21,905, which is

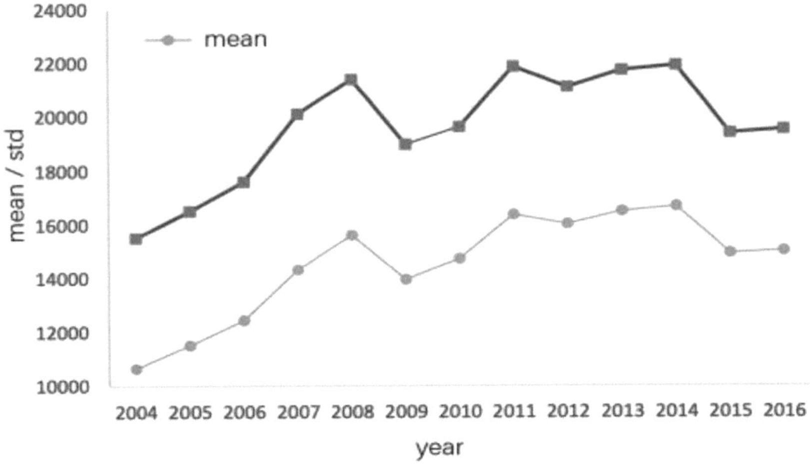

Fig. 12.16 Changes in standard deviation and mean of GDP per capita for 120 countries worldwide (constant2011PPP$)[2]

roughly in line with the absolute difference in 2011 ($21,860). However, after 2014, the absolute difference in GDP per capita began to decline again, to $19,522 in 2016.

Overall, the changes in the mean value of GDP per capita and the changes in the absolute variance (standard deviation) of the 120 countries have a certain correlation, and the fluctuations in the changes in the mean value mainly occurred during the global economic crisis in the period of 2007–2009. However, the overall situation is in a stable fluctuation state, and the fluctuations in the changes in the absolute variance are relatively more obvious.

(i) Analysis of absolute differences for 32 developed countries

The changes in absolute differences (standard deviations) between GDP per capita in 32 developed countries worldwide show similar volatility to that shown in Fig. 12.16 and also exhibit clear phasing, whereas the means of GDP per capita in developed countries generally show an upward trend similar to that seen globally (Fig. 12.17). Since the standard deviation is subject to some influence from changes in the mean during the calculation of the indicator, the increase in absolute differences can be partially attributed to the increase in GDP per capita. Therefore, to remove this partial effect on the standard deviation results, the change in GDP per capita and its relationship with the absolute difference (standard deviation) between GDP per capita is further examined.

[2] constant2011PPP$represents the dollar value adjusted to 2011 purchasing power parity (PPP), used to compare economic indicators across different years or countries while eliminating the effects of inflation and price level differences.For example:If a country's GDP in 2015 is $500 billion (constant2011PPP$), it means the country's actual economic output in 2015, adjusted to the purchasing power parity and price level of 2011, is equivalent to $500 billion in 2011 terms.

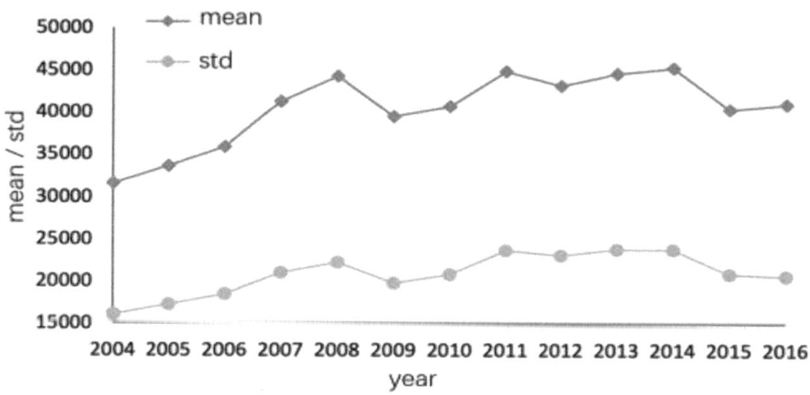

Fig. 12.17 Standard deviation and mean change in GDP per capita at purchasing power parity (PPP) for 32 developed countries

Although the trends in both the standard deviation and mean of GDP per capita in developed countries are similar to the global trend, the range of change in the mean of GDP per capita in developed countries [$31,533–$33,594] is higher than the range of change in the standard deviation [$16,020–$20,583]. The change in the absolute variance of GDP per capita in developed countries from 2004 to 2016 was $4,563. The rate of change was 28.48%, whereas the change in the mean was $2,507, a rate of change of only 6.54%.

(ii) **Analysis of absolute differences in 88 less developed countries**

Changes in the absolute differences (standard deviation) in GDP per capita among 88 less developed countries worldwide show similar volatility to that observed globally, and there are still distinct phases, with the exception that the absolute differences have been decreasing since 2012 (Fig. 12.18). However, the mean value of GDP per capita in less developed countries generally shows an upward trend similar to that shown in Fig. 12.29. Since the standard deviation is affected by changes in the mean in the calculation of the indicator, the increase in absolute differences can be partially attributed to the increase in GDP per capita. Therefore, to remove this partial influence on the standard deviation results, the change in GDP per capita and its relationship with the absolute difference (standard deviation) between GDP per capita is further examined.

The standard deviation and the change in the mean of GDP per capita in the less developed countries showed an opposite trend in 2012–2013. During this period, the mean value of GDP per capita in the less developed countries increased; however, the absolute variance is decreased, and overall, the standard deviation of GDP per capita in the less developed countries has shown an upward trend in the last 13 years, with an increase of $1,362 and a rate of change of 31.80%, which indicates that the GDP per capita in the less developed countries has been gradually increasing.

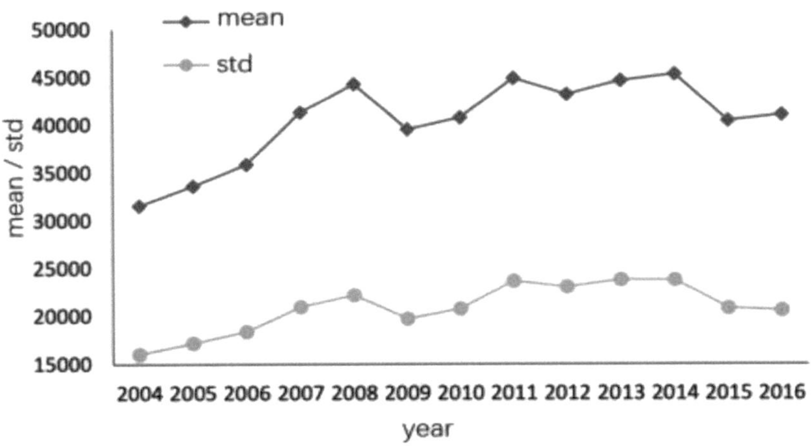

Fig. 12.18 Standard deviation and mean change in GDP per capita at purchasing power parity (PPP), 88 less developed countries

12.2.2.2 Descriptive Analysis of Differences in Total Per Capita Income

Based on the World Bank's Gross National Income (GNI) Per Capita data (https://data.worldbank.org/) for 120 countries worldwide from 2004 to 2016 and the World Bank's Atlas method (converting GNI to current US dollars using the Atlas method) and PPP method (conversion to current international currency using the PPP method), we statistically analyzed the status of differences in GNI per capita among 120 countries worldwide in terms of the mean, standard deviation, and coefficient of variation (Table 12.18).

The results of this study using the Atlas and PPP methods of analysis show that changes in the overall absolute difference (standard deviation) of GNI per capita in 120 countries worldwide show some volatility and exhibit clear phase characteristics, with GNI per capita generally showing an upward trend (Fig. 12.19). The standard deviation of GNI per capita under the two methods declined from 2008 to 2009. The absolute difference in GNI per capita under the PPP method continued to become larger after 2014, while the absolute difference in GNI per capita under the Atlas method began to decline. Since the standard deviation indicator is subject to some influence from changes in the mean during the calculation process, the increase in the absolute variance can be partially attributed to an increase in overall GNI per capita. Therefore, to remove this partial effect on the standard deviation results, a further correlation between the change in GNI per capita and its relationship with the absolute difference in GNI per capita (standard deviation) was examined.

Prior to 2014, the average GNI per capita for the 120 countries in the world increased under both methodologies, and after 2014, the average GNI per capita under the PPP methodology continued to increase, whereas the average GNI per capita under the Atlas methodology began to decline (Fig. 12.20).

Table 12.18 Descriptive statistics of gross national income per capita (GNI per capita) for 120 countries

Year	Mean value		Standard deviation		Variation coefficient	
	Atlas method	PPP	Atlas method	PPP	Atlas method	PPP
2004	9761	13,199	14,207	13,707	1.46	1.04
2005	11,178	13,899	16,296	14,363	1.46	1.03
2006	11,999	15,059	17,100	15,365	1.43	1.02
2007	13,046	16,042	18,457	16,182	1.41	1.01
2008	13,988	16,698	19,152	16,365	1.37	0.9801
2009	13,824	16,291	18,453	15,522	1.33	0.9528
2010	14,289	17,018	19,023	16,334	1.33	0.9598
2011	14,774	17,780	19,311	16,947	1.31	0.9532
2012	154,111	18,291	19,969	17,315	1.30	0.9466
2013	15,984	18,974	20,641	17,861	1.29	0.9390
2014	16,026	19,572	20,604	18,221	1.29	0.9310
2015	15,434	20,085	19,665	18,653	1.27	0.9287
2016	14,860	20,594	18,766	19,045	1.26	0.9248

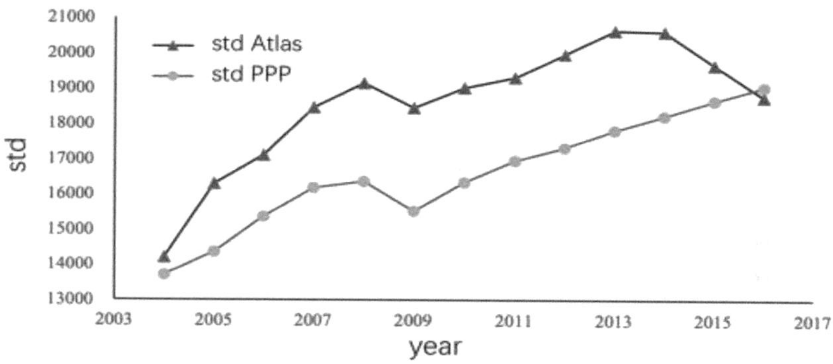

Fig. 12.19 Standard deviation change in GNI per capita for 120 countries

(i) Analysis of absolute differences for 32 developed countries

This study uses the Atlas and PPP methods, and the changes in the absolute differences (standard deviation) of GNI per capita in the 32 developed countries show some volatility, all of which exhibit more pronounced phases, with an overall upward trend in GNI per capita. Similar to the overall trend of change, the standard deviation of GNI per capita under both methods declined in 2008–2009, and the absolute difference in GNI per capita under the PPP method continued to increase after 2014, whereas the absolute difference in GNI per capita under the Atlas method began to decline. Since the standard deviation indicators are affected by changes from the mean in the calculation process, the increase in the absolute variance can be partially attributed to

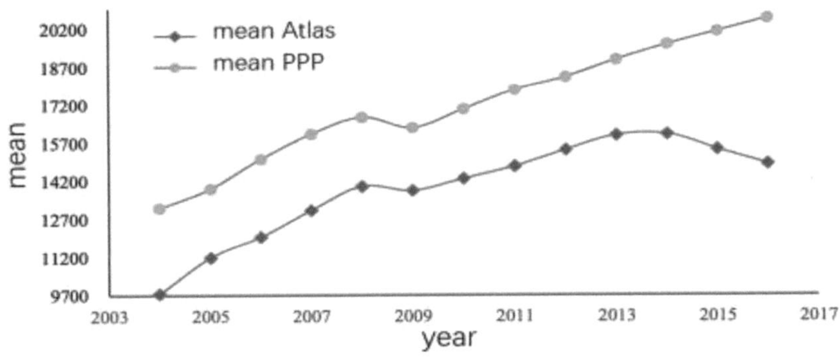

Fig. 12.20 Changes in the mean value of GNI per capita for 120 countries

the increase in overall GNI per capita. Therefore, to remove this partial influence on the standard deviation results, the change in the mean value of GNI per capita and its relationship with the absolute variance of GNI per capita (standard deviation) were further examined in relation with the change in the mean value of GNI per capita (Fig. 12.21).

The average GNI per capita for the 32 developed countries in the world increased under both methods until 2014 but experienced a significant decline in 2008–2009, and the average GNI per capita under the PPP method continued to increase after 2014, whereas the average GNI per capita under the Atlas method began to decline (Fig. 12.22).

(ii) **Analysis of absolute differences in 88 less developed countries**

The absolute variance (standard deviation) of GNI per capita for the 88 less developed countries in the world shows an overall upward trend, with a larger absolute variance under the PPP method than the mean under the Atlas method. Similar to the overall trend, the standard deviation of GNI per capita under both methods declined in 2008–2009, and the absolute difference in GNI per capita under the PPP method continued

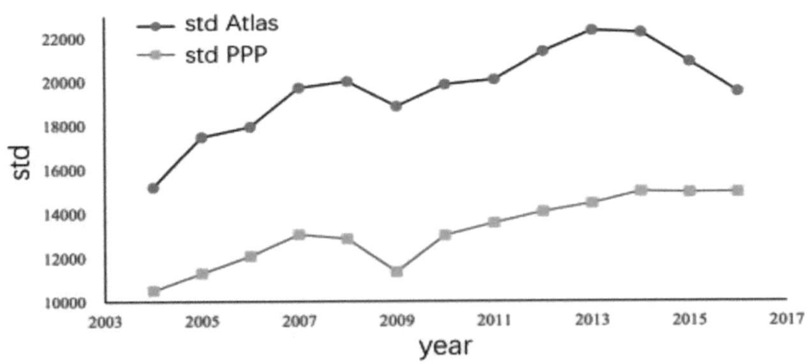

Fig. 12.21 Evolution of standard deviation of GNI per capita in 32 developed countries

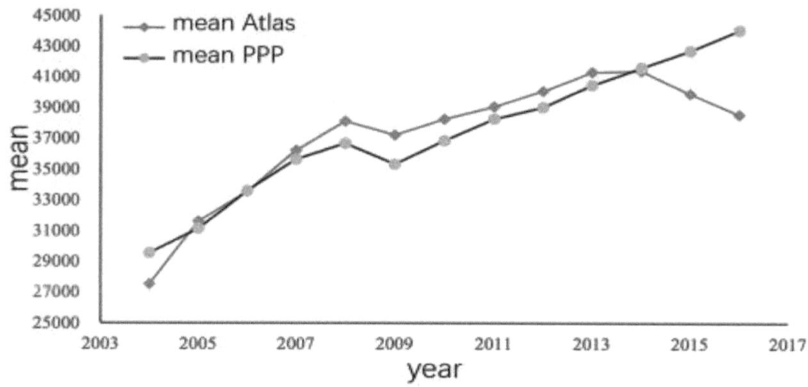

Fig. 12.22 Changes in the mean value of GNI per capita in 32 developed countries

to become larger after 2014, while the absolute difference in GNI per capita under the Atlas method began to decline (Fig. 12.23). Since the standard deviation indicator is affected by the correlation from the change in the mean during the calculation of the indicator, the increase in the absolute difference can then be partially attributed to the increase in overall GNI per capita. Therefore, to remove this partial influence on the standard deviation results, the change in GNI per capita and its relationship with the absolute difference in GNI per capita (standard deviation) was further examined in relation with the change in GNI per capita and its relationship with the absolute difference in GNI per capita (standard deviation).

Before 2014, the mean value of GNI per capita in less developed countries under both methods generally showed an upward trend, with a larger mean under the PPP method than that under the Atlas method, and after 2014, the mean value of GNI per capita under the PPP method continued to increase, whereas the mean value of GNI per capita under the Atlas method began to decline (Fig. 12.24).

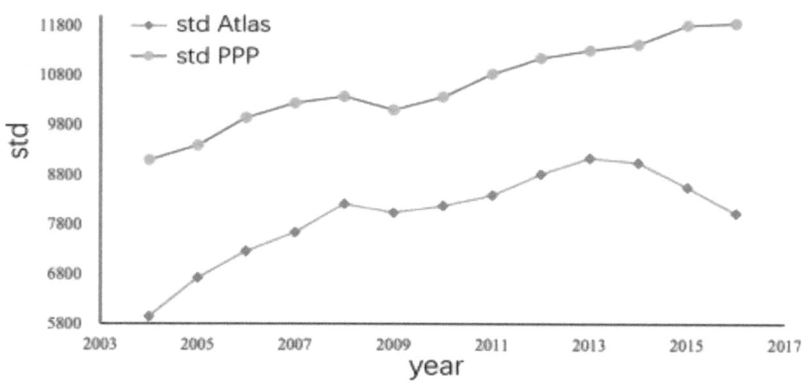

Fig. 12.23 Standard deviation change in GNI per capita for 88 less developed countries

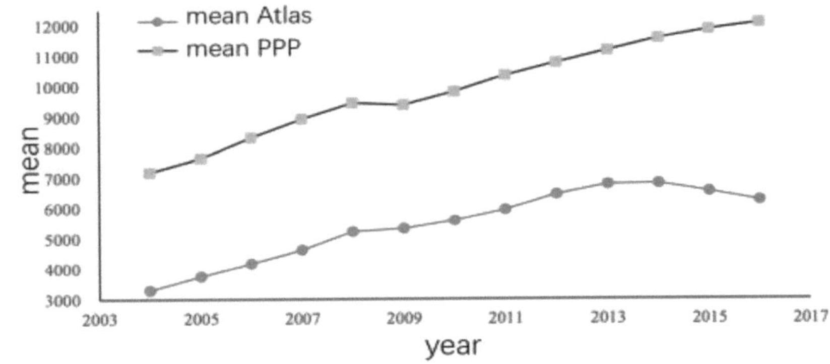

Fig. 12.24 Mean change in GNI per capita for 88 less developed countries

12.2.3 Analysis of Factors Influencing the Convergence of Regional Economic Development

12.2.3.1 Developments in Globalization and Regional Economic Development

Within the framework of the development of global regional networks and the increasing economic linkages among countries, globalization-driven capital flows, labor migration, trade activities, cultural exchanges, technology sharing, and institutional synergies have had a profound impact on the upgrading of the global economy. However, the differences in output and income among countries, and the specific manifestations of this impact between developed and less developed countries, still need further clarification. Against this backdrop, globalization has become a key force in supporting the economic growth of all countries, and its importance has become increasingly prominent. The rapid progress of science and technology, the continuous improvement of productivity, the deepening of the international division of labor, and the vigorous development of global trade and financial investment have led to the interdependence of the world's economies and more frequent economic exchanges, thus strengthening the trend of economic globalization. Globalization is reflected in the economic level and has far-reaching impacts in social fields such as culture and religion, forming a global cultural landscape with multiple intertwining of technology, finance, media, and ideology. To explore the specific impact of globalization on regional economic development in depth, we first need to analyze the intrinsic connection between the level of total globalization and regional economic development. Globalization not only includes economic globalization but also social globalization. Within economic globalization, financial globalization, has a particularly significant impact on the regional economy. While at the level of social globalization, interpersonal, cultural, and information globalization also have a significant impact on regional economic development. In this section, an in-depth analysis of the mean and standard deviation of the levels of financial, interpersonal, cultural, and information globalization for 120 countries is presented using globalization data published by the Konjunkturforschungsstelle (KOF) Swiss Economic Institute (Figs. 12.25 and 12.26).

The analysis of changes in the mean values of financial, interpersonal, cultural, and information globalization in 120 countries shows that the information globalization index is significantly higher than that of financial globalization, interpersonal,

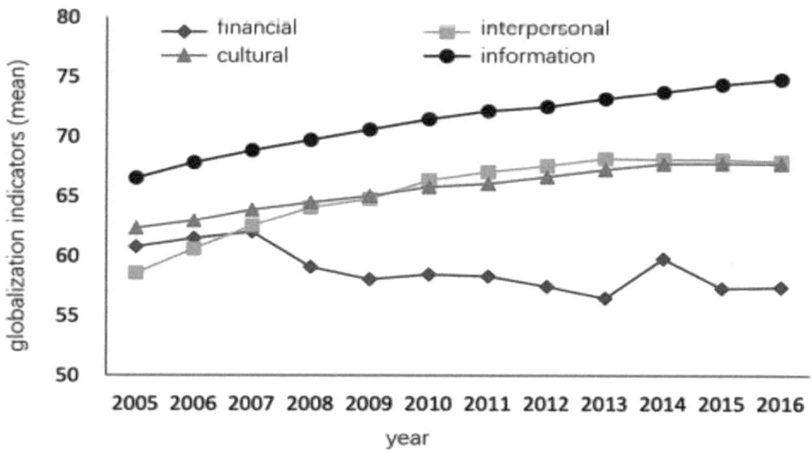

Fig. 12.25 Changes in the means of financial, interpersonal, cultural, and information globalization for 120 countries worldwide

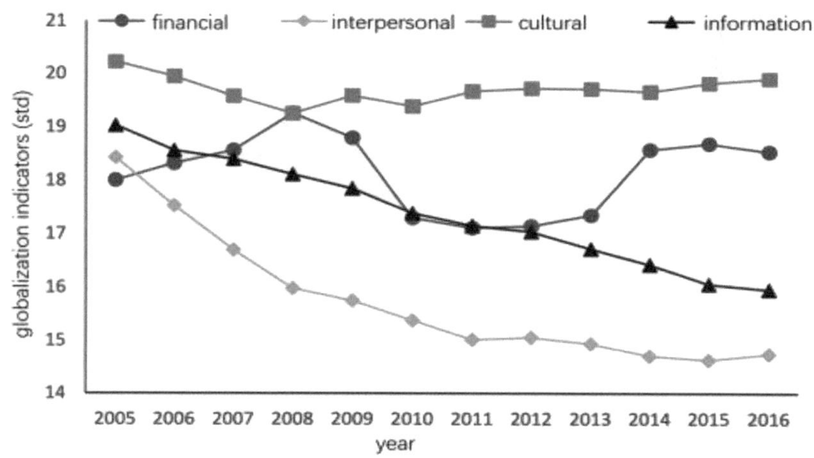

Fig. 12.26 Changes in the standard deviation of financial, interpersonal, cultural, and information globalization in 120 countries worldwide

and information globalization, of which financial globalization has the strongest fluctuating tendency. The financial globalization index showed a downward trend from 2007 to 2013 due to the impact of the global financial crisis and was significantly lower than that of the social globalization index. In 2014, the economy recovered, and financial globalization reached the highest value.

Analysis of the changes in the standard deviation of financial, interpersonal, cultural, and information globalization in 120 countries shows that the standard deviation of financial globalization in 120 countries showed an increasing trend in the period 2005–2008, indicating that the variability of financial globalization in each country is expanding. The standard deviation of cultural globalization is significantly larger than that of interpersonal and information globalization, indicating that there is more variability in cultural globalization than in interpersonal and information globalization (Fig. 12.24). During the period 2005–2016, the interpersonal and information globalization of the 120 countries showed a downward trend, indicating a decrease in the variability of interpersonal and information globalization of the 120 countries.

(i) **Analysis of economic globalization in 32 developed countries**

We analyzed changes in the mean values of financial, interpersonal, cultural globalization, and information globalization in 32 developed countries worldwide. The range of changes in the mean value of the financial globalization index for the 32 developed countries from 2005 to 2016 is [73.65,80.43], and the range of changes in the financial globalization index is low, with the lowest in 2013 and the highest in 2007. In terms of the breakdown of the social globalization index, the interpersonal globalization index is the lowest and the information globalization index is the highest (Fig. 12.27).

Changes in the standard deviation of financial, interpersonal, cultural, and information globalization in 32 developed countries worldwide. Among the detailed indicators of social globalization, the standard deviation of cultural globalization is higher than that of interpersonal and information globalization, and the standard deviation of cultural and information globalization shows a decreasing trend during 2005–2016, indicating that the difference in the level of interpersonal and information globalization of each developed country is gradually narrowing. During 2005–2016, the standard deviation of the financial globalization index of 32 developed countries was [7.42,11.11], and the standard deviation of the financial globalization index fluctuated within a high range (Fig. 12.28).

(ii) **Analysis of economic globalization in 88 less developed countries**

The changes in the mean values of financial, interpersonal, cultural, and information globalization of 88 less developed countries show that the range of changes in the mean value of the index of trade globalization of the 88 countries in 2005–2016 is [50.10, 55.33], which is lower than the range of fluctuation of financial globalization of developed countries, and the index of financial globalization fluctuates in a lower range, reaching its highest value in 2007. As for the detailed indicators of social

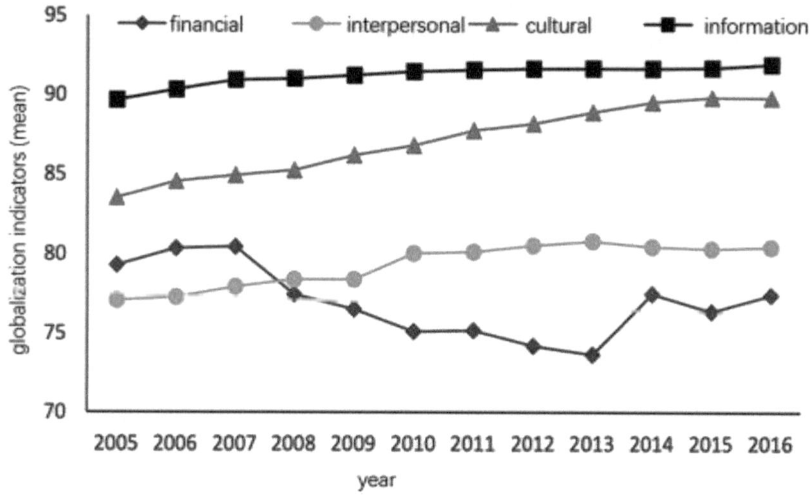

Fig. 12.27 Changes in the means of financial globalization, interpersonal globalization, cultural globalization, and information globalization in 32 developed countries

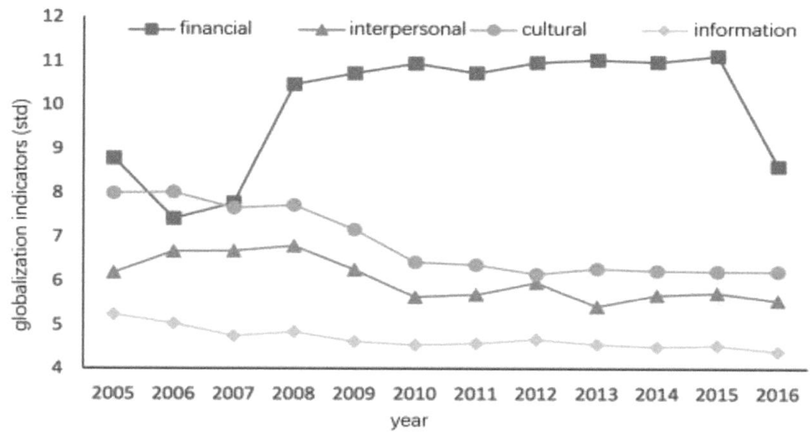

Fig. 12.28 Changes in standard deviation of financial globalization, interpersonal globalization, cultural globalization, and information globalization in 32 developed countries

globalization, the index of information globalization is significantly higher than that of cultural and interpersonal globalization (Fig. 12.29).

The analysis of the changes in the standard deviation of financial, interpersonal, cultural, and information globalization in 88 less developed countries worldwide shows that the standard deviation of the financial globalization index of the 88 less developed countries in the period 2005–2016 varied in the range of [14.65, 17.29],

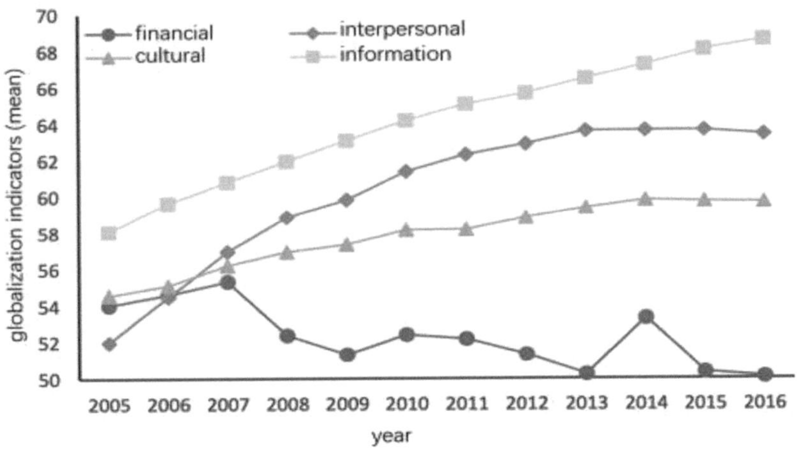

Fig. 12.29 Changes in the means of financial globalization, interpersonal globalization, cultural globalization, and information globalization in 88 less developed countries

which is significantly higher than that of developed countries, and the standard deviation of the financial globalization index fluctuates in a higher range.

As far as the detailed indicators of social globalization are concerned, the index of cultural globalization is significantly higher than that of interpersonal and information globalization, and both interpersonal and information globalization show a downward trend, indicating that the difference between the levels of cultural globalization and information globalization of less developed countries is narrowing (Fig. 12.30).

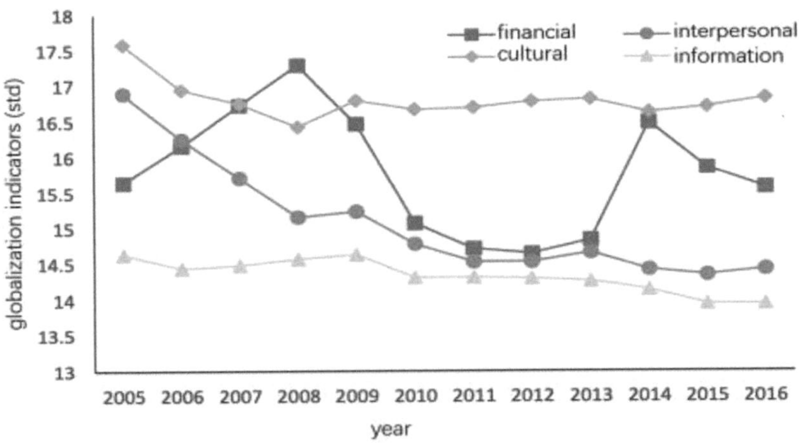

Fig. 12.30 Changes in standard deviation of financial globalization, interpersonal globalization, cultural globalization, and information globalization in 88 less developed countries

12.2.3.2 Level of Industrial Development and Regional Economic Development

The industrial structure, as a visual manifestation of the pattern of economic development and the close coupling between primary, secondary, and tertiary industries within it, has become an important indicator of the level of regional economic development. Since the Industrial Revolution, the rise of the secondary industry, especially the manufacturing industry, has made a great contribution to the total output value of the national economy. In recent years, the rapid development of the tertiary industry, especially the service sector, has played a key role in promoting the transformation of the economic development model. The economic gap between developed and less developed countries is largely reflected in the level of development of the secondary and tertiary industries. At the international level, many developed countries or regions have proposed the strategy of synergistic development based on the advanced service sector and traditional manufacturing industry in the process of economic development to promote the transformation of regional industrial structure from a "single-core-driven" mode to a "double-core-driven" mode and realize the transformation of the industrial development mode. In this way, the regional industrial structure will be transformed from a single "single-core-driven" mode to a "double-core-driven" mode, thus realizing the optimization and upgrading of the industrial structure. For the mode of collaborative industrial agglomeration development, China has attached great importance to it as an important means and path to promote the continuous optimization of the regional industrial structure and achieve the sustainable development of the regional economy. Meanwhile, the industrial transfer has a far-reaching impact on economic development through the promotion of the agglomeration of production factors. It is not only a key way to optimize the spatial distribution of production factors but also one of the important means to promote the coordinated development of the region. To further explore the extent and differences of the impact of different industrialization levels on the level of regional economic development, we analyze the relationship between the level of industrialization and the level of regional economic development.

The analysis of the changes in the industrialization level of the service sector and that of industry in 120 countries worldwide found that the industrialization level of the service sector is much higher than that of industry, and the range of fluctuations between industry and service industrialization levels during the period 2005–2016 is small (Fig. 12.31).

The analysis of standard deviation changes in the industrialization level of industry and that of services in 120 countries shows that the fluctuation range of industry industrialization level is [8.34, 11.07], and the fluctuation range of service industrialization level is [9.94, 11.26]. The fluctuation range of service industrialization level is significantly higher than that of industry industrialization level, which indicates that the influence of service industrialization level is stronger than that of industry industrialization level. After 2011, the industrialization level of industry shows a downward trend, indicating that the differences in the industrialization levels of industry in each country are narrowing (Fig. 12.32).

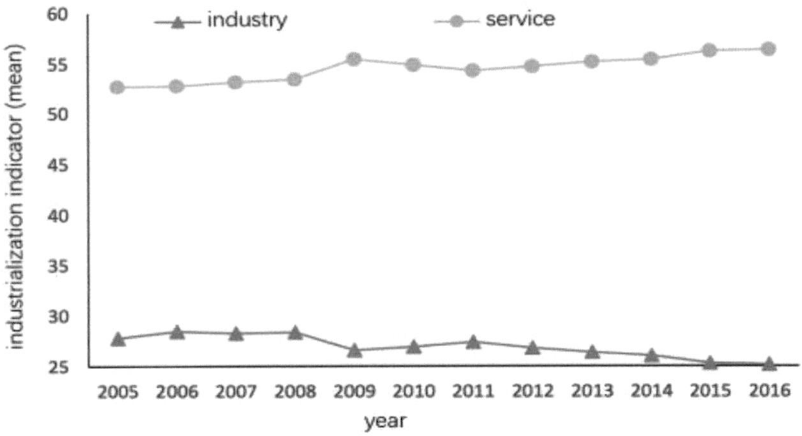

Fig. 12.31 Mean change in the industrialization level of industry and industrialization level of services

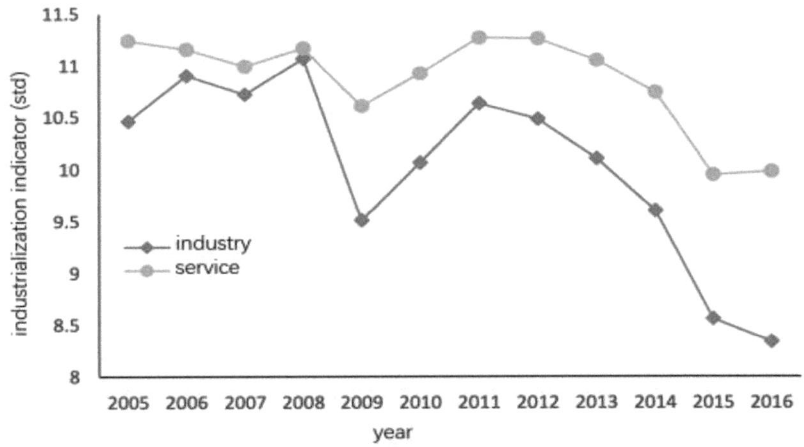

Fig. 12.32 Change in the standard deviation of industrialization level of industry and services

12.2.3.3 Human Capital Investment and Regional Economic Development

As a core factor of production, the state of labor force allocation largely reflects the level of socio-economic development of countries. Urbanization and unemployment rates are not only important indicators of this state but also profoundly affect the development process of regional economies at many levels. Employment is not only an effective input of human capital but also a key factor in guaranteeing basic income, promoting consumption, and improving the quality of life. In developed countries, the unemployment rate is usually higher, whereas in less developed countries, due to the supporting role of public welfare expenditure, the unemployment rate is relatively low, and this difference exerts a positive impetus on economic growth, showing a positive change relationship. Regarding the relationship between unemployment rate and economic growth, research conclusions are not consistent across countries. Nevertheless, the prevailing view is that there is an inverse change relationship between the two, that is, the rise in unemployment rate is often accompanied by a slowdown in economic growth. The role of urbanization development in promoting economic growth cannot be ignored. The pursuit of economic development is an important driving force for urbanization, and there is a strong unidirectional causal relationship between the urbanization process and economic growth. Urbanization can not only optimize resource allocation and improve production efficiency but also promote innovation and industrial upgrading through the aggregation effect. Given the close connection between labor force, employment, urbanization, and economic growth, we analyze the relationship between the level of unemployment rate, urbanization rate, and regional economic development to reveal the specific impact of human capital investment on regional economic development.

The analysis of the change in the mean value of the unemployment rate and the level of urbanization rate in 120 countries worldwide shows that the urbanization rate and unemployment rate fluctuated in a relatively stable range from 2005 to 2016, and the fluctuation range of the urbanization rate is [56.71, 60.38]. As indicated in the graph, the urbanization rate has been increasing, and it was the lowest in 2005, at 56.71, and highest in 2016, at 60.38. Unemployment rate fluctuates in the range of [6.59, 8.16], reaching its lowest value in 2008, at 6.59, and the highest value in 2013, at 8.16. Overall, the urbanization rate fluctuated in a higher range and the unemployment rate fluctuated in a lower range (Fig. 12.33).

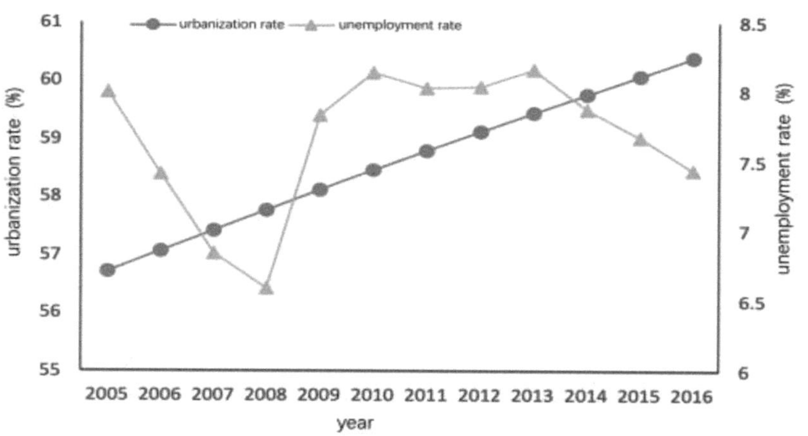

Fig. 12.33 Mean change in unemployment rate, level of urbanization rate, 120 countries worldwide

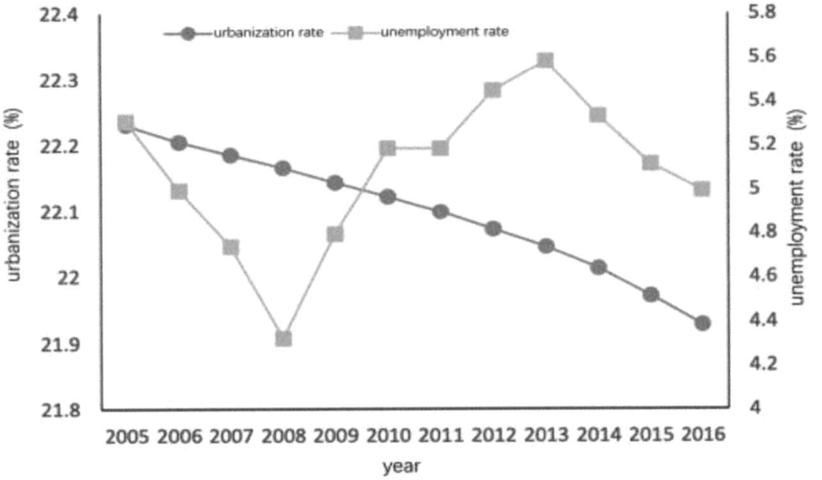

Fig. 12.34 Standard deviation change in unemployment rate and urbanization rate

The analysis of the changes in the standard deviation of the unemployment rate and the level of urbanization rate in 120 countries worldwide shows that the fluctuation range of the urbanization rate is [21.93, 22.23], the standard deviation of the urbanization rate in 2005 was 22.23, which was the maximum, and was 21.93 in 2016. During the period 2005–2016, the standard deviation of the urbanization rate showed a steady downward trend, which indicates that the differences in the level of urbanization rate across countries are narrowing. The fluctuation range of unemployment rate is [4.32, 5.58], with the standard deviation of the unemployment rate at 4.32 in 2008, the lowest value, and 5.58 in 2013, the highest value. Overall, the level of urbanization rate fluctuates in a higher range and the level of unemployment rate fluctuates in a lower range (Fig. 12.34).

12.2.4 Empirical Study of Convergence Models for Regional Economic Development

12.2.4.1 α Convergence Analysis of Regional Economic Development

(i) Analysis of relative differences in GDP per capita

Since there will be some errors in the examination of differences in economic growth by the absolute difference indicator, we use the relative difference indicator to analyze the coefficient of variation of GDP per capita in 120 countries worldwide.

As far as relative differences are concerned, the coefficient of variation of GDP per capita for the 120 countries worldwide shows an overall downward trend, especially

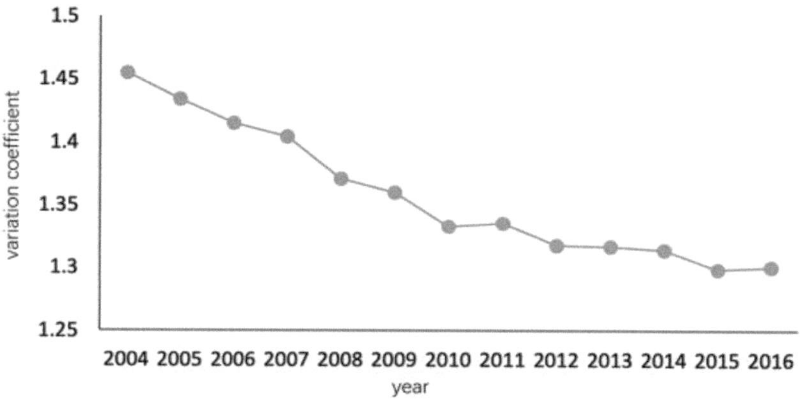

Fig. 12.35 Change in the coefficient of variation of GDP per capita for 120 countries worldwide

in the period 2004–2010, when the coefficient of variation declined from 1.45 to 1.33, and in the period 2011–2014, the relative differences still show a downward trend, with the rate of change declining somewhat from the previous eight years, from 1.34 to 1.31, and after 2014, the change in the coefficient of variation is smaller. To further observe and compare the trends in the coefficient of variation of GDP per capita in developed and less developed countries, we analyzed the trends in the coefficient of variation of GDP per capita in 32 developed countries and 88 less developed countries, respectively (Fig. 12.35).

The change in the relative variance of GDP per capita in developed countries is characterized by large fluctuations. A relatively large decline occurred in 2006–2009, with the coefficient of variation dropping from 0.51 to 0.49, which is related to the impact of the global economic crisis. The relative difference in the GDP per capita of developed countries began to increase substantially after 2009, which indicates that the economic differences of developed countries are gradually expanding. The relative difference in the GDP per capita of developed countries reached its maximum in 2012 (0.53) in 2012, and then began to decline significantly, indicating that the economic disparity of developed countries has been gradually narrowing in recent years (Fig. 12.36).

Changes in the relative variance of GDP per capita in less developed countries show similar characteristics of changes as those globally, and in general, the coefficient of variation has steadily declined year after year, from 1.40 in 2004 to 1.01 in 2016, which indicates that the economic disparities of less developed countries have been gradually narrowing in recent years (Fig. 12.37).

(ii) **Analysis of relative differences in GNI per capita**

Since the absolute difference indicator will have some errors in the examination of economic growth differences, we use the relative difference indicator to analyze the coefficient of variation of GNI per capita in 120 countries worldwide.

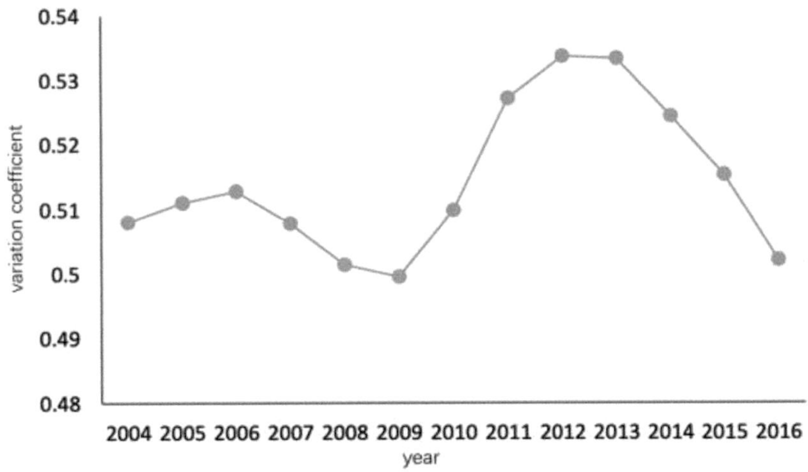

Fig. 12.36 Changes in the coefficient of variation of GDP per capita for 32 developed countries

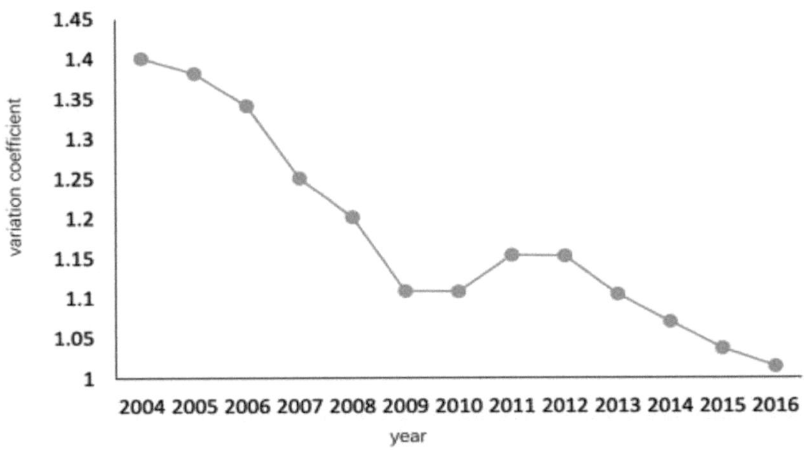

Fig. 12.37 Changes in the coefficient of variation of GDP per capita for 88 less developed countries

According to Fig. 12.38, the coefficient of variation of GNI per capita for the 120 countries shows an overall decreasing trend in terms of relative differences. In particular, between 2004 and 2009, this coefficient of variation declined at a significantly faster rate, suggesting that differences in economic development are narrowing globally. However, between 2009 and 2014, while the relative variance remained on a downward trend, the rate of decline slowed down compared to the previous five years. After 2014, the change in the coefficient of variation leveled off, indicating a gradual stabilization of changes in global economic development differences.

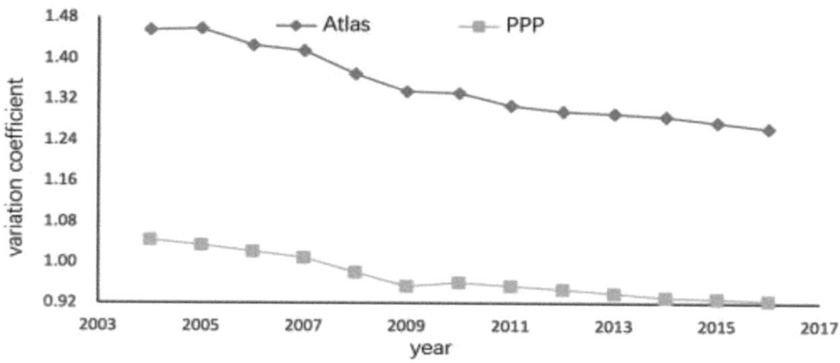

Fig. 12.38 Change in the coefficient of variation of GNI per capita for 120 countries worldwide

To explore this phenomenon in greater depth, we further analyzed the trend of the coefficient of variation of GNI per capita in 32 developed countries and 88 less developed countries. As shown in the figure, before 2014, the average value of GNI per capita in developed countries, whether calculated using the PPP or Atlas method, decreased between 2008 and 2009, but overall showed an increasing trend. Interestingly, after 2014, the mean GNI per capita under the PPP method continues to grow, whereas the mean under the Atlas method begins to decline. In terms of relative differences, the coefficient of variation of GNI per capita for both developed and less developed countries has changed relatively little overall. In particular, the coefficient of variation for developed countries showed a clear downward trend between 2007 and 2009, indicating that the differences in the economic development of developed countries decreased during this period. After 2014, this coefficient of variation continued to show a slow downward trend, further reflecting the trend of narrowing economic development differences among developed countries (Fig. 12.39).

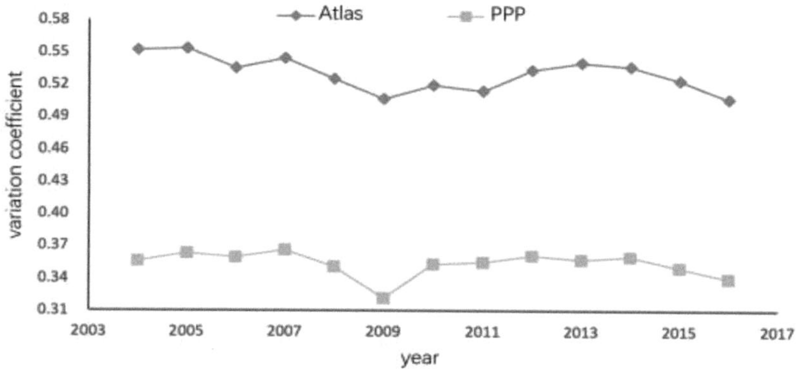

Fig. 12.39 Change in coefficient of variation of GNI per capita for 32 developed countries

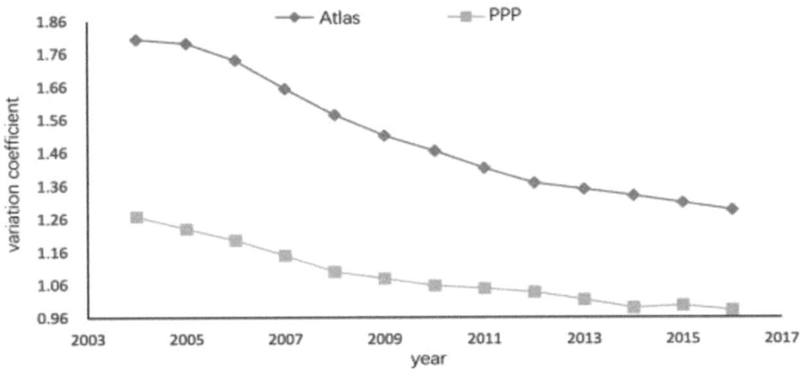

Fig. 12.40 Changes in the coefficient of variation of GNI per capita in 88 less developed countries

In terms of relative differences, the coefficient of variation of GNI per capita in less developed countries under the two methods shows an overall downward trend, with the rate of change under the Atlas method being slightly larger than that under the PPP method before 2011, and the rates of change under the two methods being essentially the same after 2011 (Fig. 12.40).

The comparison reveals that the absolute differences in per capita GNI between developed and less developed countries are generally on a downward trend, but the relative differences in per capita GNI in developed countries have changed by a smaller margin, whereas the relative differences in per capita GNI in less developed countries have changed by a larger margin.

12.2.4.2 β Convergence Analysis of Regional Economic Development

(i) **β-absolute convergence analysis**

We conducted a Hausman test on panel data for 120 countries, 32 developed countries, and 88 less developed countries globally for the period 2005–2016. The test results show that the fixed effects model is more suitable for absolute β convergence analysis; thus, we reject the original hypothesis (Prob > chi2 = 0.0000). Next, we conducted β-absolute convergence analyses for the growth rates of GDP per capita (Model I), national income per capita (Atlas method) (model and national income per capita (PPP method).

The results of the analysis show that the global economic development shows convergence under both models, namely national income per capita using the Atlas method and national income per capita using the PPP method. This means that countries with a lower level of economic development at the initial stage have relatively faster economic growth in subsequent stages, thus gradually narrowing the economic gap with developed countries. However, under the GDP per capita model, global economic development shows a divergence pattern, which means that countries with a higher initial level of economic development maintain a faster growth rate. In particular, the model with the rate of change in GDP per capita as the explanatory variable shows the largest absolute value of the coefficient of convergence (0.6992), whereas the model with the rate of change in national income per capita (Atlas method) as the explanatory variable has the highest coefficient of regression fit (0.3810) (Table 12.19). Similarly, the economic development of developed countries shows a convergence pattern. In all models, the convergence coefficients for the initial values are negative. Among them, the model with the rate of change in national income per capita (Atlas method) as the explanatory variable not only exhibits the largest absolute value of the convergence coefficient (− 0.2476), but also the highest regression fit coefficient (0.3956) (Table 12.19). Similarly, the economic development of less developed countries shows convergence, with negative convergence coefficients for the initial values. However, similar to the global and developed countries, under the GDP per capita model, the economic development of these countries shows a departure dispersion. The model with the rate of change in national income per capita (Atlas method) as the explanatory variable shows the largest absolute value of the convergence coefficient (−0.1584) and has the highest coefficient of regression fit (0.3890) of all the models (Tables 12.19).

In summary, the global regional economic development showed a certain convergence trend during 2005–2016. Notably, economic growth in developed countries converged faster than in less developed countries. These findings provide an important perspective on the dynamics of global economic development and contribute to a deeper understanding of the economic differences between different countries and their evolutionary trends.

(ii) **β-conditional convergence analysis**

In exploring the issue of β-conditional convergence, we adopt GNI per capita as a key indicator of regional economic development. In order to deeply analyze the convergence mechanism, we choose the KOF globalization index to represent the level of globalization, as well as the share of industrial value added in GDP and the share of service-sector value added in GDP to measure the level of industrialization development, and also consider the urbanization rate and unemployment rate as the control variables for the conditional convergence analysis.

By conducting the Hausman test on the panel data of 120 countries, that is, 32 developed countries and 88 less developed countries worldwide, from 2005 to 2016, we find that the fixed-effects model is more suitable for β-conditional convergence analysis (Prob > chi2 = 0.0000, rejecting the original hypothesis). Subsequently, we conducted a β-conditional convergence analysis using the growth rate of GDP per capita. Model I is the original β-absolute convergence model, whereas Models II through VIII each incorporate different control variables to explore their effects on economic development convergence. Specifically, Models II, III, and IV introduce the level of globalization as a control variable, Model V focuses on the industrialization level of industry, Model VI focuses on the industrialization level of services, Model VII examines the level of urbanization, and Model VIII considers the level of

Table 12.19 Coefficient values of β-absolute convergence analysis results (*Note* * $p < 0.10$, ** $p < 0.05$, *** $p < 0.001$)

Explanatory variables	120 countries			32 developed countries			88 less developed countries		
	GDP per capita	GNI_A per capita	GNI_P per capita	GDP per capita	GNI_A per capita	GNI_P per capita	GDP per capita	GNI_A per capita	GNI_P per capita
Constant	– 0.656 3***	1.486 6***	1.121 1***	0.6768***	2.5931***	1.33347***	– 0.7272***	1.3214***	1.0602***
Initial value	0.6992***	– 0.1668***	– 0.1173*** 1	– 0.0668***	– 0.2476***	– 0.1246***	0.0834***	– 0.1584***	– 0.1160***
R^2	0.0217	0.3810	0.2064	0.0087	0.3956	0.1296	0.0354	0.389	0.2338
F-value	0.98***	7.98***	5.20***	0.5800***	7.27***	2.03***	1.19***	8.45***	6.30***

unemployment. Model IX is a comprehensive model that contains multiple control variables such as the globalization, industrialization, and urbanization levels.

The analytical results of the nine models consistently show that global economic development shows a convergence pattern, that is, countries with a higher level of economic development at the initial stage have relatively slower economic growth, whereas countries with a lower level at the initial stage have relatively faster growth, leading to the convergence of economic growth rates. It is particularly noteworthy that in Models II, III, and IV, the effect of the level of globalization on the convergence of global economic development is positive, indicating that globalization helps promote the convergence of national economies. Further comparing Models V and VI, we find that the industrialization level of industry has a positive impact on the convergence of global economic development, whereas the industrialization level of services has a negative impact. This implies that the advancement of industrialization helps promote the convergence of the global economy, whereas the development of the service sector may have widened the economic differences between countries to a certain extent. Model IX, which combines the level of globalization and the level of industrialization, shows that the impact of the industrialization level of industry on the convergence of global economic development remains positive, whereas the impact of the industrialization level of the service sector remains negative. This result provides a new perspective for us to deeply understand the convergence mechanism of global economic development (Table 12.20).

For the 32 developed countries analyzed, we conducted a β-conditional convergence analysis of the growth rate of GNI per capita. Model I serves as a benchmark to demonstrate the original β-absolute convergence model, whereas Models II to VIII each introduce different control variables to explore their impact on the convergence of economic development. Specifically, Models II, III, and IV incorporate the level of globalization as a control variable, Model V considers the industrialization level of industry, Model VI considers the industrialization level of services, Model VII considers the level of urbanization, and Model VIII considers the level of unemployment. Finally, Model IX is a comprehensive model that covers several control variables such as the globalization, industrialization, and urbanization levels.

The results of the nine models consistently show that the economic development of developed countries also shows a convergence pattern, that is, the economic growth rates gradually converge. In Models II, III, and IV, the effect of the level of globalization on the convergence of the economic development of developed countries is negative, which suggests that the process of globalization may have exacerbated the economic differences among developed countries. Further comparing Models V and VI, we find that the industrialization level of industry has a positive impact on the convergence of economic development in developed countries, whereas the industrialization level of services has a negative impact. It is particularly noteworthy that the negative impact of the industrialization level of services is stronger than the positive impact of the industrialization level of industry in developed countries, which is reflected in the absolute value of the coefficients and the fitted regression coefficients of the model. Finally, Model IX, which combines the globalization and industrialization levels, indicates that the impact of the industrialization level of services on

Table 12.20 Estimates of conditional convergence of β for each setup model (*Note* * $p < 0.10$, ** $p < 0.05$, *** $p < 0.001$)

Explanatory variables	Model number								
	I	II	III	IV	V	VI	VII	VIII	IX
Constant	1.4866***	1.4391***	1.5240***	1.5187***	1.1137**	1.6472***	1.5713***	1.5746***	1.5085***
Initial value	– 0.1688***	– 0.1694***	– 0.1851***	– 0.1607***	– 0.1447***	– 0.1415***	– 0.1564***	– 0.1708***	– 0.1478***
Trade globalization		0.0011**							0.0007*
Human globalization			0.0018***						0.0017**
Culture globalization				– 0.0013***					– 0.0009
Industrialization level (industry)					0.0068***				0.0038***
Industrialization level (services)						– 0.0069***			– 0.0039***
Urbanization rate							– 0.0030***		– 0.0023
Unemployment rate								– 0.007***	– 0.0051***
R^2	0.3810	0.3830	0.3862	0.3843	0.4287	0.4242	0.3832	0.4066	0.4646
F-value	7.98***	7.88***	8.07***	8.06***	8.73***	9.07***	7.65***	8.77***	9.17***

the convergence of economic development in developed countries remains negative, whereas the industrialization level of industry shows a positive impact. This result provides important clues for a deeper understanding of the mechanism of convergence of economic development in developed countries (Table 12.21).

In our analysis of the 88 less developed countries, we conduct a β-conditional convergence analysis of the growth rate of GNI per capita. Model I is the original β-absolute convergence model, whereas Models II to VIII each incorporate different control variables to examine their impact on economic development. Specifically, Models II, III, and IV consider the level of globalization, Model V considers the industrialization level of industry, Model VI considers the industrialization level of services, Model VII considers the level of urbanization, and Model VIII incorporates the level of unemployment. Model IX is a comprehensive model that includes several control variables such as the level of globalization, the level of industrialization, and the level of urbanization.

After the simulation and analysis of the nine models, we find that the economic development of less developed countries generally shows a convergence trend, that is, the economic growth rate of the countries with higher initial values is relatively slow, whereas the countries with lower initial values are relatively fast, which leads to the convergence of the economic growth rate. It is particularly noteworthy that the level of globalization in Model IV has no significant effect on the convergence of economic development in less developed countries, whereas Model III shows a significant effect. Further comparing Models VI and V, we find that the industrialization level of industry has a positive impact on the convergence of economic development in less developed countries, that is, the increase in industrialization helps promote the convergence of economic growth. On the contrary, the industrialization level of the service sector has a negative impact on the convergence of economic development in less developed countries, which may mean that the development of the service sector has widened the economic differences between countries. It is worth noting that the positive effect of the industrialization level of industry is stronger than the negative effect of the industrialization level of services in less developed countries, which is reflected in the absolute value of the coefficients and the fitted regression coefficients of the model. Finally, in Model IX, which combines the level of globalization and the level of industrialization, we find that the level of globalization has a more significant effect on the convergence of regional economic development in less developed countries. Meanwhile, both the industrialization level of industry and that of services contribute significantly to the convergence of regional economic development. This result provides an important perspective to deeply understand the mechanism of economic development convergence in less developed countries (Table 12.22).

Table 12.21 Estimates of conditional convergence of β for each set model for 32 developed countries (*Note* * $p < 0.10$, ** $p < 0.05$, *** $p < 0.001$)

Explanatory variables	Model number								
	I	II	III	IV	V	VI	VII	VIII	IX
Constant	2.5931***	2.6344***	2.6375***	2.6220**	1.7204***	2.8018***	3.6482***	2.5554***	2.9659***
Initial value	− 0.2476***	− 0.2478***	− 0.2282***	− 0.2064***	− 0.1932***	− 0.1812***	− 0.2251***	− 0.2386***	− 0.1850***
Trade globalization		− 0.0005							0.0011
Human globalization			− 0.0031*						0.0012
Culture globalization				− 0.0052***					− 0.0014
Industrialization level (industry)					0.0129***				0.0013
Industrialization level (services)						− 0.0139***			− 0.0083**
Urbanization rate							− 0.0166***		− 0.0072**
Unemployment rate								− 0.0071***	− 0.0032***
R^2	0.3956	0.3959	0.4015	0.4390	0.4888	0.5059	0.4410	0.4763	0.5353
F-value	7.27***	7.23***	7.04***	8.50***	9.54***	10.44***	8.69***	8.00***	7.99***

Table 12.22 β-conditional convergence estimates for each setting model for 88 less developed countries (*Note* * $p < 0.10$, ** $p < 0.05$, *** $p < 0.001$)

Explanatory variables	Model number								
	I	II	III	IV	V	VI	VII	VIII	IX
Constant	1.3214***	1.2789***	1.3650***	1.3288***	1.0068**	1.4535***	1.3682***	1.4026***	1.3295***
Initial value	– 0.1584***	– 0.1606***	-0.1789***	– 0.1570***	– 0.1397***	– 0.1380***	– 0.1505***	– 0.1629***	– 0.1413***
Trade globalization		0.0011**							0.0006*
Human globalization			0.0020***						0.0015**
Culture globalization				– 0.0003					0.0005
Industrialization level (industry)					0.0059***				0.0038***
Industrialization level (services)						– 0.0058			0.0035***
Urbanization rate							– 0.0021		– 0.0024
Unemployment rate								– 0.006***	0.0052***
R^2	0.3890	0.3920	0.3958	0.3892	0.4295	0.4216	0.3902	0.3976	0.4552
F-value	8.45***	8.37***	8.56***	8.42***	8.90***	9.22***	7.99***	8.71***	8.87***

12.2.4.3 Analysis of the Driving Mechanisms of Regional Economic Development

Based on an in-depth analysis of the convergence trend of regional economic development, we constructed an analytical framework for the driving mechanism of global economic development centered on the spatial interaction of global regional economic development and the Cobb–Douglas production function. In this study, we selected GNI per capita as the explanatory variable and the level of globalization as the core explanatory variable, and further refined the measurement of the level of globalization, including trade, financial, interpersonal, cultural, and information globalization and other dimensions. Meanwhile, considering the key factors such as land, capital, industrial structure, and human resources, we introduce the industrialization level of industry, the industrialization level of the service sector, unemployment rate, and urbanization rate as control variables.

In terms of model setting, Models I and I focus on the effects of economic globalization, whereas Models III, IV, and V focus on the effects of social globalization, and Model VI integrates all globalization indicators.

The panel data regression results of global economic development in the context of globalization show that social globalization has a significant positive driving effect on regional economic development. In terms of economic globalization, trade globalization shows a positive effect on economic development, whereas financial globalization shows a significant negative effect. It is particularly noteworthy that the regression coefficient of social globalization in Model V is higher than the coefficient of economic globalization in Models I and II, and only social globalization significantly affects the level of global economic development in Model VI, which highlights the important position of social globalization in promoting global economic development. In addition, the level of industrialization, as a control variable, also has a significant positive contribution to global economic development. It is especially worth mentioning that the regression coefficients of the service sector are higher than those of industry in all six models, which indicates that the development of the service sector has exceeded the contribution of industry to the global economy in recent years. In terms of model fitting effects, the models with the addition of social globalization in Models III, IV, and V perform the best, further verifying the key role of social globalization in explaining global economic development (Table 12.23).

After regression analysis of the panel data on the economic development of 32 developed countries under different globalization contexts, we come to the following conclusions. First, social globalization has a significant positive impact on regional economic development, whereas the level of economic globalization has a significant negative impact on regional economic development. Further comparison reveals that the regression coefficients of social globalization in Models III, IV, and V are higher than those of economic globalization in Models I and II, especially in Model IV, where the regression coefficient of social globalization reaches the highest level, which demonstrates that social globalization significantly outweighs economic globalization in terms of its impact on economic development in developed countries. Among the control indicators examined, only the level of industrialization of the

Table 12.23 Estimates of global economic development for 120 countries (*Note* * $p < 0.10$, ** $p < 0.05$, *** $p < 0.001$)

Explanatory variables	Model number					
	I	II	III	IV	V	VI
Constant	2.8202***	3.5730***	4.1296***	3.1601***	3.1501***	4.2606***
Initial value	0.0059***					
Trade globalization		– 0.0054***				
Human globalization			0.0285***			
Culture globalization				0.0127***		
Information globalization					0.0386***	0.240***
Industrialization level (industry)	– 0.0017	– 0.0019	– 0.0073***	– 0.0026	0.0019	– 0.0035***
Industrialization level (services)	0.0184**	0.0165**	0.0071***	0.0152***	0.0098***	0.0057***
Urbanization rate	0.0791***	0.0803***	0.0442***	0.0750***	0.0381***	0.0238***
Unemployment rate	– 0.0249***	– 0.0252***	– 0.0211***	– 0.0285***	– 0.0159***	– 0.0196***
R^2	0.4436	0.4252	0.5694	0.4458	0.5660	0.5073

service sector has a significant impact on the level of economic development of developed countries, which reflects that the booming development of the service sector in recent years has played an important role in the improvement of the economic level of developed countries. Meanwhile, the regression coefficients of the unemployment rate in all six models are significantly negative, which reveals the significant impact of the employment level on the level of economic development of the developed countries, that is, the condition of the employment market is closely related to the economic development of developed countries. In addition, Models III, IV, and V with the inclusion of social globalization have the best fit, which further validates the importance and effectiveness of social globalization in explaining the economic development of developed countries. These findings provide an important basis for an in-depth understanding of the mechanism of the economic development of developed countries in the context of globalization (Table 12.24).

After regression analysis of the panel data on the economic development of 88 less developed countries under different globalization contexts, we conclude that social globalization has a significant positive effect on regional economic development in Models III, IV, V, and VI. In Model II, the regression coefficient of economic globalization is significantly negative. Meanwhile, comparing Models III, IV, and V with Models I and II, the regression coefficients of social globalization are higher than the coefficients of economic globalization, which further suggests that the impact of social globalization is more significant than that of economic globalization. As a control indicator, the level of industrialization also plays a significant positive role in promoting the economic development of less developed countries. It is particularly noteworthy that the level of industrialization of the service sector has a significantly positive impact on the economic development of less developed countries, and the regression coefficients of the service sector are higher than those of industry in all six models, which highlights the important position of the service sector in promoting the economic development of less developed countries. In addition, the unemployment rate shows significance in all six models, which indicates that the employment situation has a greater contributing power to the level of economic development of less developed countries, and the stability and activity of the job market has a non-negligible role in economic development. In Models III, IV, and V, the model with the inclusion of social globalization has the best fit, which further verifies the importance and validity of social globalization in explaining the economic development of less developed countries (Table 12.25).

12.3 Summary

This chapter discusses the complex dynamics of regional economic development, exploring the factors that shape the economic trajectories of different regions and the convergence models that seek to explain how less developed regions can catch up with their more advanced counterparts. The chapter begins by analyzing the role of human capital investment in regional economic development, emphasizing the

Table 12.24 Estimates of economic development in 32 developed countries (*Note* * $p < 0.10$, ** $p < 0.05$, *** $p < 0.001$)

Explanatory variables	Model number					
	I	II	III	IV	V	VI
Constant	7.3942***	7.8546***	6.3677***	7.3447***	3.7483***	4.5437***
Initial value	– 0.0016					– 0.0009
Trade globalization		– 0.0036*				– 0.0002
Human globalization			0.0336**			0.0269***
Culture globalization				0.0166***		0.0095***
Information globalization					0.0509***	0.0315***
Industrialization level (industry)	0.0041	0.0021	0.0078***	0.0090	0.0105	0.0130
Industrialization level (services)	0.0334***	0.03***	0.0286***	0.0318***	0.0277***	0.0243***
Urbanization rate	0.0126	– 0.0120	– 0.0072	– 0.0070	0.0005	– 0.0213***
Unemployment rate	– 0.0143***	– 0.0158***	– 0.1666***	– 0.0182***	– 0.0119***	– 0.0167***
R^2	0.1363	0.1432	0.3174	0.2189	0.2674	0.4107

Table 12.25 Estimates of economic development in 88 less developed Countries (*Note* * $p < 0.10$, ** $p < 0.05$, *** $p < 0.001$)

Explanatory variables	Model number					
	I	II	III	IV	V	VI
Constant	2.7129^{***}	3.3019^{***}	3.8280^{***}	2.6959^{***}	3.0624^{***}	8.4054^{***}
Initial value	0.0063^{***}					0.002^{**}
Trade globalization		-0.0050^{***}				-0.0052^{***}
Human globalization			0.0273^{**}			0.0171^{***}
Culture globalization				0.0130^{***}		0.0025^{*}
Information globalization					0.0370^{***}	0.0227^{***}
Industrialization level (industry)	-0.0013	-0.0012^{**}	-0.0069^{***}	-0.0024	0.0024	-0.0034
Industrialization level (services)	0.0170^{**}	0.0154^{***}	0.0063^{**}	0.0140^{***}	0.0087^{***}	0.0052^{**}
Urbanization rate	0.0839^{***}	0.0857^{***}	0.0497^{***}	0.0806^{***}	0.0427^{***}	0.0291^{***}
Unemployment rate	-0.0347^{***}	-0.0324^{***}	-0.0258^{***}	-0.0373^{***}	-0.0230^{***}	-0.0232^{**}
R^2	0.4964	0.4695	0.6030	0.4914	0.6014	0.6752

Schedules 12.1 List of developed and developing countries

Country	Continent	Classification of level of development
Albania	Europe	Developing Country
Argentina	South America	Developing Country
Australia	Oceania	Developed Country
Austrian	Europe	Developed Country
Azerbaijan	Asia	Developing Country
Burundi	Africa	Developing Country
Belgium	Europe	Developed Country
Benin	Africa	Developing Country
Burkina Faso	Africa	Developing Country
Bangladesh	Asia	Developing Country
Bulgaria	Europe	Developing Country
Bahamas	Latin America	Developing Country
Belarus	Europe	Developing Country
Belize	Central America	Developing Country
Bolivia	South America	Developing Country
Brazil	South America	Developing Country
Barbados	Latin America	Developing Country
Brunei	Asia	Developing Country
Bhutan	Asia	Developing Country
Botswana	Africa	Developing Country
Switzerland	Europe	Developed Country
Chile	South America	Developing Country
China	Asia	Developing Country
Cameroon	Africa	Developing Country
Colombia	South America	Developing Country
Comoros	Africa	Developing Country
Cape Verde	Africa	Developing Country
Costa Rica	South America	Developing Country
Cyprus	Europe	Developed Country
Czech Republic	Europe	Developed Country
Germany	Europe	Developed Country
Dominican Republic	North America	Developing Country
Denmark	Europe	Developed Country
Algeria	Africa	Developing Country
Ecuador	South America	Developing Country
Egypt	South America	Developing Country
Spain	Europe	Developed Country

(continued)

Schedules 12.1 (continued)

Country	Continent	Classification of level of development
Estonia	Europe	Developed Country
Finland	Europe	Developed Country
France	Europe	Developed Country
Gabon	Africa	Developing Country
United Kingdom	Europe	Developed Country
Papua New Guinea	Oceania	Developing Country
Greece	Europe	Developed Country
Guatemala	Central America	Developing Country
Guyana	South America	Developing Country
Honduras	Central America	Developing Country
Croatia	Europe	Developing Country
Hungary	Europe	Developing Country
Indonesia	Asia	Developing Country
India	Asia	Developing Country
Ireland	Europe	Developed Country
Iran (Islamic Republic of)	Asia	Developing Country
Iceland	Europe	Developed Country
Israel	Asia	Developed Country
Italy	Europe	Developed Country
Jamaica	North America	Developing Country
Jordan	Asia	Developing Country
Japan	Asia	Developed Country
Kazakhstan	Asia	Developing Country
Kirghizia	Asia	Developing Country
Cambodia	Asia	Developing Country
Korea, Republic of	Asia	Developed Country
Lebanon	Asia	Developing Country
Sri Lanka	Asia	Developing Country
Lesotho	Africa	Developing Country
Lithuania	Europe	Developed Country
Luxembourg	Europe	Developed Country
Latvia	Europe	Developed Country
Morocco	Africa	Developing Country
Moldova	Europe	Developing Country
Madagascar	Africa	Developing Country
Maldives	Asia	Developing Country
Mexico	North America	Developing Country

(continued)

Schedules 12.1 (continued)

Country	Continent	Classification of level of development
Mali	Africa	Developing Country
Mozambique	Africa	Developing Country
Mauritania	Africa	Developing Country
Mauritius	Africa	Developing Country
Malawi	Africa	Developing Country
Malaysia	Asia	Developing Country
Namibia	Africa	Developing Country
Nigeria	Africa	Developing Country
Nicaragua	Central America	Developing Country
Netherlands	Europe	Developed Country
Norway	Europe	Developed Country
Nepal	Asia	Developing Country
New Zealand	Oceania	Developed Country
Pakistan	Asia	Developing Country
Panama	Central America	Developing Country
Peru	South America	Developing Country
Philippines	Asia	Developing Country
Poland	Europe	Developing Country
Portugal	Europe	Developed Country
Paraguay	South America	Developing Country
Romania	Europe	Developing Country
Russia	Europe	Developing Country
Rwanda	Africa	Developing Country
Saudi Arabia	Asia	Developing Country
Sudan	Africa	Developing Country
Senegal	Africa	Developing Country
Singapore	Asia	Developed Country
El Salvador	Central America	Developing Country
Serbia	Europe	Developing Country
Slovakia	Europe	Developed Country
Slovenia	Europe	Developed Country
Sweden	Europe	Developed Country
Togo	Africa	Developing Country
Thailand	Asia	Developing Country
Tajikistan	Asia	Developing Country
Turkey	Europe	Developing Country
Tanzania	Africa	Developing Country

(continued)

Schedules 12.1 (continued)

Country	Continent	Classification of level of development
Uganda	Africa	Developing Country
Ukraine	Europe	Developing Country
Uruguay	South America	Developing Country
United States of America	North America	Developed Country
Uzbekistan	Asia	Developing Country
Vietnam	Asia	Developing Country
South Africa	Africa	Developing Country
Zambia	Africa	Developing Country
Zimbabwe	Africa	Developing Country
Dominica	North America	Developing Country
Georgia	Asia	Developing Country
Malta	Europe	Developed Country

Note (i) Blacked out are the 46 countries categorized as convergent in the Global Social Development (GSD) study. (ii) Dominica, Georgia, and Malta are categorized as converging only in the GSD study; the others are categorized as converging in the GED study

importance of education, health, and skills development in fostering productivity and growth.

It then turns its attention to industrial development, discussing how the level and structure of industrial activity can influence regional economic performance. The chapter explores the interplay between traditional and modern sectors, examining how industrialization processes can lead to structural transformation and economic diversification.

The impact of globalization on regional economic development is also considered, focusing on how trade, investment, and technological diffusion can shape regional growth patterns. The chapter examines both the opportunities and challenges presented by globalization, highlighting the need for policies that can harness its benefits while mitigating its potential risks.

The empirical section of the chapter provides a detailed analysis of the relative and absolute differences in per capita income and GNI across different regions and countries. Using convergence models, the chapter explores the extent to which less developed regions are converging toward their more developed counterparts, identifying the factors that facilitate or hinder this process. The chapter concludes by outlining policies and strategies that can support the convergence process, emphasizing the importance of targeted interventions that address the unique needs and challenges of different regions.

Exercise

1. How do the level of globalization and the level of industrialization of services affect the convergence of global economic development, according to β-conditional convergence analysis?
2. Why is it that the level of industrialization of industry has gradually become less significant in the convergence of economic development in developed countries while it can still significantly contribute to the convergence of regional economic development in less developed countries?
3. Globalization increases the differences in economic development among developed countries, but in less developed countries, the convergence of regional economic development is not significant. What are the reasons behind this phenomenon?
4. Where does China stand in terms of global economic development trends and patterns of convergence? Is it more like a developed country or a less developed country? Why?
5. Given the emergence and growth of emerging economies, how does China interact with these emerging economies? What is its role in the global economy?

Extended Reading

Dicken, Peter. (2003). Global Shift: Reshaping The Global Economic Map In The 21st Century 4th Ed.

References

Blecker, R.A., Moreno-Brid, J.C., & Salat, I. (2017). Trumping the NAFTA renegotiation: An alternative policy framework for Mexican-US cooperation and economic convergence. *Real-World Economics Review, 94*.

Borozan, D. (2017). Internal migration, regional economic convergence, and growth in Croatia. *International Regional Science Review, 40*(2), 141–163.

Coulibaly, S. K., Erbao, C., & Mekongcho, T. M. (2017). Economic globalization, entrepreneurship and development. *Technological Forecasting and Social Change, 127*, 271–280.

Dall'Erba, S., & Le Gallo, J. (2008). Regional convergence and the impact of European structural funds over 1989–1999: A spatial econometric analysis. *Papers in Regional Science, 87*(2), 219–244.

Ezcurra, R., & Rodrgíuez-Pose, A. (2013). Does economic globalization affect regional inequality? A cross-country analysis. *World Development, 52*, 92–103.

Fischer, M.M., & Stirb?ck, C. (2006). Pan-European regional income growth and club-convergence. *The Annals of Regional Science, 40*(4), 693–721.

Gohou, G., & Soumaré, I. (2012). Does foreign direct investment reduce poverty in Africa and are there regional differences? *World Development, 40*(1), 75–95.

González, P., & Figueiredo, A. (2019). Social convergence, development failures and industrial relations: The case of Portugal. In Towards Convergence in Europe. Edward Elgar Publishing.

Herman, E. (2016). The importance of the manufacturing sector in the Romanian economy. *Procedia Technology, 22*, 976–983.

Howell, J. (2005). *Tailoring globalization to reduce global poverty and inequality: A case study of the U.S.-Cambodian bilateral textile agreement*. Social Science Electronic Publishing.

Lauridsen, L. S. (2017). *New economic globalisation, new industrial policy and late-development in the 21st century: A critical analytical review*. Social Science Electronic Publishing.

Lynch, R. G. (2003). Estimates of income and income inequality in the United States and in each of the fifty states: 1988–1999. *Journal of Regional Science, 43*(3), 571–588.

M. Monfort, J. Ordó?ez, & H. Sala. (2018). Inequality and unemployment patterns in Europe: Does integration lead to (real) convergence?. *Open Economies Review, 29*(4), 703–724.

Mussini, M. (2013). A matrix approach to the Gini index decomposition by subgroup and by income source. *Applied Economics, 45*(17), 2457–2468.

Portnov, B. A., & Felsenstein, D. (2010). On the suitability of income inequality measures for regional analysis: Some evidence from simulation analysis and bootstrapping tests. *Socio-Economic Planning Sciences, 44*(4), 212–219.

Rey, S. J., & Montouri, B. D. (1999). US regional income convergence: A spatial econometric perspective. *Regional Studies, 33*(2), 143–156.

Rosenthal, S. S., & Strange, W.C. (2004). Chapter 49 Evidence on the Nature and Sources of Agglomeration Economies. In J. V. Henderson & J.-F. Thisse (Ed.), Handbook of regional and urban economics.

Royuela, V., & García, G. A. (2015). Economic and social convergence in Colombia. *Regional Studies, 49*(2), 219–239.

Yin, L., Zestos, G. K., & Michelis, L. (2003). Economic convergence in the European Union. *Journal of Economic Integration*, 188–213.

Chapter 13
International Trade, Cooperative Organizations, and Regional Development

Abstract International trade plays an active role in promoting regional economic growth. China has actively advocated and participated in numerous international organizations aimed at strengthening exchanges and cooperation with countries around the world. The diversity of its geographical location has provided rich opportunities for trade activities and economic growth. International trade can promote the optimal allocation of resources, the wide dissemination of technology, and the specialized division of labor in the economy, which in turn promotes the vigorous development of regional economies. An in-depth analysis of the impact of geopolitics and protectionism will help us better grasp the challenges faced in international trade and explore problem-solving strategies to promote closer cooperation and common development. Comparing the trade policies of different countries helps assess the effectiveness of each country in promoting international trade and regional economic growth, and then make constructive suggestions and recommendations to improve the trade environment. This chapter provides an in-depth discussion of the four dimensions of the theoretical underpinnings of international trade for geographic development, international cooperation organizations and institutions, geopolitics, and protectionism and comparative analyses of inter-country trade policies to provide a comprehensive introduction to the interrelationships and impacts of international trade, cooperation organizations, and regional development. The objectives of this chapter are as follows: 1. To master the theoretical basis of international trade in geographical development. 2. To understand international cooperation organizations and institutions. 3. To understand geopolitics, protectionism, and trade policies between different countries.

13.1 Theoretical Basis of International Trade on Geographical Development

Adam Smith (1776) first proposed the theory of absolute advantage in international trade in *The Wealth of Nations*. He believed that international trade occurred because countries produced different commodities, and disagreed with the trade view of

X. Deng and M. Song, *Development Geography*,
https://doi.org/10.1007/978-981-96-6917-2_13

rewarding output and restricting income advocated by mercantilism. The premise of the theory of absolute advantage is that two countries produce two different commodities and only use labor production factors in production. One country spends less labor time and lower cost on the production of one product; if the production of another product takes less labor time and costs less, then, the two countries will respectively have an absolute advantage in the production of goods with lower labor costs. The theory of absolute advantage holds that international trade can improve the welfare level of both countries involved. If the two countries conduct specialized production and international trade according to their respective absolute advantages, they can both benefit from international trade.

David Ricardo (1817) proposed the theory of comparative advantage in *Principles of Political Economy and Taxation*, which is very important in international trade theory. By analyzing the role of comparative advantage in international trade, Ricardo explored how to improve the overall welfare level of the world. Ricardo believed that, in international trade, if a country is in a disadvantageous position in the production of two commodities relative to another country in trade (i.e., the labor cost of the two commodities is higher than that of the other country), as long as China faces differing degrees of disadvantage in the production of the two commodities, it can still specialize in producing the commodity where its disadvantage is relatively smaller. If there are different degrees of disadvantages in the production of two commodities, then, China has a comparative advantage in the production of commodities with fewer disadvantages, based on the idea of "the two inferiors have the right to choose the lesser," and China has relatively minor disadvantages in professional production. The other country produces the commodity with the greatest production cost advantage, and the two countries can benefit from it when they conduct international trade.

Hechscher and Ohlin (1933) proposed the well-known H–O theory. Based on the theory of comparative advantage, they further explored the source of comparative advantage and proposed the international trade theory of factor endowment. As the conditions and basis for trade between countries, the assumption of H–O theory is that two countries have two commodities and two production factors of labor and capital. Heckscher and Ohlin posited that the basis for international trade between two countries is the abundance of both countries' production factor endowments. Both countries can pass if the products with a rich endowment of production factors are traded, and both countries can benefit from international trade. For example, if one country with abundant capital factors specializes in the production of capital-intensive products, and another country with abundant labor factors specializes in the production of labor-intensive products, both countries can benefit from the exchange.

Masahisa Fujita of Kyoto University in Japan, Paul Krugman of Princeton University in the United States, and Anthony J. Venables of the London School of Economics and Political Science in the United Kingdom founded spatial economics and co-authored *Spatial Economics: Cities, Regions, International Trade*, published by MIT in 1999 (Fujita et al., 1999). They studied the relationship between economic growth and regional development from the perspectives of cities, regions, and international trade, arguing that in traditional international economics, the basis of international trade is that factors of production cannot flow freely, and it is assumed that there are no transportation costs. These foundations and assumptions are quite different from reality. Thus, they add factors such as space, geography, and transportation costs to international trade. In the analysis of trade theory, international trade theory combines with regional economic theory to form a new spatial geography international trade theory. Agglomeration is the core of spatial economics. Specific industries gather in specific places. This phenomenon is the combined effect of the self-realization mechanism of cyclical accumulation, the impact of historical and accidental events, and expectations. Historical and accidental events are the source of industrial location, and the process of cyclical accumulation will fix the industry in a certain place for a long period, and the number of enterprises will increase.

Head and Mayer (2004) studied the relationship between trade costs and regional economic development. They believed that the cost of trade protection was higher than that of free trade, which reduced the degree of integration among countries, thereby affecting regional economic growth. To develop the regional economy, trade costs must be reduced. Their research findings can provide certain theoretical guidance for the development of China's regional economy. China's local protectionism is serious, and the cost of inter-regional trade is also high, which objectively affects regional economic development.

Studies by several foreign scholars have shown that trade is an important factor that affects economic growth. Classical economists have explained that international trade is conducive to the welfare growth of a country and region in the aspects of international trade division of labor, absolute advantage, comparative advantage, and factor endowments. Modern economists have studied and demonstrated the role of international trade in promoting regional economic growth from the perspectives of spatial geography, economics, monopolistic competition, institutional culture, and trade costs.

13.2 International Cooperation Organizations and Institutions

13.2.1 Belt and Road International Cooperation Summit Forum

The Belt and Road is the inheritance of the ancient Chinese Silk Road culture. The five key areas of the Belt and Road construction are policy interoperability, facility connectivity, unimpeded trade, financial integration, and people-to-people bonds, covering factors such as politics, economy, culture, and diplomacy. In many aspects, the ancient Chinese wisdom of "connecting the world" has been integrated into the construction of the Belt and Road. On May 14 and 15, 2017, the first Belt and Road International Cooperation Summit Forum was held in Beijing. The guests attending this forum included the Secretary-General of the United Nations, Chairman of the International Committee of the Red Cross, heads of three other important international organizations, and 29 foreign heads of state. From April 25–27, 2019, China held the second Belt and Road International Cooperation Summit Forum. The two forums achieved fruitful results.

13.2.1.1 Target

The Belt and Road International Cooperation Summit Forum was held to achieve the following goals:

- Comprehensively summarize the positive progress of the Belt and Road construction, show the important early harvest results, further gather cooperation consensus, and consolidate the good cooperation situation.
- Discuss important cooperation measures for the next stage, further promote the docking of development strategies of all parties, deepen partnerships, and realize joint development.
- While promoting China's economic and social development and structural adjustment, also promote international cooperation to achieve win–win cooperation.

13.2.1.2 Features

(i) No threshold.

It is not only suitable for the participation of developed countries in the West but also for that of developing countries. Accordingly, the enthusiasm of countries worldwide to participate in the construction of the Belt and Road has been enhanced. China welcomes all countries to join the construction of the Belt and Road.

(ii) Freedom and equality.

The freedom and equality created by the Belt and Road Initiative is freedom and equality in the complete sense. All countries, big or small, strong or weak, treat each other equally, negotiate on an equal footing, jointly build through consultation, and share benefits. Thus, the drawbacks of the current difficulty in joining other international organizations are eliminated. Moreover, this type of interactive voluntary participation and freedom of coming and going can make small and medium-sized underdeveloped countries feel more comfortable and respected.

(iii) It is innovative, advanced, and scientific.

The Belt and Road regional cooperation organization is innovative, advanced, and scientific, which differs from other international cooperation organizations. It is created with strong Chinese wisdom and Chinese charm.

13.2.2 Shanghai Cooperation Organization

The Shanghai Cooperation Organization was established in Shanghai, China, on June 15, 2001 by the People's Republic of China, Republic of Kazakhstan, Kyrgyz Republic, Russian Federation, Republic of Tajikistan, and Republic of Uzbekistan. The Shanghai Cooperation Organization is the first international organization named after a Chinese city. The establishment of this organization has strengthened the relationship between China and neighboring countries.

In 1996, China, Russia, Kazakhstan, Kyrgyzstan, and Tajikistan signed the "Agreement on Strengthening Trust in the Military Field in Border Areas" to strengthen the negotiation process of border area trust and disarmament, which laid the foundation for cooperation among the five countries and established the "Shanghai Five" meeting mechanism (the predecessor of the Shanghai Cooperation Organization). After the membership expansion, India and Pakistan became members of the Shanghai Cooperation Organization. The Shanghai Cooperation Organization has now become a trans-regional multilateral comprehensive organization with the largest population and area and has great potential. It makes an important contribution to maintaining regional security and stability and promoting common development and prosperity.

13.2.2.1 Purpose and Mission

- Strengthen trust among member states.
- Maintain and strengthen regional peace, security, and stability, and jointly combat terrorism, separatism and extremism, drug smuggling, illegal arms trafficking, and other transnational crimes.
- Actively cooperate in fields such as the economy, trade, environmental protection, culture, science and technology, education, energy, transportation, and finance; promote balanced development of the economy, society, and culture among regions; and continuously improve the living standards of the people living in the member states.
- Promote the establishment of a democratic, fair, and rational new international political and economic order.
- Strictly abide by the purposes and principles of the Charter of the United Nations.
- Show mutual respect for independence, sovereignty, and territorial integrity; non-interference in each other's internal affairs; and non-use or threat of use of force by each other.
- Consider all member states as equal among themselves.
- Solve all problems through negotiation rather than force.
- Pursue non-alignment, not targeting other countries and organizations, and adhere to the principles of opening to the outside world.

13.2.3 Boao Forum for Asia

Boao Forum for Asia (BFA) is an unofficial, non-profit, regular, and fixed-site international association. It held a conference in Boao Town, Qionghai City, Hainan Province on February 27, 2001. Twenty-nine countries officially announced the establishment. The headquarters of the forum is in Boao, Hainan, and will never change.

13.2.3.1 Target

- Provide a high-level dialogue platform for government officials, business leaders, and well-known scholars to discuss economic, social, and environmental issues to enhance and deepen trade and investment ties, promote the establishment of partnerships, and propose solutions to global economic or social problems.
- Enhance cross-cultural communication in Asia and the sense of social responsibility of private groups in the region.
- Create a good business culture environment to achieve the sustainable development of the region's economy.
- Foster and promote the concept of intra-regional networking mechanisms and regional strategic alliances to increase trade and investment opportunities within Asia and between Asia and the rest of the world in the globalization process.
- Provide intellectual support for human resource development, as well as for important research activities concerning the region and its relations with the rest of the world.

13.2.4 Asian Infrastructure Investment Bank

The Asian Infrastructure Investment Bank (AIIB) is an intergovernmental multilateral development institution in Asia that focuses on supporting infrastructure construction in Asian countries. The purpose of this organization is to speed up the process of economic integration, promote the interconnection of the Asian region, and strengthen the cooperation and exchanges between China and other Asian countries and regions. The AIIB is the first multilateral financial institution initiated and established by China. It is headquartered in Beijing, China, with an authorized capital of US$100 billion. The governance structure of the AIIB is divided into three layers: the council, board of directors, and management. The board of directors is the highest decision-making body of the AIIB. Each member country has one regular director and one deputy director in the AIIB; there are 12 directors on the board, including nine from within the region and three from outside the region. The management consists of a president and five vice presidents. As of July 13, 2019, a total of 100 countries have joined the AIIB.

13.2.4.1 Background

At the global level, the rise of emerging powers is one of the important backgrounds for the establishment of the Asian Development Bank (ADB). Since entering the new century, developing countries have generally realized faster growth, and emerging countries have gradually become an important force in the global economy and are playing an increasingly important role in global governance. To better play the role of emerging countries in the world economy and global financial governance, the reform of the original international financial system has become an urgent need. However, although the World Bank and the International Monetary Fund (IMF) have adopted corresponding decisions on the reform of equity weighting and voting weighting, the irrational international financial mechanism has not been fundamentally changed due to the opposition of the United States Congress.

At the regional level, the backwardness of Asian infrastructure is another important reason for the establishment of the ADB. Asia's economy accounts for 1/3 of the global economy and is the region with the highest economic vitality and growth potential in the world today, but the lack of construction funds has limited the region's infrastructure development, such as railroads, highways, bridges, ports, airports, and communications. This infrastructure lag has to some extent constrained the economic development of the Asian region.

With the rise of China's economy and its global status, China also expects to play a greater role in the international financial system and promote the reform and restructuring of the global financial governance system. The establishment of the ADB provides a platform for China to participate more actively in global financial governance and to promote infrastructure construction in the Asian region.

13.2.4.2 Main Purpose

- Promote sustainable economic development in the Asian region by investing in infrastructure and other productive areas to create more wealth.
- Strengthen cooperation with other multilateral and bilateral development agencies, promote regional cooperation and build partnerships, and jointly address various challenges.

13.2.4.3 Main Function

- Attract more public and private capital investment to promote regional economic development, especially in infrastructure and other productive areas.
- Promote the development of the regional economy and use its disposable funds for financing support. These projects can efficiently support the development of the regional economy. Pay special attention to and provide the necessary support for the needs of some underdeveloped member countries in the region.

- Encourage private capital to actively invest in areas conducive to regional economic development, especially projects or enterprises that develop infrastructure and other productive areas. When private capital cannot be reasonably invested, it will replace private capital to invest.
- Specific activities and other services to be provided to strengthen these functions.

13.2.5 World Trade Organization

The World Trade Organization (WTO) was established at the General Agreement on Tariffs and Trade (GATT) Uruguay Round Ministerial Conference in Morocco in 1994. The meeting decided to establish a more global WTO to replace GATT. The WTO is one of the most important contemporary international economic organizations and has 164 member countries. China joined the WTO in 2001. The total trade volume of member countries has reached 98% of the total global trade volume, and it has been given titled "Economic United Nations."

13.2.5.1 History

The predecessor of the WTO was the GATT, which was signed in Geneva, Switzerland, on October 30, 1947, and was temporarily applied on January 1, 1948. After its establishment on January 1, 1995, the WTO coexisted with the GATT for one year.

The WTO officially began to operate on January 1, 1995. Its headquarters was established on the banks of Lake Leman in Geneva, Switzerland, and it is responsible for maintaining and managing the world economic and trade order. Its basic principle is to achieve the goal of liberalizing the world trade economy by implementing the principles of non-discrimination, opening, and fair trade. On January 1, 1996, it formally replaced the Provisional Institutions of the GATT. Compared with the GATT, the scope of the WTO's services covers trade in goods, services, and intellectual property rights, while GATT only applies to trade in goods. In addition, the WTO has the status of a legal person and has higher authority in mediating disputes among member states.

China officially joined the WTO in 2001, and since then, the degree of opening up of the Chinese market to the outside world has reached a higher level. Since then, the World Trade Organization has continued to grow and expand, admitting more members and promoting the improvement and development of trade rules in several areas.

13.2.5.2 Purpose

- Increase the employment rate, gradually increase the income level, and improve the living standards of the people.

- Continuously expand the scale of production and trade of goods and services.
- Protect the environment and strengthen the full use of natural resources.
- Efforts should be made to ensure that developing countries, especially the least-developed countries, receive a share and benefits commensurate with their level of economic development in the growth of international trade.
- Build an integrated multilateral trading system.
- Through effective measures such as cutting tariffs, establish a dynamic, complete, and durable multilateral trading system.
- Based on the principles of openness, equality, and mutual benefit, gradually reduce the tariff and non-tariff trade barriers of each member state and eliminate the unfair treatment experienced by member states in international trade.

13.2.6 United Nations

The United Nations is an international organization established after World War II and is composed of sovereign states. The symbol of the formal establishment of the United Nations was on October 24, 1945, when the Charter of the United Nations was signed and officially entered into force in San Francisco, California, in the United States. The purpose of the United Nations is to strengthen economic and trade cooperation, international law, international security, human rights, social progress, maintenance of world peace, and ease of international tension cooperation among countries.

Headquartered in New York City, New York in the United States, the United Nations currently has 193 member states, including 39 in Asia, 54 in Africa, 28 in Eastern Europe and the Commonwealth of Independent States, 23 in Western Europe, 33 in Latin America, and 16 in North America and Oceania, including all internationally recognized sovereign states. There are also two observer states (The Holy See and Palestine).

The five permanent members of the United Nations Security Council are the United States of America, Russian Federation, United Kingdom of Great Britain and Northern Ireland, French Republic, and People's Republic of China. The United Nations plays a fairly active role in maintaining world peace, easing international tensions, and resolving regional conflicts, as well as in coordinating international economic relations and promoting economic, scientific, and cultural cooperation and exchanges among the countries of the world.

13.2.7 Asia–Pacific Economic Cooperation

Asia–Pacific Economic Cooperation (APEC) is the highest-level intergovernmental economic cooperation mechanism in the Asia–Pacific region. Its official advisory body is the Alliance of Pacific Rim Universities. This organization was established to promote the liberalization of regional trade and investment and strengthen the economic, technical, and cultural cooperation among member countries. Economic cooperation, development, trade, and investment in various regions within APEC are promoted by holding several forums.

13.2.7.1 Background

The symbol of the establishment of APEC was the first ministerial meeting of the Asia–Pacific Economic Cooperation Conference in 1989. In 1991, China officially joined APEC as a sovereign state, and Chinese Taipei and Hong Kong, China, as regional economies. APEC has a total of 21 members. In October 2001, the APEC meeting was held in China for the first time, in Shanghai, and the second APEC meeting in China was held in 2014 in Beijing. As of September 2019, APEC has 21 full members and three observers.

13.2.7.2 Purpose

The purpose of APEC is to:

- Support sustainable economic growth and prosperity in the Asia–Pacific region and build a dynamic and harmonious Asia–Pacific family.
- Defend free and open trade and investment, accelerate the process of regional economic integration, and encourage economic and technical cooperation.
- Ensure people's security and promote a sound and sustainable business environment.

13.3 Geopolitics and Protectionism

13.3.1 Geopolitical Theories

13.3.1.1 Sea Power Theory

Alfred Thayer Mahan was an American naval officer and historian. After some research on maritime hegemony and the development of the British navy, he proposed a theory to explain historical development with sea dominance. To elaborate on the theory of sea power, he also wrote the book *The Influence of Sea Power on History 1660–1783* which was officially published in 1890. The ability to act freely is a necessary condition for a country to become a great power.

13.3.1.2 Land Power Theory

Halford Mackinder is the founder of British geopolitics. In 1904, he published the paper "The Geographical Hub of History," marking the establishment of the theory of land rights. He posited that the hub of world politics exists in the hub zone, and the center of Eurasia is the hub of the world. In 1919, he revised the "hub zone" to the "heartland" of the "world island" and collectively referred to the continents of Asia, Europe, and Africa as the"world island". The "Heartland" theory holds that controlling Eastern Europe is equivalent to controlling the Heartland, controlling the Heartland is equivalent to controlling the World Island, and controlling the World Island is equivalent to controlling the entire world.

Nicholas Spykman was an American geopolitical scientist. In the 1940s, after the end of World War II, he proposed the "rimland" theory based on Halford Mackinder's "heartland" theory. Through the study of history, he came to believe that the economic aggregate and population of the fringe areas exceed those of the heartland. Moreover, the two world wars took place in the fringe areas. Controlling the fringe areas is equivalent to controlling the Eurasian continent, and controlling Eurasia is tantamount to controlling the entire world.

13.3.1.3 Sky Power Theory

Italian general Giulio Douhet was the first to propose the theory of air power. He believed that the shortest distance between two points is the "sky road," and that air power determines the outcome of war. In 1921, Douhet published the book *Superiority of the Air*, in which he noted that a country should develop a powerful air force, strive for the supremacy of the air, gain the strategic initiative, and destroy the enemy's strategic targets through the air force, to successfully defeat enemy nations and wear down their willpower.

In the 1950s, the theory that the Arctic region is very important for the United States to compete for air supremacy was proposed. This theory was developed based on analyzing the important role played by the air force in military strategy and the overlapping areas controlled by the air forces of the United States and the Soviet Union, known as emptiness.

13.3.1.4 The Case

In World War II, the Nordic countries of Denmark and Norway were of notable strategic importance because of their unique geographic location. Denmark was situated at the mouth of the Baltic Sea and controlled the gateway to and from the Baltic Sea for all the powers. Norway, on the other hand, was located on the eastern shore of the North Sea, across the sea from the British mainland. For Britain, control of these two countries meant that Nazi Germany could be blockaded inside the Baltic Sea, thus weakening its maritime power. However, Germany was also well aware of the strategic value of these two countries, and Hitler's decision to occupy Denmark and Norway in 1940 was aimed at countering Britain's maritime superiority, securing the North Sea route, and controlling access to and from the Baltic Sea. This German action highlighted the important influence of geographic location in geopolitics on the strategy and outcome of war.

In addition, during World War II, Japan's aggressive behavior in Asia also fully embodied the application of geopolitical theories. For example, the "Military Aircraft Dedication Campaign" launched by Japan in Weihai was an aggressive strategy based on geopolitical considerations, aiming to realize the purpose of "war by war" by plundering resources and controlling key areas. These cases show that geopolitical theories were widely applied during World War II. Based on their geographic location, resource conditions, and relations with neighboring countries, countries formulated and implemented a series of strategies and tactics to seek advantages and victory in the war. These cases not only demonstrate the real-life application of geopolitical theories but also provide us with important references for an in-depth understanding of geopolitical patterns and war strategies.

13.3.2 Trade Protectionism

13.3.2.1 Definition

Trade protectionism refers to the policy of restricting the import of foreign goods in the process of a country's foreign trade to protect the goods produced in China from the competition of foreign goods in the domestic market. Proposals and policies for commodities offer various benefits to increase their international competitiveness.

Trade protectionism mainly restricts foreign goods from entering the domestic market through the adoption of both tariff and non-tariff measures. The former mainly prevents many foreign goods from entering the domestic market by imposing high import tariffs, while the latter includes quantitative restriction measures and other non-tariff measures that hinder trade. Quantity restriction measures are manifested in quotas, import licenses, automatic export restrictions, and quantitative foreign exchange controls. Other non-tariff measures include technical barriers to trade, animal and plant inspection and quarantine measures, customs valuation, rules of origin, and investment management measures such as local content, trade balance, and domestic sales requirements.

13.3.2.2 Features

(i) The means of trade protection have gradually shifted from tariff to non-tariff measures. A series of resolutions formed through international trade and tariff negotiations have substantially reduced tariff levels; therefore, countries have adopted non-tariff measures to implement trade protectionist policies, and more than half of the world's total trade is subject to various non-tariff restrictions. These measures are highly flexible, concealed, and restrictive.

(ii) Trade protection policies are increasingly targeted, such as reducing restrictions on the import of industrial products, but the restrictions on agricultural products are still very strict. However, the restrictions on the import of different commodities in industrial products are also different.

(iii) Implement different trade protection policies for different countries and regions. If one country is a strong competitor of another country, strict trade protection policies will be implemented. For other countries, trade protection policies can be appropriately relaxed to protect the products produced by their own domestic enterprises, continuously develop and expand enterprises, and enhance the international competition of enterprise force.

(iv) The impact on the global economy is multifaceted: trade protectionism leads to higher trade barriers and restricts the free flow of international trade, thereby reducing global trade and weakening the stability of global supply chains and markets. In addition, it can lead to higher commodity prices, increased inflationary pressures, and the disruption of global cooperation. These effects all point to the significant negative impact of trade protectionism on the global economy.

13.3.2.3 Protection Purpose

The main purpose of implementing trade protection policies includes:

(i) Protection of domestic industries: A central purpose of trade protectionism is to protect domestic industries from foreign competition. By restricting imports, domestic firms can develop and increase their competitiveness in a relatively less competitive environment, thereby maintaining or improving their position in domestic and foreign markets.

(ii) Protection of employment opportunities: Protectionist trade policies help protect employment opportunities in China. Restrictions on imports can reduce competition from foreign products in the domestic market and encourage domestic enterprises to increase production and employ more domestic workers, thus maintaining domestic employment stability.

(iii) Protection of national security: Sometimes trade protectionism is also used to protect China's strategic industries and key technologies. Some countries may restrict imports of specific products or technologies to ensure self-sufficiency and maintain national security in emergency situations.

(iv) Protection of intellectual property rights: Trade protectionism can also be used to protect the intellectual property rights of national firms. Restrictions on imports can reduce the inflow of infringing products and ensure that the innovations of domestic firms are protected and can be economically rewarded accordingly.

13.3.2.4 The Case

Post-Brexit Trade Strategy. Since the UK's referendum decision to leave the EU in 2016, it has begun to renegotiate trade agreements with the EU to mitigate the economic impact of Brexit and to protect its industries. This protectionist approach to trade, while intended to safeguard the UK's economic and industrial interests, has also created uncertainty and additional trade costs for UK businesses and consumers.

13.4 A Comparison of Trade Policies Between Countries

A government's foreign trade policy generally refers to the guidelines for import and export trade activities within a certain period, formulated based on China's political and economic interests, and its basis is China's development goals. The laws, rules, regulations, and measures implemented on import and export trade in a certain period are the concentrated expression of a country's foreign trade policy. It is an important part of a country's foreign policy and general economic policy. Foreign trade policies are mainly divided into two major policies, namely free trade and trade protection (Clemens & Williamson, 2004). In international trade, one should choose which international trade policy to use according to the overall international trade pattern at that time and the closeness of its relationship with the world's economic powers.

This book discusses trade policy options in capitalist countries as well as in developing countries.

13.4.1 Foreign Trade Policy of Major Capitalist Countries

In the historical progress of human society, the fifteenth and sixteenth centuries were the gestation period of capitalist production relations. Because commercial capital was dominant in society at that time, the theory of mercantilism was gradually formed. The theory of mercantilism argues that the wealth of a country can be increased as follows. First, a large amount of mining of gold, silver, and other minerals is performed to gain profits from other countries' trade and increase wealth accumulation. Therefore, mercantilists generally advocate state intervention by economic activities, through the encouragement of raw material imports, and the implementation of protectionist policies and measures such as protective tariffs, the rapid development of international trade, and incentives for rewarding exports and restricting imports. This policy has been widely used in Britain, France, Italy, Spain, and other countries where capitalist production relations came into being at an early point in history.

Mercantilism is the earliest foreign trade policy of capitalist countries. This foreign trade protection policy played a greater role in the period of primitive capital accumulation in capitalist countries and promoted the development of capitalist productive forces. However, the implementation of the mercantilism policy also brought many negative effects. The mercantilism policy brutally destroyed the economies of some Eastern countries, whereas the follow-up foreign economic policies of these old capitalist countries had a great impact. Later, capitalist countries generally established tariff barriers and launched overseas wars, which were also related to the implementation of mercantilism policies to a certain extent.

In the middle of the seventeenth century, feudal production relations gradually disintegrated, and social productive forces were liberated. Correspondingly, capitalist production relations also rapidly spread and developed. The bourgeois revolution that broke out in England in 1640 also started the entirety of human society moving from the era of feudalism to the era of capitalism. In the mid-to-late eighteenth century, industrial revolutions took place in European countries one after another, establishing the capitalist large-scale machine industry, and ultimately establishing the capitalist mode of production. At the same time, the United States and France were successively victorious in their bourgeois revolutions. Capitalism then entered a stage of free competition. At this time, the theory and policy of mercantilism could no longer adapt to the rapid development of capitalist productivity because the power of industrial capital far exceeded that of commercial capital. Continuing to adopt mercantilism would only hinder the continued development of capitalist productivity (Fujita et al., 1999); therefore, a policy of free trade replaced that of trade protectionism and is in an important position.

As the first country to implement a free trade policy, the United Kingdom took the lead in completing the first industrial revolution as early as the 1830s and 1940s and entered the era of large-scale machine industry. The development of British productivity has made the contradiction between the small domestic market and the continuous development of productivity increasingly prominent. Adam Smith proposed the idea of developing free trade and "regional division of labor theory" against trade protectionism. Later, David Ricardo developed "comparative interest theory." Their theories became an important theoretical reference for Britain's free trade policy and provided a theoretical basis for its colonial expansion. Britain took full advantage of a century of capitalist free competition to build a strong industrial base and a monopoly in free trade. In contrast to Britain, rising stars such as Germany, the United States, and France are relatively backward economically. Therefore, they oppose free trade policies, adopt state intervention in the economy, and use protectionist policies to protect and develop their own immature national industries. Friedrich List is a representative of infant industries theory and has argued that the productivity of wealth is more important than wealth. He also noted that Britain's policy of free trade is intended to hinder the development of other countries' productivity; therefore, for the development of Germany's productivity, strict protective tariffs should be implemented to limit the entry of low-priced foreign goods into China, temporarily sacrificing some. Only in this way can we promote the development of international trade and the shipping industry, strengthen our national defense, improve our cultural level, increase our own productivity, and maintain our own industrial independence to compete with Britain in the world market. This viewpoint also once played a certain role in promoting the opposition to the rapid expansion of the UK and the development of China's relatively backward capitalist national industry.

At the end of the nineteenth century and beginning of the twentieth century, liberal capitalism entered the stage of monopoly capitalism. Although great differences remain among the major capitalist countries in terms of economic strength, level of productivity development, status, and role in the world political situation, they share the same basic economic characteristics of imperialism. Large monopoly organizations and financial capital rely on their controlling positions in international trade to combine commodity and capital output, and through the international division of labor, world markets, and capital tend to internationalize, and finally form the capitalist world economic system. Further, to monopolize the domestic market and seize foreign markets, various imperialist countries have intervened in the economy one after another and implemented stricter protectionist trade policies—super-protection trade policies. Various contradictions inherent in imperialism are intensifying. The era of "free trade" has passed, and countries have generally adopted incentives as well as restrictions on imports and other new protective trade measures and established tariff barriers, which have undoubtedly deepened the contradictions among imperialist countries and made the struggle for the world market more acute.

The above discussion shows that protective trade policy runs through the history of international trade development, and no pure protection trade or free trade policy has ever existed. The banner of "free trade" is generally upheld by countries whose economic strength occupies an absolute advantage in the world economy. Therefore, the current international trade situation is beneficial to China, and forcing other countries to implement free trade is of great significance to exert their favorable position in the world economy. Countries with relatively inferior economic strength will implement protective trade policies with different degrees of relaxation. Thus, whether it is trade protectionism or free trade policy, the premise is to protect a country's economic development and trade interests.

13.4.2 Foreign Trade Policies of Developing Countries

13.4.2.1 Foreign Trade Policies of Developing Countries

Several scholars have proposed their own views on foreign trade policies implemented by developing countries and conducted in-depth explorations. Some scholars have analyzed the foreign trade policies implemented by developing countries in different historical stages around a certain international trade theory. Others have systematically sorted out the trade policies implemented by developing countries. Mustafa (2020) explored the previous question and noted that the growth of infant industries requires trade protection for a certain period. Therefore, the governments of developing countries use measures such as quotas, tariffs, and subsidies to protect their own infant industries. However, it is crucial for this protection to be time-bound and flexible to adapt as the industry matures. The use of a combination of protection tools such as tariffs, quotas, subsidies, and strategic partnerships can aid in fostering the competitiveness of infant industries. The ultimate goal should not only be to shield domestic industries but also to propel them toward international competitiveness, thereby contributing to overall economic growth.

To analyze the latter issue, several scholars have conducted research. Geller (2023) explored the theoretical basis of trade policy in detail and discussed the foreign trade policy of developing countries. He holds that the theoretical basis of trade policy in developing countries can be anchored in theories such as dependency theory, export-led growth theory, and institutional economics theory. These theories provide frameworks for understanding how trade agreements can impact economic development. Tahirović (2020) found that, under the background of trade globalization, developing countries can catch up with the economies of developed countries by implementing free trade policies. By opening up their markets and cooperating with international trade institutions, developing countries can harness the benefits of international trade flows, ultimately contributing to their growth and economic development. Nuralifia and Yasin (2023) divided the trade policies of developing countries into "import substitution" and "export orientation." Developing countries often choose between these strategies based on factors such as balance of payments, foreign exchange availability, and economic independence goals. Both strategies play crucial roles in industrial development, with some countries, including Indonesia, initially opting for import substitution before transitioning toward export-oriented policies.

13.4.2.2 The Practice of Foreign Trade Policy in Developing Countries

(i) Theory of comparative advantage

The United States, the European Union, Japan, and Hong Kong are China's four major trading partners. Compared with these countries or regions, the ratio of capital to labor in China is quite different. The total exports of these four major trading partners account for more than 70% of China's total exports, and the proportion of exports with more than 100 other countries and regions is less than 30%. Based on the theory of comparative advantage, this shows that China has established foreign trade. From the perspective of China's trade structure, general and processing trade each account for approximately 50%, of which refined processing and high-tech products account for a small proportion, and textiles and general mechanical and electrical products account for a relatively high proportion. Most of China's current export products are still labor-intensive products with low added value. Because China's protection of patented designs is not sufficiently strict, it does not monopolize many means of production, and production does not generate large-scale returns. Therefore, China currently does not have the conditions to form an oligopoly in the international market, and thus fails to grasp the pricing power.

(ii) Infant industry protection theory

The theory of infant industry protection in developing countries was first proposed more than 200 years ago, and after the end of World War II, it was widely used in the post-war development of developing countries. A typical example of the successful application of this theory is the “Asian Tigers.”

(iii) Union of developing countries

To strive for, maintain, and consolidate their legal and equal status and rights in international trade, a vast number of developing countries have united and adopted a series of necessary policy measures with three main aspects. First, to protect the rights and interests of exporting countries of primary agricultural and sideline products, several international commodity agreements have been signed. Second, these countries aim to boycott developed countries’ trade protectionism against industrial products exported by developing countries. Third, they have strengthened economic and trade cooperation among them and built a regional trade group. Examples of developing countries uniting mainly include the establishment of the Organization of the Petroleum Exporting Countries (OPEC) in 1960, the struggle for the rights of the “Group of 77” in 1963, and the establishment and development of the ASEAN Free Trade Area (CAFTA).

13.4.2.3 Suggestions on Foreign Trade Policy of Developing Countries Under the Background of Financial Crisis

(i) Optimizing export structure: Developing countries should make efforts to optimize the structure of their export products, promote industrial upgrading, and increase the added value of their products. In addition, they should actively explore new export markets, reduce their dependence on a single market and lower their risks.
(ii) Strengthening trade protection: In the context of the financial crisis, trade protectionism may rise. Developing countries should persist in opposing and resisting trade protectionism, and at the same time strengthen their own trade protection measures to protect domestic industries and market stability.
(iii) Promoting trade liberalization: Despite the pressure of trade protectionism, developing countries should actively promote the process of trade liberalization, reduce tariffs and non-tariff barriers through multilateral trade negotiations and regional trade cooperation, and promote the free flow of goods and services.
(iv) Strengthening foreign exchange management: The financial crisis may trigger volatility in the foreign exchange market, and developing countries should strengthen foreign exchange management, stabilize exchange rates, and prevent excessive capital outflows. In addition, they should actively accumulate foreign exchange reserves to cope with possible payment risks.
(v) Promoting technological innovation: Developing countries should increase their investment in technological research and development and innovation to improve the technological content and competitiveness of their products.

Through technological innovation, they can promote industrial upgrading and enhance their position in the global value chain.

(vi) Strengthening international cooperation: In the face of the global financial crisis, developing countries should strengthen cooperation with other countries and regions to jointly meet the challenges. Through international cooperation, resources, technologies, and markets can be shared to achieve win–win development.

13.4.3 Trade Policy and Trade Environment Coordination

Free trade and trade protectionism are the focus of a long-standing debate in the field of international economics. Relevant empirical literature on trade policy indicates that a country's optimal trade policy should be closely related to the international trade environment in which China is located at that time. According to Clemens and Williamson, when the international trade environment favors protectionism, the optimal trade policy should be a protection trade policy; however, when the international trade environment favors free trade, the optimal trade policy should be a free trade policy. Given the post-financial crisis era, to properly protect and develop their own economies, all countries have adopted a certain degree of protection trade policy. Judging from the above analysis, China should take appropriate protective trade measures. Simultaneously, the adjustment of trade policy should also be combined with China's actual national conditions. At present, China's foreign exchange reserves are relatively abundant. Therefore, the trade policy that needs to be adopted under such circumstances should gradually dilute the policy of encouraging exports, foreign direct investment, and processing trade. The original positive trade policy should be changed to a more neutral foreign economic and trade strategy and policy to balance the relationship between promoting domestic economic development and expanding openness to the outside world.

13.4.4 Recent Trade Policies in China

(i) Promoting the construction of free trade zones: China actively participates in the construction of free trade zones around the globe and has signed free trade agreements with several countries and regions to promote trade liberalization and facilitation. The signing of these free trade agreements has helped reduce tariff and non-tariff barriers and promote the free flow of goods and services.

(ii) Implementation of import diversification strategy: To reduce its dependence on a single market, China has implemented an import diversification strategy and actively explored new import channels and markets. This helps to increase the variety and quantity of imported goods to meet the needs of the domestic market and also helps to promote the balanced development of global trade.

(iii) Optimizing export structure: China is committed to optimizing its export structure and promoting the transformation of export products in the direction of high technology and high value-addedness. Improving product quality and value-addedness will enhance the competitiveness of export products, and China's position in global trade will be upgraded.
(iv) Strengthening the protection of intellectual property rights: China has strengthened the protection of intellectual property rights, cracked down on infringement, and maintained a fair and competitive trade environment. This helps to enhance China's reputation and image in global trade and attract more foreign investment and technology into the Chinese market.
(v) Promoting cross-border e-commerce: With the rapid development of e-commerce, China has actively promoted the development of cross-border e-commerce to provide small and medium enterprises (SMEs) with more opportunities to participate in international trade. Building a cross-border e-commerce platform and improving the logistics system and other measures have lowered the threshold and cost for SMEs to participate in international trade.

13.5 Summary

Studies by many foreign scholars have shown that trade is an important factor affecting economic growth. Classical economists have explained that international trade is conducive to the welfare growth of a country and region with respect to international trade division of labor, absolute advantage, comparative advantage, and factor endowments. Modern economists have studied and demonstrated the role of international trade in promoting regional economic growth from the perspectives of spatial geography, economics, monopolistic competition, institutional culture, and trade costs.

To strengthen exchanges and cooperation between China and other countries, maintain world peace and stability, and promote global prosperity and development, China has advocated the establishment of some international organizations such as the Belt and Road International Cooperation Summit Forum, AIIB, Boao Forum for Asia, and Shanghai Cooperation Organizations. China has also joined some international organizations, such as the WTO, United Nations, and Asia–Pacific Economic Cooperation.

Geopolitics is divided into sea power theory, land power theory, and air power theory. Sea power theory posits that the ability to act freely on the sea is a necessary condition for a country to become powerful. Land power theory is divided into "heartland" theory and "periphery" theory. The former argues that controlling the heartland is equivalent to controlling the entire world, whereas the latter posits that controlling the fringes is equal to controlling the entire world. Air power theory argues that air power determines the outcome of war. Geographical factors will affect the nation, China, and the world, and then have a certain impact on a country's national culture, character traits, overall strength, and foreign policy.

Trade protectionist policies mainly include two types of measures—tariff barriers and non-tariff barriers—to restrict foreign goods from entering the domestic market. In recent years, trade protectionism has increasingly been characterized by trade protection shifting from tariff to non-tariff measures. The pertinence is becoming increasingly stronger, strengthening the distinctions between trade protection policies implemented by different countries and regions.

The foreign trade policies implemented by capitalist countries have gone through different stages of mercantilism, free trade, and monopoly; however, protection trade policies have always run through them, and there has never been a pure free trade or protection trade policy. Developing countries mainly implement comparative advantage theory, infant industry protection theory, and joint foreign trade policies. In the future, China needs to continue to fully utilize its comparative advantages and more effectively combine trade and industrial policy.

Exercise

1. Briefly describe the theoretical basis of international trade for geographical development.
2. What are the international cooperation organizations and institutions?
3. What are the objectives of the establishment? What are their characteristics?
4. Briefly summarize the relevant theories of geopolitics.
5. Briefly summarize trade protectionism.
6. What are the differences in trade policies between capitalist countries and developing countries?
7. What are the recommendations for the foreign trade policies of developing countries in the context of a financial crisis?
8. Briefly summarize the recent trade policies practiced by China.

Extended Reading

Krugman, P. R., & Obstfeld, M. (1998). International Economics: Theory and Policy (3rd ed.). Addison-Wesley.

Smith, A. (1776). An Inquiry into the Nature and Causes of the Wealth of Nations. W. Strahan and T. Cadell.

Trade policies in developing countries (English). World Bank reprint series; no. REP 1 Washington, D.C.: World Bank Group. http://documents.worldbank.org/curated/en/827391468138588144/Trade-policies-in-developing-countries

References

Clemens, M.A., & Williamson J.G. (2004). Why did the Tariff–growth correlation change after 1950?. *Journal of Economic Growth, 9*(1).

Fujita, M., Krugman, P., & Venables, A. J. (1999). *The spatial economy: Cities, regions, and international trade.* MIT Press.

Geller, C. (2023). Trade policy and economic development: A study of trade agreements' effects on developing countries. *Journal of Modern Law and Policy, 3*(3), 64–76. https://doi.org/10.47941/jmlp.1555

Head, K., & Mayer, T. (2004). Chapter 59—The empirics of agglomeration and trade. In J. V. Henderson & J.-F. Thisse (Eds.), *Handbook of regional and urban economics* (Vol. 4, pp. 2609–2669). Elsevier. https://doi.org/10.1016/S1574-0080(04)80016-6

Heckscher, E. F., & Ohlin, B. (1933). *Interregional and international trade.* Harvard University Press.

Mustafa, D. I. (2020). Infant industries protectionism: The case of automobile industry in Malaysia. *International Journal of Business and Economics Research, 9*(2), Article 2. https://doi.org/10.11648/j.ijber.20200902.12

Nuralifia, D. F., & Yasin, M. (2023). Persaingan Industri dan Subtitusi Industri Terhadap Perkembangan Industri. *Jurnal Riset Ekonomi Dan Akuntansi, 1*(4), 219–225. https://doi.org/10.54066/jrea-itb.v1i4.990

Ricardo, D. (1817). *On the principles of political economy and taxation.* John Murray.

Smith, A. (1776). An inquiry into the nature and causes of the wealth of nations. W. Strahan and T. Cadell.

Tahirović, H. (2020). The importance of improving free trade in developing countries. *Ekonomski Izazovi, 9*(17), 28–44. https://doi.org/10.5937/EkoIzazov2017028T

Chapter 14
International Balance of Payment, Financial Stability, and Regional Development

Abstract Financial stability is fundamental to national economic development, and countries generally recognize the far-reaching impact of the external financial environment on national financial systems and, therefore, actively participate in international financial institutions to jointly maintain global financial stability. This chapter systematically analyzes the role of international financial stability on regional development, balance-of-payments deficits and trade wars, the impact of the global financial crisis on developing countries, and the modern financial system and its evolution process from the following four perspectives to provide a comprehensive picture of the relationship between the balance of payments, financial stability, and regional development. The objectives of this chapter are as follows: 1. To master the role of international financial stability on regional development. 2. To understand the balance of payment deficits and trade wars. 3. To become familiar with financial system reform and policies to promote regional development.

14.1 Role of International Financial Stability on Regional Development

14.1.1 Formation of International and Regional Financial Stability

14.1.1.1 International Financial Stability

International finance is an important part of international economics, which studies all types of financial activities carried out by various countries, regions, and economic entities based on currency and the relationships among different national currencies. The Mundell–Fleming model of two countries shows that the economies of two countries can be transmitted internationally through income, relative price, and interest rate mechanisms. With the financial globalization process and acceleration of international capital flows, the stability of the global financial market affects the

X. Deng and M. Song, *Development Geography*,
https://doi.org/10.1007/978-981-96-6917-2_14

internal and external equilibrium of each region. In particular, the international financial crisis has caused the global economy to fall into recession, affecting economic growth. Countries are increasingly paying attention to the impact of the external financial environment on their national financial system, and an increasing number of countries are joining international financial institutions to maintain international financial stability.

14.1.1.2 Formation of Regional Financial Stability

As an important part of national economic development, financial stability can promote the efficiency of the transformation from savings to investment, improve the optimal allocation of social resources, and enhance a country's ability to resist risks. Financial stability is a dynamic process that involves the reallocation and adjustment of financial elements within and between regions. In a regional financial system, the differences in regional resource allocation determine the differences in regional financial factors, financial institutions, and other aspects of regional financial development, thereby affecting regional economic development (Laulajainen, 2003). However, the differences in regional economic development levels and investment preferences affect the allocation of financial resources by regional financial institutions according to the regional background and financial environment. Regional economic development attracts the layout and agglomeration of financial institutions and businesses in the region, forming the process of financial agglomeration and showing a pattern of spatial gradient development in the financial industry. Financial agglomeration accelerates the utilization of regional resources, produces the Marshall effect, and forms positive externalities, and the flow and diffusion of regional financial resources have an impact on the marginal regional economy (Fig. 14.1).

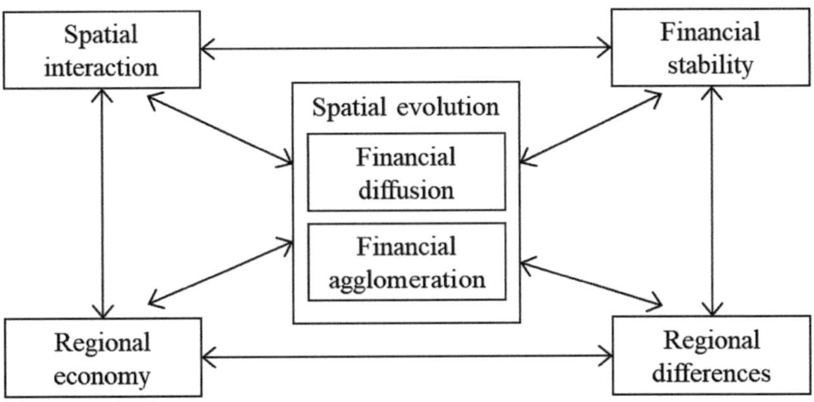

Fig. 14.1 Formation mechanism of regional financial stability

14.1.2 Promotion of Financial Development to Regional Development

Finance is the core of the modern economy and an important force promoting scientific and technological innovation, which provides important support for innovation-driven social and economic development through financial activities, effective allocation of resources, mobility, and activation of credit resources. The role of the financial system should be fully utilized to promote the development of the real economy, guide enterprises to transform and upgrade in the direction of "science and technology" and "innovation," and promote regional high-quality development.

14.1.2.1 Providing Capital Support for Regional Development

As an effective tool for resource allocation, finance provides capital support for regional development by promoting social savings and transforming them into investments. This can be used to allocate idle funds from the supply side to the demand side through financial intermediaries or financial markets, improve the investment level of society, and accelerate the flow of funds between surplus and deficit.

14.1.2.2 Promoting the Adjustment and Optimization of Regional Industrial Structure

Regional industrial structure can be adjusted and upgraded mainly through two methods. One is incremental development, which promotes the development of original industries to emerging industries by expanding capital investment. The other is to integrate inefficient and efficient industries through asset reorganization, enrich industrial structures, improve industrial chains, and optimize industrial layouts. Relying on the development of the financial market, it provides financial support for adjusting industrial structure and mobilizing external resources. It is further necessary to absorb high-quality resources from home and abroad, meet the needs of regional high-quality enterprises and industries for financing, and enhance the upgrading capacity of regional industrial development and structural adjustment.

14.1.2.3 Stimulating the Vitality of Scientific and Technological Innovation in Enterprises

Scientific and technological innovation is the core content of industrial upgrading and development, and financial development provides important support for the mass entrepreneurship and innovation trends. The emergence and improvement of financial services effectively promote tens of thousands of small and micro enterprises to actively innovate and stimulate the vitality of scientific and technological innovation.

The financial market solves the financial support needed for technological innovation from the root cause. Further, scientific and technological innovation has many uncertain factors. The financial system has the function of risk identification, and the financial market can identify risky projects by analyzing and selecting scientific and technological innovation projects. In addition, finance promotes the exchange of technological information between different industrial entities through foreign direct investment, multinational corporations' investment, or mergers and reorganization, which is conducive to the upgrading of industrial technology.

14.1.2.4 Providing the Basis for Endogenous Economic Growth

Endogenous economic growth refers to achieving sustained economic growth without relying on the external environment. The financial system provides financial support for the market economy and convenience for capital flow and improves the efficiency and economic benefits of capital use. The profit-seeking nature of financial investment urges the market to screen projects and enterprises with poor prospects, no profits, or even losses, to optimize the rational allocation of regional resources.

14.1.2.5 Conducive to the Maintenance of Social Stability

Financial security is an important part of national security; a financial crisis will gradually lead to a social crisis. High risk in the financial market indicates potential problems in the operation of any financial institution, which leads to public concern. The herd effect causes that panic to spread, which will affect the development of other industries. If remedial measures are not taken promptly, the financial crisis will further spread to impact social stability. Tocqueville stated that revolutions tend to take place at the inflection point between economic prosperity and recession, rather than from a process of sustained economic decline.

14.1.3 Impact of International Financial Stability on Regional Development

14.1.3.1 Promoting International Investment

With the development of international finance, global resources can flow freely. International financial stability means that financial institutions in the international financial system are sound and are able to resist risks, the international financial market maintains a normal credit order, and the security of international capital circulation is guaranteed. The relevant subjects of international investment pursue the maximization of interests. In the environment of international financial stability, international investors can reduce decision-making risk through factors such as interest and exchange rates to promote the effective implementation of international investment.

14.1.3.2 Creating a Stable Environment for International Trade

The stability of international finance is beneficial for the economic subjects of international trade to predict future market situations and make appropriate decisions according to the change trends in exchange rates and other factors. To accomplish this, it is necessary to promote the rationalization and scientization of regional foreign trade, turn the trade operation into a virtuous circle, promote the improvement of export quality and the transformation of foreign trade mode, and promote the high-quality development of the regional manufacturing industry.

14.1.3.3 Improving the Efficiency of Regional Financial Industry

The transnational development of financial institutions promotes the exchange of information technology between countries, forms demonstration and technology diffusion effects and other functions, optimizes the innovative development of regional financial technology, and improves regional financial efficiency. However, financial institutions will expand their business abroad, which increases their competitiveness with local financial institutions. It is necessary to motivate financial institutions to improve their viability by improving service quality, reducing costs, improving financial efficiency, and carrying out financial innovation, to improve the overall efficiency of the regional financial system.

14.2 Balance of Payments Deficit and Trade War

14.2.1 Theoretical Basis of the Balance of Payments Deficit

14.2.1.1 Meaning of the Balance of Payments Deficit

The balance of payments is a comprehensive reflection of a country's international capital flows, which provides an important basis for national macro-control. Through analysis of the current situation and causes of an imbalance in payments and its impact on China, we can regulate domestic development and the policy of opening up to the outside world, promote the internal and external balance of the national economy, and achieve high-quality national economic development.

(i) International balance of payments

Balance of payments refers to the systematic record of all economic transactions between the residents of a country and the residents of other countries within a certain period. The balance of payments reflects the inflow and outflow of a country's international funds, and the international income is the inflow of funds, which includes three points. First, the balance of payments must be an economic transaction between residents and non-residents of a country, on the contrary, it does not belong to the reflection of the balance of payments. Second, the content reflected in the balance of payments is based on economic transactions, which refers to the transfer of economic value from one economic unit to another. Third, the balance of payments is a concept of flow. The balance of payments reflects economic transaction activities within a certain period; thus, it belongs to the flow value.

(ii) Deficit in the balance of payments

A balance of payments deficit means that the net balance of payments in a country is negative; thus, net exports are less than net capital inflows. In the balance of payments account, the amount of the credit item is greater than that of the debit item. The items of the balance of payments mainly include current and capital accounts, which are spontaneous transactions. If spontaneous trading cannot be balanced, financial authorities carry out regulatory trading, using domestic gold, foreign exchange, and other international capital reserves for adjustment, or through financing to foreign central banks and international financial institutions to compensate for the gap between income and expenditure caused by spontaneous transactions.

14.2.1.2 Caliber of the Balance of Payments

The balance of payments deficit is generally the difference in the total, and the adjustment of the balance of payments deficit is often analyzed according to the structural caliber of the balance of payments. According to different calibers, the balance of payments is divided into four main categories, as described below (Fig. 14.2).

(i) Balance of trade

The balance of trade is the net amount between the import and export of a country's goods over a certain period. When the net value is positive (i.e., exports are larger than imports), it is a trade surplus; otherwise, it is a trade deficit. Trade balance is an integral part of the balance of payments and cannot represent the balance of payments of a country as a whole. For some countries, the total amount accounts for most of the total balance of payments, which reflects the actual competitiveness and foreign exchange earning ability of a country's industrial market.

(ii) Balance on current account

Current account income and expenditure includes a country's income and expenditure on goods and services, income, and current transfer income and expenditure, in which

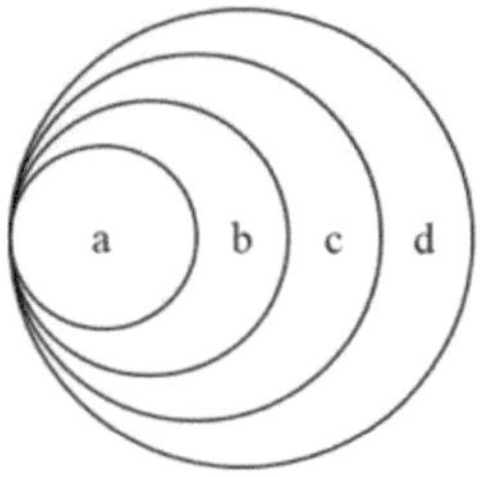

a. balance of trade (income andexpenditure on tangible goods)
b. balance on current account = a +revenue and expenditure of intangiblegoods, income and current transfer
c. basic balance = b+ long-termcapital flows
d. overall balance = c+ short-termcapital flows

Fig. 14.2 Schematic diagram of balance of payments of various calibers. ***Data source* Jiang Boke. New edition of international finance (in Chinese). Fudan University Press. 2012

goods and services income and expenditure is the main body of current account income and expenditure. It reflects a country's import and export situation and the comprehensive competitiveness of the primary, secondary, and tertiary industries. Therefore, its index has become an important basis by which a country can adjust its balance of payments policy.

(iii) Basic balance

The basic balance is the sum of the current account balance and the long-term capital account difference. Long-term capital account refers to long-term current financial assets with a repayment period of more than one year in securities investment, direct investment, and other investment accounts. It reflects the status and strength of a country in international economic exchanges and is an important indicator for countries with large long-term capital flows to observe and judge the balance of payments.

(iv) Overall balance

The overall balance is based on the basic difference, plus the difference between the short-term capital account difference and the net error and omission item. This index measures the pressure of a country's international capital reserves, comprehensively reflects the surplus or deficit of the balance of payments, and is an indicator to comprehensively measure and analyze the balance of payments.

14.2.1.3 Reasons for the Formation of the Balance of Payments Deficit

(i) Monetarism

Monetarists believe that the balance of payments deficit is purely a monetary phenomenon caused by an excess supply of goods, resulting in the increase of monetary demand less than the increment of domestic credit level, the loss of foreign exchange reserves, and the formation of the balance of payments deficit. Under the fixed exchange rate system, if the domestic demand for money is too large, the local currency exchange rate rises relative to foreign countries, and the inflow of funds

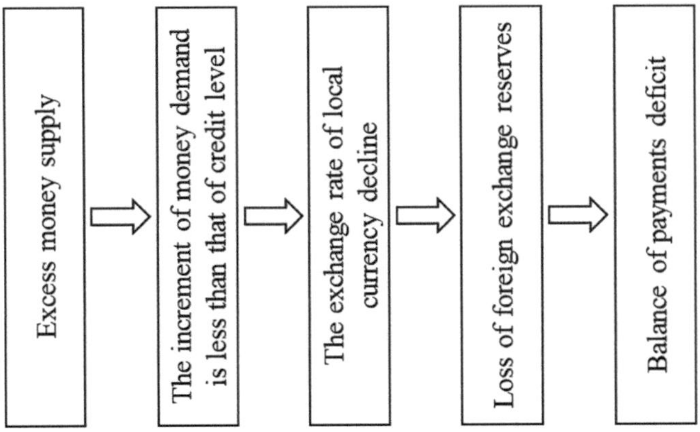

Fig. 14.3 Formation of balance of payments deficit

forms a foreign exchange surplus. In the case of constant demand for money, fluctuations in the domestic credit level are reflected in the changes in foreign exchange reserves, and a high level of credit creation will lead to a loss of foreign exchange reserves and the formation of a foreign exchange deficit (Fig. 14.3).

(ii) Structuralism

The early structuralists formed a current account deficit based on the belief that the product structure did not adapt to changes in the global market. A balance of payments deficit can occur for three main reasons. First, in the competition of a country's industrial structure in the global market, the income elasticity of its export demand is small, resulting in a trade deficit over time. Second, the growth of market economy is slow, the competitiveness of products is relatively weak, and product exports decline. Third, low efficiency leads to high export costs that restrict export profits and create a low balance of payments deficit.

(iii) Domestic absorption

Absorption theory, also known as expenditure analysis, has been proposed by the International Monetary Fund using the Keynesian income and expenditure formula, which considers the impact of domestic total income and expenditure on the balance of payments. Its basic theory is as follows:

According to Keynesian macroeconomics, the relationship between national income and national expenditure is:

National income (Y) = national expenditure (E).

In closed economic conditions:

National expenditure (E) = consumption (C) + investment (I).

In an open economy:

National expenditure (E) = consumption (C) + investment (I) + [export (X)-import (M)].

By moving the identity above, you can obtain the following:

$$X - M = Y - (C + I) \tag{14.1}$$

where $X - M$ is the trade balance, which can be regarded as the balance of payments. $C + I$ is the total domestic expenditure, which represents the total national income spent in China, called domestic absorption (A).

14.2.1.4 Impact of the Formation of the Balance of Payments Deficit

(i) Advantages of balance of payments deficit

The balance of payments deficit is not always negative. Imports in international trade can increase effective supply while also promoting the adjustment of domestic industrial structure. As a result, there will be a short-term account deficit in the balance of payments.

(ii) Disadvantages of the balance of payments deficit

If the local currency exchange rate is floating down, the balance of payments deficit will lead to the decrease of foreign exchange supply in the domestic foreign exchange market, increase in demand, floating of the domestic currency exchange rate, flight of international capital, and an increase in cost. If the government uses foreign exchange reserves and buys local currencies to stabilize the exchange rate, it will deplete reserves and reduce the supply of local currency. It will further lead to a rise in domestic interest rates and the devaluation of the local currency. If the government intervenes by tightening fiscal and monetary policies, the balance of payments deficit will affect economic development speed.

Crowding out domestic production leads to increased unemployment. The balance of payments deficit leads to excessive imports of foreign merchandise trade, crowding out domestic production and occupying domestic resources, resulting in an increase in domestic unemployment and a slowdown in economic development. Through the multiplier effect of foreign trade, we can obtain the following:

$$\Delta Y = K[\Delta I + (\Delta X - \Delta M)] \tag{14.2}$$

where ΔY is national income, ΔI is the increment of investment and consumption, $(\Delta X - \Delta M)$ is the increment of net export, and K is the multiplier of foreign trade. Under the condition that the marginal propensity to consume remains unchanged, the increase in foreign trade will increase the income and consumption of relevant departments through the industrial chain of the national economy, which will eventually have a doubling effect on national economic growth and income.

Regarding entering a debt crisis, to pay a large balance on a payment's deficit, a country will incur foreign debt from international financial organizations or developed countries for a long period. The heavy interest burden causes a country to fail to repay its debts, which may lead to a debt crisis or even a financial crisis.

14.2.2 Analysis of the United States Trade Deficit

The persistent current account deficit in the United States has always been the main manifestation of the imbalance in the United States' international trade (Siddiqui, 2019). This problem has lingered in the United States and become increasingly serious. As an economic power, the trade deficit of the United States has a far-reaching impact on the world economy and has become the focus of governments and scholars.

14.2.2.1 Summary of the United States' Trade Deficit

(i) A long-standing huge trade deficit

Since the 1970s, the United States has experienced a persistent trade deficit. According to the data released by the US Bureau of Economic Analysis, the US trade deficit continues to expand (Fig. 14.4).

(ii) Trade deficit in goods and trade surplus in services

The United States is well-known to be a global importer of goods and exporter of services. In the foreign trade of the United States, the export of goods and services

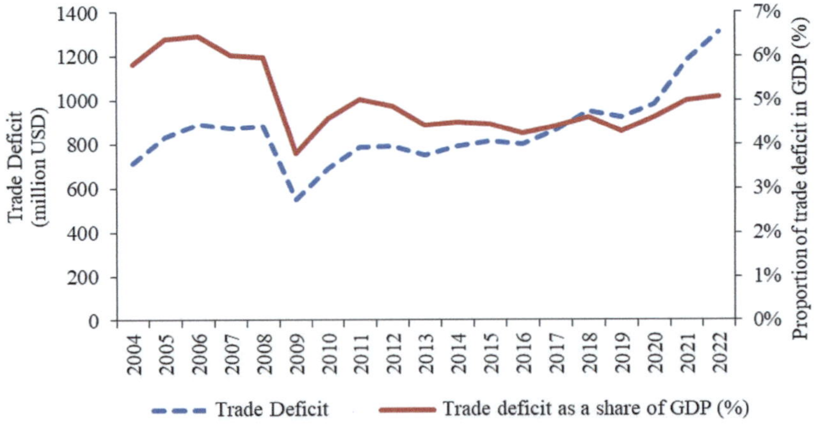

Fig. 14.4 The size of the US trade deficit and the proportion of the US trade deficit to GDP. **Data source* U.S. Bureau of Economic Analysis

continues to rise. The foreign trade of goods continues to run at a deficit, whereas the foreign trade of services continues to have a surplus. The merchandise trade deficit of the United States shows a continuous growth trend (Fig. 14.5).

(iii) Regional distribution of foreign trade deficit is uneven

As Fig. 14.6 shows, the US trade deficit is unevenly distributed among its trading partners. The export trading partners of processing trade with lower added value are the main sources of the US trade deficit, and China and Mexico are typical representatives.

The trade deficit of the United States with trading partners processing lower value-added products is rapidly rising. The lower value-added processing trade exports represented by the United States to China and Mexico, and the higher value-added high-end manufacturing exports represented by Germany and Japan (Li, 2004), show

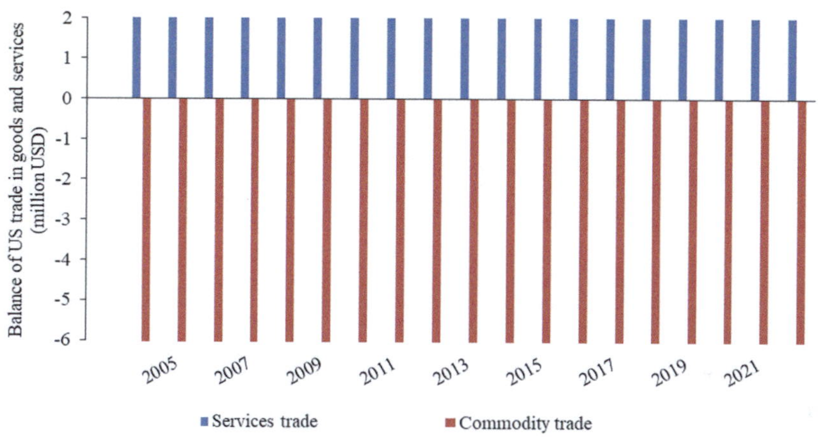

Fig. 14.5 Balance of US trade in goods and services. *Data source: U.S. Bureau of Economic Analysis

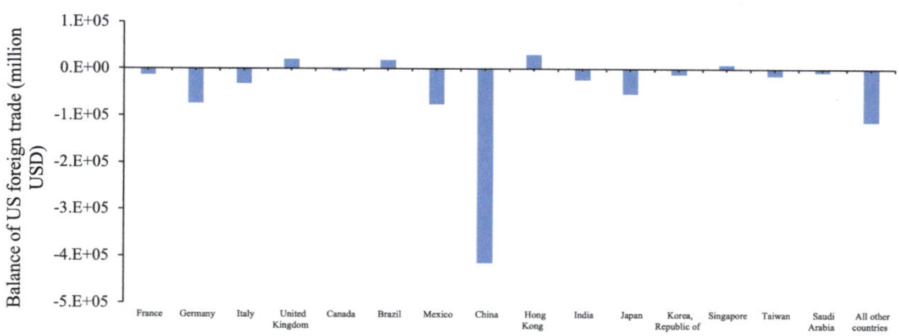

Fig. 14.6 Schematic diagram of US foreign trade balance in 2018. **Data source* U.S. Bureau of Economic Analysis

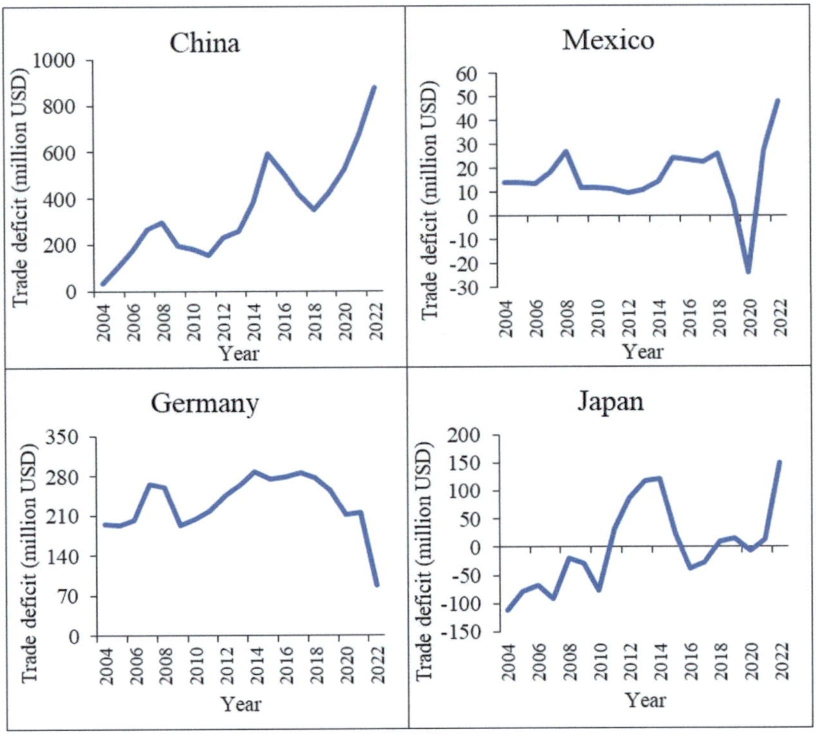

Fig. 14.7 Comparison of the trade deficit between the United States and four countries. **Data source* U.S. Bureau of Economic Analysis

that the trade deficit caused by the United States to the lower value-added processing trade exports continues to increase at a faster rate (Fig. 14.7).

14.2.2.2 Causes of the US Trade Deficit

(i) An economic structure with a low savings rate, high investment, and high consumption

According to the macroeconomics identity, the domestic output is Y = C + I + G + (X−M) from the perspective of expenditure and Y = C + S + T from the income point of view. When the national income is balanced, X−M = (S−I) + (T−G), that is, the net export of foreign trade is equal to the sum of the difference between personal savings and investment and the difference between government expenditure and taxes.

In terms of personal consumption and savings, long-term personal savings in the United States are low; however, consumer spending is very high. Since 2008, personal savings have been much smaller than personal consumption expenditure. In 2018, personal savings increased by 5.16%, whereas personal consumption expenditure increased by 17.41%.

Total domestic investment in the United States has continued to rise, growing by 46.44% from 2008 to 2018. Domestic investment continues to increase, exceeding savings, resulting in an investment–savings gap and an increasing negative value (S-I).

The US government implements a positive fiscal policy, resulting in a fiscal deficit. Since the 1980s, the US government has been implementing an expansionary fiscal policy and sharply reducing personal income tax. In the twenty-first century, the Bush administration implemented further tax cuts. In 2017, the Trump administration implemented tax reforms and large-scale tax cuts to entice multinationals and other companies to return to the United States and expand investment (Table 14.1).

The long-term savings rate in the US shows the opposite trend to that of net trade exports (Fig. 14.8).

(ii) As an international currency, the United States dollar faces the Triffin Dilemma

After World War II, the Bretton Woods system implemented the international monetary system centered on the US dollar, which strengthened its dominant position in the international monetary system and made it the standard and savings currency worldwide. The US dollar accounts for the majority of the global foreign exchange

Table 14.1 Changes in US domestic consumption and savings from 2008 to 2018 (in billion yuan)

Year	GDP	Personal savings	Personal consumption expenditure	Total domestic investment	Government recurrent revenue	Government recurrent expenditure
2008	14,712.8	544.9	9976.3	2477.6	4125.0	4977.0
2009	14,448.9	666.5	9842.2	1929.7	3696.6	5286.8
2010	14,992.1	740.9	10,185.8	2165.5	3933.2	5565.7
2011	15,542.6	849.8	10,641.1	2332.6	4130.6	5647.7
2012	16,197.0	1107.6	11,006.8	2621.8	4312.2	5673.6
2013	16,784.9	801.4	11,317.2	2826	4834.5	5737.8
2014	17,527.3	970.8	11,822.8	3044.2	5054.4	5896.7
2015	18,224.8	1048.8	12,284.3	3223.1	5288.2	6078.5
2016	18,715.0	958.8	12,748.5	3178.7	5335.4	6259.2
2017	19,519.4	1030.9	13,312.1	3370.7	5481.7	6454.5
2018	20,580.2	1210.4	13,998.7	3628.3	5537.7	6786.6
2019	21,521.40	4828.75	14,417.61	774.97	945.67	1198.18
2020	21,322.95	6265.35	14,206.23	816.47	1016.08	1468.85
2021	23,594.03	6038.25	16,042.96	826.78	1188.57	1517.41
2022	25,744.11	4979.41	17,511.75	876.76	1234.68	1542.93
2023	27,360.94	5559.95	18,570.64	992.51	1240.66	1608.14

* *Data source* U.S. Bureau of Economic Analysis

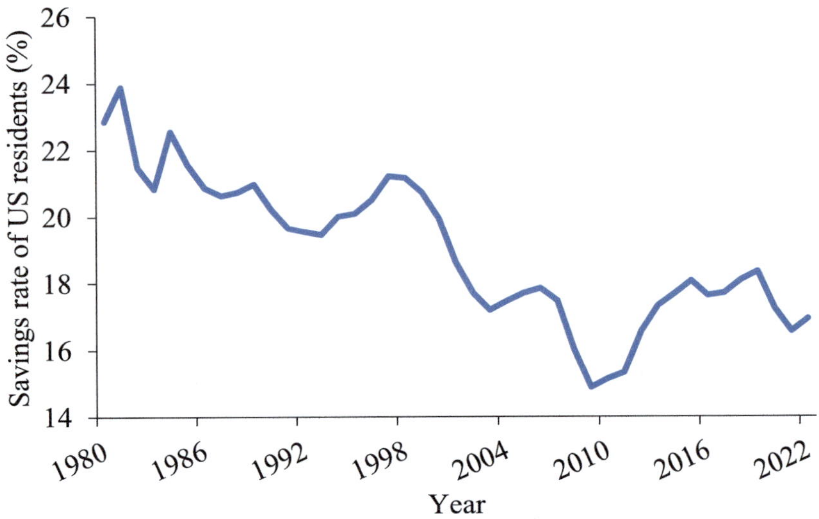

Fig. 14.8 The savings rate of US residents from 1965–2017. * *Data source* U.S. Bureau of Economic Analysis

reserves. Although the US dollar reserves have fluctuated since 2010, they have basically remained above 60%, reaching 61.68% in 2018. With the development of international trade, the demand for foreign exchange reserves increases, and the demand for the US dollar increases accordingly; thus, the trade deficit continues to form (Fig. 14.9).

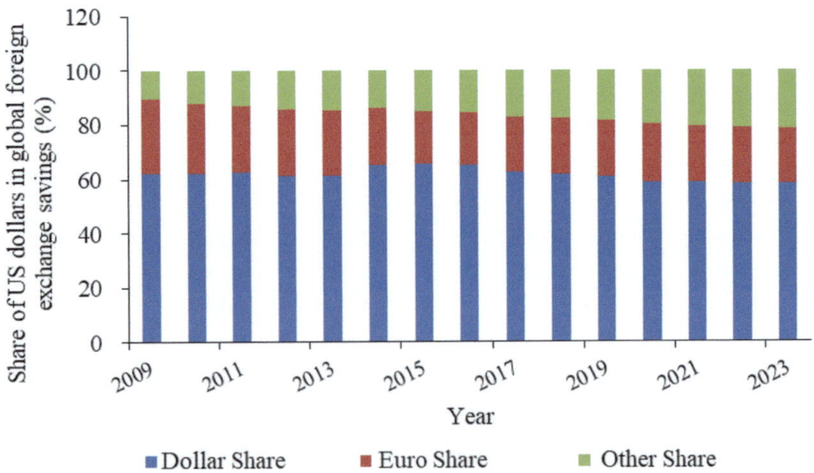

Fig. 14.9 Share of US Dollars in global foreign exchange savings from 2000 to 2018. * *Data source* International Monetary Fund

(iii) Influence of the United States foreign trade policy

US foreign trade policy is the main factor affecting its trade deficit. The choice of trade policy is not only the response of the world's international and domestic political situation but also how the United States government has chosen to deal with external shocks (Nanto, 2011; Irwin, 2023). The United States mainly focuses on the manufacture of high-tech industries; however, after the September 11 terrorist attacks, the United States began to implement export control policies on high-tech products and military and civilian supplies. Subsequently, the United States had a deficit in high-tech products, and the industries with export advantages did not fully utilize them.

(iv) Influence of other factors

The United States imports some textile and mechanical and electrical products with low technological content that create serious pollution abroad, focuses on the production of high-tech industries, and imports a large number of low-end, value-added manufactured goods. It has not only made up for the domestic demand for basic consumer goods and national economic development but also caused a supplementary domestic trade deficit.

14.2.2.3 Impact of the United States Trade Deficit

(i) Impact of the US trade deficit on the United States

The historical data show that the long-term international trade deficit did not affect the status of the United States as a global trading power, nor did it lead to an economic recession or financial crisis. However, Trump began a trade war on the grounds of trade protection, believing that the huge trade deficit affects the international competitiveness of US industry and weakens the global economic strength of the United States (Fig. 14.10).

(ii) Impact of the US trade deficit on the world

The US trade deficit is one of the main channels for exporting liquidity to the world while meeting domestic demand. As the global currency, the US dollar provides both domestic money and an international monetary supply for the world. Owing to the loose fiscal policy of the United States, the monetary supply continues to increase, which not only meets the domestic currency demand but also provides liquidity to the world through the trade deficit.

14.2.3 Impact of the International Trade War

International trade not only promotes the development of the domestic market but also leads the competition to the domestic economy. With the increasing speed of the

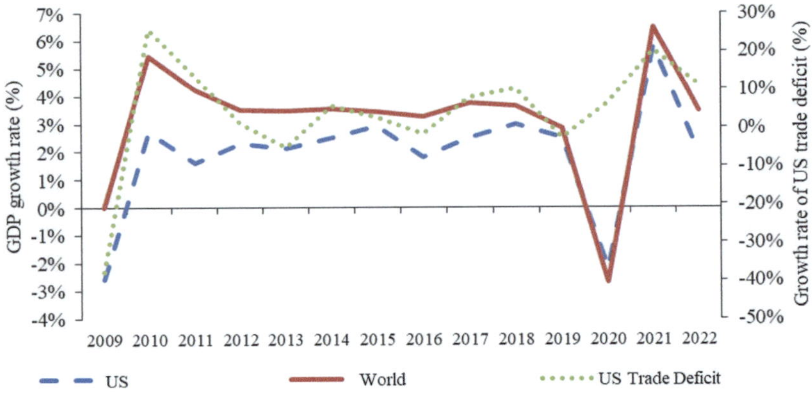

Fig. 14.10 Change trend in GDP growth rate and trade deficit between the world and the United States from 2009 to 2018. **Data source* International Monetary Fund

globalization of international trade, trade frictions among countries become increasingly frequent. Owing to the mutual flow of global resources and the investment and financing in international finance, the economic ties between countries are growing closer. Trade wars not only represent friction between two countries but also affect other countries that have trade relations with the countries experiencing trade friction (Obstfeld and Taylor, 2017), and even affect the stability of the global economy.

14.2.3.1 Trade War Threatens Global Trade Stability

(i) Violation of the basic principles of international trade

International trade means that countries export superior products to the world through their own resource endowments and industrial advantages and carry out international division of labor to improve global well-being. In contrast, a trade war will restrict the entry of goods from other countries through tariff and non-tariff barriers to maintain a country's own weak industries. As the industrial chain system constructed by the global international division of labor in the twenty-first century has been formed, the economic ties between countries growing closer, and the occurrence of a trade war will violate the basic principles of international trade. All countries and multinational corporations will be affected by disrupting the global industrial and value chains and affecting the cost, prices, and circulation of goods on a global scale.

(ii) Trade frictions caused by unilateral trade protectionism

With the development of the world, the global economy still faces greater potential risks. The occurrence of a trade war aggravates the tension in global trade and affects global economic recovery. After World War II, the United States actively built a free trade system and signed the General Agreement on Tariffs and Trade, forming a global trading system centered on the United States.

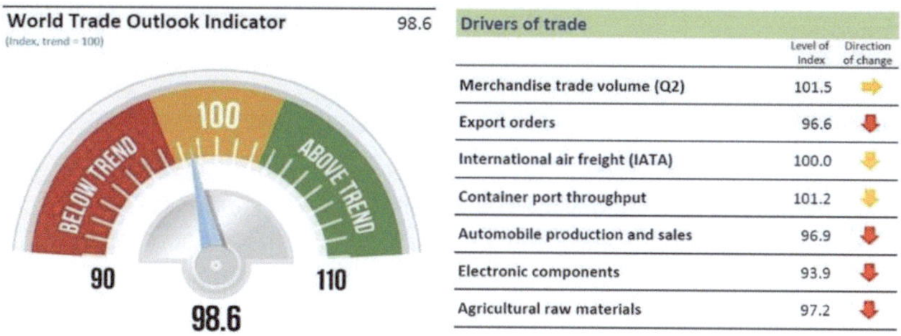

Fig. 14.11 World trade outlook indicator (WTOI) for the first quarter of 2019. **Data source* World Trade Organization

14.2.3.2 Current Tense Trade Situation

In May 2019, the WTO released its latest first-quarter World Trade Outlook Indicator (WTOI) of 96.3 (Fig. 14.11), the lowest level since 2010. Recently, influenced by the continuous global tariff policy implemented by the United States, the WTO stated that the current international trade situation is tense and global trade growth is expected to slow down.

14.2.4 Sino-US Trade War

In recent years, the US trade deficit has grown increasingly large, and the pressure of trade protectionism has shifted from the original free trade policy to a unilateral trade policy. In particular, after former President Trump took office, he focused on implementing a unilateral trade policy and erected barriers to trade and investment in China on the grounds of a huge trade deficit, creating bilateral trade frictions between China and the United States (Cerutti et al., 2019), which escalated into a trade war.

14.2.4.1 Course of Occurrence of the Incident

(i) Beginning of trade friction

After Trump took office, he advocated trade protectionism and implemented trade sanctions and investigations successively. On August 4, 2017, the United States Trade Representative was authorized to launch a 301 investigation against China in accordance with the Trade Act of 1974 to create trade friction with China. This was followed by a survey of steel and aluminum products (a 232 investigation) and a full tax of 25% and 10%, respectively. China's Ministry of Commerce decided to extend anti-dumping duties and strictly examine the imports of hydroquinone from the United States and Japan, thus intensifying trade frictions between China and the United States.

(ii) Trade war begins

On March 23, 2018, the United States imposed tariffs on about 60 billion yuan of Chinese imports to the United States based on the "301" survey, restricted Chinese investment in the United States, and imposed 25% tariffs on aerospace, information and communications technology, and other high-end technology products. In response, China took countermeasures, planning to impose tariffs on some US imports, and Sino-US trade war began.

(iii) Negotiation and more negotiation

On May 3 and 4, 2018, US Treasury Secretary Mnuchin led a delegation to China for the first round of consultations on economic and trade issues. The two sides exchanged views on expanding US exports to China, bilateral trade in services, and resolving tariff and non-tariff measures and reached a consensus in some areas. Repeated consultations between China and the United States followed; however, on May 29, ten days after the issuance of the joint statement, the United States overturned the consensus of the consultation, saying that it would continue with the plan to increase tariffs, which brought the negotiation process between the two countries to a standstill and re-escalated the dispute.

(iv) Trade disputes escalate again

On June 15, 2018, the United States introduced a 25% tariff on the total value of Chinese products worth US$50 billion, and the tariff system Committee of the State Council of China decided to impose a tariff countermeasure of the same scale and intensity on the United States. Subsequently, the United States successively issued a list of goods taxed at US$16 billion and U$200 billion against China. China then also released a list of US$34 billion, US$16 billion, and US$60 billion taxed goods, and the Sino-US trade dispute reached a white-hot stage (Figs. 14.12 and 14.13).

(v) Compromise between the two sides in the trade war

The deterioration of Sino-US trade war gradually had a negative impact on the economies of both countries, and China and the United States held many consultations. On December 1, 2018, the heads of state of China and the United States reached a consensus on Sino-US economic and trade issues during the G20 Argentina summit and signed a 90-day interim extension agreement (Table 14.2).

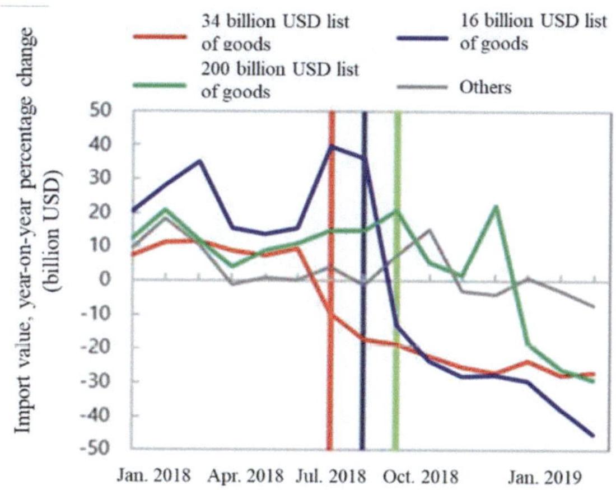

Fig. 14.12 Imports of three additional groups of listed goods from China levied by the United States. **Data source* Cerutti et al. The impact of Sino-US trade tensions. (International Monetary Fund, 2019.5.23)

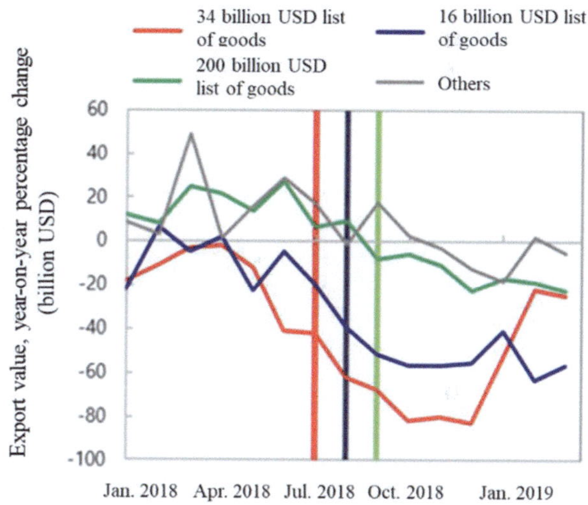

Fig. 14.13 Exports from the United States among the three additional groups of listed commodities levied by China. **Data source* Cerutti et al., The impact of Sino-US trade tensions. (International Monetary Fund, 2019.5.23)

14.2.4.2 Main Reasons the United States Launched a Trade War Against China

(i) The widening US trade deficit with China

The US has a huge trade deficit with China. Since the 1970s, the United States has followed a trade deficit trend showing increased volatility. In the total trade deficit of the United States, the deficit in goods between the United States and China is increasing annually (Fig. 14.14).

The United States has an industrial structure dominated by the production of high-tech products; however, it has issued the Export Control list to China, restricting the export of 20 categories of high-end technology products, such as aero-engines, to

Table 14.2 Timetable for the course of Sino-US trade war

Stage	Date	Initiator	Event
Beginning of trade friction	2017.08.14	United States	Trump signed an executive memorandum authorizing the United States trade representative to launch an investigation and review of China in accordance with section 301 of the Trade Act of 1974
	2017.08.18	United States	The United States officially launched the "301 investigation" against China
	2018.01.23	United States	Trump signed a bill imposing tariffs of 30% and 50% on all imported solar cells and modules, as well as large washing machines, which was to be implemented on February 7
	2018.02.23	United States	The US Department of Commerce released a safety investigation into US steel and aluminum products (the 232 investigation)
	2018.03.08	United States	A comprehensive tax is imposed on imported steel and aluminum products at 25% and 10%, respectively
	2018.03.15	China	China's Ministry of Commerce decided to extend the anti-dumping duties on imports of photographic paper from the European Union, the United States, and Japan, to be implemented on March 23, and planned to repeatedly review the imports of hydroquinone from the United States and Japan
Trade war begins	2018.03.23	United States	Based on the "301 investigation," tariffs were imposed on imports of approximately US$60 billion of Chinese goods, restrictions on Chinese investment and mergers and acquisitions in the United States, and tariffs on aerospace, information and communication technology, machinery, and other products
		China	China planned to impose tariffs on some US imports in response to the "List of discontinued concession products for US imports of steel and aluminum products 232," and Sino-US trade war began

(continued)

Table 14.2 (continued)

Stage	Date	Initiator	Event
	2018.04.02	China	The Customs Tariff Commission of the State Council of China announced the decision to impose tariffs equivalent to about US$3 billion on 7 categories and 128 imported goods from the United States
	2018.04.03	United States	Based on the "301" survey, the United States announced a 25% tariff on imports originating in China, involving approximately US$50 billion in Chinese exports to the United States
	2018.04.04	China	China decided to impose a 25% tariff on 14 categories of products equivalent to approximately US$50 billion, including soybeans imported by the United States
	2018.04.16	United States	The US Department of Commerce banned American companies from selling components to ZTE in the next seven years, sparking a science and technology war in Sino-US trade war
	2018.04.17	China	China's Ministry of Commerce announced an anti-dumping investigation against sorghum imported from the United States
Negotiation and more negotiation	2018.05.03–2018.05.04	United States	China and the United States held the first round of consultations on economic and trade issues in Beijing and reached a consensus in some areas
	2018.05.15–2018.05.18	China	Chinese Vice Premier Liu went to the United States for the second round of economic and trade consultations, where the two sides reached an agreement on the economic and trade framework and told the media that they were willing to "suspend" the trade war
	2018.05.29	United States	Ten days after the release of the joint statement, the United States overturned the consensus of the consultation, saying that it would continue with its plan to impose tariffs

(continued)

Table 14.2 (continued)

Stage	Date	Initiator	Event
	2018.06.02–2018.06.03	China	Chinese Vice Premier Liu and US Secretary of Commerce Ross completed a new round of consultations on trade issues between the two countries at the Diaoyutai State Guesthouse in Beijing
Trade war escalates again	2018.06.15	United States	The United States introduced a 25% tariff on Chinese products worth US$50 billion
	2018.06.16	China	The Customs Tariff Commission of the State Council of China decided to impose a 25% tariff on approximately US$50 billion
	2018.07.06	United States	The United States formally imposed a 25% tariff on Chinese goods worth US$34 billion on the first list
		China	China issued a list of additional tariffs on 545 US goods worth approximately US$34 billion
	2018.07.10	United States	The United States announced an additional 10% levy on Chinese imports of US$200 billion
	2018.08.02	United States	The United States stated it would raise tariffs on imported goods worth US$200 billion to 25%
	2018.08.03	China	In accordance with the law, China imposed tariffs on imports of approximately US$60 billion from the United States at different tax rates
	2018.08.07	United States	The Office of the Trade Representative announced a 25% tax rate on Chinese goods worth US$16 billion, which was to be implemented on the 23rd of the same month
	2018.08.08	China	The Customs Tariff Commission of the State Council decided to impose tariffs on imports originating in the United States of approximately US$16 billion, which was formally implemented on the 23rd of the same month

(continued)

Table 14.2 (continued)

Stage	Date	Initiator	Event
	2018.08.22–2018.08.23	China	Wang Shouwen, Vice Minister of Commerce and Deputy China International Trade Representative, led the Chinese delegation to exchange economic and trade issues with the United States in Washington, DC
	2018.09.18	United States	The White House imposed a 10% tariff on Chinese imports worth US$200 billion beginning September 24th
		China	The Customs Commission of the State Council issued a circular imposing a tariff of 10 or 5% on US$60 billion worth of goods
Both sides compromise and negotiate	2018.12.01	United States	The heads of state of China and the United States reached a consensus on Sino-US economic and trade issues during the G20 summit in Argentina, and Trump signed a 90-day interim extension agreement
	2018.12.05	United States	China and the United States hold the eleventh round of consultations
-	From 2019 to the present	United States	China and the United States have held many consultations to discuss trade balance, technology transfer, intellectual property protection, and non-tariff barriers

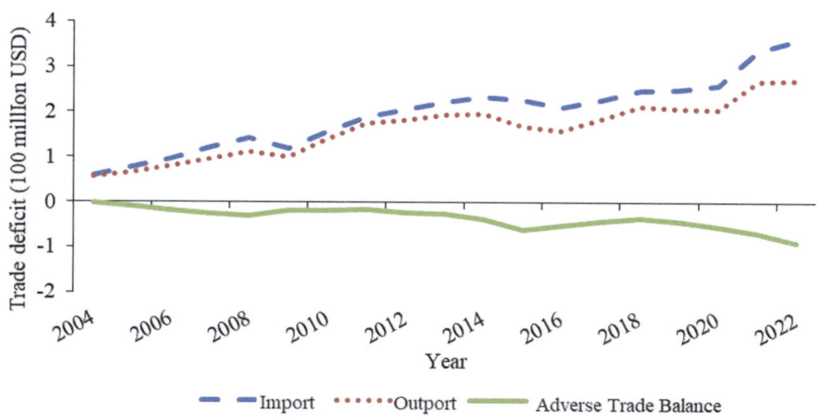

Fig. 14.14 US trade deficit with China in goods and services from 2004 to 2022. *Data source* U.S. Bureau of Economic Analysis

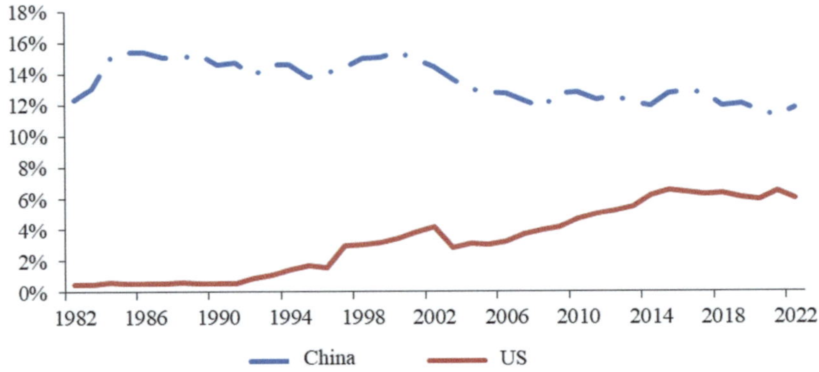

Fig. 14.15 Ratio of Sino-US trade in goods and services to the world total between 1982–2022. **Data source* United Nations Conference on Trade and Development (UNCTAD)

China. This restricted the export of industries with advantages; however, a large number of low-end value-added manufactured products were imported to China, resulting in a huge trade deficit.

(ii) Strengthen the maintenance of the international dominant position of the United States

The export volume of Sino-US trade in goods and services is the result of a comparison of the international status of Sino-US trade in goods and services. The global pattern of Sino-US trade in goods and services shows that, for quite some time, the proportion of US exports of goods and services worldwide has been much higher than that of China (Fig. 14.15).

(iii) Concerns in the United States about the hollowing out of domestic industries

In the 1950s, the United States adjusted the industrial structure of domestic natural resources, environmental protection, and other factors; transferred steel, textile, and other traditional industries to Japan, Germany, and other regions; and focused on the development of technology-intensive industries domestically. In the 1980s, some industries, including automobiles and electronics were transferred abroad, and only the high-end technology industry was retained in China, and the financial industry was vigorously developed. The domestic economy is "virtualized," and the industrial structure is gradually "hollow."

14.2.4.3 Impact of Sino-US Trade War

(i) Consumers and producers in China and the United States have suffered the most

The occurrence of Sino-US trade war will certainly have an impact on the economic development of both China and the United States, restricting the industrial development of the goods on the tax list issued by both sides, of which consumers and producers in China and the United States will experience the greatest losses. The increase in tariffs will lead to a rise in prices, the tariff costs borne by importers will be passed on to consumers, and consumers will become the bearers of the final tariff costs.

(ii) Limited impact on trade balance between China and the United States

Since the establishment of Sino-US trade barriers, the import and export balance between China and the United States has not changed much. The establishment of trade barriers between China and the United States will have an impact on the trade deficit, but the long-term trade deficit of the United States is not only the main manifestation of the global international monetary status but also the impact of the global international division of labor pattern. Therefore, the occurrence of a trade war has a limited impact on the trade balance between China and the United States.

(iii) Serious effect the global economic and trade patterns

The tariffs imposed by China and the United States will not only have a significant impact on countries but expand to all trade between China and the United States and the global economic and trade patterns. The impact of unilateral trade protection will have a more far-reaching negative impact on the global economy; especially, the increase in trade barriers will affect the global international division of labor and disrupt the global supply chain. Reducing domestic competition also affects the spread of new technologies and the exchange of ideas, leading to weakened global production efficiency.

14.3 Impact of the Global Financial Crisis on Developing Countries

14.3.1 Evolution of the Global Financial Crisis

14.3.1.1 Occurrence of the Subprime Crisis

The subprime crisis refers to the subprime mortgage crisis. Owing to the sharp increase in defaults and tightening of credit problems in the subprime housing credit industry in the United States, subprime lenders became bankrupt, investment funds were forced to close, and the stock market fluctuated violently, resulting in a financial crisis. A subprime mortgage is a type of subprime mortgage loan, which is made by commercial banks and other lending institutions to individuals with poor credit unstable sources of income. This type of loan not only has a high return but also has a substantial amount of credit risk.

The trigger of the subprime mortgage crisis in the United States was the economic recession caused by the "Dot-com Bubble Burst" in March 2000. The terrorist attacks on September 11, 2001 aggravated the recession of the American economy and led to a spread in market panic. The low-interest rate policy for four consecutive years has contributed to the continued rise in the core consumer price index and the rapid recovery of the US economy. As a result, the real estate market was booming and house prices were rising rapidly. However, with the implementation of the easing policy, US inflation rose faster than expected, and the Federal Reserve increased interest rates 17 times in a row to reduce inflation and tighten liquidity. The growth rate of house prices began to slow or even reverse. After peaking in 2006, house prices began to fall, and the use of subprime loans for financing was no longer effective.

14.3.1.2 Formation of the United States Financial Crisis

(i) Subprime crisis to Bond Market crisis

In the United States, with the development of asset securitization, investment banks packaged and securitized the subprime loan assets sold by credit institutions into new financial derivatives and trade in the capital market. Owing to the asset securitization of subprime loans, the occurrence of the subprime crisis also meant that more financial institutions were involved. The rise of subprime default rate led to an increase in the credit risk of subprime loans, and the credit rating of subprime bonds was downgraded, resulting in panic selling. The market value of asset-backed securities made by securitization based on subprime loans shrunk (Croke et al., 2005), resulting in serious losses in the hedge and investment funds that had bought the securities.

(ii) The deepening of financial crisis

The subprime mortgage crisis spread to the financial market, large financial institutions closed down and went bankrupt, and the financial crisis deepened further. As commercial banks and other financial institutions reported losses in succession, the Federal Reserve was forced to intervene. Bear Stearns declared bankruptcy in March 2008, followed by Lehman Brothers, the fourth-largest investment bank, on September 15, 2008. The Federal Reserve mistakenly determined that the collapse of Lehman Brothers would not cause a systemic crisis and refused to provide the company with a bail out. Subsequently, large financial institutions such as Merrill Lynch, the third largest investment bank; American International Group, the largest insurance company; and Washington Mutual Bank filed for bankruptcy, were acquired cheaply, or were taken over. Market confidence completely collapsed, panic continued to spread, and the subprime crisis evolved into a full-scale financial crisis.

(iii) The financial crisis shifts to the economic crisis

Owing to the impact of the financial crisis, the real economy of the United States began to show weakness. The credit crunch increased the rate of overdue defaults, problems grew with a large number of junk bonds and credit default swaps (CDSs), the US real estate market showed a steep decline (Fig. 14.16).

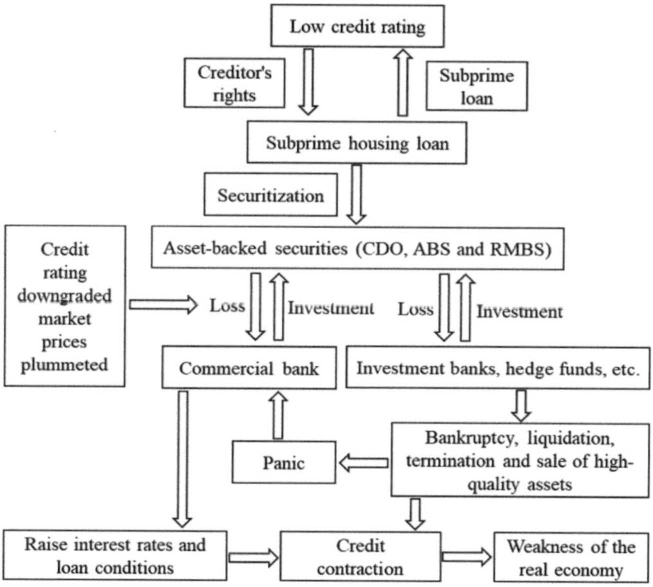

Fig. 14.16 Evolution of the American financial crisis. *Note* CDO: collateralized debt obligation; ABS: asset-backed securitization; RMBS: residential mortgage-backed securities

14.3.1.3 Spread of the Global Financial Crisis

With the collapse of major financial institutions in the United States, the financial crisis ran out of control, further spreading worldwide and becoming a global financial crisis. Owing to the relevance of the world economy, banks in Europe, Asia, and other places with close ties with the United States were affected, capital difficulties occurred, related businesses showed a large-scale contraction, global liquidity dried up, and global stock markets were hit hard. Several countries also faced a serious financial crisis because of the global financial crisis. Global market confidence collapsed, the Icelandic government declared bankruptcy because of heavy debt, and Lithuania, Latvia, and Estonia faced the threat of a collapse of the national banking system.

14.3.2 *Impact of the Global Financial Crisis on Developing Countries*

14.3.2.1 Multidimensional Impact of the Global Financial Crisis on Developing Countries

(i) Import and export international trade has dropped sharply

As a result of the financial crisis, the demand for commodities in developed countries decreased, whereas the exports of developing countries also plummeted. The impact of the financial crisis led to weakness in the real economy and a decline in market demand. Global international trade has declined since the global financial crisis in 2008 (Table 14.3).

(ii) The economic growth rate has fallen sharply

The impact of the global financial crisis on the economies of developing countries was mainly caused by international trade exports and investment. The United States and other developed countries were affected by the financial crisis, resulting in a decline in domestic demand for consumable products such as resources and energy. This led to a significant decline in production efficiency and economic growth in developing countries that rely on commodity production and exports to drive economic growth (Cömert & Ugurlu, 2015) (Table 14.4).

(iii) The shortage of investment funds leads to the second round of effects

The previous round of economies in developing countries came largely from investment. As a result of the financial crisis, global investors pulled their money out of the financial markets. The real economy lacked funding sources, its own available funds were limited, and many ongoing projects could not be carried out because of the rupture of the capital chain. Global market demand plummeted, prices fell, completed projects were ineffective, funds could not be returned, and profits could not be obtained.

Table 14.3 Growth rate of trade in goods and services of developing countries in 2007–2010

	2007 (%)	2008 (%)	2009 (%)	2010 (%)
Global international trade				
Commodity export	15.61	15.18	– 22.25	21.87
Service export	19.54	12.33	– 10.58	8.94
Goods and services	16.66	14.59	– 19.90	19.07
International trade of developing countries				
Commodity export	16.46	18.87	– 20.57	28.61
Service export	22.10	15.35	– 9.86	21.33
Goods and services	17.70	18.47	– 19.03	27.01

* *Data source* United Nations Conference on Trade and Development (UNCTAD)

Table 14.4 Average annual GDP growth rate of developing countries in 2006–2010

	2006 (%)	2007 (%)	2008 (%)	2009 (%)	2010 (%)
Developing countries	7.50	7.88	5.26	2.86	7.62
Developing countries: Africa	5.74	5.99	5.55	3.41	5.32
Developing countries: America	5.44	5.69	4.05	– 1.87	5.94
Developing countries: Asia	8.65	9.06	5.71	4.63	8.55
Developing countries: Australia	2.80	5.62	0.32	2.04	5.78

* *Data source* United Nations Conference on Trade and Development (UNCTAD)

(iv) An increase in unemployment

Due to the market downturn, many enterprises stopped production or even closed down, resulting in a sharp increase in the unemployment rate. Particularly in the developing countries, which mainly have labor-intensive industries, many small and medium-sized intensive enterprises closed down under the effects of the financial crisis; workers lost their jobs, and the unemployment rate increased.

14.3.2.2 Reasons for the Far-Reaching Impact of the Global Financial Crisis on Developing Countries

(i) Underdeveloped industrial economic structure

In developing countries, the tertiary industry is underdeveloped and the financial system is imperfect, which is mainly devoted to the development of the secondary industry based on the manufacturing industry. The outbreak of the financial crisis in the United States had the greatest impact on developing countries in the real economy field. Most developing countries have limited domestic markets, are overly dependent on external demand to drive economic growth, and rely heavily on developed countries. When the financial crisis broke out, the development of emerging economies was hit hard by a decline in demand from developed countries and a sharp drop in exports from developing countries. Industries in developing countries are mainly labor-intensive, with low knowledge content, added value, and profit margins, and were therefore greatly affected by the global financial crisis.

(ii) Lower income level of residents

The per capita income of residents in developed countries is relatively high, and their consumption is not only to meet clothing, food, housing, and transportation needs but also tends to include entertainment and luxury consumption. In contrast, the majority of residents in most developing countries tend to be used to meet their daily living expenses, forming well-off consumption. When the financial crisis leads to a decline in the income of residents in developed countries, the most important thing is the decline in luxury and enjoyable consumer goods. However, if residents in developing countries lose their jobs and their incomes decline, their daily tipping supplies will be greatly affected.

(iii) Industrial development mainly depends on international capital investment

Affected by the financial crisis, financial institutions in developed countries needed to maintain their own survival and ensure capital adequacy to resist further effects of the financial crisis. International investment fell sharply and even fled developing countries. Industrial projects in developing countries that relied on capital inflows were suspended or even stopped production, and many enterprises went bankrupt and closed down. In addition, for developing countries that rely on international industrial transfer, the international capital investment with multinational corporations as the carriers also withdrew under the impact of the financial crisis, causing industrial transfer to stagnate, and national industrial development lacked the promotion of external factors.

14.3.3 Impact of the Global Financial Crisis on China

14.3.3.1 Macroeconomic Situation of China Before the Financial Crisis

(i) Import and export trade and domestic economic growth

Since China joined the WTO in 2001, the export volume of trade in goods and services has continued to increase, and the contribution rate of net exports of goods and services to GDP growth has increased in recent years, showing that China's import and export trade plays a substantial role in supporting GDP growth, whereas external demand has a great impact on economic growth (Fig. 14.17).

(ii) Status of obtaining foreign direct investment

Since the reform and opening up, the total amount of foreign direct investment received by China has continued to rise. In 1992, Comrade Deng Xiaoping inspected China's southern region and delivered an important speech, clearly opening China wider to the outside world and boldly utilizing foreign capital to promote China's economic development (Fig. 14.18).

(iii) Situation of domestic industrial structure

China's industries are mainly secondary and tertiary industries, and the secondary industry accounts for the highest proportion of GDP. Since 2001, the value-added proportion of the primary industry in GDP has decreased annually. Although the value-added proportion of the tertiary industry in GDP continues to increase, it is relatively low compared with that of the secondary industry (Fig. 14.19).

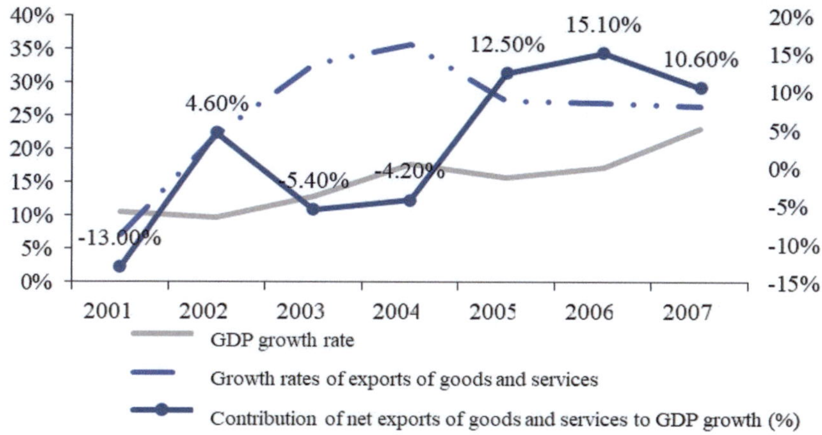

Fig. 14.17 Growth rate of China's exports of GDP, goods, and services in 2001–2007. **Data source* National Bureau of Statistics of China

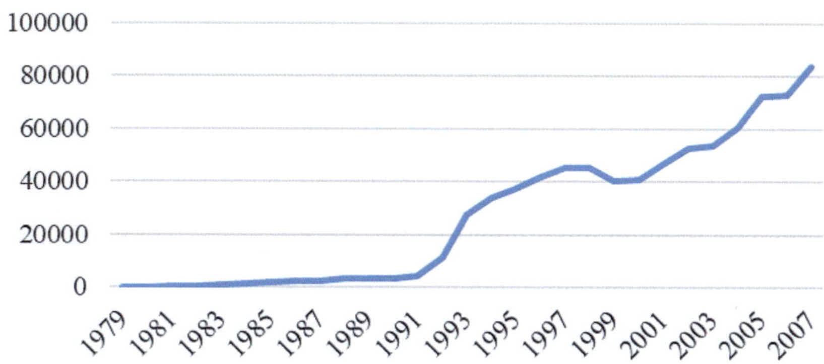

Fig. 14.18 Foreign direct investment in China from 1979 to 2007. **Data source* United Nations Conference on Trade and Development (UNCTAD)

14.3.3.2 Impact of the Financial Crisis on China

(i) Impact on import and export trade

The impact of the financial crisis on China's economy was mainly reflected through the economic interaction among countries as a result of international trade. Therefore, the occurrence of the financial crisis led to a decline in external demand, and China's import and export trade therefore continued to decline.

(ii) Impact on the economy of various industries

Affected by the financial crisis, global oil prices fell, and domestic petrochemical industries, along with other industries, were affected. Global liquidity dried up, market demand tightened, and a large amount of capital fled China, leading

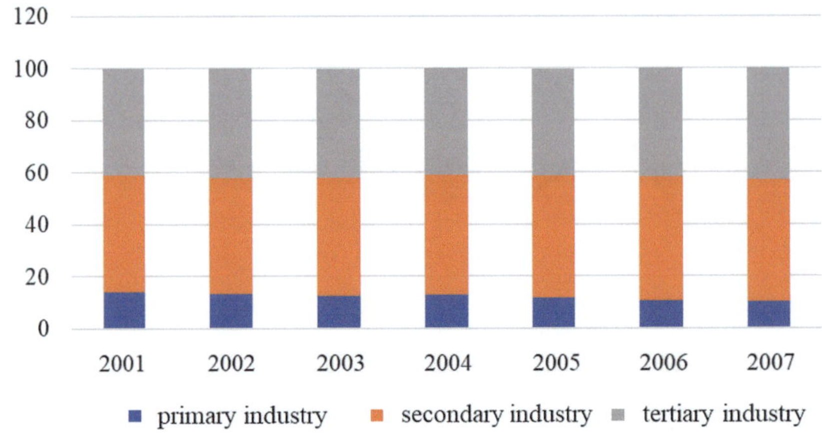

Fig. 14.19 Proportion of added value of China's industries to GDP in 2001–2007. **Data source* National Bureau of Statistics of China

to setbacks in the operation of industrial manufacturing enterprises. As a result, the GDP growth rate of the secondary and tertiary industries declined significantly (Table 14.5).

(iii) Impact on the financial market

With the development of China's reform and opening up, China's stock market continues to rise, many times to refresh the history of the best establishment. Prior to the financial crisis, in October 2007, the closing price of China's Shanghai Composite Index reached 5030.25 yuan. In the global financial crisis, the global stock market plummeted, and the Chinese stock market was no exception (Fig. 14.20).

The financial crisis not only depressed the securities market but also dealt a blow to transactions in the capital market. From the perspective of real estate transactions, owing to the impact of the global financial crisis, the monthly cumulative growth rate of commercial housing sales in China began to decline annually and to show negative growth in February 2008 (Fig. 14.21).

(iv) Impact on social stability and employment

A certain transmission mechanism exists for the impact of the financial crisis on social stability, and the high risk of the financial market determines whether problems occur in the operation of any financial institution, which can lead to public panic. The herd effect will cause panic to spread rapidly, which will affect the operation of other financial institutions. If remedial measures are not taken on time, the financial crisis will further spread to impact the stability of the social environment. The income of residents will decline, and low-income people will experience the greatest loss. In response to the international financial crisis, China implemented several interest rate cuts in 2008 (Table 14.6).

Table 14.5 Changes in GDP in 2007–2009

Year	Gross domestic product		Primary industry		Secondary industry		Tertiary industry	
	Absolute value (100 million yuan)	Year-on-year growth rate	Absolute value (100 million yuan)	Year-on-year growth rate	Absolute value (100 million yuan)	Year-on-year growth rate	Absolute value (100 million yuan)	Year-on-year growth rate
2007	270,232.3	14.20	27,788	3.50	126,633.6	15.10	115,810.7	16.10
2008	319,515.5	9.70	32,753.2	5.20	149,956.6	9.80	136,805.8	10.50
2009	349,081.4	9.40	34,161.8	4.00	160,171.7	10.30	154,747.9	9.60

* *Data source* WIND Economic Database

Fig. 14.20 Shanghai Composite Index in 2007–2009. *Data source* TongHuaShun iFinD

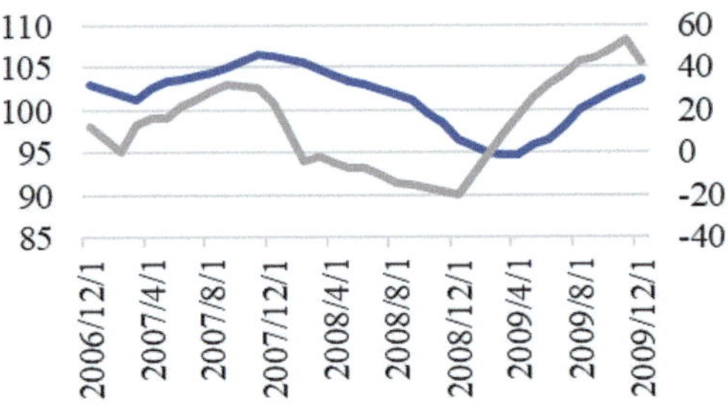

Fig. 14.21 Real Estate Development in China. *Data source* TongHuaShun iFinD

Table 14.6 Interest rate cuts implemented by China in 2008

Time	Pre-adjustment	Post-adjustment (%)
Deposit reserve ratio		
September 25, 2008	17.5%	16.5
October 15, 2008	16.5%	16
December 5, 2008	Large financial institutions: 16%	15
	Small and medium-sized financial institutions: 16%	14
December 25, 2008	Large financial institutions: 15%	14.5
	Small and medium-sized financial institutions: 15%	13.5
One-year deposit and loan interest rate		
September 16, 2008	One-year loan 7.47%	6.21
October 9, 2008	One-year loan 6.21%	6.12
	One-year deposit 4.14%	3.87
October 30, 2008	One-year loan 6.12%	6.03
	One-year deposit 3.87%	3.60
November 27, 2008	One-year loan 6.03%	5.04
	One-year deposit 3.6%	2.52
December 23, 2008	One-year loan 5.04%	4.86
	One-year deposit 2.52%	2.25

* *Data source* Li Q. China's monetary policy in response to the international financial crisis (in Chinese). The Chinese market, 2010, 35–38

In addition to interest rate tools to stimulate consumption and increase corporate investment, the People's Bank of China can use open market operations to increase monetary liquidity. In September 2008, the People's Bank of China increased the money supply by reducing the frequency of issuing bills, buying back government bonds, promoting the entry of monetary funds into the market, and improving money liquidity. Simultaneously, China implemented a loose credit policy, increased the credit scale of banks to individuals and small and medium-sized enterprises, increased the credit line, and provided financial support for the development of small and medium-sized enterprises.

14.3.3.3 Effect of Policy Implementation

After implementing pro-active fiscal, loose monetary, and continuous opening-up policies, China's economic growth rate has rebounded from 9.17% in 2009 to 18.25% since 2010. China's economic growth level has rapidly improved, and resident enterprises have been well developed. The household consumption index rose to 1026.1 from 934.3 in 2008, the consumption level of Chinese residents has improved, and investor confidence has increased. Furthermore, China's foreign trade has also rapidly increased, with exports of goods and services rising from 15.71% in 2009 to 30.77% in 2010. China's economy is improving overall, and the real economy is out of the recession.

With the implementation of national policies, the global economy began to recover. Panic has eased, expectations of an economic downturn have decreased, confidence in the capital markets has gradually increased, and global stock markets have gradually improved.

14.4 Policies to Reform the Financial System and Promote Regional Development

14.4.1 Modern Financial System and Its Evolution

The financial system is a synthesis of financial assets, intermediaries, and trading markets. The economic activities between various departments at domestically and internationally must be realized through finance, and finance is gradually formed in the process of serving society. The financial systems of various countries form different frameworks according to their own national conditions, provide support for the development of the national economy, and become the core content of the modern economic system.

14.4.1.1 Composition and Operation of the Modern Financial System

(i) Elements of the modern financial system

The modern financial system is characterized by marketization, stability, openness, and science and technology, and carries out highly market-oriented resource allocation to improve allocation efficiency. Constructing a modern financial system is the only way to achieve rapid social and economic development. Views differ on the composition the modern financial system. Some believe that the modern financial system includes three elements: the financial sector, financing model, and corporate governance and regulatory system. However, others hold the view that it is composed of five elements: money circulation, financial market, financial intermediary, financial instrument, and regulatory framework.

(ii) Operation of modern financial system

The financial system consists of various subsystems, including financial regulation, financial institution, financial market, financial supervision, and financial environment systems. In modern economic activities, residents, enterprises, financial institutions, governments, and other economic entities must use financial institutions and financial markets as carriers to perform financial activities such as asset liquidation and payment, financing and equity refinement, and asset allocation. As financial activities have strong externalities (Luo, 2022), once financial risks occur, they will have serious adverse effects on the activities of social and economic entities (Fig. 14.22).

14.4.1.2 Evolution of Modern Financial System

The modern financial system formed following the occurrence of the "French Mississippi Bubble" and "British South Sea Bubble" at the end of the eighteenth century. The different responses of the two countries created a watershed moment, resulting in two different financial systems: bank-led and stock market-led. In response to the bubble incident, both Britain and France strictly regulated the stock market; however, Britain lifted the restriction at the beginning of the nineteenth century, whereas France did not abolish it until the end of the twentieth century.

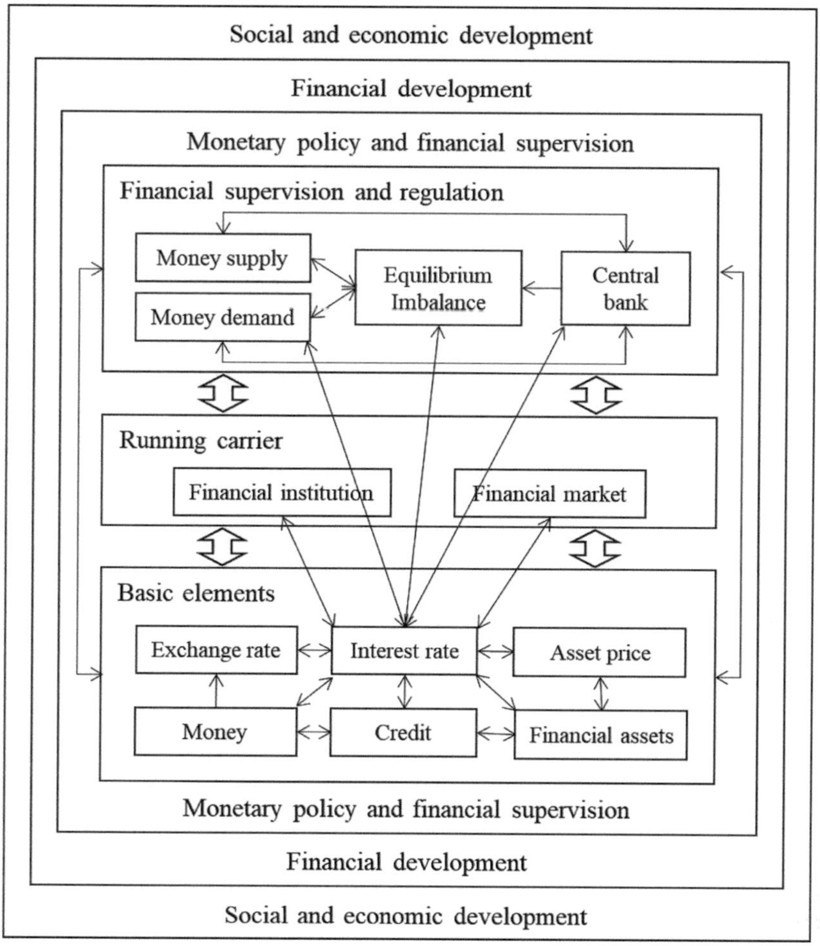

Fig. 14.22 Operation chart of the modern financial system. * *Data source* Li Jian. Finance (essential Edition) (in Chinese). Beijing: Higher Education Press, 2016–11

14.4.2 Financial System Reform and Policies to Promote Regional Development in the United States

14.4.2.1 The Course of Financial System Reform in the United States

The United States has a major economic impact on the world, and its market economy is highly developed. The American financial system developed earlier, and after more than 200 years of evolution, the United States has built a relatively complete financial system, and its financial market development, financial supervision, and financial instrument innovation have become the benchmark for financial reform in other countries. The American financial system plays an important role in promoting American economic development and industrial structural adjustment, laying the foundation for the hegemonic position of the United States in the global economy, and its financial system has an important influence on the operation of the world's economy.

(i) Formation stage of the financial system (1782–1933)

The beginning of the American banking industry originated with the establishment of the North American Bank in 1782. Then the state bank management system was established, which was supervised and managed by the state government. State banks had the right to issue currency and carried out mixed operation. In 1863, to unify the issuance of national currency, the United States Congress passed the National Bank Act, which stipulated that only federally registered banks had the right to issue currency, forming a two-line bank management system with the coexistence of national and state banks. In 1914, to optimize the management of the banking system, the Federal Reserve system of the United States was formally established and became the real central bank of the United States, exercising the functions of monetary policy formulation and banking supervision.

(ii) The stage of tightening the financial system (1933–1980)

Since World War I, the economy of the United States has developed rapidly, as have the banking and securities markets. The US government advocates laissez-faire liberalism and does not interfere with the market economy, which leads to the prevalence of speculation in the US financial market, and economic prosperity promotes the upsurge of speculators. In 1933, after former President Roosevelt took office, he adopted the theory of government intervention advocated by Keynes, and formulated and implemented a series of "New Deal" policies to optimize the American financial system. In 1935, the Banking Act was passed, which strengthened the rights of the Federal Reserve Bank, concentrated the supervision functions of the central bank, made use of open market operations for macro-control, and expanded the monetary and credit management capacity of the federal government.

(iii) The stage of liberalization of the financial system (1980–2008)

A stricter financial management system provides order maintenance for the financial market and prevents financial risks; however, it also limits the development and the market and financial institutions. After the end of World War II, economic growth continued to decline in the US, the Bretton Woods system disintegrated, and US savings institutions fell into crisis in 1980. The United States Government began to deregulate financial markets and sequentially issued four legal provisions for deposit-taking institutions. In 1980, the Depository Institutions Deregulation and Monetary Control Act was implemented, which abolished the restrictions on deposit and loan interest rates, expanded the sources of funds for deposit institutions, and required all deposit institutions to uniformly pay deposit reserves. In 1982, the Garn-St. Germain Depository Institutions Act was promulgated, which further relaxed the restrictions on the business of deposit institutions, expanded the scope of their debt business, allowed for the establishment of money market deposit accounts, and abolished Regulation Q. In 1989 and 1990, the Financial Institutions Reform, Recovery and Enforcement Act and the Federal Deposit Insurance Corporation Improvement Act were successively implemented to address the savings and loan association crisis and recapitalize bank insurance funds.

(iv) The stage of perfection of financial supervision (2008–present)

In 2008, owing to the innovative development of financial institutions and financial derivatives, the American economy was virtualized and economic development became severely separated from the real economy. In addition to the dereliction of duty in financial supervision, the collapse of Fannie Mae and Freddie Mac, and the subprime mortgage crisis, the five major investment banks in the United States either went bankrupt, were acquired cheaply, or were transformed into bank holding companies. With the outbreak of the crisis, the spread of terror, and the lack of confidence in the market, the financial crisis in the United States evolved into a global financial crisis. In response to the crisis, the US government actively introduced policies to restore market confidence and introduced financial regulatory reform measures to repair loopholes in the US regulatory system. Under the Federal Reserve are financial institutions that protect consumers and supervise the high pay of corporate executives.

14.4.2.2 Policies to Promote Regional Development

(i) Financial policies to promote regional development

In 1914, the Federal Reserve was formally established with the power to formulate and implement monetary policy. Currently, the monetary policy implemented by the Federal Reserve regulates the money supply through traditional monetary policy tools such as open market operations and deposit reserves, and non-traditional policy tools such as forward-looking guidance and quantitative easing.

Based on the implementation of traditional monetary instruments, before the subprime crisis, the United States adjusted the federal funds rate and implemented loose or tight monetary policy mainly through traditional methods. In 1997, a financial crisis occurred in Asia, and to avoid the impact of recessions in other countries on the United States, the Federal Reserve cut interest rates three times from September to November 1998 to increase market liquidity. From 1999 to February 2000, the Federal Reserve raised interest rates four times in a row to prevent excess liquidity and reduce the danger of excess inflation.

Regarding the innovation of non-traditional monetary instruments, since the subprime crisis, the traditional monetary policy has failed and cannot effectively regulate the macro-economy; therefore, the Federal Reserve began to implement the non-traditional monetary policy of quantitative easing. Non-traditional monetary policies such as quantitative easing can effectively help maintain financial market stability, provide sufficient liquidity, and reduce market interest rates.

After 2014, the US monetary policy normalized and began to gradually withdraw from the quantitative easing policy, and combined with the guidance of market public expectations, cautiously raised interest rates. After the economic recovery of the United States, the monetary policy has been steadily adjusted, the federal funds rate has gradually increased, price stability and balanced economic development have been maintained, and the maximization of employment has been promoted. By 2018, after a sustained rate hike, the federal funds rate rebounded to 2.4%, ending the zero-interest rate policy.

(ii) Fiscal policy to promote regional development

Regarding the tax reforms of the Reagan administration, before 1981, the United States Government implemented an expansionary fiscal policy to expand government spending while reducing tax revenue, which to some extent boosted economic growth. However, the scale of the government deficit was expanding and the economy was experiencing "stagflation." After former President Reagan took office, he launched an economic recovery plan, the core content of which was a tax reduction plan. To address the problem of excessive nominal income caused by inflation, inflation was linked to the tax system, income tax was slashed to stimulate economic demand, and tax breaks for investment were introduced to increase economic growth and boost employment. In 1982 and 1985, to reduce fiscal pressure, the United States Congress passed the Tax Equity and Fiscal Responsibility Act and the Social Security Amendment Act respectively, which significantly increased taxes and offset the effects of the reduction in personal income tax in 1981.

Owing to the impact of the subprime crisis in 2008, the Obama administration implemented a structural fiscal policy with a larger fiscal deficit. The United States Government implemented the 787 billion economic stimulus package, strengthened infrastructure construction, and supported the development of the renewable energy industry. Taxes were cut further and social welfare spending was increased to help low-and middle-income level and unemployed people and narrow the gap between the rich and the poor. In addition, large-scale acquisitions of problematic financial assets and guarantees for financial institutions were performed to restore market confidence.

14.4.3 Financial System Reform and Policies to Promote Regional Development in Japan

14.4.3.1 The Course of Financial System Reform in Japan

The reform of Japan's financial system has a long history, and the process of financial liberalization is tortuous. Over the course of financial development, we continue to summarize experiences and lessons and constantly improve the financial system.

Currently, the financial system plays an irreplaceable role in the development of the Japanese economy, particularly for major industries, and promotes the sustainable development of the national economy.

(i) The initial stage of the financial system (1868–1966)

Japan's financial system was built earlier. From the Meiji Restoration in 1868 to the outbreak of World War II, Japan established a relatively sound currency management mechanism and financial system. In 1885, Japan implemented a cabinet system and established the Ministry of Tibet as an executive office responsible for financial management and monetary and banking supervision. In 1931, Japan formed a financial system with banks as the dominant model. After World War II ended in 1945, Japan's economy was in urgent need of recovery, and there was a shortage of capital for democratic progress, which could not meet the needs of economic development. Therefore, the Japanese government established policy-oriented financial institutions to provide capital for national economic investment.

(ii) The stage of promoting financial liberalization (1966–2008)

In 1966, Sumita Satoshi, director of the Banking Bureau of the Ministry of Tibet of Japan, proposed financial reform, implemented efficient administration, introduced a competition mechanism, carried out moderate self-competition, and established a free competition market. In the mid- and late 1960s, with the development of the Euro-dollar market, global finance experienced the phenomenon of internationalization. Japan has also correspondingly expanded foreign exchange trading and made foreign exchange investments, and the financial market has become internationalized. In 1987, Japan combined banks with securities to issue and circulate commercial paper, breaking the boundary between financial institutions and liberalizing financial business. However, the bubble of the Japanese real estate market burst in 1990, and the drawbacks of the financial system gradually emerged, which led to the collapse of the market economy, and the Japanese stock market fell into a bear market that lasted for decades. In the late 1990s, Japanese financial institutions had large-scale non-performing debts. Japan reformed its financial system and issued the Reform of the Financial System-the Revival of the Tokyo Market in 2001. Japan decided to accelerate financial liberalization, reform the original separate management system into mixed operation, reform the financial investment and financing system, standardize market trading rules and market systems, abolish the foreign exchange banking system, and promote financial innovation.

(iii) The stage of functional return of policy-oriented financial institutions (2008–present)

In 2008, with the outbreak of the global financial crisis, policy-oriented financial institutions provided a great deal of financial support to the national economy. The Japanese Government is aware of the importance of policy-oriented financial institutions to economic development; therefore, in 2009, through the Japanese Policy Investment Banking Law, the full privatization of policy-oriented financial institutions was postponed for four years. Then, in March 2011, a major earthquake in

Japan triggered a tsunami, resulting in a leak at the Daiichi nuclear power plant, which resulted in heavy losses of Japanese property. The real economy was hit, the financial market was affected, and the stock market plummeted. Policy-oriented financial institutions provided financial assistance to deal with the earthquake and help the quake-hit areas restore economic development. In May 2011, the newly revised Japanese Policy Investment Banking Law and the Commercial and Industrial Combination Central Treasury Law were issued, delaying the privatization process and proposing privatization reforms to maintain the government's shareholding and financial base and the integrity of business operations.

14.4.3.2 Policies to Promote Regional Development Over the years

(i) Financial policies to promote regional development

In 1990, Japan's real estate bubble burst, the economy declined for a long time, and the stock market fell into a bear market that lasted over a decade. The Bank of Japan repeatedly cut interest rates to the "zero-interest rate floor." In 2001, the Japanese monetary authorities began to innovate monetary policy, implementing the untraditional monetary policy of quantitative easing for the first time worldwide. This was aimed to reactivate the zero-interest rate policy, provide liquidity to the market through open market operations, change the monetary policy operating tool from unsecured overnight lending rates to reserve account balances, and increase the purchase of long-term government bonds.

In 2018, when the global financial crisis broke out, Japan's GDP growth slowed, and the real GDP growth rate was only −1.04%; thus, the Japanese government decided to use quantitative easing again. In December 2018, interest rates were cut and government bond purchases were expanded. In 2010, a 35-trillion-yen asset purchase plan was released. In October 2012, the Bank of Japan set up an unlimited loan support program to provide support loans to financial institutions. The policy of quantitative easing continues to be implemented and coexists with the zero-interest rate policy.

On April 4, 2013, Japan began to implement quantitative and qualitative easing, changing the unsecured overnight lending rate into a base currency, significantly increasing the loan support program and the purchase of long-term treasury bonds, and increasing the use of base money. However, the implementation of the policy did not achieve the desired results: export and credit growth declined, and the 2% core inflation target was not met. In September 2016, the Bank of Japan began to build a new monetary policy framework, adjust the operational index of the quantitative and qualitative easing policy to the regulation of long-term and short-term interest rates, innovate the interest rate transmission mechanism, and return the intermediary indicators from quantity to interest rates. This aimed to regulate the yield curves of different terms, create a stable and relaxed environment for the market, and promote economic growth. Under the new monetary policy framework, interest rates and quantitative and qualitative dimensions were combined to create a loose monetary policy environment to restrain the rise of long-term interest rates to a certain extent.

(ii) Fiscal policy to promote regional development

After the end of World War II, the Japanese economy urgently needed to recover to prevent inflation while promoting industrial production and construction. In 1946, Japan advocated Keynesianism, strengthened the government's intervention in the economy, increased government investment, and carried out an active fiscal policy of "production first." Focusing on the development of industries, price adjustment subsidies were set up for the coal, iron and steel, and other industries, accounting for 8.90% of the total expenditure in that year. The Revitalization Financial Treasury, a financial institution established at that time, prioritized providing loan support to key industries. In 1948, the Japanese economy recovered rapidly, especially the coal and iron and steel industries, whose GDP recovered to 91.60% and 48.20% of that before the war, respectively.

After the asset bubble burst, the Japanese economy fell into a long recession. The proactive fiscal policy of the Japanese government not only failed to restore the economy but led to a substantial increase in the level of government debt, which accounted for 125% of the total value of GDP in 2000. By 2001, the Koizumi Cabinet called for structural reform, emphasizing that "without reform, there would be no development," insisting on reducing government expenditure, simplifying government functions, and building an efficient and convenient small government. In 2002, the "Basic Principles for Economic and Financial Operation and Structural Reform 2002" was issued to reduce fiscal expenditure, reform the distribution of tax sources between the central and local governments, promote the improvement of the social security system, and carry out tax reduction and reform to stimulate economic growth. In 2006, the Abe cabinet continued to promote the fiscal policy of structural reform, reduce government expenditure, simplify government departments, improve the tax system, and promote the soundness of Japan's financial system.

Regarding the fiscal policy of Abenomics, owing to the impact of the 2008 financial crisis, Japan had to change its original fiscal system to a proactive fiscal policy. To stabilize the hearts and minds of the people and stimulate economic growth, the Japanese government issued a series of large-scale financial rescue plans, which, to a certain extent, alleviated the impact of the global financial crisis and led to a short-term economic recovery; however, the overall economy was in a recession. To revive the Japanese economy and deal with austerity, Shinzo Abe launched the "three arrows" of Abenomics after becoming prime minister in 2012: loose monetary policy, flexible fiscal policy, and industrial policy of structural reform. Among them, the mobile fiscal policy is to adopt a larger fiscal deficit and increase the construction of public infrastructure to stimulate economic development.

14.4.4 *Financial System Reform and Policies to Promote Regional Development in China*

14.4.4.1 The Course of Financial System Reform in China

Over the past 40 years of China's reform and opening up, the reforms of China's financial and economic systems have been coordinated to jointly promote the development of the economic system. The reform of China's financial system stems from breaking the "unified" banking monopoly system and establishing an independent financial system. This aimed to gradually improve the establishment of banks and other financial institutions, strengthen the role of financial supervision, and build a market-oriented and international modern financial system based on reform and opening up. The reform of China's financial system has gone through four stages: initial construction of the financial system, reform of marketization and legalization, deepening transformation of financial institutions, and reform of marketization and internationalization of the financial system.

(i) Preliminary construction stage of the financial system (1978–1993).

Regarding breaking the "Great Unification," with the reform and opening up as the starting point, China began to reform its economic system. In China, the financial system has been constantly optimized, gradually changing from the original "unified" financial system, and initially constructed as a dual banking system and a diversified financial institution system. In 1979, the Bank of China was separated from the People's Bank of China to be responsible for the operation and management of foreign exchange funds and finance. In September 1983, the State Council clearly stipulated that the People's Bank of China exercises the functions of the central bank. In 1984, the People's Bank of China exercised the functions of the central bank alone, gradually established industry, agriculture, and construction, took charge of the industrial and commercial credit and savings business, and initially established a secondary banking system.

Further, China diversified its reform of financial organizations. In 1980, the People's Insurance Company of China resumed its domestic insurance business, and then set up China Pacific Insurance and other national insurance companies, such as Shanghai Volkswagen Insurance, accelerating the pace of the diversification of the insurance market. In 1986, comprehensive banks such as the Bank of Communications were established. At the end of 2017, the Shanghai Stock Exchange and Shenzhen Stock Exchange were established successively, the Chinese stock market was initially established, and trust companies and securities companies were set up, which greatly enriched China's financial market.

(ii) Marketization and legalization stage of the financial system (1993–2003).

China has worked to show marketization in the financial system and introduce competition and market mechanisms. Cross-business exists among banks to promote cross-competition. In 1994, China established three major policy banks—the China Development Bank, Export–Import Bank of China, and Agricultural Development Bank of

China—transforming the four major state-owned specialized banks into state-owned commercial banks. In May 1998, the People's Bank of China allowed eight foreign banks to enter the global interbank lending market to operate RMB business. In 2002, China introduced the qualified foreign institutional investor system to increase the openness of China's securities market, and foreign funds began to engage in limited participation in the securities market. In 2003, the goal of market-oriented interest rate reform was established in the third Plenary session of the 16th CPC National Congress, and the market mechanism of determining interest rates by market supply and demand was established and improved.

With respect to legalization in the management mechanism, to standardize the operation and development of financial institutions, legal provisions needed to be developed to standardize risk management and strengthen financial market supervision. In 1995, the Law of the People's Republic of China on Commercial Banks and Law of the People's Republic of China on Negotiable Instruments were promulgated, followed by the Regulations on the Monetary policy Commission of the Peoples Bank of China. The implementation of these legal provisions established the legal management of China's financial system and gave full play to the important position of the central bank. To strengthen the supervision and management of the financial market, the China Securities Regulatory Commission, China Insurance Regulatory Commission, and China Banking Regulatory Commission were set up after 1992, forming a separate supervision mode of "one line and three associations."

(iii) Deepening transformation stage of financial institutions (2003–2012).

To optimize the operating efficiency and overall efficiency of commercial banks, it is necessary carry out the enterprise transformation of banks and promote the development of the banking industry. In January 2004, the State Council of China decided to start a pilot joint-stock reform of state-owned banks. Bank of China Co., Ltd. and China Construction Bank Co., Ltd. were established, followed by non-tradable shares, and the major banks publicly issued shares and were listed on the stock market. In July 2010, the successful listing of Agricultural Bank shares on the Shanghai Stock Exchange and the Hong Kong Stock Exchange marked the end of the shareholding reform of state-owned commercial banks.

To expand the authority of operation and increase the openness of the financial system, in 2004, the People's Bank of China chartered China Construction Bank to engage in financial derivatives trading and expand the banking business. In 2005, the promulgation of the "Administration of Pilot Projects for the Establishment of Fund Management Companies by Commercial Banks Procedures" established that commercial banks can participate in the establishment of fund management companies, and the banking operation in China uses a diversified mixed operation model. In July 2005, the People's Bank of China reformed the exchange rate system and began implementing a floating exchange rate system based on market supply and demand. The original single exchange rate system pegged to the US dollar was changed to improve the flexibility of China's exchange rate management and determine the market-oriented orientation of the RMB exchange rate. In 2006, the operating authority of foreign banks was expanded, and the restrictions on the regions and customers of RMB business were lifted. Subsequently, the State Council agreed to the establishment of the China Financial Futures Exchange to further deepen the reform process of the capital market in China. By 2010, the stock index futures of China's financial futures exchange were also officially listed, the capital market system had been further improved, and the function of the capital market had been further brought fully into play.

(iv) Marketization and internationalization stage of the financial system (2012–present)

The domestic and international economic and financial situation is constantly changing, and China continues to promote the reform and opening up of the financial system. Exchange rate fluctuations were relaxed in 2019 to meet market demand. In 2013, the Shanghai Free Trade Zone was established, and a pilot project of highly open finance was implemented. Subsequently, the "Shanghai-Hong Kong Stock Connect" and the "Shenzhen-Hong Kong Stock Connect" were launched, and in 2016, the RMB formally joined the International Monetary Fund (IMF) Special Drawing Rights (SDR) basket, increasing the openness of the financial system. To improve the marketization of the financial system, the loan interest rates of financial institutions were liberalized in 2013, and the deposit interest rate ceiling was lifted in 2015. In 2017, China's A-share market was included in the MSCI[1] Emerging Markets Index and MSCI Global Index, and the financial stock market was internationalized. The Measures for Liquidity Risk Management of Commercial Banks were issued in 2018, stipulating new liquidity risk supervision indicators, optimizing liquidity risk management, and addressing the problems brought about by the marketization of interest rates and the deepening of financial innovation.

14.4.4.2 Policies to Promote Regional Development Over the years

(i) Financial policies to promote regional development

Financial policy is a general term for the monetary and credit policies adopted by a country's government or the central bank to adjust the money supply. The central bank of China mainly uses monetary policy to act on operational indicators through monetary policy tools, causing changes in intermediary indicators, such as total credit and market interest rates. In turn, it affects price stability, full employment, balance of payments, economic growth, and financial stability.

With the development of China's economy and the improvement of the socialist economic system, the implementation of monetary policy has played an important role in macroeconomic regulation and control. The implementation of monetary policy in China began in 1984 and established the statutory deposit reserve as a policy

[1] MSCI stands for Morgan Stanley Capital International. It is a globally renowned financial services company that specializes in providing stock market indices and related analytical tools. The MSCI Emerging Markets Index is one of the key indices published by the company, designed to measure the overall performance of stock markets in emerging market countries.

tool to control the scale of credit in the form of loans. Subsequently, innovations have been implemented to set up policy tools for re-lending/rediscounting and open market operations, and to adjust the base currency and market liquidity.

With the complication of the economic situation, the "quantitative" monetary policy implemented in China was gradually invalidated. The central bank began to pay attention to the implementation of "structural" monetary policy and improve the regulation and control ability of monetary policy. Since the financial crisis, there has been a double surplus in China's international trade, resulting in a substantial net inflow and persistent excess liquidity. The central bank's active regulation and control capacity is limited, and in 2013, it formally used the new standing lending facility, an innovative monetary policy tool, to enhance the central bank's ability to regulate and control credit funds. From 2014 to 2017, the central bank locked in reserve demand by raising the required reserve ratio, using pledged supplementary lending and medium-term lending facility to supply reserves to construct structural liquidity shortages and promote liquidity balance.

(ii) Fiscal policy to promote regional development

China has adhered to the path of building a socialist system with Chinese characteristics, promoting the construction of a market economy, actively implementing fiscal policies, and enacting macro-control in the light of national conditions. Since the reform and opening up, China's fiscal policy tools have been gradually improved and become the main means of macroeconomic regulation and control. Currently, the main tools of China's fiscal policy are national debt, taxation, public expenditure, and government investment. Since 1998, China has been alternately implementing steady and active fiscal policy to eliminate or reduce the impact of economic cyclical fluctuations and promote stable economic development.

14.5 Summary

Finance is the core of the national economy, and international trade is a part of international finance. Countries export advantageous products to the rest of the world through their own resource endowments and industrial advantages; carry out international division of labor; build global value chains, supply chains, and industrial chains; obtain their own interests; and ultimately, promote global well-being. The economies of various countries transmit to and influence each other through the income, relative price, and interest rate mechanisms generated by international trade. Therefore, the stability of international finance is particularly important for the steady economic development of all countries.

The global financial crisis in 2008 led to a decline in the economies of various countries and decreases in household income. Low-income households experienced the greatest losses; thus, developing countries were hit hard by this crisis. However, both developing and developed countries experienced harsh economic setbacks, and all countries have participated in rescuing the market and adopted active monetary and fiscal policies. Affected by the financial crisis, the disadvantages of the financial system of various countries have been gradually revealed, and financial system reform has been carried out across several countries worldwide. To strengthen the supervision and inspection of the financial system, the United States implemented a stricter regulatory system reform. To restore the national economy, Japan paid more attention to the functions of financial institutions. As the financial systems of developing countries were not perfected at that time, they were less affected by the financial crisis; however, they all adopted loose monetary and active fiscal policies to stimulate market demand and economic growth.

Exercise

1. Explain the role of international financial stability in regional development and discuss the possible impact of financial stability in one country on other countries in the context of globalization.
2. Describe the dynamic process of the formation of regional financial stability and analyze how financial agglomeration affects the development of regional economies.
3. Analyze the possible positive and negative effects of a balance of payments deficit on a country's economy and discuss the causes of the US trade deficit and its impact on the global economy.
4. Discuss the history of financial system reforms in the United States and Japan and how these reforms have affected economic development and regional stability in both countries.
5. In the face of a financial crisis, discuss how monetary and fiscal policies can be coordinated to achieve rapid economic recovery, and analyze China's policy response as an example.
6. In a globalized economy, discuss how a country's financial policy affects other countries, especially small open economies, and discuss the importance of international cooperation in financial policymaking.

Extended Reading

Bajo-Rubio O., Díaz-Roldán C. 2015. Economic growth and the balance of payments constraint: the case of the Spanish regions,1998-2009. Acta Oeconomica, 65(4):617-628.

BOP Analysis Group State Administration of Foreign Exchange. 2019. Report on China's balance of payments in 2018. State Administration of Foreign Exchange

Ferda H. 2012. Balance-of-payments constrained growth: the case of Turkey. MPRA Paper, (7):1-18.

Hummels D., Ishii J., Kei-M. Y. 2001. The nature and growth of vertical specialization in world trade. Journal if International Economics, 54:75-96.

International Monetary Fund. 2019. World Economic Outlook: Challenges to Steady Growth. Washington.

References

Cerutti E., Gopinath G., & Mohammed A. (2019). The impact of Sino-US trade tensions. International Monetary Fund.

Cömert, H., & Ugurlu, E. (2015). The impacts of the 2008 global financial crisis on developing countries: The case of the 15 most affected countries (ERC Working Paper 1509). ERC–Economic Research Center, Middle East Technical University. https://econpapers.repec.org/paper/metwpaper/1509.htm

Croke, H., Kamin, S. B., & Leduc, S. (2005). Financial market developments and economic activity during current account adjustments in industrial economies. Discussion Paper, 827 (Board of Governors of the Federal Reserve System, February).

De Seversky, A. P. (1950). *Air power: Key to survival.* Simon and Schuster.

Douhet, G. (1921). The command of the air. Translated by Dino Ferrari, 1942.

Irwin, D. A. (2023). Trade policy in historical political economy. In J. A. Jenkins & J. Rubin (Eds.), *The oxford handbook of historical political economy* (p. 0). Oxford University Press. https://doi.org/10.1093/oxfordhb/9780197618608.013.24

Laulajainen, R. (2003). In *Financial geography: A banker's view.* Routledge.https://doi.org/10.4324/9780203987346

Li, S. (2004). An analysis of US trade deficit: Growth inertia and sustainability. *Contemporary Finance and Economics, 239*(10), 83–87.

Luo, D. (2022). Research on financial systemic risk assessment algorithm based on risk data fuzzy cluster analysis. In *2022 2nd International conference on networking, communications and information technology (NetCIT)* (pp. 216–219). https://doi.org/10.1109/NetCIT57419.2022.00060

Mackinder, H. J. (1904). The geographical pivot of history. *The Geographical Journal, 23*(4), 421–437.

Mackinder, H. J. (1919). *Democratic ideals and reality: A study in the politics of reconstruction.* Constable and Company, Ltd. Retrieved from Internet Archive.

Mahan, A. T. (1890). *The influence of sea power upon history, 1660–1783.* Little, Brown, and Company.

Nanto, D. K. (2011). The global crisis: Analysis and policy implications. *Journal of Current Issues in Finance, Business and Economics, 4*(1), 43–150.

Obstfeld M., & Taylor A. (2017). *International monetary relations: Taking finance seriously.* National Bureau of Economic Research.

Siddiqui, K. (2019). The US economy, global imbalances and recent development: A critical review. *Istanbul Journal of Economics/İstanbul İktisat Dergisi*, 175–205. https://doi.org/10.26650/ISTJECON2019-0027

Spykman, N. J. (1942). *America's strategy in world politics: The United States and the balance of power*. Harcourt, Brace and Company.

MIX
Papier aus verantwortungsvollen Quellen
Paper from responsible sources
FSC® C105338

If you have any concerns about our products,
you can contact us on
ProductSafety@springernature.com

In case Publisher is established outside the EU,
the EU authorized representative is:
Springer Nature Customer Service Center GmbH
Europaplatz 3, 69115 Heidelberg, Germany

Printed by Libri Plureos GmbH
in Hamburg, Germany